Diversity and Applications of New Age Nanoparticles

Suneev Anil Bansal
Department of Mechanical Engineering, Bharat Institute of Engineering and Technology, India

Virat Khanna
Department of Mechanical Engineering, MAIT, Maharaja Agrasen University, India

Nilanthy Balakrishnan
School of Chemical and Physical Sciences, Keele University, UK

Pallav Gupta
Department of Mechanical Engineering, A.S.E.T., Amity University, India

IGI Global
PUBLISHER of TIMELY KNOWLEDGE

A volume in the Advances in Chemical and Materials Engineering (ACME) Book Series

Published in the United States of America by
IGI Global
Engineering Science Reference (an imprint of IGI Global)
701 E. Chocolate Avenue
Hershey PA, USA 17033
Tel: 717-533-8845
Fax: 717-533-8661
E-mail: cust@igi-global.com
Web site: http://www.igi-global.com

Library of Congress Cataloging-in-Publication Data

Names: Bansal, Suneev, 1982- editor.
Title: Diversity and applications of new age nanoparticles / Suneev Bansal, Virat Khanna, Nilanthy Balakrishnan, and Pallav Gupta, editors.
Description: Hershey, PA : Engineering Science Reference, [2023] | Includes bibliographical references and index. | Summary: "This book is focused on emerging and developing field of new age nanoparticles, offering cutting edge knowledge on Nanotechnology Materials and applications"-- Provided by publisher.
Identifiers: LCCN 2022039754 (print) | LCCN 2022039755 (ebook) | ISBN 9781668473580 (h/c) | ISBN 9781668473597 (s/c) | ISBN 9781668473603 (eISBN)
Subjects: LCSH: Nanoparticles.
Classification: LCC TA418.78 .D63 2023 (print) | LCC TA418.78 (ebook) | DDC 620/.5--dc23/eng/20221004
LC record available at https://lccn.loc.gov/2022039754
LC ebook record available at https://lccn.loc.gov/2022039755

This book is published in the IGI Global book series Advances in Chemical and Materials Engineering (ACME) (ISSN: 2327-5448; eISSN: 2327-5456)

British Cataloguing in Publication Data
A Cataloguing in Publication record for this book is available from the British Library.

All work contributed to this book is new, previously-unpublished material. The views expressed in this book are those of the authors, but not necessarily of the publisher.

For electronic access to this publication, please contact: eresources@igi-global.com.

Advances in Chemical and Materials Engineering (ACME) Book Series

J. Paulo Davim
University of Aveiro, Portugal

ISSN:2327-5448
EISSN:2327-5456

Mission

The cross disciplinary approach of chemical and materials engineering is rapidly growing as it applies to the study of educational, scientific and industrial research activities by solving complex chemical problems using computational techniques and statistical methods.

The **Advances in Chemical and Materials Engineering (ACME) Book Series** provides research on the recent advances throughout computational and statistical methods of analysis and modeling. This series brings together collaboration between chemists, engineers, statisticians, and computer scientists and offers a wealth of knowledge and useful tools to academics, practitioners, and professionals through high quality publications.

Coverage

- Numerical Techniques
- Sustainable Materials
- Fracture Mechanics
- Wear of Materials
- Multifuncional and Smart Materials
- Computational methods
- Ceramics and glasses
- Biomaterials
- Statistical Methods
- Metallic Alloys

IGI Global is currently accepting manuscripts for publication within this series. To submit a proposal for a volume in this series, please contact our Acquisition Editors at Acquisitions@igi-global.com or visit: http://www.igi-global.com/publish/.

The Advances in Chemical and Materials Engineering (ACME) Book Series (ISSN 2327-5448) is published by IGI Global, 701 E. Chocolate Avenue, Hershey, PA 17033-1240, USA, www.igi-global.com. This series is composed of titles available for purchase individually; each title is edited to be contextually exclusive from any other title within the series. For pricing and ordering information please visit http://www.igi-global.com/book-series/advances-chemical-materials-engineering/73687. Postmaster: Send all address changes to above address.

Titles in this Series

For a list of additional titles in this series, please visit: http://www.igi-global.com/book-series/advances-chemical-materials-engineering/73687

Polymer-Inorganic Nanostructured Composites Based on Amorphous Silica, Layered Silicates, and Polyionenes
Kostyantyn M. Sukhyy (Ukrainian State University of Chemical Engineering, Ukraine) and Elena A. Belyanovskaya (Ukrainian State University of Chemical Engineering,Ukraine)
Engineering Science Reference • © 2023 • 304pp • H/C (ISBN: 9781799897286) • US $260.00

Development, Properties, and Industrial Applications of 3D Printed Polymer Composites
R. Keshavamurthy (Dayananda Sagar College of Engineering, India) Vijay Tambrallimath (Manipal Institute of Technology, Manipal Academy of Higher Education, Manipal, India) and J. Paulo Davim (University of Aveiro, Portugal)
Engineering Science Reference • © 2023 • 319pp • H/C (ISBN: 9781668460092) • US $270.00

The Classes of Higher Dimensional Polytopes in Chemical, Physical, and Biological Systems
Gennadiy Vladimirovich Zhizhin (Russian Academy of Natural Sciences, Russia)
Engineering Science Reference • © 2022 • 366pp • H/C (ISBN: 9781799883746) • US $215.00

Advances in the Modelling of Thermodynamic Systems
Elhoucine Essefi (University of Gabes, Tunisia) and Ibtissem Jendoubi (Faculty of Sciences of Bizerte, Tunisia)
Engineering Science Reference • © 2022 • 309pp • H/C (ISBN: 9781799888017) • US $270.00

Handbook of Research on Tribology in Coatings and Surface Treatment
Amirhossein Pakseresht (FunGlass – Centre for Functional and Surface Functionalized Glass, Alexander Dubček University of Trenčín, Slovakia) and Omid Sharifahmadian (FunGlass – Centre for Functional and Surface Functionalized Glass, Alexander Dubček University of Trenčín, Slovakia)
Engineering Science Reference • © 2022 • 470pp • H/C (ISBN: 9781799896838) • US $315.00

Advanced Manufacturing Techniques for Engineering and Engineered Materials
R. Thanigaivelan (Muthayammal Engineering College, India) N. Rajan (Vinayaga Mission Kripananda Variyar Engineering College, India) and T.G. Argul (St. Mary's Engineering College, India)
Engineering Science Reference • © 2022 • 351pp • H/C (ISBN: 9781799895749) • US $270.00

Handbook of Research on Green Synthesis and Applications of Nanomaterials
Rajni Garg (Rayat Bahra University, India) Rishav Garg (Galgotias College of Engineering and Technology, India) and Nnabuk Okon Eddy (Department of Pure and Industrial Chemistry, University of Nigeria, Nsukka, Nigeria)
Engineering Science Reference • © 2022 • 569pp • H/C (ISBN: 9781799889366) • US $325.00

IGI Global
PUBLISHER of TIMELY KNOWLEDGE

701 East Chocolate Avenue, Hershey, PA 17033, USA
Tel: 717-533-8845 x100 • Fax: 717-533-8661
E-Mail: cust@igi-global.com • www.igi-global.com

Table of Contents

Detailed Table of Contents

Ferrites–the electro-ceramic materials have been considered highly important magnetic materials for more than half a century. The crystal structure of ferrimagnetic oxides falls into one of the following categories: spinel, magnetoplumbite(hexagonal), garnet, and orthoferrite. The development of spinel ferrite nanoparticles (SFNPs) has become more popular owing to their fascinating properties, such as their potential adsorption properties, superparamagnetic behavior, high stability, high value of resistivity, saturation magnetization, coercivity, low power losses, and ease of functionalization. These advantageous features make them suitable for use in various applications. This chapter aims to discuss the structure, general properties, and some recent applications of spinel ferrites. Commonly used synthesis techniques for nanocrystalline ferrites are discussed. The applications of nano-structured ferrites like antimicrobial activity, gas sensors, and dye degradation are discussed.

Green synthesised nanoparticles are a new source of inspiration for clinical research, as this method uses natural materials, such as plant-derived materials, for synthesis. Green nanoparticle synthesis using plant extracts is simple, inexpensive, chemicals, resulting in biologically active shape- and size-dependent products with no contaminants or byproducts. Silver nanoparticles (AgNPs) have unique physiochemical properties and promising biological applications. To assess AgNPs' functionality, they must be characterized. Natural compounds found in plant extracts reduce and stabilise AgNPs. Green synthesis is more environmentally friendly than conventional methods, but many questions remain. To understand the plant-mediated process, many studies are needed. This proposed book chapter examines the synthesis, characterization, and bio activities of green-synthesised AgNPs. The goal is to provide an overview of green AgNP synthesis and their bioactivity assessment.

Smart nanomaterials are grabbing more significant attention. Photocatalysts are catalysts that can be stimuli by light with an appropriate wavelength. Photocatalysis is deemed to be a promising approach for the utilization of photo energy and has been extensively studied for many processes. Therefore, the synthesis of nanomaterials with tailored-made catalytic characteristics, are of great importance. For applications using visible light and hence solar energy, the modification of photocatalysts can occur via the formation of heterojunction nanocomposites by inhibiting the recombination of charge carriers. These applications are driven by solar energy, which is conducive to the sustainable development of energy resources with no impact on the environment. This chapter will include recent enhancements in the smart nanomaterials for photocatalytic applications, especially in the fields of removing environmental pollutants, self-cleaning surfaces, water splitting for hydrogen production to provide clean fuel resources, selective alcohol oxidation and CO2 reduction in the environment.

Neha Sharma, Keshav Mahavidyalaya, University of Delhi, India
Prashant Kumar, Keshav Mahavidyalaya, University of Delhi, India

A cluster of silicate minerals known as garnet is widely used as abrasives and gemstones, has now found an application as smart nonmaterial. The garnet possesses a unique feature of having different chemical composition with similar crystal form and physical properties. Yttrium iron garnet (YIG) also popularly known as iron yttrium oxide, is an example of lab grown garnet. It can have chemical composition of Y3Fe5O1 or Y3Fe2(FeO4). The YIG exhibits a ferrimagnetic nature, with a Curie temperature of 560 K. The chapter focuses on the research around YIG with a mention on fabrication methods and applications.

Seitkhan Azat, Al-Farabi Kazakh National University, Kazakhstan
Ulan Zhantikeyev, Al-Farabi Kazakh National University, Kazakhstan
Kydyr Askaruly, Al-Farabi Kazakh National University, Kazakhstan
Kalampyr Bexseitova, Al-Farabi Kazakh National University, Kazakhstan
Yerbolat Sailaukhanuly, Al-Farabi Kazakh National University, Kazakhstan
Kainaubek Toshtay, Al-Farabi Kazakh National University, Kazakhstan
Zhandos Tauanov, Al-Farabi Kazakh National University, Kazakhstan
Darkhan Ybyraiymkul, Al-Farabi Kazakh National University, Kazakhstan

Rice, which provides a major source of food for billions of people, covers 1% of the surface of the earth. Rice husk (RH), the outer covering of paddy rice, is an important byproduct during the milling process, and residue ash is generated after the burning of RH that is called rice husk ash (RHA), which is a primary waste material in the agricultural industry. The main components in RH are lignin, cellulose, and hemicellulose, which are generally named lignocellulose. RH also contains ca. 15 to 28 wt% of silica. The high content of silica in RH presents opportunities for the preparation of value-added silicon-based materials. Since the 1970s, various silicon-based materials, including silica, silicon carbide, silicon nitride, silicon tetrachloride, zeolite, and silicates, have been successfully synthesized using RH as the silicon source. This field of research has been significantly advanced and expanded in the past decade spurred by the global attention on sustainable and renewable resources.

Cadmium telluride (CdTe) is a binary II-VI direct band gap semiconducting material. Cadmium telluride is a promising electrochemical and photovoltaic material for thin film solar cells. It shows both p and n type conductivity. A nanocrystalline thin film of CdTe was deposited on nickel plate by chemical bath deposition method containing 0.01M Cadmium acetate and 0.02M tellurium dioxide. The structural, compositional, and optical analysis were studied by X-ray diffraction (XRD), energy-dispersive X-ray spectroscopy (EDS), scanning electron microscope (SEM), fourier transform infrared spectroscopy (FTIR) and photoluminescence (PL).The diffraction peak observed at 2=23.59° with (111) plane indicate the crystalline phase of CdTe film. The average crystalline size is measured to be 7nm.Compositional analysis reveals the presence of both Cd and Te elements. The EDS spectroscopy shows the ratio of Cd and Te is1:2 in case of as deposited film. The photoluminescence peak is observed at 471nm.

The need for electric vehicles and other portable devices globally, advancements in charging technologies, are inflection points in the evolution of upcoming batteries. The proper engineering of electrodes are essential for stability, lifetime and cyclability. Several materials have high theoretical capacity i.e Na, Al, Li are desirable for next generation fast charging batteries. However there are some drawbacks such as low columbic efficiency, dendrite growth and volume expansion restricting their practical applications. We organized many scheme and strategies for the engineering of these metal electrodes and electrolytes. The effect of many composite structure were found to improve the stability and cyclability. The carbon derivatives specially graphene, graphene oxide, carbon nanotube's, carbon fiber, graphite and their composites were studies and found to be the most suitable candidates for making the anode composite for better electrochemical performance of fast charging batteries.

In last few decades, nanomaterials have gained enormous attention in the scientific industry due to their tunable physico-chemical and biological properties with enhanced performance over their bulk counterparts. In particular, nanoparticles have been extensively investigated for their usefulness in X-ray and gamma-ray shielding applications. Various elements and compounds, with high atomic numbers and effective atomic numbers respectively, have the potential to form nanoparticles that offer remarkable enhancement in the shielding performance. Composites, obtained by doping different nanoparticles into structural matrices (concrete, glass, or polymers), not only possess striking thermo-mechanical properties but also are effective shielding materials to replace conventional lead shields. This review is an attempt to throw light on various aspects of nanoparticles and their influence on shielding effectiveness. The authors also summarize the experimental findings so as to highlight the potential underlying the radiation-matter interaction mechanism in nanostructured systems.

This chapter provides an exhaustive review of piezoelectric smart materials and their applications for commercialization with the addition of recent research and breakthroughs. Piezoelectric smart materials convert electrical signals into mechanical stresses and vice versa. They are remarkably used in energy harvesting from different processes and diverse human activities. Piezoelectric materials are convenient, readily available, adaptable, and have a speedy recurrence reaction time. Smart piezoelectric actuators embedded into adaptive structures are significant. Recent progress in their applications is evident in the versatile smart wings present in an aircraft, since they correlate with the overall effectiveness. In addition, a wide range of industrial applications, for example, diesel fuel injectors, quick responding solenoids, optical lenses, mechanical technology, vibrational damping, etc., utilize piezoelectric actuators. Further, this chapter aims to address the latest improvements in piezoelectric materials in connection to the realm of nanotechnology.

Structural health monitoring (SHM) is widely used to examine the structural integrity of a structure to improve safety considerations, minimize maintenance cost, and avoid sudden breakdowns that might occur under various loadings under service conditions. Fiber reinforced polymer (FRP) composites are widely used in aerospace, automotive, marine, energy, infrastructure, armor, and biomedical applications due to their high specific stiffness and strength, high degree of dimensional and thermal stability, and good resistance to corrosion. Thus, SHM of the damages developed in FRPs during service conditions is very crucial. However, traditionally used sensors for SHM in composites provide limited sensitivity and might negatively affect the structural integrity. Therefore, a self-sensing approach by evaluating the electrical resistance change (ERC) of FRPs has been successfully used by many researchers. The purpose of this chapter is to describe the ERC techniques using various nanoparticles such as graphene, carbon nanotube, expanded graphite, and carbon black.

Preface

Every industry, including engineering, medical, food safety, transportation, energy, and environmental science, is being revolutionised by new age nanoparticles and being made better. It is necessary to review the developments and difficulties in the field to ensure that companies fully capitalise on the prospects offered by new age nanoparticles.

Keeping this in mind, the book *Diversity and Applications of New Age Nanoparticles* discusses the creation of novel materials, synthesis techniques, and cutting-edge research in this area while taking into account recent advancements and applications of nanoparticles. This authoritative reference work is perfect for business professionals, computer scientists, policymakers, engineers, pharmacists, medical professionals, researchers, scholars, practitioners, instructors, and students. It covers important topics like antibiotics, thin films, battery technologies, and composites.

The book presents voice of editors and author from across the world. Everyone has attempted to present cutting edge latest research in the file of nanoparticles. There are thirteen chapter in the book catering various filed of nanoparticles such as eco-friendly and sustainable development, intelligent sensing using nanoparticles, nano-sized splines ferrites and its applications, bioactivity investigation of silver nano-particles, photo catalytic application, futuristic scope of Yttrium iron garnet, synthesis of nanoparticles from bio-waste, CdTe thin film development, carbon nanoparticles for energy applications, radiation shielding applications, commercialisation of piezoelectric nanoparticles, new age nanoparticles for health monitoring and development of quantum dots from waste.

Finally, all editors are grateful to team of IGI-Global team. The editors are grateful to all authors, across the world, for their outstanding contributions not only to this book, but also to the Development & Advancement Of Nanoparticles.

ORGANIZATION OF THE BOOK

Chapter 1: Application of Nanotechnology for Eco-Friendly and Sustainable Economic Development

Nanotechnology is the study of sub-micron particles (nm). Nanotechnology is today's most advanced technology. Nanotechnology has a favourable impact on sustainable development and global warming/climate change. Recycling garbage and removing pollutants from soil and water are two further uses. Bio-based nanoparticles have replaced synthetic chemicals and hazardous metals in industry. The increasing digitalization of the economy is driving demand for ecologically sustainable 0D, 1D, and 2D nanomaterial microchips. Nanoparticles have a high surface-to-volume ratio and can communicate different properties. Nanotechnology can help achieve Zero Hunger, Good Health and Well-Being, Clean Water and Sanitation, Affordable and Clean Energy, Responsible Consumption and Production, and Climate Action.

Chapter 2: Intelligent Nanoparticles for Antibiotics Sensing – Nanoparticles for Sensing

Nanosensors are acquiring expanding consideration because of the need to identify and quantify synthetic and actual properties in hard to arrive at natural and modern frameworks that are in the nanoscale locale. The overviews of different sensors based on nanomaterials, arranged into three expansive categories: mechanical, electromagnetic, and optical nanosensors. Nanotechnology has led to the synthesis of nanostructures having unique properties like morphology, conductance, and optical properties. Nanostructures provide increased surface for reactivity and transduction signal due to increase in surface area to volume ratio. Herein we are discussing different nanomaterials; composites which is applied for exceedingly selective detection of analytes. The detecting ideas and their comparing benefits are talked about concerning their applications.

Chapter 3: Nano-Structured Spinel Ferrites and Their Applications as Antimicrobial Agents, Gas Sensors, and Dye Adsorbents

Ferrites–the electro-ceramic materials have been considered highly important magnetic materials for more than half a century. The crystal structure of ferrimagnetic oxides falls into one of the following categories: spinel, magnetoplumbite (hexagonal), garnet, and orthoferrite. The development of spinel ferrite nanoparticles (SFNPs) has become more popular owing to their fascinating properties, such as their potential adsorption properties, superparamagnetic behaviour, high stability, high value of resistivity, saturation magnetization, coercivity low power losses, and ease of functionalization. These advantageous features make them suitable for use in various applications. This chapter aims to discuss the structure, general properties, and some recent applications of spinel ferrites. Commonly used synthesis techniques for nanocrystalline ferrites are discussed. The applications of nano-structured ferrites like antimicrobial activity, gas sensors, and dye degradation are discussed.

Chapter 4: Plant Extract-Based Silver Nanoparticles and Their Bioactivities Investigations

Green synthesised nanoparticles are a new source of inspiration for clinical research, as this method uses natural materials such plant-derived materials for synthesis. Green nanoparticle synthesis using plant extracts is simple, inexpensive, chemicals, resulting in biologically active shape- and size-dependent products with no contaminants or by-products. Silver nanoparticles (AgNPs) have unique physiochemical properties and promising biological applications. To assess AgNPs' functionality, they must be characterized. Natural compounds found in plant extracts reduce and stabilise AgNPs. Green synthesis is more environmentally friendly than conventional methods, but many questions remain. To understand the plant-mediated process, many studies are needed. This proposed book chapter examines the synthesis, characterization, and bio activities of green-synthesised AgNPs. The goal is to provide an overview of green AgNP synthesis and their bioactivity assessment.

Chapter 5: Smart Nanomaterials for Photo-Catalytic Applications

Smart nanomaterials are grabbing more significant attention. Photocatalysts are catalysts that can be stimuli by light with an appropriate wavelength. Photocatalysis is deemed to be a promising approach for the utilization of photo energy and has been extensively studied for many processes. Therefore, the synthesis of nanomaterials with tailored-made catalytic characteristics, are of great importance. For applications using visible light and hence solar energy, the modification of photocatalysts can occur via the formation of heterojunction nanocomposites by inhibiting the recombination of charge carriers. These applications are driven by solar energy, which is conducive to the sustainable development of energy resources with no impact on the environment. This chapter will include recent enhancements in the smart nanomaterials for photocatalytic applications, especially in the fields of removing environmental pollutants, self-cleaning surfaces, water splitting for hydrogen production to provide clean fuel resources, selective alcohol oxidation and CO2 reduction in the environment.

Chapter 6: Yttrium Iron Garnet (YIG) – A Nanomaterial for Tomorrow

A cluster of silicate minerals known as garnet is widely used as abrasives and gemstones, has now find an application as smart nonmaterial. The garnet possesses a unique feature of having different chemical composition with similar crystal form and physical properties. Yttrium iron garnet (YIG) also popularly known as iron yttrium oxide, is an example of lab grown garnet. It can have chemical composition of Y3Fe5O1 or Y3Fe2(FeO4). The YIG exhibits a ferrimagnetic nature. with a Curie temperature of 560 K. The chapter focuses on the research around YIG with a mention on fabrication methods and applications.

Chapter 7: Synthesis Nanoparticles of SiO2 From Rice Husk and Its Industrial Application

Rice, which provides a major source of food for billions of people, covers 1% of the surface of the earth. Rice husk (RH), the outer covering of paddy rice, is an important by-product during the milling process and residue ash is generated after the burning of RH and the ash is called rice husk ash (RHA), which is a primary waste material in the agricultural industry. The main components in RH are lignin, cellulose, and hemicellulose, which are generally named lignocellulose. RH also contains ca. 15 to 28 wt% of silica. The high content of silica in RH presents opportunities for the preparation of value-added silicon-based materials. Since the 1970s, various silicon-based materials, including silica, silicon carbide, silicon nitride, silicon tetrachloride, zeolite, and silicates, have been successfully synthesized using RH as the silicon source. This field of research has been significantly advanced and expanded in the past decade spurred by the global attention on sustainable and renewable resources.

Chapter 8: Development and Characterization of As-Deposited CdTe Thin Films in a Non-Aqueous Medium

Cadmium telluride (CdTe) is a binary II-VI direct band gap semiconducting material. Cadmium telluride is a promising electrochemical and photovoltaic material for thin film solar cells. It shows both p and n type conductivity. A nanocrystalline thin film of CdTe was deposited on nickel plate by chemical bath deposition method containing 0.01M Cadmium acetate and 0.02M tellurium dioxide. The Structural, Compositional and Optical analysis were studied by X-ray diffraction (XRD), Energy-Dispersive X-ray Spectroscopy (EDS), Scanning Electron Microscope (SEM), Fourier Transform Infrared Spectroscopy (FTIR) and Photoluminescence (PL).The diffraction peak observed at 2=23.59° with (111) plane indicate the crystalline phase of CdTe film. The average crystalline size is measured to be 7nm.Compositional analysis reveals the presence of both Cd and Te elements. The EDS Spectroscopy shows the ratio of Cd and Te is1:2 in case of as deposited film. The Photoluminescence peak is observed at 471nm.

Chapter 9: Effect of Carbon-Based Composites in Fast-Charging Battery Technologies

The need for electric vehicles and other portable devices globally, advancements in charging technologies, are inflection points in the evolution of upcoming batteries. The proper engineering of electrodes is essential for stability, lifetime and cyclability. Several materials have high theoretical capacity, i.e., Na, Al, Li, are desirable for next generation fast charging batteries. However, there are some drawbacks such as low columbic efficiency, dendrite growth and volume expansion restricting their practical applications. We organized many schemes and strategies for the engineering of these metal electrodes and electrolytes. The effect of many composite structures was found to improve the stability and cyclability. The carbon derivatives specially graphene, graphene oxide, carbon nanotube's, carbon fibre, graphite and their composites were studies and found to be the most suitable candidates for making the anode composite for better electrochemical performance of fast charging batteries.

Chapter 10: Nanoparticles as Fillers in Composites for X-Ray and Gamma-Ray Shielding – A Review

In last few decades, nanomaterials have gained enormous attention in scientific industry due to their tunable physico-chemical and biological properties with enhanced performance over their bulk counterparts. In particular, nanoparticles, have been extensively investigated for their usefulness in X-ray and gamma-ray shielding applications. Various elements and compounds, with high atomic number and effective atomic number respectively, have the potential to form nanoparticles that offer remarkable enhancement in the shielding performance. Composites, obtained by doping different nanoparticles into structural matrices (concrete, glass or polymers), not only possess striking thermo-mechanical properties but also are effective shielding materials to replace conventional lead shields. This review is an attempt to throw light on various aspects of nanoparticles and their influence on shielding effectiveness. We also summarize the experimental findings so as to highlight the potential underlying the radiation-matter interaction mechanism in nanostructured systems.

Chapter 11: Piezoelectric Smart Materials and Commercialization

This chapter provides an exhaustive review of piezoelectric smart materials and their applications for commercialization with the addition of recent research and breakthroughs. Piezoelectric smart materials convert electrical signals into mechanical stresses and vice versa. They are remarkably used in energy harvesting from different processes and diverse human activities. Piezoelectric materials are convenient, readily available, adaptable, and have a speedy recurrence reaction time. Smart piezoelectric actuators embedded into adaptive structures are significant. Recent progress in their applications is evident in the versatile smart wings present in an aircraft since they correlate with the overall effectiveness. In addition, a wide range of industrial applications, for example, diesel fuel injectors, quick responding solenoids, optical lenses, mechanical technology, vibrational damping, etc., utilize piezoelectric actuators. Further, this chapter aims to address the latest improvements in piezoelectric materials in connection to the realm of nanotechnology.

Chapter 12: Structural Health Monitoring by Electrical Resistance Change Method – Using Nanoparticles

Structural health monitoring (SHM) is widely used to examine the structural integrity of a structure to improve safety considerations, minimize maintenance cost, and avoid sudden breakdowns that might occur under various loadings under service conditions. Fiber reinforced polymer (FRP) composites are widely used in aerospace, automotive, marine, energy, infrastructure, Armor, and biomedical applications due to their high specific stiffness and strength, high degree of dimensional and thermal stability, and good resistance to corrosion. Thus, SHM of the damages developed in FRPs during service conditions is very crucial. However, traditionally used sensors for SHM in composites provide limited sensitivity and might negatively affect the structural integrity. Therefore, a self-sensing approach by evaluating the electrical resistance change (ERC) of FRPs have been successfully used by many researchers. The purpose of this chapter is to describe the ERC techniques using various nanoparticles such as graphene, carbon nanotube, expanded graphite, and carbon black.

Chapter 13: Synthesis and Characterization of Carbon-Based Quantum Dot From Rice, Sugar, and Aloe-Vera

Carbon quantum dots (CQDs), a novel class of zero-dimensional carbon nanostructures with diameters below 10 nm, have recently come into existence. Here, we have synthesized and analysed carbon-based quantum dots from aloe-vera, rice and sugar. The characterization of these quantum dots is done via UV-Visible spectrophotometer. The quantum dots which are made from chemicals are toxic and are harmful to both humans and animals. However, the carbon quantum dots (CQD) which we have synthesized in this research work are bio- compatible, less toxicity and has the property of high aqueous dispersibility. CQDs are being used in various fields of applications such as bio-imaging, medical diagnosis, bio-sensing, chemical sensing, photo catalysis, and photovoltaic devices etc. This study helps the scientific community to understand the nature of carbon quantum dots and gives a brief idea of suitable materials that create CQD.

Suneev Anil Bansal
Department of Mechanical Engineering, Bharat Institute of Engineering and Technology, India

Virat Khanna
Department of Mechanical Engineering, MAIT, Maharaja Agrasen University, India

Nilanthy Balakrishnan
School of Chemical and Physical Sciences, Keele University, UK

Pallav Gupta
Department of Mechanical Engineering, A.S.E.T., Amity University, India

Acknowledgment

The editors would like to acknowledge the help of all the people involved in this project and, more particularly, to all the authors and the reviewers that took part in the review process of this book. Without their support, this book would not have become a reality. Our sincere thanks to all the authors of chapter's who contributed their time and expertise to this book. All the Editors are extremely thankful to almighty God.

Dr. Virat Khanna is greatly indebted to his wife Anisha and his daughter Avishka who helped to keep him sane during the whole journey by providing a motivating and supporting environment at home.

Dr. Pallav Gupta is also thankful to his beloved mother (Beena Gupta), caring wife (Dr. Ritu Agrahari) and beloved son (Saahas Gupta) for the support provided during the course of Editing this book.

Chapter 1
Application of Nanotechnology for Eco-Friendly and Sustainable Economic Development

Sumanta Bhattacharya
https://orcid.org/0000-0003-2563-2787
NALSAR University of Law, India & Maulana Abul Kalam Azad University of Technology, India

Arkadyuti Seth
Indira Gandhi National Open University, India

ABSTRACT

Nanotechnology is the study of sub-micron particles (nm). Nanotechnology is today's most advanced technology. Nanotechnology has a favorable impact on sustainable development and global warming/ climate change. Recycling garbage and removing pollutants from soil and water are two further uses. Bio-based nanoparticles have replaced synthetic chemicals and hazardous metals in industry. The increasing digitalization of the economy is driving demand for ecologically sustainable 0D, 1D, and 2D nanomaterial microchips. Nanoparticles have a high surface-to-volume ratio and can communicate different properties. Nanotechnology can help achieve Zero hunger, good health and well-being, clean water and sanitation, affordable and clean energy, responsible consumption and production, and climate action.

INTRODUCTION

Since the beginning of agricultural practices, humans have been utilizing the natural resources. During the industrial revolution, the propensity of this unchecked affliction to Mother Nature reaches its height. The first industrial revolution introduced mechanization, which was quickly advanced by mass manufacturing in the second industrial revolution and digitalization in the third. However, the so-called modernization of the economy via rapid industrialization has not yet addressed the fundamental issues

DOI: 10.4018/978-1-6684-7358-0.ch001

facing human society, such as ensuring access to food, reducing poverty, providing high-quality health-care, and eliminating inequality. Furthermore, the intensive use of natural resources has negative effects on biodiversity and environmental sustainability. A sizeable portion of the global population is currently afflicted with serious health problems and complex genetic disorders. The scarcity of wholesome foods is threatened by a sharp increase in agricultural output brought on by a rising tendency for climate change, specifically the occurrence of either excessive rainfall or extreme drought. Given that a larger percentage of the population is still dependent on agriculture, particularly in emerging countries, recession is thus becoming a common phenomenon in the agricultural sector and is leading to enormous economic loss. The industry must play a significant role in the development of eco-friendly technology to address these critical environmental and social concerns. 4.0. The United Nations' member nations adopted 17 sustainable development goals (SDG) in 2015, and they are expected to be completed by 2030.

One of the most revolutionary discoveries of the 20th century was nanotechnology, which reveals the power of minute particles found in matter. Nanoparticles come in a variety of shapes and hues, each having a unique appearance. For instance, at a wavelength of 20 nanometers, the colours of gold (Au), platinum (Pt), silver (Ag), and palladium (Pd) are red, yellowish grey, and black, respectively. The substance exhibits elevated physical and chemical properties at the nanoscale, or between 1 and 100 nm, which leads to cutting-edge changes in all economic sectors. Agriculture, water treatment, medical science, the cosmetics industry, the automotive, aerospace, telecommunication, defence, waste management, biotechnology, the food processing industry, genetics, information technology, and other areas of human civilization are just a few of the many fields where nanotechnology is revolutionizing society. As a result of the internal potential of nanoparticles, natural chemicals are more effective at overcoming current environmental issues, resulting in the eco-friendly and sustainable economic development of society. Cosmetics, customized drug delivery systems, the treatment of chronic diseases like cancer, and faster and more effective faster and more efficient faster If the microprocessors in batteries use less power, the batteries can keep more charge and operate for longer. The use of nanotechnology has a substantial positive impact on both achieving the goals of sustainable development and addressing the challenges posed by global warming and climate change. Other uses of this technique include recycling waste and removing contaminants from soil and water. Due to the increasing frequency of harsh weather patterns and the rising global food need, indoor farming systems are swiftly gaining popularity as an alternative agriculture technique. Nanomaterials play a vital role in the manufacture of an effective nutrition solution, which is crucial for soilless agricultural operations. Bio-based nanoparticles have become an environmentally friendly and sustainable substitute for synthetic chemicals and dangerous metals in the industrial sector. The demand for environmentally friendly microchips built on 0D, 1D, and 2D nanomaterials is rapidly increasing as a result of the increasing digitalization of different aspects of the economy. This is particularly true in the time period after the COVID standard. Additionally, nanoparticles are crucial in the creation of bioplastics, an environmentally friendly replacement for conventional plastics. Nanotechnology is the science of creating nanoparticles of different chemical compositions, shapes, and sizes and employing them to enhance human wellness. Because of their greater surface-to-volume ratio and ability to communicate different features, nanoparticles can be utilized in specialized applications. Different cutting-edge nanotechnology methods point us in the direction of achieving sustainable development objectives.

Nanotechnology and Sustainable Economic Growth

Quantitative and qualitative expansion of the economy are both components of what is meant by "sustainable economic development." Human development factors such as food security, equality, poverty alleviation, quality health services, etc. have, since the last decade of the 20th century A.D., become an integral component of the global economy alongside growth factors like gross domestic product, per capita income, etc. The deterioration of environmental factors has a direct impact on the economy's quantitative and qualitative factors, so they too are linked to economic growth. Many of the Sustainable Development Goals (SDGs) are directly affected by the various breakthroughs in nanotechnology. These include, but are not limited to, achieving zero hunger, good health and well-being, clean water and sanitation, affordable and clean energy, responsible consumption and production, climate action, and many more.

Zero Hunger

Due to worsening quality and quantity in the agricultural sector, achieving food security at zero hunger levels has become an important challenge in the modern era. Biotic and abiotic stress caused by global warming and climate change are having a cumulative negative effect on agricultural output, which is on a downward trend. These days, the world has to deal with the negative effects of both overeating and undereating on agricultural output, both in terms of quantity and quality. Moreover, soil fertility is diminished by the widespread use of chemical-based fertilizers. From soil preparation to post-harvest packaging, nanotechnology increases the sustainability and productivity of agriculture.

Potential nonabsorbents, nanometals and polymer-encapsulated nanometals, can remove metallic toxicants from the soil by forming a stable chelating complex with them. Through photolytic degradation, the oxygen molecules are transformed into Reactive Oxygen species (ROS) when semiconductor nanoparticles are applied to seeds, speeding up the germination process and allowing for more rapid plant growth. The ability of nanoparticles to transport genetic materials into plant cells has the potential to boost plant growth and make plants more resistant to biotic and abiotic stress. Micro and macronutrients in the soil could be transported by nanoparticles. (Kaliannan et al, 2019)

Incorporating nanomaterials into the production of fertilizers has been shown to increase the effectiveness of said fertilizers, which in turn has improved soil health and increased crop yield. Soil fertility can be increased and toxicants removed from the soil with the help of organic fertilizers enhanced with nanotechnology. These materials have the potential to replace chemical fertilizers in agriculture while reducing environmental impact and improving crop nutrition. The remediation of soil toxicants, induction of reactive oxygen species accumulation, stimulation of growth-promoting phytohormone activity, chlorophyll synthesis, adsorption of toxic contaminants, etc. are all ways in which nano fertilizers boost agricultural output. Doping biochars with nanometals improves soil fertility by increasing the soil's capacity to sequester carbon. (Nanogaki, 2006)

The photocatalytic degradation of oxygen molecules and the reduction of harmful contaminants to innocuous compounds all contribute to the accumulation of reactive oxygen species (ROS) in nanoparticles of aluminum, zinc, copper, nickel, titanium, silver, and gold, as well as their oxides. (Ghidan and Al Antary, 2019) Photolysis of oxygen molecules and large inorganic contaminants are both aided by nanometal doping of organic fertilizers. Enclosing nanoparticles in biopolymers like chitin, lignin, etc., improves their ability to adsorb toxic metals, leading to more efficient remediation of toxicants in the medium. Polymer-encased nanometals have been shown to be effective biosensors for the detection of trace amounts of toxic heavy metals in soil, such as mercury, cadmium, palladium, chromium, etc. Examples of effective sensors for mercury, arsenic, and lead detection include gold nanometal coated with chitosan and silver nanometal coated with lignin. (Malini et al, 2022)

Plants' ability to absorb water and nutrients is enhanced when polymeric nanomaterials are incorporated into their root systems. It also causes the plant to produce leaves with a larger surface area, which improves its photosynthetic efficiency. An increase in the rate of chlorophyll synthesis is a direct result of the nanoparticle-induced activity of the carbonic anhydrase enzyme in plants. The polymeric shell-like structure has strong adhesive properties and can prevent nutrient loss by forming a barrier around plant roots. Starch molecules in plant cell walls can have their polysaccharide bonds broken by the nanoparticles, allowing the plant to develop more quickly. (Prasad et al, 2017).

Figure 1. Application of nanotechnology towards achievement of "Zero hunger"

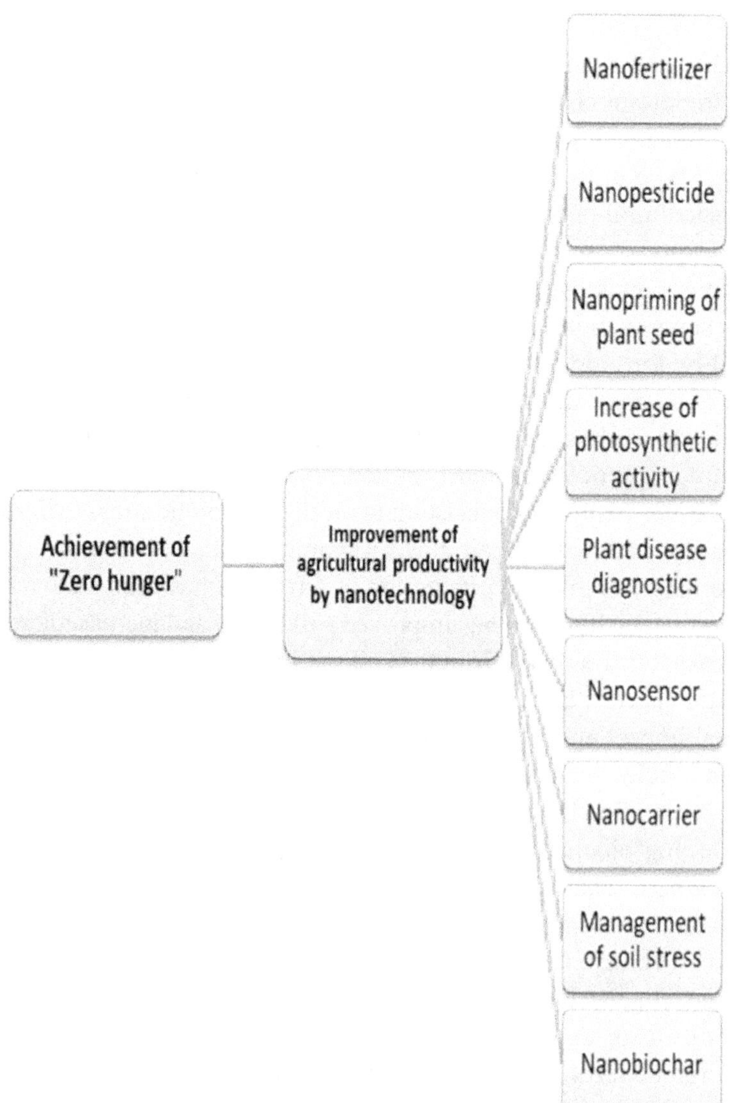

Good Health and Well-Being

Nanoparticles are commonly utilised as therapeutic molecule carriers due to their increased chemical activity and potential to overcome tissue barriers. This is because nanoparticles can deliver therapeutic molecules more effectively. Specifically tailored nanomaterials are made in order to transfer medicine molecules, proteins, plasmids, antibodies, oligonucleotides, fluorophores, ligands, polymers, radio-isotopes, tissue engineered products, and so on to the human body and have an effect on target cells. Nanoparticles are able to recognise a particular cell of interest as another potential target. Cantilever (a biosensor-based nanomechanical system), nanoshells (tiny spherical shells), dendrimers (nanopolymeric dendritic structures that carried drug molecules), liposomes (phospholipid nanostructures that carried drug molecules), quantum dots (tiny light-emitting nanocrystals that readily enter into small human cells and also facilitate magnetic resonance imaging of the body while emitting light after being activated), and dendrimers are all examples. They are also able to easily differentiate the target cells from other cells because of the presence of distinctive membrane characteristics. These nanoparticles are commonly utilised as chemotherapeutic agents because of their efficient target selection qualities and controlled drug release capabilities. These qualities and capabilities help prevent the destruction of healthy cells while chemotherapy is being administered. (Idrees, 2015; Anjum et al., 2021)

Figure 2. Nanotechnology and health

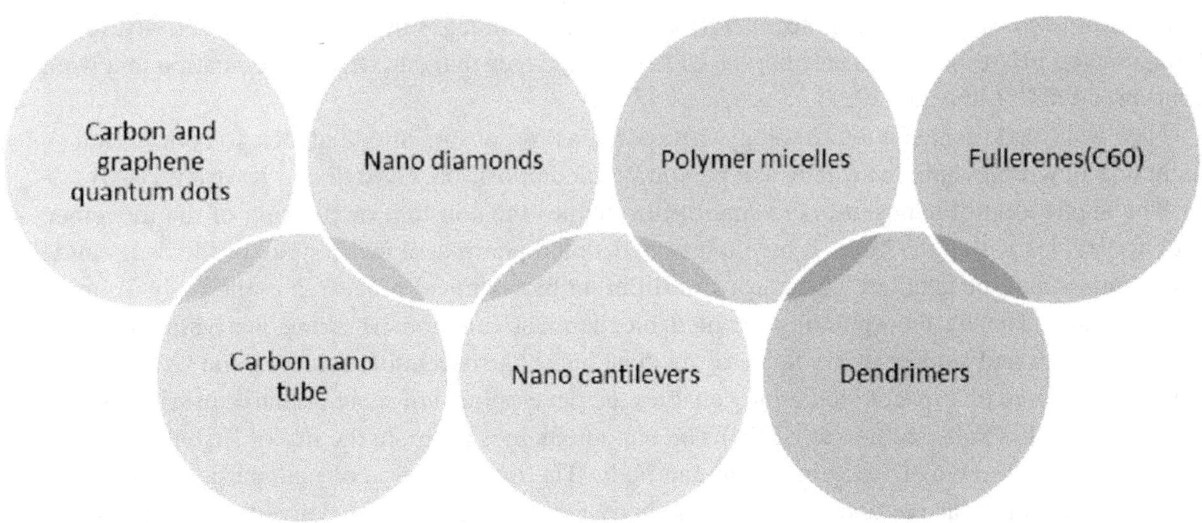

As the second leading cause of mortality in the modern period, cancer is a complicated disease with a multitude of side effects, including toxicity, complexity, low therapeutic selectivity, and the potential for cancer resistance that can be treated using nanomaterials. Numerous nanoparticle types, including carbon quantum dots, carbon nanotubes, and nanodiamonds, have special features that can be exploited to target particular malignancies. Several nanopills are currently undergoing clinical testing, but more could be created and employed in the next wave of cancer therapies. (Jin et al., 2021)

In order to create smart pills, nano-based electronic devices that perform advanced imaging, sensing, and drug delivery are being added to pharmaceutical pills. It enables medical professionals to monitor the drug delivery process within the human body and research the mechanism underlying pharmacological action. An enhanced smart pill called an "Atmo Gas Capsule" monitors the activity of gaseous chemicals in the human body. Advanced nanotechnology is employed to create "Smart Sensor Capsules," which are given with vaccines to track their effectiveness in the body. Additionally, it can be taken orally rather than intravenously. This is a ground-breaking development in the realm of medical science study. In order to reduce the risk of infection, nanopatch vaccines use nanoparticles to deposit the vaccines on immune cells found in the skin. The necessity for vaccine refrigeration is removed, making the vaccination process simpler. (Talegaonkar et al., 2015)

Nanoflares are specially designed nanomaterials used in the treatment of cancer. The presence of cancer cells in the blood stream is efficiently detected. When the target genetic sequence is discovered, it can bind with the genetic target in cancer cells and produce light. (Halo et al., 2014)

When implanted into a human body, nanobots, the cutting-edge robots created with nanotechnology and robotics, perform surgery in miniature. It efficiently restores intracellular structures, which heal a specific ailment within the human body. By altering the target genome sequence, DNA-based nanobots are also manufactured for the eradication of genetic disorders. Nanomaterials can effectively transport genetic material throughout the body. (Bhat, 2014)

Nanofibers, which are appropriate polymers with a high surface area to volume ratio, increased loading efficiency, and enhanced penetration activity, are used to create "Smart Bandages" for use in surgery and adamantine wound dressing. Because nano molecules are more reactive, they improve the healing effectiveness of the bandages. In addition to clotting agents, antibiotics, and sensors (for the detection of infection signals), these implanted nanofibers in smart bandages increase healing activity. Recently, India created pH-responsive nanofiber-based smart bandages that can deliver medication at a wound's appropriate pH. (Liu et al., 2021)

New gear made using nanotechnology incorporates tiny sensors into the fabric to collect vital signs, including blood pressure, heart rate, and perspiration composition. (Avdyli and Kastrati, 2020)

The application of nanosensors to smart pills enables the constant monitoring of the movement of drug molecules within the body. It facilitates the detailed pharmacokinetic study of the drugs and also helps understand the detailed mechanisms of different biochemical processes in the body. It improves bioinformatic data by incorporating detailed biochemical functions, making it a valuable asset for future research and studies in the fields of medical and pharmaceutical sciences and technology. The nanosensor-based pharmacokinetic study enables the development of more effective medicines through further research. (Yellepeddi et al., 2019) The nanomedicines promote the use of herbal medicines by increasing their in vivo efficacy in the human body. The incorporation of nanoparticles increases the bioavailability of bioactive components in the body and also ensures the delivery of products to their target. The development of nanophytomedicine is an emerging field of medical science that deals with the development of medicines from phytochemicals and nanomaterials. (Hesari et al., 2021) Moreover, the nanocarriers also enable the effective delivery of specially engineered genetic materials into the body for the effective treatment of genetic diseases.

Figure 3. Application of nanotechnology in good health and well being

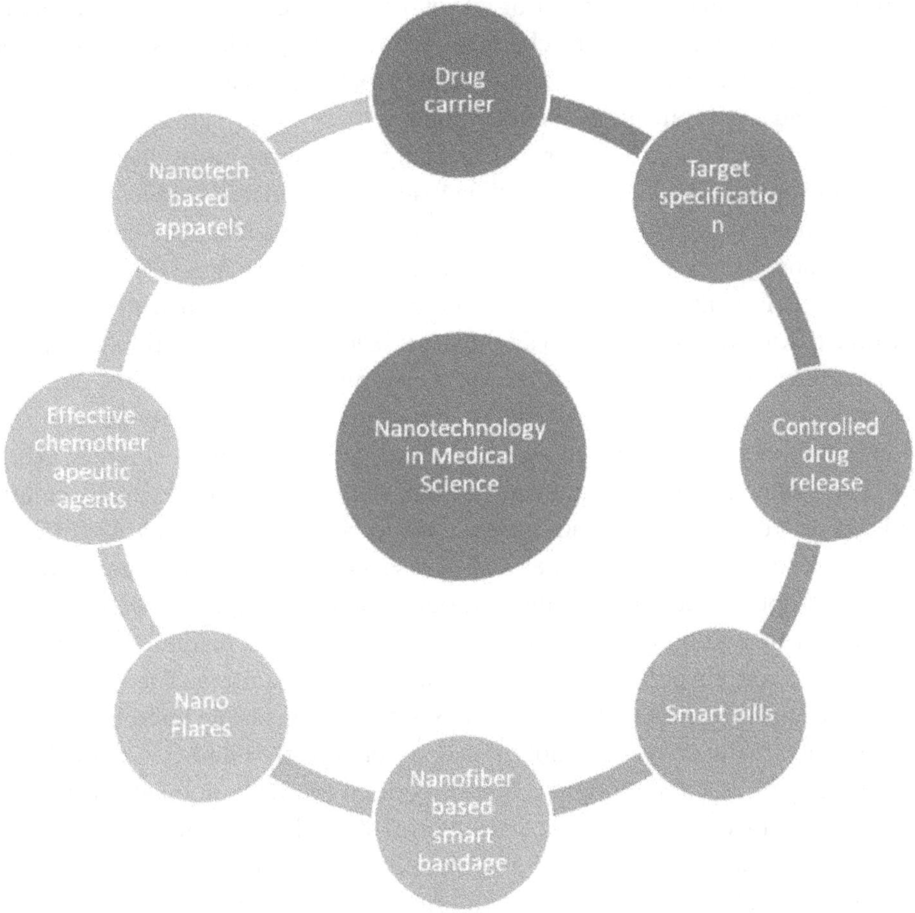

Clean Water: For both humans and animals to continue to exist, water is a material that is essential. Contrarily, industrially handled water frequently contains a variety of contaminants, such as fluoride, dyes, heavy metals, pharmaceutically active substances, pesticides, and others. Metal oxide and zero-valent metal nanoparticles are two examples of nanoporous materials that have recently gained attention as potentially useful new platforms for the elimination of pollutants. The numerous nanoporous materials that are being used to remove emerging pollutants from the environment will be the main topic of this special conference call. The secret to success in achieving all of the aforementioned objectives is the use of nanotechnology. Researchers from a range of scientific and technological fields have been allowed to explore previously uncharted area thanks to nanoporous materials. Due to their exceptional surface-area-to-volume ratio, distinctive electrical and chemical properties, quantum size effects, high scalability, and tunable surface functionalities, nanoporous materials have recently attracted enormous interest as an emerging platform for environmental remediation. A number of treatment technologies, such as nanoporous materials, have sparked this interest. Due to their impermeability through nanopores, membranes created from nanoporous materials, specifically nanofibers, are crucial in the remediation of micro-sized pollutants.

Due to their high scalability and prevalence of the relevant qualities in this field, nanoparticles are frequently used for the sustainable treatment of water. The enhanced physical and chemical properties of nanomaterials, such as their high surface charge to volume ratio, optical susceptibility, photocatalytic ability, antibacterial potential, magnetic properties, etc., make it easier to remove contaminants from water in an efficient and environmentally friendly manner.

Because nanoscale materials have higher surface charge to volume ratios than macroscale materials, they accumulate more charges on their surfaces than macroscale materials, which helps water pollutants adhere to them. The internal composition, external structure, surface area, and quantity of sorption sites that are present in nanomaterials all affect their ability to absorb substances. Different kinds of metallic oxide nanoparticles, such as titanium oxide and zinc oxide, metallic nanoparticles (gold, silver, etc.), nanostructured mixed oxides (indium(III)-tin(IV) binary mixed oxides), magnetic nanoparticles (hematite, magnetite, etc.), carbonaceous nanomaterials (carbon nanotube, carbon nanofiber, etc.), and polymeric nanomaterials (PLGA, PLA, PVA nanoparticles; dendrimers Mercaptobutyric acid, L-glutathione, -thio—(propionic acid), hepta (ethylene glycol), and meso-2,3-dimarcaptosuccinic acid are ligands that can be added to nanomaterials to speed up the process of heavy metal adsorption by creating stable co-ordinate complexes between the metals and ligands. Due to favourable hydrophobic effect, -interaction, hydrogen bonding, electrostatic interaction, etc., the presence of functional groups in the nanostructure further boosts the adsorption potential of nanomaterials. (Gehrke et al, 2015; Maheswari, 2015)

Null Value The low ionization potential of the valence electrons is potentially reduced by nanometals. They transform the water's pollutants into less dangerous substances. The zero valent nanometals reduce the phosphate units in the microorganisms' DNA, which causes the microbes' cell structure to be destroyed. Additionally, many enzymes' thiol groups are effectively interacted with by silver nanoparticles, preventing the synthesis of important proteins required for the biological survival of microbes in aqueous medium. (Sukhnove et al, 2004; Sylvester et al, 2007; Theron et al, 2010)

The inclusion of nanometals in biological polymeric structures such as polyether sulfone, cellulose acetate, carboxymethyl acetate, poly(vinylidene fluoride), poly(ether sulfone), poly(amide-imide), etc. increases the reduction potential of the nanometals. Additionally, the biopolymers boost the nanocomposite's hydrophilicity, which makes it easier for bacteria and organic contaminants to bind to it. Due to the production of co-ordination complexes that are easily washable from the aqueous medium, the co-ordination action of aromatic moieties increases the adsorption potential of nanostructured polymers towards heavy metals. (Tiede, 2008; Tiraferri, 2011; Trivedi and Axe, 2000)

The band gap of nanoparticles comprised of semiconductors and metallic oxides allows for the photolytic destruction of pollutants. Due to its excellent photocatalytic activity, photostability, cost-effectiveness, chemical, and biological stability in aqueous media, titanium oxide has been shown to be a viable photocatalyst nanoparticle. In the presence of nanophotocatalysts, the photolysis of contaminants with high molecular weight, such as chlorinated organic compounds, polycyclic aromatic hydrocarbons, dyes, phenols, arsenic, cyanides, and heavy metals, produces less harmful products with lower molecular weights. In the presence of UV radiation, photoelectrons are excited, resulting in the formation of reactive oxygen species and hydroxyl ions, which damage microbial cells' structures and impair their ability to function. The band gap of nanometallic oxides can be lowered to enable photo excitation in the visible range by doping them with metals like silver, cadmium, celenium, and tin as well as non-metals like nitrogen, phosphorus, sulphur, carbon, and graphene oxide. (Vecitis et al, 2010; Vecitis et al, 2011; Vikesland and Wiggintron, 2010)

Micro contaminants cannot pass through the nanopores in the nanostructured membrane, which makes it easier to separate them from water. A fantastic opportunity to increase membrane permeability is provided by the inclusion of functional nanomaterials. Additionally, nanofiber membranes are created using electrospinning, which makes it possible to remove microparticles from the aqueous phase with a high rejection rate and minimal fouling. Hydrophilic metal oxide nanoparticles, anti-microbial nanoparticles, and photocatalytic nanoparticles are combined to create specially tailored nanocomposite membranes. An example of an adjuvant nanocomposite membrane is thin film membrane, which was created by combining nano-zeolites, carbon nanotubes, silver nanoparticles, and titanium oxide nanoparticles. (Westerhoff et al, 2006; Westerhoff et al, 2009)

Figure 4. Application of nanotechnology in purification of water

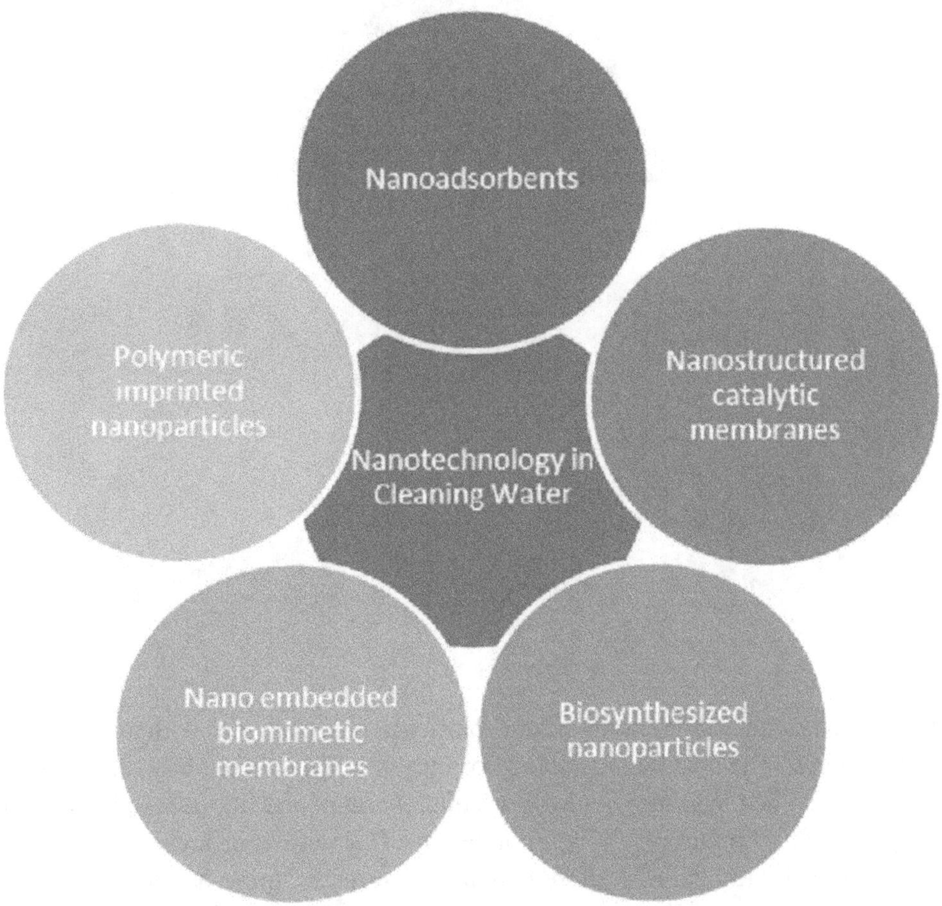

Affordable and Clean Energy: In order to reduce carbon footprints, the world is moving toward using renewable energy sources in place of conventional, non-renewable energy sources. Additionally, in order to address the concerns with the energy crisis, some countries, particularly developing ones, are adopting regulations that make it easier to produce energy at a lower cost. The current supply of fossil fuels is insufficient to meet the soaring energy demand. On the other hand, the key issues facing the currently available renewable energy sources are their lower rate of productivity, reduced ability to fulfil the current energy demand, and increased infrastructure costs. A sustainable solution to this issue is provided by nanotechnology.

Figure 5. Nanotechnology and energy sector

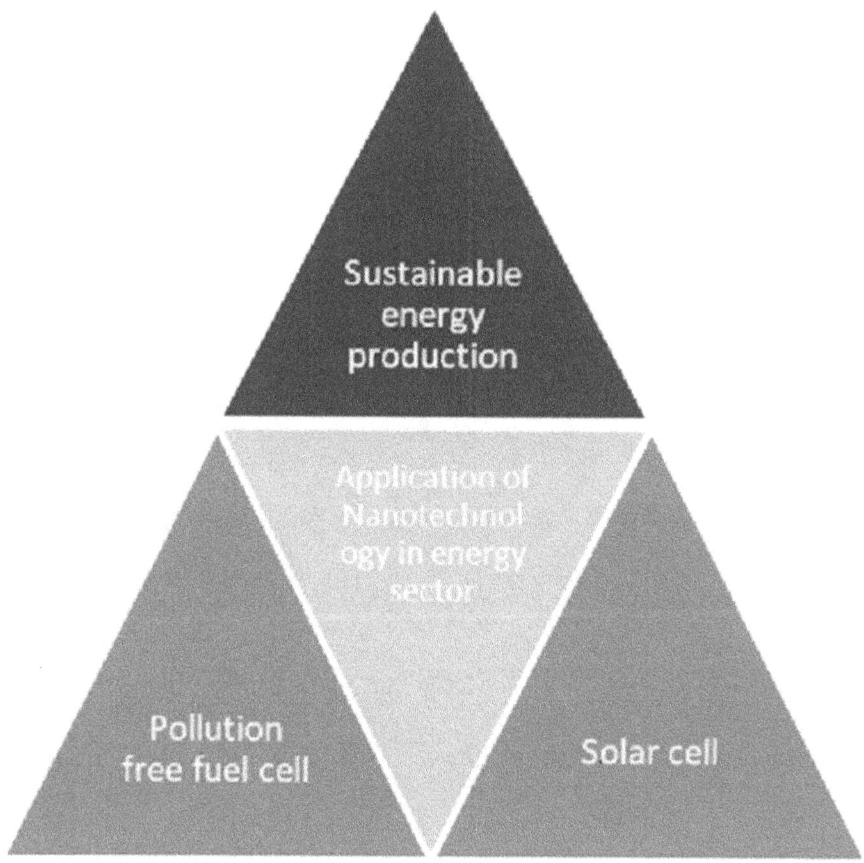

The most widely used renewable energy source, particularly in countries with tropical and subtropical climates, is solar energy. The cost and efficiency of currently available solar panels, however, cannot keep up with demand. When making photovoltaic cells, titanium and titanium oxide nanoparticles can be employed in place of traditional silicon semiconductors. Due to their increased surface area to volume ratio and exposure to solar radiation on more conducting surfaces, these nanomaterials are capable of absorbing more photon particles. The use of carbon nanotubes, fullerenes, and quantum dots produces lighter, less expensive, and more effective solar cells. When a photon strikes a nanomaterial like lead selenide, additional electrons are released into the atmosphere.

Additionally, windmill efficiency is increased by the use of nanotechnology. Making stronger and lighter windmill blades using epoxy containing carbon nanotubes increases electricity production. Nano paints can be used to extend the life of wind turbines.

Nanotechnology also improves the productivity of geothermal energy. In conventional geothermal energy production, cold fluids are applied to naturally hot rocks present at about 1500 metres beneath the earth's surface. The hot fluid is then collected and used to produce electricity. Nanotechnology is now making geothermal energy more effective by allowing the production of energy closer to the surface. The application of nanoparticles improves the fluid's heat-holding capacity as well.(Echiegu, 2016)

Nanomaterials accelerate the production of bioenergy from microalgae, lignocellulosic mass, and wastewater. The performance of biogas, biodiesel, bioethanol, and biohydrogen can be enhanced with the help of nanotechnology. Nanomaterials are an important and effective component for the manufacture of electrodes and membranes during the production of biofuel. The nano-conjugate enzymes also play important roles in the production of bioenergy through the use of both natural and synthetic substrates. (Khoo et al., 2020) Nanobiochar is a new approach towards the production of bioenergy in a sustainable manner from agricultural wastes. This process optimises the production of greenhouse gases due to the combustion of carbon-based products in the controlled presence of oxygen. (Ramanayaka et al., 2020)

Figure 6. Nanotechnology in affordable and clean energy

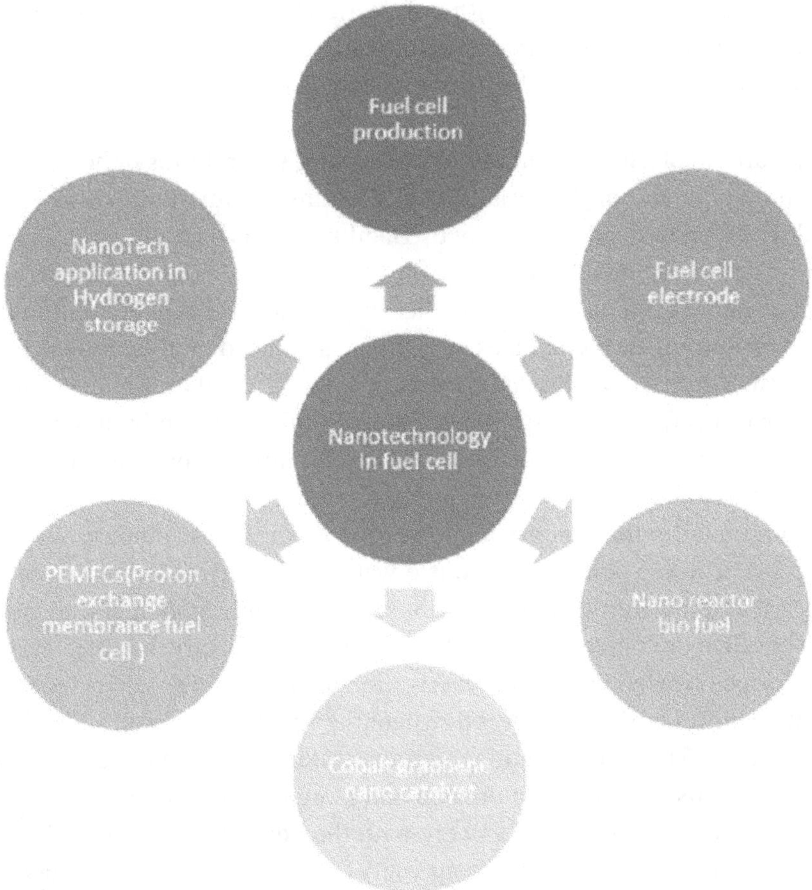

Nanotechnology in Pollution Control and Waste Management: Nanotechnology plays significant role in controlling the water and air pollution. Different specially engineered nanomaterials like carbon nanotube, dendrimers, nanoscale zeollites, bimetallic particles and metallic oxides are largely used for the remediation of contaminants from the environment. The zero valent nanoparticles like silver, zinc, copper, iron act as potential reducing agents that used to reduce the harmful contaminants to less harmful products. They remediate microorganisms from water. They can even reduce the phosphate unit present in DNA molecule and destroy the genetic material of toxic organisms.

Different nanoparticles and nanocomposits facilitate photocatalytic degradation of contaminants. Metal oxide nanoparticles viz. titanium oxide, zirconium oxide doped with nitrogen facilitates the photolysis of heavy metals present in waste water. Titanium oxide with Silver nanoparticles proved to be more effective for the removal of chromium from waste water. Zinc Oxide doped with palladium facilitates removal of E. coli bacteria from waste water. Apart from that, polychlorinated biphenyls, azo dyes, halogenated aliphatic compounds, organochlorine pesticides, halogenated herbicides and nitro aromatics are removed through the application of nanocatalysts. (Eskandarinezhad et al, 2021)

The nanostructured catalytic membranes facilitate physical separation of micro contaminants, degradation of organic pollutants, inefficiency of microorganisms, and prevention of biofouling. The filter synthesized by titanium oxide along with polymeric membranes viz. cellulose acetate, polyvinyl fluoride, polyethylene sulfonate, chitosan etc. exhibit more efficiency towards microbial inactivation and physical separation of micromolecules from aqueous solution due to the presence of nanopores. (Khalith et al, 2022)

The state-of-the-art nanotechnological approaches bring several biosynthesized nanoparticles into the forefront that show adjuvant anti-bacterial potential against the microbial contamination in the water. Silver nanoparticles synthesized extracellularly by Bacillus cereus bacteria, magnesium oxide nanoparticles, cellulose acetate fibers embedded with silver nanoparticles exhibit bioactivity against the gram positive and gram-negative bacteria, virus and other microbial pathogens present in the waste water.

The preparation of biomimetic membranes by bringing advanced modification in the protein channels of biological membranes with nano-fabrication techniques facilitates the desalinization as well as removal of other inorganic toxicants from the waste water. These biomimetic membranes have high permeability and chemical stability due to the presence of synthetic biomimetic analogues like aquaporins, carbon nanotubes etc. (Sharma et al, 2022)

The polymers imprinted with nanoparticles are one of the latest innovations of nanotechnology facilitating adsorption of micro pollutants of waste water especially generated from health sector. The nanoparticles here use as template molecules that linked with suitable monomers having functional groups by covalent, semi-covalent or non-covalent bonds which provide specific binding sites to the adsorbents. The polymers are formed by free radical polymerization techniques. (Zarejousheghani et al, 2021)

The contamination in groundwater reserves threatens the agriculture as well as drinking water resources in the drought prone areas. The contamination of groundwater with arsenic is a common phenomenon in different developing nations. The different organic and inorganic contaminants as well as heavy metals of the groundwater can be remediated effectively with the help of specially engineered nanoparticles and nanocomposites. The nano Zero Valent Irons (nZVI) are proved to have the potentiality to remove the contaminants from groundwater reserves with 80% efficiency. The typical core shell structure of nZVI with the presence of iron or iron oxide in cores exhibits dual properties of adsorbents and reducing agents towards the contaminants. Apart from that, the super magnetic property of iron oxide nanoparticles facilitates the removal of 99% arsenic from groundwater reserves. The implication of nZVI with carboxymethyl cellulose-carbo-iron alginate polymeric structure remediates the contaminants from groundwater with 100% efficiency. The emulsification of nZVI with oil liquid membrane prepared by biodegradable oil and food grade surfactants facilitates the adsorption of organic contaminants due to the hydrophobic nature of the membrane. The synthesis of Bi-metallic Nanoparticles by the incorporation of metals like platinum, gold, nickel, palladium etc. with iron nanoparticles facilitates easy removal of trichloroethane from the groundwater. (Mehnidiratta, 2013)

Titanium oxide, zinc oxide, magnetite, tungsten oxide nanoparticles are used for the treatment of nitrogen oxides, carbon dioxide, and other greenhouse gases generated from vehicles and industries. They act as semiconductor catalyst that facilitates the chemical conversion of the toxic gases into innocuous gaseous products. Different metal nanocatalysts like gold are used for the prevention of pollution at the source. Nanoparticles are used as potential filters in the vehicles and industrial sector for purification of air. Nanowires and nanotube based sensors can be used as potential sensors and detection of the quality of pollution. Gold nanoparticles are mostly used as the remediator of air pollutants due to their hemispherical shape and presence of effective perimeter interfaces around the particles. Carbon monoxide present in the smoke adsorbs on the edge of the gold surface and oxygen adsorbs on the support surface. After that both reacts in the interface region. (Abdelbasir et al, 2020)

Nanometals can be synthesized effectively from the waste products especially from the electronic wastes. The carbon, lead, zinc, copper, lead, palladium etc. are some metals that can be extracted effectively from the e-waste in nanoscale. These can be extracted physically through grinding and milling and chemically by treating with strong acids like sulphuric acid, nitric acid, hydrochloric acid etc. The reduction with sodium borohydrate is the most widely acceptable process of the extraction of nanometals from solid wastes. Nanoparticles can also be synthesized from the agricultural wastes that can again be used for the remediation of contaminants from the field. Sugar cane bagasse and leaves, bamboo leaves, egg shells, walnut shells, wheat straw, coconut shells, banana peel are some potential starting materials for the synthesis of nanometals like silver, silicon, gold etc. (Aswathi et al, 2022)

Figure 7. Nanotechnology and fuel cell

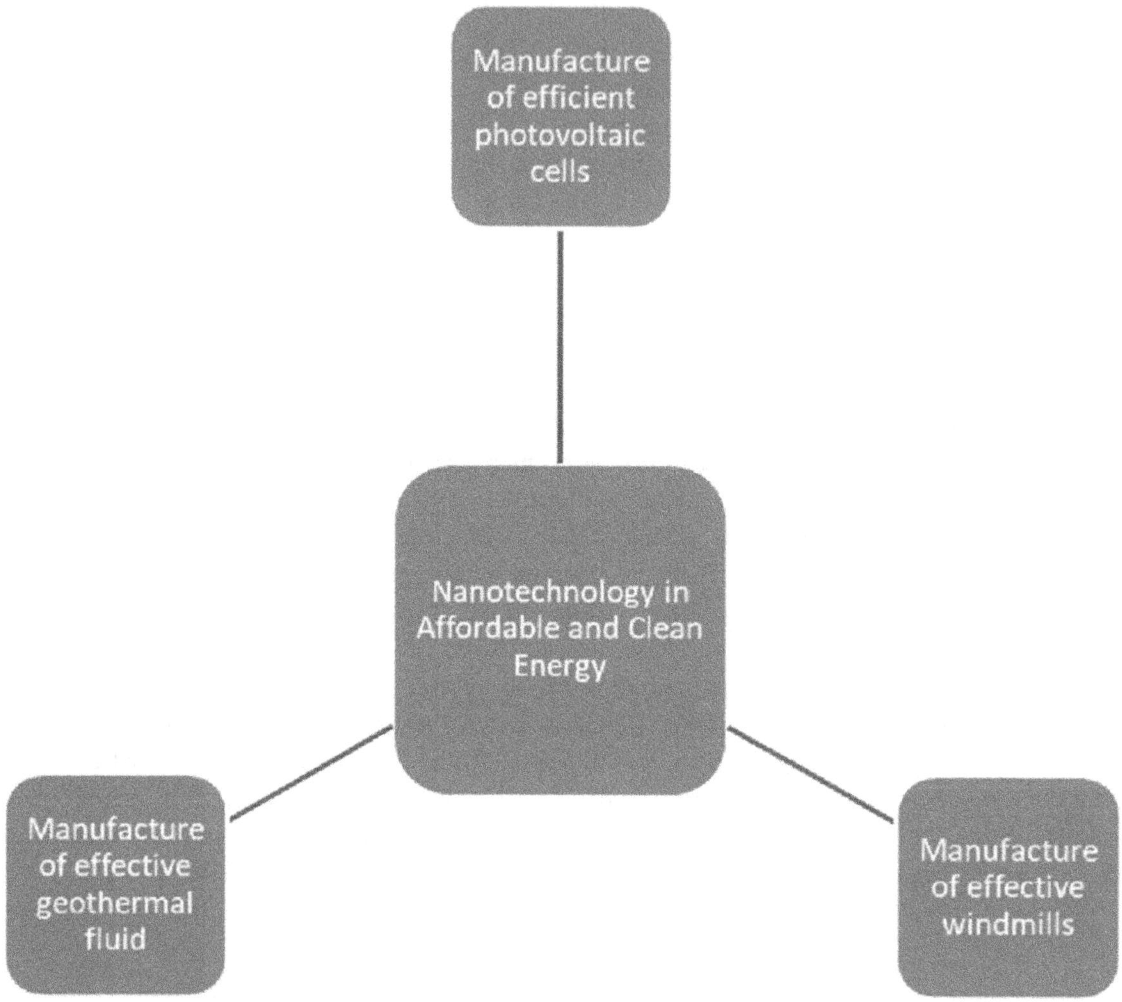

APPLICATION OF NANOTECHNOLOGY IN TEXTILE

Hydrophobicity, antibacterial properties, conductivity, antiwrinkle properties, antistatic behaviour, and light guidance and scattering are just some of the functions that can be nanoengineered into textiles. Incorporating nanotechnology allows for these qualities to be attained without sacrificing breathability or texture. (Nanografti, n. d.; Shishoo, 2004)

Basically, nanosilver particles realise positively charged ions that stop bacterial cells' functions, which pursue a powerful ability to kill bacteria and fungi and prevent the nasty odours caused by them. Nanofabric is a prime example of textile engineering that provides numerous benefits such as superhydrophobicity, odour and moisture elimination, increased elasticity and strength, and bacterial resistance. Although nanotechnology in the textile industry reflected important characteristics such as dye capability enhancement, self-cleaning fabrics, inhebit microbes, blocking UV rays, and frame resistance, (Shah et al., 2022)

Wrinkle-free treatment and antistatic finishes provide a luminous future for nanotechnology in the textile industry.

Nanotechnology plays a significant role in the improvement of the efficiency of geotextile products. Geotextile is an emerging field of textile technology that enables the use of specially engineered fabrics in geotechnical applications such as the construction of roads, railway embankments, coastal protection structures, and earth dikes. It facilitates improvements in the performance of different functions like filtration, separation, reinforcement, and stabilisation of soil layers. The addition of nanoparticles to the polyester enables the preparation of nonwoven geotextiles that enhance the toxicological remediation capabilities of the materials. The nanomaterials also increase the water permeability of the geotextile products through the development of sustainable laminar geotextile composites with nanoporous structural features. (Jeon, 2015)

Figure 8. Application of nanotechnology in Textiles

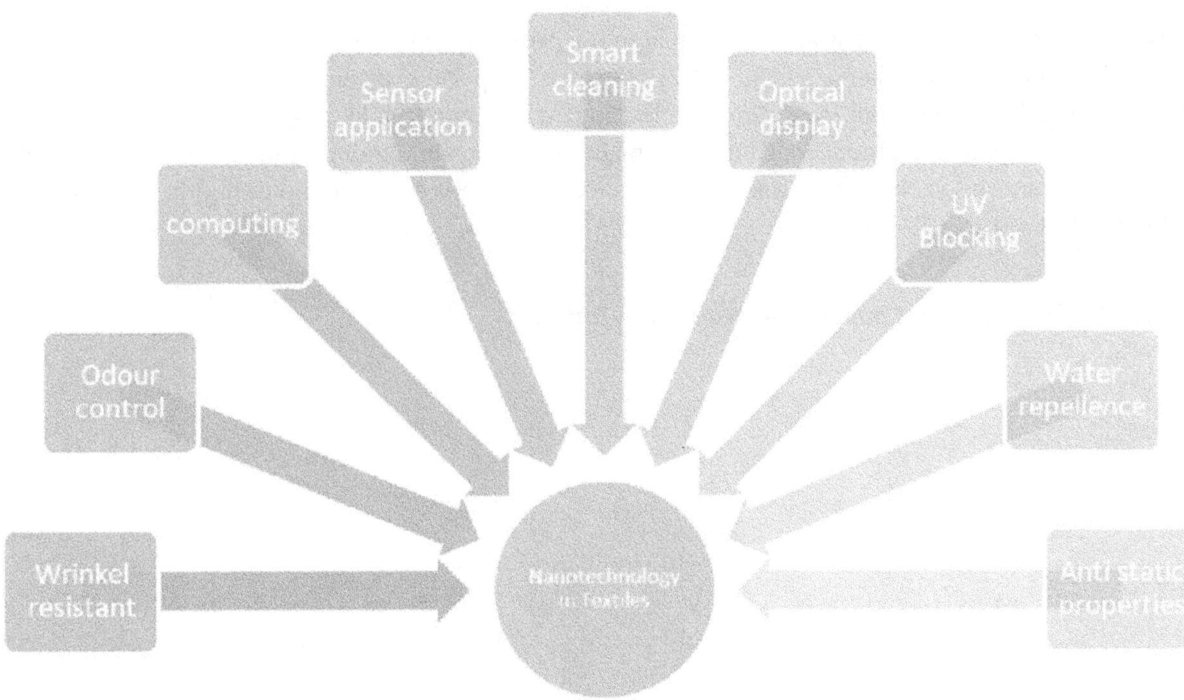

Current Uses of Nanotechnology and its Impact in Society

Due to quantum effects, the unique shape, size, surface characteristics, and inner structure of nanoparticles enable their application in a variety of industrial fields such as electronics and information technology, health care, energy, transportation, and so on. It plays a significant role in the smart infrastructural development of society. According to a report by Allied Market Research, an international market research organization, released in 2020, the global market value of nanotechnology in 2020 was $75.8 billion, which is expected to experience a major boom and reach 36.4% in 2030. (Divyanshi, T., 2022)

Figure 9. Market value of nanotechnology in 2nd decade of 21st century A.D.

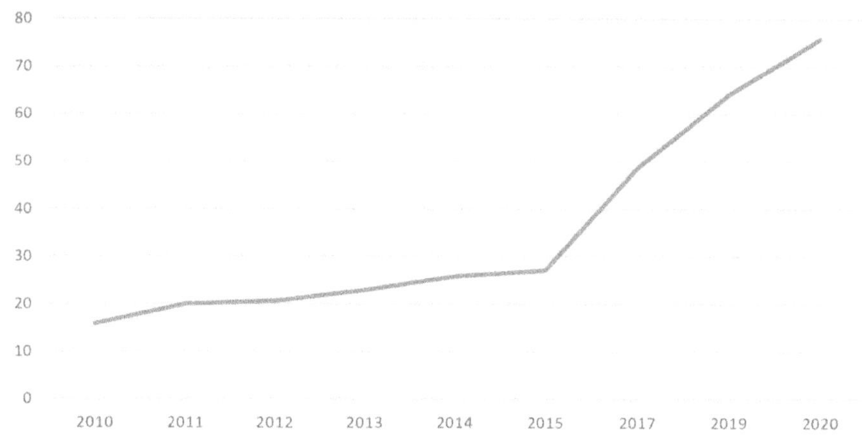

Nanotechnology applications and research are now primarily concentrated in the developed nations of America and Europe. (Santos et al., 2022) However, as per the recent report released by Insider Monkey, a US-based financial institution, the leading Asian countries like Japan and South Korea are coming out on top, holding the second and third places in nanotechnology patents in 2022, respectively. Developing countries such as India and South-East Asian nations are also improving their capabilities in this field. (Fredrick, T.; 2022)

Figure 10. Top 5 countries having nanotechnology patents

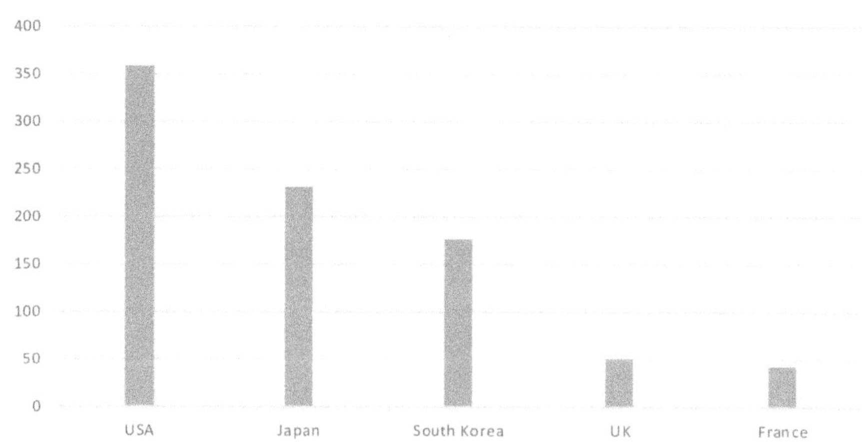

The nanoparticles are able to store more energy than their macro counterparts due to quantum effects. It is now widely used in the electronic industries to create portable and efficient gadgets with increased storage capacity. The MRAM (Magnetic Random Access Memory) enables computers and mobile phones to boot quickly due to its highly efficient nanoscale magnetic tunnel junction. It is able to store the data by changing the magnetic state of the material due to its improved magnetic characteristics. (Ohno et al., 2014) Nanomaterials like quantum dots are used for improving the displays of televisions, laptops, mobile phones, and other electronic gadgets. The improved optical characteristics of the quantum dots facilitate the production of more vibrant colours. The semiconductor nanomembranes are mostly used in the manufacture of smartphone displays. The nanomaterials enable the manufacture of stretchable, flexible, bendable, rollable, and foldable electronics products that are currently used at large scale in the aerospace, defence, medical, and aerospace sectors. The photovoltaic nanochips can be used in textile products that have large-scale applications in the defence and pharmaceutical sectors. (Xuan and Qian, 2017) Nanomaterials are also used as a protective mechanism in various electrical and electronic devices on a large scale.The nanocoatings are used on keyboards and monitors as a preventive measure against microbes and liquids. (Chaudhary et al., 2018) Nano-based copper suspensions are a sustainable and environmentally friendly alternative to lead-based solder and other hazardous materials commonly used for assembling electronics products.

Nanobiotechnology has effective applications in the modernised treatment of acute genetic diseases and cancer. The gold nanoparticles are largely used as probes for the detection of the sequences of nucleic acids in cells, which facilitates the determination of the genetic sequences responsible for the rare diseases. The nanoparticles are used in the manufacture of advanced medical equipment, facilitating more effective diagnosis. (Anjum et al., 2021)

The application of nanomaterials enhances the synthesis of fuel from petroleum products by improving the catalysis process. The scrubbers made by nanoparticles enable effective purification of toxicants from the gaseous effluent generated in an industry. The use of carbon nanotubes in the preparation of electrical wires reduces the resistance, leading to a more effective flow of current than with normal wires. It optimises the cost of power consumption. The solar panel prepared by nanomaterials enables the effective storage of solar energy, leading to an increase in the efficiency of solar energy resources. The use of epoxy containing carbon nanotubes increases the efficiency, strength, and longevity of wind mills. (Sethi et al, 2011; Muzammil et al, 2019)

Figure 11. Current applications of nanotechnology

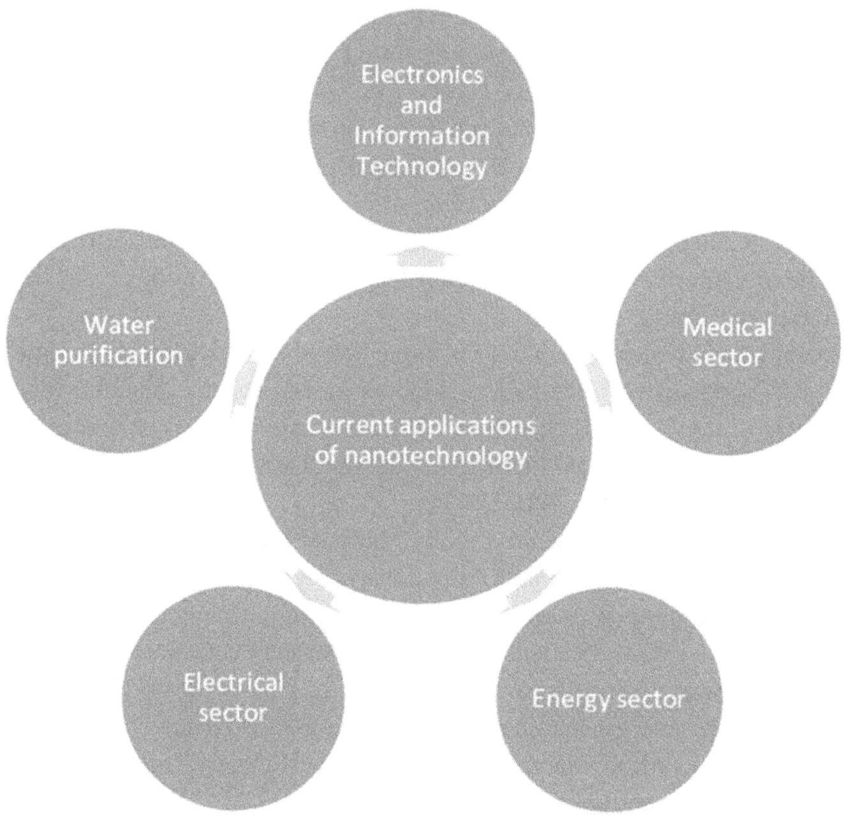

FUTURE PROSPECTIVE OF NANOTECHNOLOGY

Modern transport policy is moving towards the adoption of electric vehicles by replacing fossil fuel-based vehicles in order to reduce the carbon footprint. Nanotechnology enables the manufacture of highly efficient, portable, energy-saving, long-lasting batteries that are capable of storing energy for a long timeanotechnology enables the manufacture of highly efficient, portable, energy-saving, long-lasting batteries that are capable of storing energy for a long time. Industry 4.0 and society 5.0 bring automation into every aspect of life, which increases the demand for new, sophisticated, flexible electronic gadgets. Nano-based electronic chips and batteries are gaining more importance in this sector due to their greater efficiency than their macro counterparts. Long-lasting textile products with self-healing capabilities have emerged as critical components of the medical industry. The nano-enabled pharmaceutical products ensure effective targeted and non-targeted drug molecule delivery. Nanosensors attached to drug molecules aid in the study of pharmacokinetics as well as the detailed study of therapeutic and physiological mechanisms. It facilitates the development of more accurate and precise bioinformatic data, leading to the prosperity of medical, pharmaceutical, and biotechnological research. The agricultural sector is facing a paradigm shift from conventional practices. The introduction of smart technologies like artificial intelligence, machine learning, the Internet of Things, agricultural drones, sensor networks for integrated pest management, cellular data communication systems, telematics positioning technology, big data analytics, integrated land morphology studies, cloud computing, etc. revolutionises the agricultural sector. Nanotechnology has a wide scope of applications in this sector to promote a smart agricultural industry as it increases the demand for smart electrical and electronic equipment in the agricultural sector as well. Moreover, the organic farming system in the agricultural sector is increasing day by day in order to promote green and sustainable agriculture. However, the productivity of organic farming is still too low to replace chemical fertilizer. Nanotechnology comes forward to prepare more effective organic fertilisers by incorporating specially engineered nanomaterials in the organic fertilizers. The growth of agricultural crops under current climatically stressed conditions becomes a challenge in modern agriculture. The application of nanofertilizers facilitates the recovery of crops from biotic and abiotic stress conditions by increasing the water and nutrient uptake capacity of plant roots. It also facilitates the increase of photosynthesis rates in plants by ameliorating the photosynthetic pigments present in plants. Nanotechnology enables the production of green energy, which is being continuously adopted by most nations in order to achieve sustainable development goals. The nanobiosynthesis approaches promote the production of bioenergy from agricultural wastes. The nano-enabled biochar increases the carbon sequestration rate in agricultural fields. Moreover, nanotechnology plays an important role in the smart city and smart village programmes adopted by developing nations. The developing nations are also focusing on large-scale infrastructural development based on the different components of Industry 4.0 in order to increase the quality of the citizens' livelihoods. Nanotechnology needs to be an important component of the developmental policies and programmes of the developing nations to increase their growth as well as the human development index. Due to the recent outbreak of the COVID 19 pandemic, digitalization is reaching even the most economically disadvantaged segments of society, including the majority of the informal sectors. The smart innovations in the field of nanotechnology will enable the manufacture of electronic gadgets at a lower cost. It will also reduce the operational costs of installing internet facilities, allowing digitalization to reach a larger population of society and promoting both the quantitative and qualitative aspects of development.

Table 1. Future prospective in nanotechnology

Sector	Uses
Electronics and IT	• Increase in production of cost effective, long lasting electrical and electronic devices in order to meet the demand of increasing digitalization. • Development of new sophisticated nano chips. • Improvement of the optical display of electronics gadgets.
Textile sector	• Promotion of nano based smart bandages in healthcare sector for exhibiting effective self-healing characteristics. • Introduction of specially engineered nonwoven geotextiles for quality infrastructural development in land transportation sector. • Application of nanofibers for quality enhancement of textile products.
Healthcare sector	• Development of nano based smart pills to ensure precise delivery of drug molecules. • Promotion of nanophytomedicines for improvement and diversification of health treatment sector. • Use of nanosensors with smart pills for detailed pharmacokinetic study and development of bioinformatic database. • Development of smart medical textile products with nanotechnology.
Agricultural sector	• Promotion of nanofertilizers in agricultural field. • Performance enhancement of organic farming practices. • Sustainable biotic and abiotic stress management. • Development of highly equipped electronic devices for the promotion of precision farming. • Effective integrated pest management. • Carbon sequestration and improvement of soil health through nanobiochar.
Energy sector	• Enhancement of green energy outcome through the development of effective equipment like nano enabled solar panels and windmills. • Increased performance of bioenergy production from agricultural wastes.
Remediation of environmental pollution	• Sustainable treatment of waste water. • Quality waste management through smart cleaning, sensor application, computing, odour control, wrinkle resistance. • Reduction of green house effect through UV blocking.

CONCLUSION

Modern science and technology have made significant contributions to achieving sustainable societal development. It plays an important role in promoting a number of sustainable development goals. Almost every nation is now putting a special emphasis on eco-friendly growth and development, which allows for a wide range of research in the field of nanotechnology. However, there are several limitations to nanotechnology. Many nanometals also exhibit negative consequences for the environment. The biosynthesis of nanomaterials from plant-based extracts is being promoted in order to avoid the negative impacts of nanotechnology on the environment. It is evident that further research is required in this field in order to foster the eco-friendly and sustainable economic development of society through nanotechnology.

REFERENCE

Abdelbasir, S. M., McCourt, K. M., Lee, C. M., & Vanegas, D. C. (2020). Waste-Derived Nanoparticles: Synthesis Approaches, Environmental Applications, and Sustainability Considerations. *Frontiers in Chemistry*, *8*, 782. doi:10.3389/fchem.2020.00782 PMID:33110911

Anjum, S., Ishaque, S., Fatima, H., Farooq, W., Hano, C., Abbasi, B. H., & Anjum, I. (2021). Emerging Applications of Nanotechnology in Healthcare Systems: Grand Challenges and Perspectives; MDPI. *Pharmaceuticals*, *14*(8), 707. doi:10.3390/ph14080707 PMID:34451803

Aswathi, V. P., Meera, S., Ann Maria, C. G., & Nidhin, M. (2022). *Green synthesis of nanoparticles from biodegradable waste extracts and their applications: a critical review; Nanotechnology for Environmental Engineering*. Springer.

Avdyli, M., & Kastrati, E. (2020). Aspects of the use of nanotechnology and nanomedicine in medicine. *International Journal of Engineering and Innovative Technology*, *10*(3), 10–12.

Bhat, A. S. (2014). Nanobots: The future of Medicine. *International Journal of Engineering and Management Sciences*, *5*(1), 44–49.

Choudhary, D., Patil, T., Raichurkar, P., & Daberao, A. (2018). Review on Nanotechnology & Its Application in Coating Industry. *International Journal of Scientific and Engineering Research*, *09*(4).

Divyanshi, T. (2018). Nanotechnology Market By Type (Nanosensor and Nanodevice) and Application (Electronics, Energy, Chemical Manufacturing, Aerospace & Defense, Healthcare, and Others). *Global Opportunity Analysis and Industry Forecast, 2021-2030; Report Code: A04928*. Allied Market Research. https://www.alliedmarketresearch.com/nanotechnology-market

Echiegu, E. A. (2016). Nanotechnology as a Tool for Enhanced Renewable Energy Application in Developing Countries (Editorial). *Journal of Fundamentals of Renewable Energy J and Applications*, *6*(6). doi:10.4172/2090-4541.1000e113

Eskandarinezhad, S., Khosravi, R. K. R., Amarzadeh, M., Mondal, P., & Fiho, F. J. C. M. (2021). Application of different Nanocatalysts in industrial effluent treatment: A review. *Journal of Xomposites and Compounds*, *2*(5). Advance online publication. doi:10.52547/jcc.3.1.5

Fredrick, T. (2022). 5 Most Advanced Countries in Nanotechnology. *Insider Monkey*. https://www.insidermonkey.com/blog/5-most-advanced-countries-in-nanotechnology-1098326/5/

. Gehrke, I.; Geiser, A.; Somborn-Schulz, A. (2015) *Innovations in nanotechnology for water treatment; Nanotechnology, Science and Applications,* 1-17. Dovepress. doi:10.2147/NSA.S43773

Ghidan, A. Y., & Al Antary, T. M. (2019). *Applications of Nanotechnology in Agriculture*. IntechOpen.

Halo, T. L., McMahon, K. M., Angeloni, N. L., Xu, Y., Wang, W., Chinen, A. B., Malin, D., Strekalova, E., Cryns, V. L., Cheng, C., Mirkin, C. A., & Thaxton, C. S. (2014, December 02). NanoFlares for the detection, isolation, and culture of live tumor cells from human blood. *Proceedings of the National Academy of Sciences of the United States of America, 111*(48), 17104–17109. doi:10.1073/pnas.1418637111

Hesari, M. Mohammadi, P., Khademi, F., Shackbaei, D., Momtaz, S., Mosaefi, N., Farzaei, M. H., & Abdollahi, M. (2021). Current advances in the use of Nanophytomedicine Therapies for Human Cardiovascular diseases. *International Journal of Nanomedicine, 16*, 3292–3315.

Idrees, M. (2015). Role of nanotechnology in Medical Sciences: A Review. *International Journal of Innovative Drug Discovery*, *5*(1), 14–24.

Jeon, H. Y. (2015). *Nanotechnology and formulations of modeling of hydraulic permeability for nonwoven geotextiles*. Intechopen. doi:10.5772/61997

Jin, C., Wang, K., Oppong-Gyebi, A., & Hu, J. (2020). Application of Nanotechnology in Cancer Diagnosis and Therapy - A Mini-Review. *International Journal of Medical Sciences, 17*(18), 2964–2973. doi:10.7150/ijms.49801

Kaliannan, D., Palaninaicker, S., Palanivel, V., Mahadeo, M. A., Ravindra, B. N., & Jae-Jin, S. (2019). February). A novel approach to preparation of nano-adsorbent from agricultural wastes (Saccharum officinarum leaves) and its environmental application; Springer. *Environmental Science and Pollution Research International, 26*(6), 5305–5314. doi:10.100711356-018-3734-z

. Khalith, S. B. M.; Karuppannan, S. K.; Dowlath, M. J. H.; Raiyan, G. I. D.; Arunachalam, K. D. (2022) Nanostructured Catalytic membranes for water filtration; Nano-Bioremediation: Fundamentals and Applications. *Micro and Nano Technologies*, 389-412

Khoo, K. S., Chia, W. Y., Tang, D. Y. Y., Show, P. L., Chew, K. W., & Chen, W.-H. (2020, February 17). Nanomaterials utilization in Biomass for Biofuel and Bioenergy production; MDPI. *Energies, 13*(4), 892. doi:10.3390/en13040892

LiuX.XuH.ZhangM.YuD. G. (2021);Electrospun Medicated Nanofibers for Wound Healing. MDPI. *Membranes (Basel), 11*(10), 770. doi:10.3390/membranes11100770

Madan, T., Pandey, S., & Talegaonkar, S. (2015). Nanotechnology: A smart drug delivery tool in modern healthcare. *Journal of Chemical and Pharmaceutical Research, 7*(6), 257–264.

Maheswari, U. B., Sivakumar, V. M., & Thirumarimurugan, M. (2019). Synthesis, characterization of modified nanoadsorbents and its application in removal of Zn2+ ions from battery effluent. *Environmental Chemistry and Ecotoxicology, 1*, 2–11. doi:10.1016/j.enceco.2019.05.001

Malini, S., Raj, K., & Madhumathy, S. (2022). El-Hady, Khalid M.; Islam, S.; Dutta, M. (2022) Bioinspired Advances in Nanomaterials for Sustainable Agriculture; p. 1-11; Hindwai. *Journal of Nanomaterials, 8926133*. doi:10 doi:.1155/2022/8926133

Mehndiratta, P., Jain, A., Srivastava, S., & Gupta, N. (2013). Environment Pollution and Nanotechnology. *Environment and Pollution*, *2*(2), 2013. doi:10.5539/ep.v2n2p49

. Muzammil, W. K.; Rahman, M. M.; Fazlizan, A.; Ismail, M. A. (2019) Nanotechnology in Renewable Energy: Critical Reviews fir Wind Energy. *Nanotechnology: Applications in Energy, Drug and Food*, 49-71. doi:10.1007/978-3-319-99602-8_3

Nanogaki, H. (2006). Seed Germination- The Biochemical and Molecular Mechanism. *Breeding Science*, *56*(2), 93–105. doi:10.1270/jsbbs.56.93

Nanografti (n.d.). Use of Nanotechnology in Textile Industry & Its future. *Nanografiti*. https://nanografi.com/blog/use-of-nanotechnology-in-textile-industry-and-its-future/#:~:text=Nanotechnology%20is%20used%20in%20the,are%20explained%20shortly%20among%20them

Ohno, H., Endoh, T., Hanyu, T., Ando, Y., & Ikeda, S. (2014). Spin-transfer-torque magnetoresistive random access memory (STT-MRAM) technology. *Advances in Non-volatile Memory and Storage Technology,* 455-494; doi:10.1533/9780857098092.3.455

Prasad, R., & Bhattacharyya, A. Nguyen, Quang D. (2017) Nanotechnology in Sustainable Agriculture: Recent Developments, Challenges, and Perspectives. *Frontiers in Microbiology*. doi:10 doi:.3389/fmicb.2017.01014 PMID:28676790

Ramanayaka, S., Vithanage, M., Alessi, D., & Liu, W. J. (2021). Nanobiochar: Production, properties, and Multifunctional Applications. *Environmental Science. Nano*, *7*(11). doi:10.1039/D0EN00486C

Santos, J., Calpa, L. R. S., & Souza, F. G. Jr. (2022). *Main countries contributing on Nanotechnology*. Qeios., doi:10.32388/4MZGZR

Sethi, V. K., Pandey, M., & Shukla, P. (2011). Use of Nanotechnology in Solar PV Cell. *International Journal of Chemical Engineering and Applications*, *2*(2), 77–80. doi:10.7763/IJCEA.2011.V2.79

Shah, M. A., Pirzada, B. M., Priece, G., Shibiru, A. L., & Qurashi, A. (2022). Applications of nanotechnology in smart textile industry: A critical review. *Journal of Advanced Research, 38*, 55–75. doi:10.1016/j.jare.2022.01.008

Sharma, V., Borkute, G., & Gumfekar, S. P. (2022). Biomimetic nanofiltration membranes: Critical review of materials, structures, and applications to water purification. *Chemical Engineering Journal*, *433*(3), 133823. doi:10.1016/j.cej.2021.133823

Shishoo, R. (2004). The Impact of Nanotechnology in the Textile Industry: Market Potential (Paper Presentation). EuroNanoTex Conference, Barcelona.

Sukhanova, A., Devy, M., Venteo, L., Kaplan, H., Artemyev, M., Oleinikov, V., Klinov, D., Pluot, M., Cohen, J. H. M., & Nabiev, I. (2004). Biocompatible fluorescent nanocrystals for immunolabeling of membrane proteins and cells. *Analytical Biochemistry*, *324*(1), 60–67. doi:10.1016/j.ab.2003.09.031 PMID:14654046

. Sylvester, P., Westerhoff, P., Mooller, T., Badruzzaman, M., Boyd, O. (2007) A hybrid sorbent utilizing nanoparticles of hydrous iron oxide for arsenic removal from drinking water. *Environmental Engineering Science 24* (1), 104e112.

. Theron, J., Cloete, T.E., de Kwaadsteniet, M. (2010) Current molecular and emerging nanobiotechnology approaches for the detection of microbial pathogens. *Critical Reviews in Microbiology 36* (4), 318e339.

. Tiede, K., Boxall, A.B.A., Tear, S.P., Lewis, J., David, H., Hassellov, M. (2008) Detection and characterization of engineered nanoparticles in food and the environment. *Food Additives and Contaminants, 25* (7), 795e821.

. Tiraferri, A., Vecitis, C.D., Elimelech, M. (2011) Covalent binding of single-walled carbon nanotubes to polyamide membranes for antimicrobial surface properties. *Acs Applied Materials & Interfaces 3*(8), 2869e2877.

Trivedi, P., & Axe, L. (2000). Modeling Cd and Zn sorption to hydrous metal oxides. *Environmental Science & Technology*, *34*(11), 2215–2223. doi:10.1021/es991110c

Vecitis, C. D., Schnoor, M. H., Rahaman, M. S., Schiffman, J. D., & Elimelech, M. (2011). Electrochemical multiwalled carbon nanotube filter for viral and bacterial removal and inactivation. *Environmental Science & Technology*, *45*(8), 3672–3679. doi:10.1021/es2000062 PMID:21388183

Vecitis, C. D., Zodrow, K. R., Kang, S., & Elimelech, M. (2010). Electronic-structure-dependent bacterial cytotoxicity of single-walled carbon nanotubes. *ACS Nano*, *4*(9), 5471–5479. doi:10.1021/nn101558x PMID:20812689

Vikesland, P. J., & Wigginton, K. R. (2010). _ Nanomaterial enabled biosensors for pathogen monitoring e a review. *Environmental Science & Technology*, *44*(10), 3656–3669. doi:10.1021/es903704z PMID:20405829

Westerhoff, P., De Haan, M., Martindale, A., & Badruzzaman, M. (2006). Arsenic adsorptive media technology selection strategie. *Water Quality Research Journal of Canada*, *41*(2), 171–184. doi:10.2166/wqrj.2006.020

Westerhoff, P., Moon, H., Minakata, D., & Crittenden, J. (2009). Oxidation of organics in retentates from reverse osmosis wastewater reuses facilities. *Water Research*, *43*(16), 3992–3998. doi:10.1016/j.watres.2009.04.010 PMID:19450863

Xuan, X., & Qian, S. (2017). Editorial for the Special Issue on Micro/Nano-Chip Electrokinetics; MDPI. *Micromechanics*, *8*(145), 1–2. doi:10.3390/mi8050145

Yellepeddi, V., Joseph, A., & Nance, E. (2019). Pharmacokinetics of nanotechnology-Based Formulations in Pediatric Populations. *Advanced Drug Delivery Reviews*, *151-152*, 44–45. doi:10.1016/j.addr.2019.08.008

Zarejousheghani, M., Rahimi, P., Borsdorf, H., Zimmermann, S., & Joseph, Y. (2021). Molecularly imprinted Polymer- Based Sensors for Priority Pollutants; MDPI. *Sensors (Basel)*, *21*(7), 2406. doi:10.339021072406 PMID:33807242

Chapter 2
Intelligent Nanoparticles for Antibiotics Sensing

Anam Munawar
University of Health Sciences, Lahore, Pakistan

Ahsan Riaz
https://orcid.org/0000-0002-9739-2519
University of Health Sciences, Lahore, Pakistan

Zulfiqar Ali Buzdar
https://orcid.org/0000-0001-6218-603X
University of Health Sciences, Lahore, Pakistan

Muhammad Jawad
University of Health Sciences, Lahore, Pakistan

ABSTRACT

Nanosensors are acquiring expanding consideration because of the need to identify and quantify synthetic and actual properties in hard to arrive at natural and modern frameworks that are in the nanoscale locale. The overviews of different sensors based on nanomaterials, arranged into three expansive categories, are: mechanical, electromagnetic, and optical nanosensors. Nanotechnology has led to the synthesis of nanostructures having unique properties like morphology, conductance, and optical properties. Nanostructures provide increased surface for reactivity and transduction signal due to increase in surface area to volume ratio. Herein, the authors are discussing different nanomaterials; composites which is applied for exceedingly selective detection of analytes. The detecting ideas and their comparing benefits are talked about concerning their applications.

DOI: 10.4018/978-1-6684-7358-0.ch002

Figure 1.Graphical abstract is presenting the depiction nanosensors and their working principle.

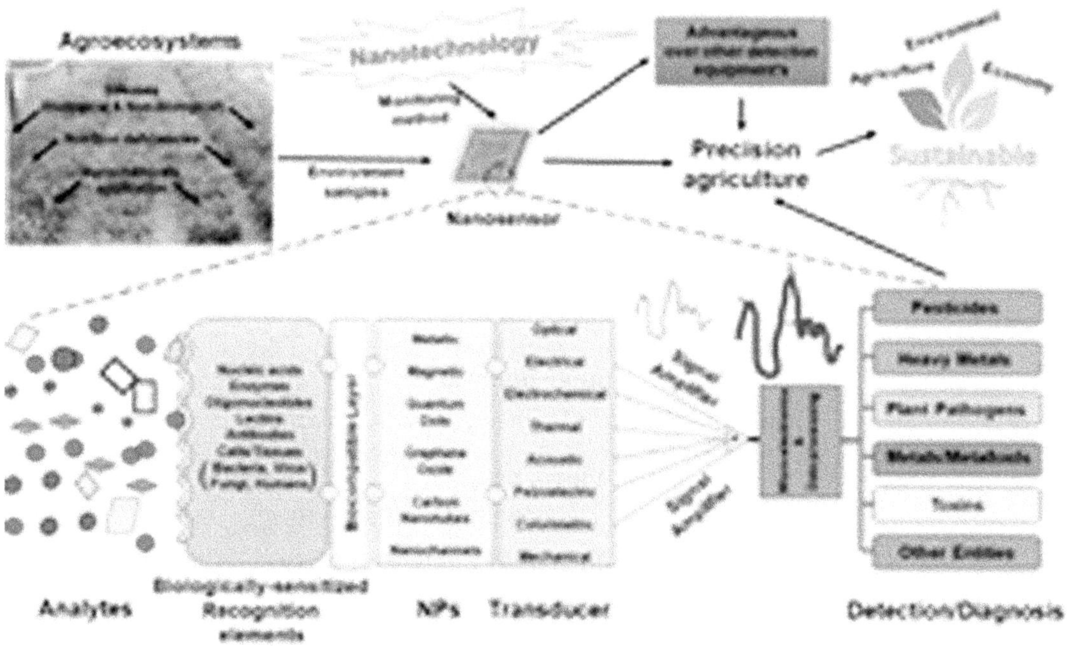

INTRODUCTION

Most articles based on nanosensors are centered around a precise kind of sensors, for example, optical nanosensors, nanobiosensors, and attractive nanosensors, with numerous specialized subtleties (Lew et al., 2020). In this boo chapter we are presenting an outline for different nanosensors, having similitudes and central contrasts between different classifications. The main point discussed in book chapter is to give an outline, that is reasonable for novices to understand the developing significance in this era (Nißler et al., 2020).

Nanosensors are detecting gadgets with something like one of their detecting aspects being not more noteworthy than 100 nm (Liu et al., 2020). In the area of nanotechnology, instruments called nanosensors are used for (a) checking of substance and physical peculiarities in locales hard to reach (J. Liu, L. Pan, et al., 2020), (b) identifying cell organelles having biochemicals (Akgönüllü, Yavuz, & Denizli, 2020), and (c) estimating nanoscopic particles in climate (Wu et al., 2020). An inquiry based on the conditions "nanosensor(s)" and "nano-sensor(s)" showing up in the nanotechnology related topics representing a developing pattern in nanosensor research, as obvious from the subsequent distribution record (Luo et al., 2020). Obviously, a far more prominent number can be anticipated assuming that a total catchphrase search is accomplished to incorporate different nanosensor distributions (Kulkarni et al., 2020). The development in logical consideration is normally trailed by mechanical advancement.

Despite the fact that sensing system have a prolonged history, generally the domain of nanosensors is new (B. Li et al., 2020). The different nanosensors can be inexactly assembled into three general classes of nanosensors:

(I) vibrational nanosensors,

(ii) mechanical nanosensors, and
(iii) electromagnetic or potentially optical nanosensors,

NANOSENSORS

Nanosensors are based on nanotechnology- sensing that action actual amounts and generate signs to identify and examine those amounts. Nanosensors are small portable devices having an affectability even at nanoscale (Kozawa et al., 2020). Different nanosensors are accessible for numerous application. Nanosensors possess a huge scope of utilizations, from space to biotechnological wellbeing, drug conveyance/screening, diagnostics, sanitation handling, and security (Liu, Jin, Cheng, Zhao, & Wang, 2020; Wu et al., 2020). At present, nanosensors are created utilizing different techniques, including atomic self-get together, hierarchical lithography, and base up gathering.

In spite of the fact that nanosensors show various sorts of estimation for various things, they all offer fundamental parts. The work process of a sensor is as per the following: restricting with the particular target, creating a sign by the communication among nanosensor and biocomponents, handling of the sign by using the concerned measurements (Al-Turjman, 2020). Nanosensors are profoundly specific and explicit. They can have improved particularity because of their usefulness with synthetic and organic atoms at similar scale as normal natural cycles.

Having both specificity and selectivity, there are numerous other benefits like cheap, fast response, small in size makes it a suitable candidate for high performance and applications. For real time sensing of different analytes nanosensors are best. Because of the enlisted benefits nanosensors are better as compared to conventional diagnostic procedures for example chromatography and spectroscopy. These conventional methods are time consuming, expensive, required expert persons to obtain results (Graboski, Martinazzo, Ballen, Steffens, & Steffens, 2020).

Single-layered nanomaterials like nanowires and nanotubes can be used in the fabrication of nanosensors, can act as transducers for the transmission of signals. Nanomaterials are unique materials with high surface area to volume ratio and small size, displaying huge sign varieties with the limiting of targeted molecules. The working of sensing device relies on either the applied nanomaterial is radioactive or fluorescent substance, or label free detection. Nanosensors are critical parts inside each advanced control framework, for instance in vehicles, security frameworks, observing, and diagnostics (Latha et al., 2020).

IMPORTANT ELEMENTS OF A SENSOR

A nanosensor is composed of different important elements: input device, transducer and data processing unit. Type of nanosensors change according to the transduction approach. When an analyte comes into contact with the nanomaterial, the transducer delivers a quantifiable result, like a current, change in mass, or an adjustment of shading (Akgönüllü et al., 2020; Sun et al., 2019). Sensors can be utilized to gauge or distinguish a huge assortment of physical, substance, and natural amounts, including proteins, microorganisms, synthetics, gases, light force, movement, position, sound and numerous others (Fakher, Rahmanian, & Hosseini-Hashemi, 2019).

HOW NANOSENSORS WORK

Nanosensors work by changing over the noticed material or a cycle into electrical signs which would then be able to be broke down. Most nanosensors work by estimating electrical changes in the sensor materials (nanomaterials) (Prajapati, Benedict, & Bhat, 2019). For instance, assuming an atom of nitrogen dioxide interacts with a synthetic carbon tube nanosensor, it will take an electron from the nanotube (nanomaterial), which will make the nanotube less conductive (V. K. Khanna, 2019). The nanosensor will see this and convey electrical messages. Nanosensors identify changes from the outer collaborations and convey to the next nanocomponents (Hüseynli, Çimen, Bereli, & Denizli, 2019).

FABRICATION METHODS

This section focuses on investigating and evaluating the pros and drawbacks of various nanomaterial production techniques (Tsao et al., 2019). Figure 2 depicts the two alternative ways of nanomaterial fabrication.

Figure 2.

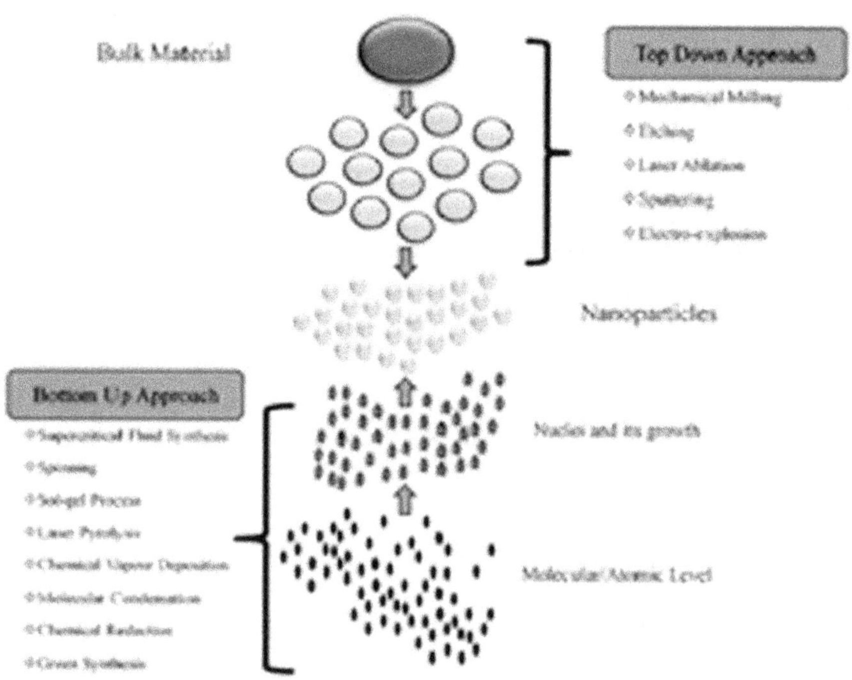

I. Top-Down Methods

a. Mechanicanosynthetic Methods

Mechanical methods are the most cost-effective approaches to mass-produce nanomaterials. From all these methods, the most basic technique is ball milling. Through mechanical attrition, ball milling creates nanostructures, that involves transferring kinetic energy from a material that is being reduced to a grinding medium (Wang, Zhang, Wang, Zhao, & Yan, 2019; Zhou et al., 2019).

A large-scale procedure, consolidation and compaction is, in which nanoparticles are "reassembled" to create materials with improved characteristics. This method can be used to create metallic alloys. Industry makes extensive use of top-down mechanical technologies (Munawar et al., 2019).

Thermal methods are a broad area, and we've tried to narrow it down to those that supply heat to a fabrication process. Electrospinning is one of these methods for creating nano-thread materials. High-energy processes necessitate a large amount of input energy, either in the form of electricity, heat, or solar power (Li, Wang, Wang, & Zhao, 2019). First method used for the production of carbon nanotubes was arc discharge that could be controlled. Solar flux and laser ablation are also effective. The issue is quality control and potential upgrade. This category includes plasma techniques. In high-energy circumstances, plasmas form (high potential bias, etc.) (Barbillon, 2019).

With the probable exception of solar flux systems, where sunlight is readily available, the challenge with this and other high-energy methods is scaling potential. Chemical fabrication procedures from the top down are always simple to scale up, and many, like as anodizing, are common industrial operations (Wang et al., 2019). Lithographic methods, as we all know, are top-down methods capable of creating micron-sized features, while being requiring intensive energy and exclusive facilities and equipment. Lithography has been used to create printed circuit boards and computer boards for decades (Szperlich & Toroń, 2019). Miniaturization will be a costly endeavor in the future because different powerful resources (shorter wavelength and high energy electron beams sources), as well as support infrastructure and equipment, would be essential. Nanoimprint lithography (NIL) is a type of lithography; however it isn't as we know it. It's more like template synthesis than anything else. To generate a pattern, first a desired analyte molecule is surrounded by monomers and cross linkers further imprinted on the surface of gel like polymer material (Nehra, Dilbaghi, Hassan, & Kumar, 2019). The template stamp, like the imprinted material, is made from the top down. Latex spheres are used to create a templated matrice in nanosphere lithography. As a result, these procedures can also be referred to as template processes.

II. Bottom-up methods

To make nanomaterials, bottom-up approaches begin from a molecule or atoms. Gas phase method, chemical vapour deposition produces nanostructure materials by reacting reactive elements with a catalyst or pre-templated surface (Shahbazi & Zare-Dorabei, 2019). CVD is the most cost-effective way to make carbon nanotubes. A catalyst based on Fe, Ni or Co placed at the bottom and different carbon sources for example methane and acetylene (in gaseous form) passed through the chamber. The catalyst particle forms nanotubes after it decomposes into carbon. An industrial process, atomic layer deposition that may coat any material to produce a thin film or monolayer, regardless of its size.

Other bottom-up industrialised processes include molecular beam epitaxy and molecular chemical vapour deposition (MOCVD) (Yadegari et al., 2019). There are numerous liquid phase approaches as well. All of the self-assembly and synthesis takes place in the liquid phase. Liquid phase approaches are cost-effective and scalable. Nanomaterials can be made in a variety of techniques, including electrodeposition and electroless deposition (dots, clusters, colloids, rods, wires, thin films). Making nanomaterials by anodizing aluminium to create a porous oxide structure is a simple process. Any substance generated within the porous structure is a nanomaterial. Porous membranes are the ultimate template in many aspects (Dewangan, Khan, Shrivas, & Sahu, 2019). The scene has been set for a new generation of nano bottom-up approaches. Different new techniques and approaches are high-throughput and low-cost. One of the disadvantages is the establishing of wide-ranging directive. The two latest methods are nanosphere lithography and nanolithography (dip pen method) (Cecchini et al., 2019; Lupan, Chow, & Adelung, 2019).

III. Sol-Gel Method

A multipurpose soft chemical method, sol-gel procedure used for the fabrication of glass materials, metal oxides and ceramics. Monolayer coatings, spherical shaped or ultra-fine powders, inorganic microporous membranes and ceramic fibres are only some of the ceramic and glass materials available. In the sol-gel technique, inorganic organometallic salts or metal alkoxides are commonly used as precursors. In this process a number of reactions take place for example hydrolysis and multiple condensation on the precursors to generate a sol or colloidal suspension (Zopf et al., 2019).

The sol-gel method entails converting a structure "sol" from a liquid (usually colloidal) to "gel" a solid phase, followed procedures include drying and calcination at various temperatures to produce metal oxide nanopowder. The shape, morphology, and textural qualities of the finished materials can all be controlled using the sol-gel process. In comparison to high-temperature techniques, the Sol-gel approach has numerous advantages, including the ability to generate metastable materials, superior purity, and product compositional uniformity at a low temperature (Parashar, Shukla, & Singh, 2020).

Furthermore, the precursor's chemical transformation at molecular scale to produce an ultimate oxidic network, this process has an impact on particle morphology. Using the sol-gel approach, several research groups have previously reported on the production of iron oxide and alumina nanoparticles (Perveen et al., 2020). The advantages linked with this approach include: at atomic scale the final product possess homogeneity, generate porosity to increase the surface area, molecular sizes can be adjusted and processing temperature is low (Moradnia, Fardood, Ramazani, Osali, & Abdolmaleki, 2020).

Furthermore, using high purity reagents, it is possible to synthesis complicated composition materials and produce greater purity products. The sol-gel procedure can produce high-quality films with a micron thickness that are challenging to achieve with physical deposition techniques (Wong et al., 2020). Furthermore, complicated composition materials can be synthesised, as well as coatings for multifaceted geometries. Initial components for the synthesis of a sol are generally metal organic complexes or inorganic metal salts, that are transformed into the sol through hydrolysis and polycondensation processes (Ganesan, Hariram, Vivekanandhan, & Muthuramkumar, 2020).

Further processing of the sol allows for the creation of a variety of ceramic materials. For making of thin films, two main procedures are in use, dip-coating and spin-coating. If a sol is poured in a mould then it will transform into a moist gel. After heat treatment and drying the gel is converted in the form of a glass materials or dense ceramic. Under supercritical conditions, low density aerogel layers with highly porosity is generated, when the liquid part of wet gel is extracted. From a sol, ceramic fibres can be pulled if the viscosity has been adjusted to a sufficient range. Precipitation, spray pyrolysis, and emulsion methods are used to create ultra-fine and homogeneous ceramic powders (Hjiri, 2020).

IV. Hydrothermal Method

Usually an autoclave used for the hydrothermal synthesis, which is a pressurised vessel where the reaction takes place in mixture liquid solution. Vapour saturation pressure is created in autoclave, because of the temperature increased above the water's boiling point. Hydrothermal synthesis is commonly utilised to generate metal oxide nanoparticles, which can be easily obtained by hydrothermally treating peptized metal precursor precipitates with water. Different factors can be adjusted additives, solvent characteristics, pressure, reaction temperature, solution composition, this hydrothermal process is used to alter different features of a nanomaterials for example surface chemistry, gain morphology, particles size, aging and crystalline phase (Hariharan et al., 2020).

V. Gas Phase Method

Gas phase technologies are ideal for the thin films manufacture. Physical or chemical methods can be performed to create gas phase synthesis. CVD (chemical vapour deposition) is an industrially common technology for coating huge regions in a small time span. When a precursor undergoes to decomposition or chemical reaction, in a gaseous form produce metal oxide. Physical vapour deposition (PVD) is another thin film production process (Skotadis, Aslanidis, Kainourgiaki, & Tsoukalas, 2020). Physical vapour deposition (PVD) technique is almost same as chemical vapour deposition (CVD), only one change is in the physical state of precursors/materials. i.e. In PVD the precursor used in solid form while in CVD the raw material used in liquid state. This method is atomistic in nature and does not involve many chemical reactions (Coots, Gandhi, Onakoya, Chen, & Larriba-Andaluz, 2020). Various strategies for removing growing materials from precursor have been developed. The thickness of deposits' might range from a few angstroms to many millimetres. These techniques may be categorized in two classes: sputtering and evaporation. The produced materials are thermally separated from the raw substances during evaporation. Sputtering is a process in which gaseous ions dislodge atoms or molecules from a solid target (plasma) (Hammad et al., 2020).

TYPES OF NANOSENSORS

Sensors have been categorised based on a variety of factors. The most frequent method of classifying sensors is based on their transducing mechanism (piezoelectric, electrical, thermal, mass, optical, etc.), recognition principle (molecular recognition, DNA and enzymatic), or uses (food, medical diagnosis, environmental and industrial). This section focuses on features which distinguish nanosensors and determine the mechanism of action. Different types of nanosensors created to far, with an emphasis on environmental applications, and highpoint the linkage exist in the kind of nanoparticles and monitored property (Abdel-Karim, Reda, & Abdel-Fattah, 2020; Deng, Zhao, Liu, Liu, & Sun, 2020; Farmani, Soroosh, Mozaffari, & Daghooghi, 2020).

Different transduction approaches are discussed in this section, which are now being employed in the establishment of nanodevices, nanosensors and nanoelectrodes. We concentrate on major nanoforms, such as nanorods, nanotubes, and nanoparticles. The experimental setup is briefly described, in figure 3 and the key benefits and different key factors of nanodevices are discussed. We introduce and explain the current state of nanosensor technology and its applications in many domains (Shafiq, Anjum, Hano, Anjum, & Abbasi, 2020).

Figure 3.

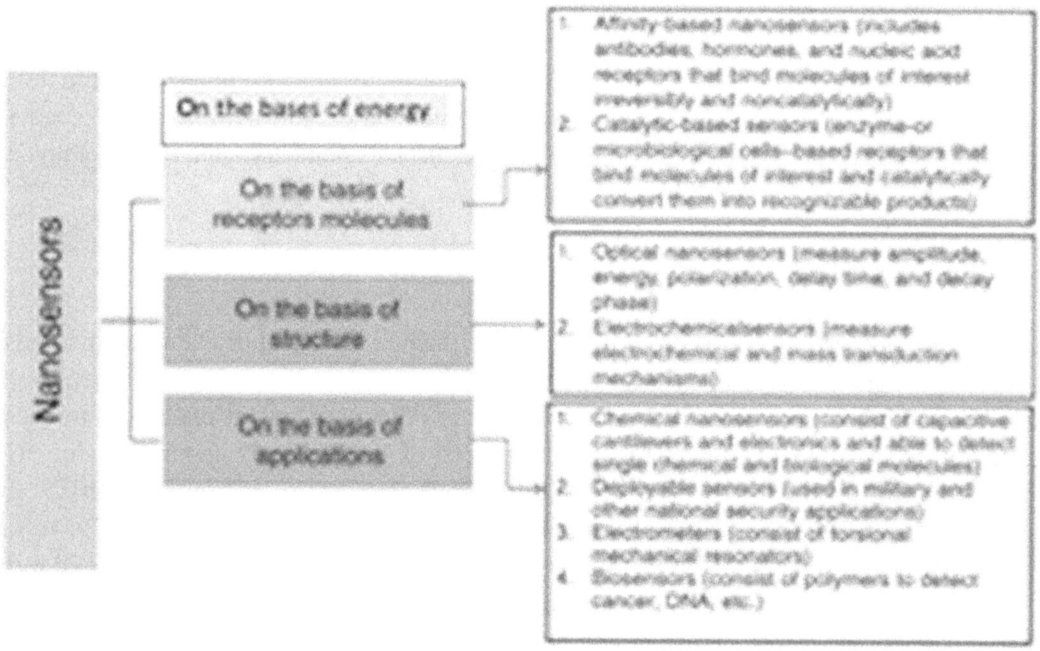

I. Piezoelectric Sensors

We will discuss piezoelectric sensors in this section. Piezoelectric sensors have a piezoelectric transduction element. They are termed active sensors since they do not require any external energy to generate an output signal (Su, Han, Kong, Nantung, & Lu, 2020).

A piezoelectric's sensitivity is defined as the open circuit voltage it creates as a result of applied stress. The sensitivity of the generated signal must be high enough to be detected above the surrounding noise. In practice, the resulting signal is weak and must be amplified with a suitable charge or voltage amplifier.

In this instance, due to their superior dielectric and piezoelectric properties, lead oxide-based compounds have been the dominant materials for piezoelectric transducers since the 1950s. Lead-containing items are now considered a health hazard as a result of growing public knowledge of health risks linked with lead and recent changes in environmental legislation. As a result, in addition to the current active efforts, additional research activity on Pb free piezoelectric materials is envisaged in the future (Guan, Xu, & Gong, 2020).

II. Electrochemical Sensors

Electrochemical sensors (first debuted in the 2nd part of twentieth century) are currently used from different ways for business settings. Electrochemical devices are distinguished for different reasons: for signal acquisition the electron usage, that is assumed as a clean technology for methodical applications because no extra substances are generated; miniaturisation in a small portable sensor (detection of analytes at micro level); cost effective: robust analysis, which has allowed these methods to become more widely used (for example commercially available glucose sensor) (Karimi-Maleh, Karimi, Alizadeh, & Sanati, 2020).

Furthermore, the advancement in electrochemical detection devices promotes the advancement of several methods for example chromatographic approach. Electrochemical devices, which are linked to nanotechnology, are getting more exact, selective, particular, and sensitive as depicted in figure 4.

Figure 4.

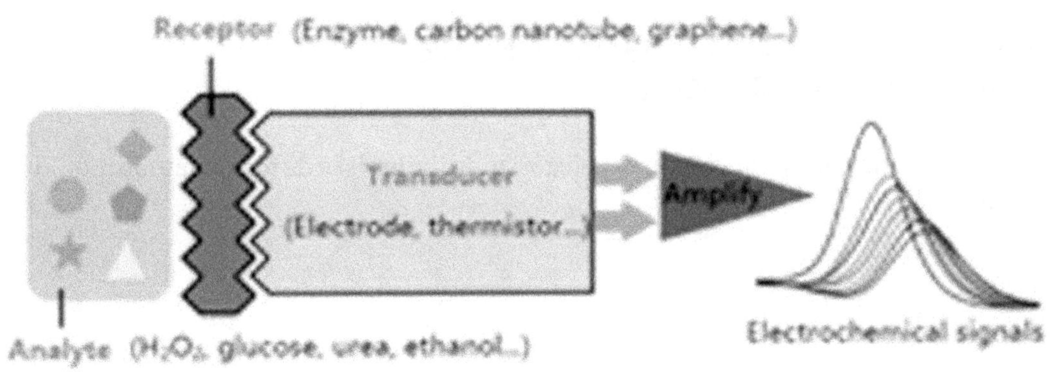

III. Conductance Based Sensors

Sensing based on conductance has a huge prospective to enable cost effective, small in size, super sensitive sensors having integrated microelectronics. Sensors based on field-effect-transistor (FET) is a smart technique used for label-free signal detection, analysing, and resultant, for realising home-care or point-of-care detection approaches. The external surface of semiconductor device mounted with a layer of molecular receptor has been used in a variety of investigations to produce conductance-based sensors (Wang, Corredor Garcia, Davidson, & Nichols, 2020).

Because receptor molecules may be made for having considerably increased binding affinities for intended analytes rather than other interfering substances present in solution, they can be used to produce extremely selective sensing.

IV. Optical-Based Nanosensors

The receptor is the part of the body that receives optical nanosensors are made up of nanoscale or nanostructured sensor materials that react differently to optical frequencies of electromagnetic excitation.

There are two primary working principles to consider. On the one hand, nanoscale materials can change the cross section of direct light-matter interactions such as absorption or scattering by raising or lowering it locally. The optical characteristics of the nanoscale sensor material, on the other hand, can vary as a result of situations or processes occurring in the sensor material's proximity (Jouyban & Rahimpour, 2020; Venkatesan et al., 2020). The change in optical characteristics can thus be investigated from the macroworld without making contact.

V. Colorimetric Sensors

Colorimetric sensors have grown in popularity and application as a result of its accessibility, ease of use, and low cost, as well as their sensitive and selective response to a variety of analytes. Colorimetric sensors are optical sensors that change colour in response to various stimuli. Single stimulus can be defined as modifications produce in the chemical or physical background. In result, the design of a colorimetric sensor is defined by a change in an environment property that needs to be detected (Yan et al., 2020).

CLASSIFICATION OF NANOSENSORS ON THE BASIS OF THEIR APPLICATIONS

Figure 5 is presenting that nanosensors can be categorize on the basis of different transduction approaches and their usage as well.

Figure 5.

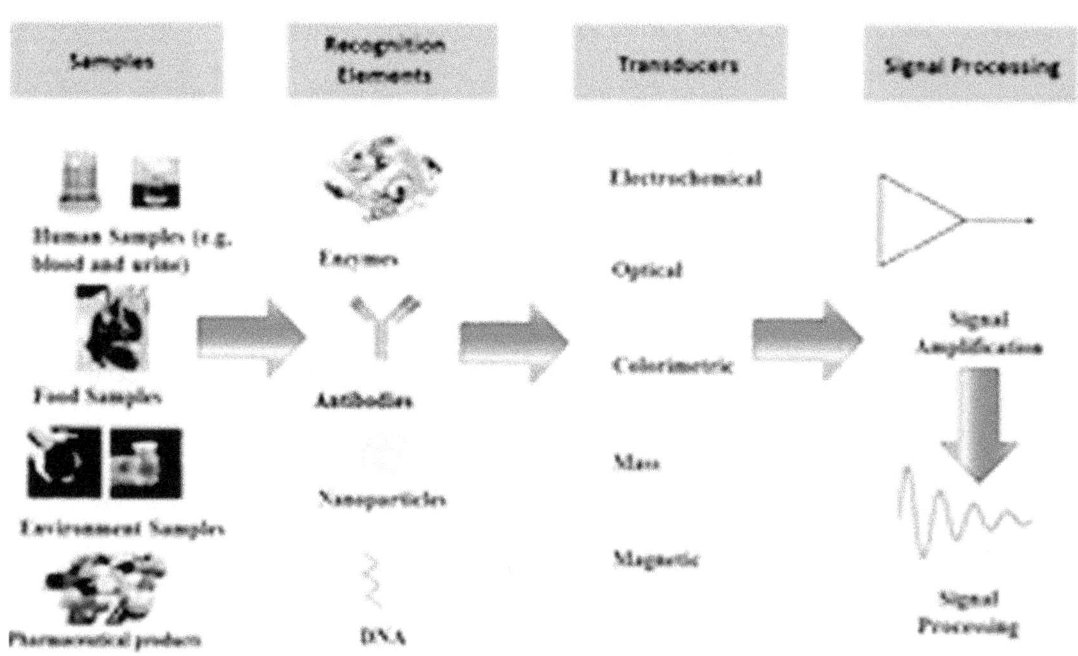

Deployable Nanosensors

This type is used by the national security or military purposes for example Sniffer STAR. It has a portable, small chemical detection system for different anaytes that uses nanoparticles to gather samples and a tiny electromechanical sensor to determine concentration (Abdel-Karim et al., 2020).

I. Chemical Nanosensors

This form of analysis can be used to look at a single chemical or molecule. Some features, for example different ion concentrations and pH, were measured using a variety of optical chemical nanosensors (Chang et al., 2020).

II. Electrometers

Mechanical element of a devices gets coupe charge by using a gate electrode, detecting electrode and mechanical resonator (Corbin et al., 2020).

III. Biosensors on a Chip

Microfluidic biosensors (biosensors-on-chip) are required to develop a cheap and reliable point-of-care detection. The combination of biosensor and microfluidic techniques allows biological and chemical constituents to be combined on a single surface, allowing for new sensing uses like unprecedented accuracies, real time detection, disposability, portability, and detection of multiple desired molecules by using a single sensing system. This technique was utilised to detect nucleic acids (cfNAs), which are found in high concentrations in cancer patients blood sample as depicted in figure 6 (Choi, 2020).

Figure 6.

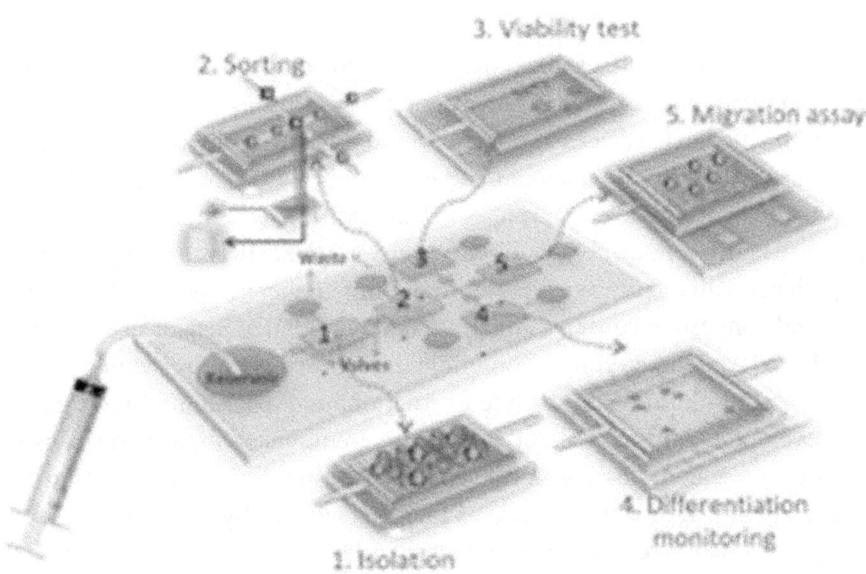

Figure 7.

APPLICATIONS

Accordingly, more than one sensor type is generally accessible to quantify an amount of interest. Each detecting choice presents its own remarkable arrangement of benefits and inconveniences. These should be made an appearance the setting of the utilization case or application being carried out to figure out which detecting innovation ought to be chosen (Sha & Bhattacharyya, 2020). Contingent upon the particular application, this can include an intricate combination of contending factors that should be thoroughly considered. Factors, for example, the sensor type, equipment choices, type of nook, and the application convention ought to be painstakingly thought of. Allow us now to momentarily check out the vital difficulties and necessities for our application areas (Fu & Ma, 2020).

I. Healthcare

Sensor applications in the medical services space range from physiological checking, for example, pulse, to screening applications, like blood natural chemistry, to falls hazard assessment. Sensors and applications used in the medical services area are commonly evolved by organizations executing a particular clinical use case requiring endorsement and enlistment with a suitable administrative body, like the US FDA (Khan, Civas, Cetinkaya, Abbasi, & Akan, 2020). In the home and local area, telehealth, telemonitoring, and mHealth (or portable wellbeing) sensor applications empower remote checking and the executives of patients with ongoing sicknesses, including diabetes, persistent obstructive pneumonic infection (COPD), and congestive cardiovascular breakdown (CHF). Sensor use in clinics and essential medical services offices zeros in additional on clinical diagnostics and screening applications, for example, point of care testing, level of electrolyte estimation, and examining blood gas fixations. There is additionally a developing business sector for over-the-counter symptomatic sensors that perform cholesterol observing, pregnancy testing, food sensitivity testing, and DNA testing (Pramanik et al., 2020). While frequently not rigorously analytic as far as exactness, by and large these over-the-counter sensors can convey demonstrative outcomes, which can help decision-production before looking for formal clinical mediation and care.

Key to the multiplication of sensors in medical services has been the improvement of minimal expense microsystem sensor advances coupled, at times, with minimal expense, low-power microcontrollers (MCUs) and radios. These gadgets have empowered the advancement of little structure factor, dependable, vigorous, exact, low-power sensor arrangements (Jin et al., 2020; Singh & Yadava, 2020). Some key application spaces of sensors in clinical medical care are:

Diagnosis: Biochemical and optical sensors are utilized for purpose in care observing and detection applications, containing tissue and blood examination. Biosensors can be utilized to recognize bacterial contamination, medications, chemicals, and levels of proteins in natural examples (Lee et al., 2020).

Physiological: Nanosensors in this classification are utilized to quantify key physiological marks of wellbeing, like ECG/EKG and circulatory strain. Noncontact thermometers include IR sensors (Chatterjee et al., 2020).

Imaging: Minimal expense CCD and ultrasound sensors are utilized for clinical imaging. Brilliant pills can be utilized for digestive imaging (Saylan & Denizli, 2020).

II. Agriculture and Food

Food handling is vital issue in food industry and farming since it is straightforwardly identified with the impact of food on the human wellbeing. Ongoing food handling occurrences, and general wellbeing worries about manufactured food added substances and compound deposits in food have driven the need to foster quick, touchy, and solid strategies to identify those food dangers. An option is given in the fast improvement of nanosensors which enjoy benefit to recognize food parts in a simple and speedy way (Shawon, Hoque, & Chowdhury, 2020). Connecting nanosensors with current Information and Communication Technologies (ICTs) empowers novel and online ways for various parts location went with high precision.

Different sorts of nanosensors are being created to meet the various necessities in food assessment (nanosensors for identification of outer and interior conditions in food bundling, carbon nanotubes based electrochemical sensors for location of cations, anions and natural mixtures in food, different aptamers for discovery of pesticides, anti-infection agents, substantial metals, microbial cells, and poisons) as depicted in figure 8 (Yang & Duncan, 2021).

Figure 8.

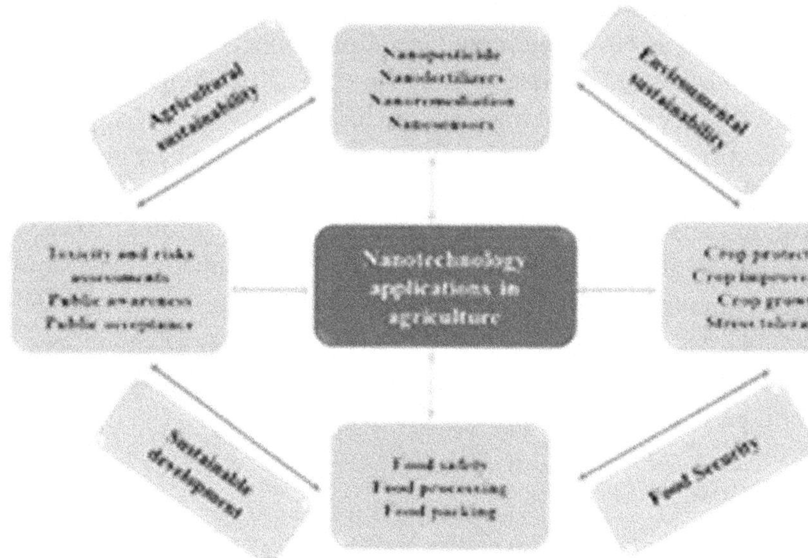

III. Environment

Expanded urbanization, serious farming techniques, industrialization, requests for power, and environmental change have altogether affected our capacity to keep a perfect climate. Foreign substances and contaminations from a huge assortment of sources can influence our air, water, soil, and surrounding climate. These reach from synthetic contaminations, organic blossoms, weighty metals, gases, and bacterial microbes, to encompassing clamor sources (Abdel-Karim et al., 2020).

Security of human wellbeing and environments is of the greatest need, requiring fast, delicate, strong, and adaptable sensor arrangements that are fit for identifying toxins, regularly at exceptionally low focuses and on a boundless premise. While numerous arrangements are as of now accessible for ecological observing, there are on-going tensions to foster new sensor advancements that are much more precise and more delicate, to drive compelling, versatile arrangements. This will empower enhancements continuously navigation, detecting granularity, and consistence checking (Cheng et al., 2020).

Many investigations particularly for bacterial pollution actually need in-situ examining with research center-based examination of tests. Severe conventions should be clung to in the testing system, especially for managed checking; if not the honesty of the examples and results can be compromised (Kumar & Guleria, 2020). Sensors, which work in situ, can resolve these issues however bring their own arrangement of difficulties; especially identifying with sensor fouling, affectability steadiness, precision, and natural impacts. Development in the utilization of sensors for these applications is as of now restrained by huge mechanical boundaries that should be survived. Progress towards supportable execution, equivalent to a research facility, should likewise be shown and kept up with in sensor networks dispersed over a wide topographical region as shown in figure 9 (Shiravand et al., 2020; Wu et al., 2020).

Figure 9.

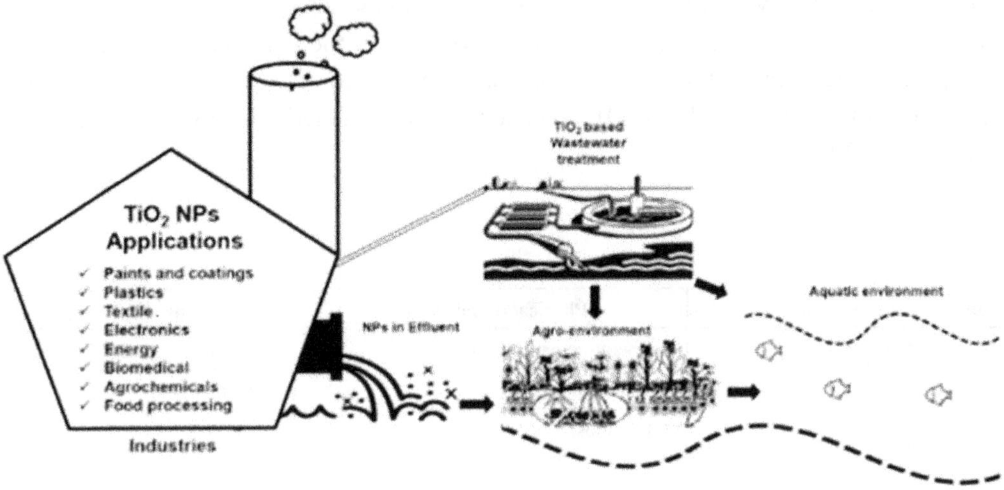

IV. Technological Advancement

Nanotechnology has devised new ways in the advancement and technology due to their working at the nanoscale. These advancements in technology lead towards the development of extremely sensitive biosensors and electronic devices that are used frequently in agriculture, food and medical sciences. At present high volume of studies are conducted for the development of these biosensors and electronic devices by utilizing the peculiar nature through which nanoparticles interact with biomolecules at nanoscale (Prakash, Parveen, Mishra, & Kaushik, 2020).

Nanosensors based devices manipulate the data at nano-scale and these devices provides high volume to size ratio that is useful for sensing the minimal signal and it is converted into the information that is used for the data analysis. Nanosensors are providing the basis for the development of new generation technologies. In the past nanosensors based devices have utilizes their low sensing abilities at large but now range is more increased and improved. Due to which these nanosensors are now not limited of working at nanoscale but they can also be a part of more complex and large electrical instruments for the sensing and detection of target materials (Xu et al., 2020).

Their importance lies in their abilities to find out the minute details and unique life process that are happening inside every cell such as the working of our biomolecules (proteins, lipids, carbohydrates and nucleic acids). These details cannot be scene at macro level of detection that makes the nano particles so much important in today's world of science and technology. They are providing their useful contribution in daily life processes that leads towards better technology with every passing day in our life (Keçili, Büyüktiryaki, & Hussain, 2019).

Nanotechnology in combination with biotechnology has leads towards the many useful contribution in serving mankind. They have linked our healthcare, environment, and technology by producing the rapid diagnostic health trackers that are not only less expensive but also very much safe for our environment. Previously used sensing instruments required an external light source for comparison and measurement of particular frequencies that the specific particle contains. Now these restrictions are eliminated as nanosensors can work without external illumination and are capable of detecting the small size similar particles. This is achieved by emitting signal at very low power and detecting their radiating energies at various frequencies (Pinnaduwage, Boiadjiev, Hawk, & Thundat, 2003).

The physical properties of materials change quite abruptly at the nanoscale as compared with their properties on macro level. For example, melting points of bull gold are 1064 °C as compared with 4 nm gold that melts at 850 °C (Zhu et al., 2020). Nanosensors are divided into three main categories as in table 1 (Zhu et al., 2020).

Table 1. Types of Nanosensors

Biological nanosensors	Physical nanosensors	Chemical nanosensors
Antibody type interactions	Temperature nanosensors	Chemical composite type
Enzymatic type interactions	Pressure nanosensors	Molecular type
DNA type interactions	Flow nanosensors	

Biological type of nanosensors utilizes the fluorescence technology and consists of receptor and a transducer. Fluorescence is used as transduction technology it has high sensitivity and relatively easy to use. Activated nanoparticles are used for synthesis of sensing probes that will detect the protein inside of the cell that results in the real time measurement to show the intracellular details (Voke, Pinals, Goh, & Landry, 2021).

Physical type sensors record the signals of analytes that is evaluated after their sensing. Physical sensors mostly work on physical parameters like pressure, temperature, force and velocity (Javaid, Haleem, Singh, Rab, & Suman, 2021).

Chemical nanosensors are most useful in medical fields in the form of diagnostic devices. Moreover they can be successfully used in environmental analysis, pollution monitoring and different chemicals and reagents preparations (Mousavi, Hashemi, Zarei, Amani, & Babapoor, 2018).

CONCLUSION

The terms transducer and sensor are generally utilized reciprocally, and the two gadgets are epitomized in sensor frameworks for a given application. As portrayed in this part, an incredible assortment of detecting strategies is accessible for estimating physical, synthetic, and organic amounts of interest. We have momentarily laid out the different strategies—mechanical, optical, semiconductor, and biosensing and enjoy featured their separate benefits and hindrances. We presented our key detecting spaces and depicted how the different sensor innovations can be used inside these areas to resolve issues related with expanded urbanization, worldwide maturing, contamination, and environmental change. The job of guideline in driving sensor applications for checking air and natural consistence was additionally laid out. At last, sensor attributes were characterized to work on the method involved with perusing datasheets and looking at execution qualities of various sensors.

CONFLICT OF INTEREST

There is no conflict of interest.

REFERENCES

Abdel-Karim, R., Reda, Y., & Abdel-Fattah, A. (2020). Nanostructured materials-based nanosensors. *Journal of the Electrochemical Society*, *167*(3), 037554. doi:10.1149/1945-7111/ab67aa

Akgönüllü, S., Yavuz, H., & Denizli, A. (2020). SPR nanosensor based on molecularly imprinted polymer film with gold nanoparticles for sensitive detection of aflatoxin B1. *Talanta*, *219*, 121219. doi:10.1016/j.talanta.2020.121219 PMID:32887120

Al-Turjman, F. (2020). Intelligence and security in big 5G-oriented IoNT: An overview. *Future Generation Computer Systems*, *102*, 357–368. doi:10.1016/j.future.2019.08.009

Barbillon, G. (2019). Fabrication and SERS performances of metal/Si and metal/ZnO nanosensors: A review. *Coatings*, *9*(2), 86. doi:10.3390/coatings9020086

Cecchini, R., Gajjela, R. S., Martella, C., Wiemer, C., Lamperti, A., Nasi, L., Lazzarini, L., Nobili, L. G., & Longo, M. (2019). High-Density Sb2Te3 Nanopillars Arrays by Templated, Bottom-Up MOCVD Growth. *Small*, *15*(37), 1901743. doi:10.1002mll.201901743 PMID:31222940

Chang, M., Morgan, G., Bedier, F., Chieng, A., Gomez, P., Raminani, S., & Wang, Y. (2020). Recent Advances in Nanosensors Built with Pre-Pulled Glass Nanopipettes and Their Applications in Chemical and Biological Sensing. *Journal of the Electrochemical Society*, *167*(3), 037533. doi:10.1149/1945-7111/ab64be PMID:34326553

Chatterjee, S., Saha, S., Barman, S. R., Khan, I., Pao, Y.-P., Lee, S., Choi, D., & Lin, Z.-H. (2020). Enhanced sensing performance of triboelectric nanosensors by solid-liquid contact electrification. *Nano Energy*, *77*, 105093. doi:10.1016/j.nanoen.2020.105093

Chen, K., Chou, W., Liu, L., Cui, Y., Xue, P., & Jia, M. (2019). Electrochemical sensors fabricated by electrospinning technology: An overview. *Sensors (Basel)*, *19*(17), 3676. doi:10.339019173676 PMID:31450877

Cheng, C., Zhang, R., Wang, J., Zhang, Y., Wen, C., Tan, Y., & Yang, M. (2020). An ultrasensitive and selective fluorescent nanosensor based on porphyrinic metal–organic framework nanoparticles for Cu 2+ detection. *Analyst (London)*, *145*(3), 797–804. doi:10.1039/C9AN02231G PMID:31894762

Choi, J. R. (2020). Development of point-of-care biosensors for COVID-19. *Frontiers in Chemistry*, *8*, 517. doi:10.3389/fchem.2020.00517 PMID:32574316

Coots, J., Gandhi, V., Onakoya, T., Chen, X., & Larriba-Andaluz, C. (2020). A parallelized tool to calculate the electrical mobility of charged aerosol nanoparticles and ions in the gas phase. *Journal of Aerosol Science*, *147*, 105570. doi:10.1016/j.jaerosci.2020.105570

Corbin, J. C., Moallemi, A., Liu, F., Gagné, S., Olfert, J. S., Smallwood, G. J., & Lobo, P. (2020). Closure between particulate matter concentrations measured ex situ by thermal–optical analysis and in situ by the CPMA–electrometer reference mass system. *Aerosol Science and Technology*, *54*(11), 1293–1309. doi:10.1080/02786826.2020.1788710

Deng, J., Zhao, S., Liu, Y., Liu, C., & Sun, J. (2020). Nanosensors for diagnosis of infectious diseases. *ACS Applied Bio Materials*, *4*(5), 3863–3879. doi:10.1021/acsabm.0c01247 PMID:35006812

Dewangan, P. K., Khan, F., Shrivas, K., & Sahu, V. (2019). Determination of uranium in environmental sample by nanosensor graphene quantum dots. *Journal of Radioanalytical and Nuclear Chemistry*, *320*(3), 757–763. doi:10.100710967-019-06512-x

Dorj, U.-O., Lee, M., Choi, J.-y., Lee, Y.-K., & Jeong, G. (2017). The Intelligent Healthcare Data Management System Using Nanosensors. *Journal of Sensors*, *2017*, 2017. doi:10.1155/2017/7483075

Fakher, M., Rahmanian, S., & Hosseini-Hashemi, S. (2019). On the carbon nanotube mass nanosensor by integral form of nonlocal elasticity. *International Journal of Mechanical Sciences*, *150*, 445–457. doi:10.1016/j.ijmecsci.2018.10.051

Farmani, A., Soroosh, M., Mozaffari, M. H., & Daghooghi, T. (2020). *Optical nanosensors for cancer and virus detections Nanosensors for Smart Cities*. Elsevier.

Fu, Y., & Ma, Q. (2020). Recent developments in electrochemiluminescence nanosensors for cancer diagnosis applications. *Nanoscale*, *12*(26), 13879–13898. doi:10.1039/D0NR02844D PMID:32578649

Ganesan, V., Hariram, M., Vivekanandhan, S., & Muthuramkumar, S. (2020). Periconium sp.(endophytic fungi) extract mediated sol-gel synthesis of ZnO nanoparticles for antimicrobial and antioxidant applications. *Materials Science in Semiconductor Processing*, *105*, 104739. doi:10.1016/j.mssp.2019.104739

Graboski, A. M., Martinazzo, J., Ballen, S. C., Steffens, J., & Steffens, C. (2020). *Nanosensors for water quality control Nanotechnology in the Beverage Industry*. Elsevier.

Guan, X., Xu, B., & Gong, J. (2020). Hierarchically architected polydopamine modified BaTiO3@ P (VDF-TrFE) nanocomposite fiber mats for flexible piezoelectric nanogenerators and self-powered sensors. *Nano Energy*, *70*, 104516. doi:10.1016/j.nanoen.2020.104516

Hammad, M., Hardt, S., Mues, B., Salamon, S., Landers, J., Slabu, I., Wende, H., Schulz, C., & Wiggers, H. (2020). Gas-phase synthesis of iron oxide nanoparticles for improved magnetic hyperthermia performance. *Journal of Alloys and Compounds*, *824*, 153814. doi:10.1016/j.jallcom.2020.153814

Hariharan, D., Thangamuniyandi, P., Christy, A. J., Vasantharaja, R., Selvakumar, P., Sagadevan, S., & Nehru, L. (2020). Enhanced photocatalysis and anticancer activity of green hydrothermal synthesized Ag@ TiO2 nanoparticles. *Journal of Photochemistry and Photobiology. B, Biology*, *202*, 111636. doi:10.1016/j.jphotobiol.2019.111636 PMID:31739259

Hjiri, M. (2020). Highly sensitive NO 2 gas sensor based on hematite nanoparticles synthesized by sol–gel technique. *Journal of Materials Science Materials in Electronics*, *31*(6), 5025–5031. doi:10.100710854-020-03069-4

Hüseynli, S., Çimen, D., Bereli, N., & Denizli, A. (2019). Molecular imprinted based quartz crystal microbalance nanosensors for mercury detection. *Global Challenges*, *3*(3), 1800071. doi:10.1002/gch2.201800071 PMID:31565367

Javaid, M., Haleem, A., Singh, R. P., Rab, S., & Suman, R. (2021). Exploring the potential of nanosensors: A brief overview. *Sensors International*, *2*, 100130. doi:10.1016/j.sintl.2021.100130

Jin, H., Yu, J., Lin, S., Gao, S., Yang, H., Haick, H., & Liu, Y. (2020). Nanosensor-based flexible electronic assisted with light fidelity communicating technology for volatolomics-based telemedicine. *ACS Nano*, *14*(11), 15517–15532. doi:10.1021/acsnano.0c06137 PMID:33141556

Jouyban, A., & Rahimpour, E. (2020). Optical sensors based on silver nanoparticles for determination of pharmaceuticals: An overview of advances in the last decade. *Talanta*, *217*, 121071. doi:10.1016/j.talanta.2020.121071 PMID:32498884

Karimi-Maleh, H., Karimi, F., Alizadeh, M., & Sanati, A. L. (2020). Electrochemical sensors, a bright future in the fabrication of portable kits in analytical systems. *Chemical Record (New York, N.Y.)*, *20*(7), 682–692. doi:10.1002/tcr.201900092 PMID:31845511

Keçili, R., Büyüktiryaki, S., & Hussain, C. M. (2019). Advancement in bioanalytical science through nanotechnology: Past, present and future. *Trends in Analytical Chemistry*, *110*, 259–276. doi:10.1016/j.trac.2018.11.012

Khan, T., Civas, M., Cetinkaya, O., Abbasi, N. A., & Akan, O. B. (2020). *Nanosensor networks for smart health care Nanosensors for Smart Cities*. Elsevier.

Khanna, P., Kaur, A., & Goyal, D. (2019). Algae-based metallic nanoparticles: Synthesis, characterization and applications. *Journal of Microbiological Methods*, *163*, 105656. doi:10.1016/j.mimet.2019.105656 PMID:31220512

Khanna, V. K. (2019). *Nanosensors: physical, chemical, and biological*. CRC Press.

Kozawa, D., Cho, S.-Y., Gong, X., Nguyen, F. T., Jin, X., Lee, M. A., & Schacherl, J. (2020). A fiber optic interface coupled to nanosensors: Applications to protein aggregation and organic molecule quantification. *ACS Nano*, *14*(8), 10141–10152. doi:10.1021/acsnano.0c03417 PMID:32667777

Kulkarni, D. R., Malode, S. J., Prabhu, K. K., Ayachit, N. H., Kulkarni, R. M., & Shetti, N. P. (2020). Development of a novel nanosensor using Ca-doped ZnO for antihistamine drug. *Materials Chemistry and Physics*, *246*, 122791. doi:10.1016/j.matchemphys.2020.122791

Kumar, V., & Guleria, P. (2020). Application of DNA-Nanosensor for Environmental Monitoring: Recent Advances and Perspectives. *Current Pollution Reports*, 1-21.

Latha, A. V., Ayyappan, M., Kallar, A. R., Kakkadavath, R. V., Victor, S. P., & Selvam, S. (2020). Fluorescence imaging of nitric oxide in living cells using o-phenylenediamine-rhodamine based polymeric nanosensors. *Materials Science and Engineering C*, *108*, 110463. doi:10.1016/j.msec.2019.110463 PMID:31923983

Lee, M. A., Wang, S., Jin, X., Bakh, N. A., Nguyen, F. T., Dong, J., & Jones, K. K. (2020). Implantable Nanosensors for Human Steroid Hormone Sensing In Vivo Using a Self-Templating Corona Phase Molecular Recognition. *Advanced Healthcare Materials*, *9*(21), 2000429. doi:10.1002/adhm.202000429 PMID:32940022

Lew, T. T. S., Koman, V. B., Silmore, K. S., Seo, J. S., Gordiichuk, P., Kwak, S.-Y., & Lee, M. A. (2020). Real-time detection of wound-induced H 2 O 2 signalling waves in plants with optical nanosensors. *Nature Plants*, *6*(4), 404–415. doi:10.103841477-020-0632-4 PMID:32296141

Li, B., Pan, W., Liu, C., Guo, J., Shen, J., Feng, J., & An, T. (2020). Homogenous magneto-fluorescent nanosensor for tumor-derived exosome isolation and analysis. *ACS Sensors*, *5*(7), 2052–2060. doi:10.1021/acssensors.0c00513 PMID:32594744

Li, D., Wang, N., Wang, F., & Zhao, Q. (2019). Boronate affinity-based surface-imprinted quantum dots as novel fluorescent nanosensors for the rapid and efficient detection of rutin. *Analytical Methods*, *11*(25), 3212–3220. doi:10.1039/C9AY00787C

Liu, C., Jin, Z., Cheng, L., Zhao, H., & Wang, L. (2020). Synthesis of nanosensors for autonomous warning of damage and self-repairing in polymeric coatings. *Nanoscale*, *12*(5), 3194–3204. doi:10.1039/C9NR09221H PMID:31967166

Liu, J., Li, F., Wang, Y., Pan, L., Lin, P., Zhang, B., & Ko, G. (2020). A sensitive and specific nanosensor for monitoring extracellular potassium levels in the brain. *Nature Nanotechnology*, *15*(4), 321–330. doi:10.103841565-020-0634-4 PMID:32042163

Liu, J., Pan, L., Shang, C., Lu, B., Wu, R., Feng, Y., & Xiong, Z. (2020). A highly sensitive and selective nanosensor for near-infrared potassium imaging. *Science Advances*, *6*(16), eaax9757. doi:10.1126ciadv.aax9757 PMID:32494594

Luo, X., Han, Y., Chen, X., Tang, W., Yue, T., & Li, Z. (2020). Carbon dots derived fluorescent nanosensors as versatile tools for food quality and safety assessment: A review. *Trends in Food Science & Technology*, *95*, 149–161. doi:10.1016/j.tifs.2019.11.017

Lupan, O. I., Chow, L., & Adelung, R. (2019). Nanosensors: Current Status and Perspectives.

Moradnia, F., Fardood, S. T., Ramazani, A., Osali, S., & Abdolmaleki, I. (2020). Green sol–gel synthesis of CoMnCrO 4 spinel nanoparticles and their photocatalytic application. *Micro & Nano Letters*, *15*(10), 674–677. doi:10.1049/mnl.2020.0189

Mousavi, S. M., Hashemi, S. A., Zarei, M., Amani, A. M., & Babapoor, A. (2018). Nanosensors for chemical and biological and medical applications. *Med Chem (Los Angeles)*, *8*(8), 2161-0444.1000515.

Munawar, A., Ong, Y., Schirhagl, R., Tahir, M. A., Khan, W. S., & Bajwa, S. Z. (2019). Nanosensors for diagnosis with optical, electric and mechanical transducers. *RSC Advances*, *9*(12), 6793–6803. doi:10.1039/C8RA10144B PMID:35518460

Nehra, M., Dilbaghi, N., Hassan, A. A., & Kumar, S. (2019). Carbon-Based Nanomaterials for the development of sensitive nanosensor platforms Advances in Nanosensors for Biological and Environmental Analysis. Elsevier.

Nißler, R., Bader, O., Dohmen, M., Walter, S. G., Noll, C., Selvaggio, G., Groß, U., & Kruss, S. (2020). Remote near infrared identification of pathogens with multiplexed nanosensors. *Nature Communications*, *11*(1), 1–12. doi:10.103841467-020-19718-5 PMID:33239609

Parashar, M., Shukla, V. K., & Singh, R. (2020). Metal oxides nanoparticles via sol–gel method: A review on synthesis, characterization, and applications. *Journal of Materials Science Materials in Electronics*, *31*(5), 3729–3749. doi:10.100710854-020-02994-8

Perveen, R., Shujaat, S., Qureshi, Z., Nawaz, S., Khan, M., & Iqbal, M. (2020). Green versus sol-gel synthesis of ZnO nanoparticles and antimicrobial activity evaluation against panel of pathogens. *Journal of Materials Research and Technology*, *9*(4), 7817–7827. doi:10.1016/j.jmrt.2020.05.004

Pinnaduwage, L., Boiadjiev, V., Hawk, J., & Thundat, T. (2003). Sensitive detection of plastic explosives with self-assembled monolayer-coated microcantilevers. *Applied Physics Letters*, *83*(7), 1471–1473. doi:10.1063/1.1602156

Prajapati, C. S., Benedict, S., & Bhat, N. (2019). An ultralow power nanosensor array for selective detection of air pollutants. *Nanotechnology*, *31*(2), 025301. doi:10.1088/1361-6528/ab44fd PMID:31530742

Prakash, J., Parveen, A., Mishra, Y. K., & Kaushik, A. K. (2020). Nanotechnology-assisted liquid crystals-based biosensors: Towards fundamental to advanced applications. *Biosensors & Bioelectronics*, *168*, 112562. doi:10.1016/j.bios.2020.112562 PMID:32919227

Pramanik, P. K. D., Solanki, A., Debnath, A., Nayyar, A., El-Sappagh, S., & Kwak, K.-S. (2020). Advancing modern healthcare with nanotechnology, nanobiosensors, and internet of nano things: Taxonomies, applications, architecture, and challenges. *IEEE Access: Practical Innovations, Open Solutions*, *8*, 65230–65266. doi:10.1109/ACCESS.2020.2984269

Rainbow, J., Sedlackova, E., Jiang, S., Maxted, G., Moschou, D., Richtera, L., & Estrela, P. (2020). Integrated electrochemical biosensors for detection of waterborne pathogens in low-resource settings. *Biosensors (Basel)*, *10*(4), 36. doi:10.3390/bios10040036 PMID:32294961

Saylan, Y., & Denizli, A. (2020). *Virus detection using nanosensors Nanosensors for Smart Cities*. Elsevier.

Sha, R., & Bhattacharyya, T. K. (2020). MoS2-based nanosensors in biomedical and environmental monitoring applications. *Electrochimica Acta*, *349*, 136370. doi:10.1016/j.electacta.2020.136370

Shafiq, M., Anjum, S., Hano, C., Anjum, I., & Abbasi, B. H. (2020). An overview of the applications of nanomaterials and nanodevices in the food industry. *Foods*, *9*(2), 148. doi:10.3390/foods9020148 PMID:32028580

Shahbazi, N., & Zare-Dorabei, R. (2019). A novel "off-on" fluorescence nanosensor for sensitive determination of sulfide ions based on carbon quantum dots and gold nanoparticles: Central composite design optimization. *Microchemical Journal*, *145*, 996–1002. doi:10.1016/j.microc.2018.12.022

Sharma, P., Pandey, V., Sharma, M. M. M., Patra, A., Singh, B., Mehta, S., & Husen, A. (2021). A Review on Biosensors and Nanosensors Application in Agroecosystems. *Nanoscale Research Letters*, *16*(1), 1–24. doi:10.118611671-021-03593-0 PMID:34460019

Shawon, Z. B. Z., Hoque, M. E., & Chowdhury, S. R. (2020). Nanosensors and nanobiosensors: Agricultural and food technology aspects. Nanofabrication for Smart Nanosensor Applications. Elsevier.

Shiravand, G., Badiei, A., Goldooz, H., Karimi, M., Ziarani, G. M., Faridbod, F., & Ganjali, M. R. (2020). A Fluorescent g-C3N4 nanosensor for detection of dichromate ions. *Current Analytical Chemistry*, *16*(5), 593–601. doi:10.2174/1573411014666180627150248

Singh, P., & Yadava, R. (2020). *Nanosensors for health care Nanosensors for Smart Cities*. Elsevier.

Skotadis, E., Aslanidis, E., Kainourgiaki, M., & Tsoukalas, D. (2020). Nanoparticles Synthesised in the Gas-Phase and Their Applications in Sensors: A Review. *Applied Nanoscience*, *1*(1), 70–86.

Su, Y.-F., Han, G., Kong, Z., Nantung, T., & Lu, N. (2020). Embeddable piezoelectric sensors for strength gain monitoring of cementitious materials: The influence of coating materials. *Engineered Science*, *11*(8), 66–75. doi:10.30919/es8d1114

Sun, C., Pan, L., Zhang, L., Huang, J., Yao, D., Wang, C.-Z., Zhang, Y., Jiang, N., Chen, L., & Yuan, C. (2019). A biomimetic fluorescent nanosensor based on imprinted polymers modified with carbon dots for sensitive detection of alpha-fetoprotein in clinical samples. *Analyst (London)*, *144*(22), 6760–6772. doi:10.1039/C9AN01065C PMID:31617507

Szperlich, P., & Toroń, B. (2019). An ultrasonic fabrication method for epoxy resin/SbSI nanowire composites, and their application in nanosensors and nanogenerators. *Polymers*, *11*(3), 479. doi:10.3390/polym11030479 PMID:30960463

Tsao, Y.-H., Husain, R. A., Lin, Y.-J., Khan, I., Chen, S.-W., & Lin, Z.-H. (2019). A self-powered mercury ion nanosensor based on the thermoelectric effect and chemical transformation mechanism. *Nano Energy*, *62*, 268–274. doi:10.1016/j.nanoen.2019.05.032

Ur Rahim, H., Qaswar, M., Uddin, M., Giannini, C., Herrera, M. L., & Rea, G. (2021). Nano-enable materials promoting sustainability and resilience in modern agriculture. *Nanomaterials (Basel, Switzerland)*, *11*(8), 2068. doi:10.3390/nano11082068 PMID:34443899

Venkatesan, M., Veeramuthu, L., Liang, F.-C., Chen, W.-C., Cho, C.-J., Chen, C.-W., Chen, J.-Y., Yan, Y., Chang, S.-H., & Kuo, C.-C. (2020). Evolution of electrospun nanofibers fluorescent and colorimetric sensors for environmental toxicants, pH, temperature, and cancer cells–A review with insights on applications. *Chemical Engineering Journal*, *397*, 125431. doi:10.1016/j.cej.2020.125431

Voke, E., Pinals, R. L., Goh, N. S., & Landry, M. P. (2021). In planta nanosensors: Understanding Biocorona formation for functional design. *ACS Sensors*, *6*(8), 2802–2814. doi:10.1021/acssensors.1c01159 PMID:34279907

Wang, M., Zhang, M., Wang, Y., Zhao, R., & Yan, S. (2019). Fano resonance in an asymmetric MIM waveguide structure and its application in a refractive index nanosensor. *Sensors (Basel)*, *19*(4), 791. doi:10.339019040791 PMID:30781360

Wang, S., Corredor Garcia, J. L., Davidson, J., & Nichols, A. (2020). Conductance-Based Interface Detection for Multi-Phase Pipe Flow. *Sensors (Basel)*, *20*(20), 5854. doi:10.339020205854 PMID:33081178

Wang, Z., Hao, Z., Yu, S., De Moraes, C. G., Suh, L. H., Zhao, X., & Lin, Q. (2019). An ultraflexible and stretchable aptameric graphene nanosensor for biomarker detection and monitoring. *Advanced Functional Materials, 29*(44), 1905202. doi:10.1002/adfm.201905202 PMID:33551711

Wong, C. W., San Chan, Y., Jeevanandam, J., Pal, K., Bechelany, M., Abd Elkodous, M., & El-Sayyad, G. S. (2020). Response surface methodology optimization of mono-dispersed MgO nanoparticles fabricated by ultrasonic-assisted sol–gel method for outstanding antimicrobial and antibiofilm activities. *Journal of Cluster Science, 31*(2), 367–389. doi:10.100710876-019-01651-3

Wu, H., Nißler, R., Morris, V., Herrmann, N., Hu, P., Jeon, S.-J., Kruss, S., & Giraldo, J. P. (2020). Monitoring plant health with near-infrared fluorescent H2O2 Nanosensors. *Nano Letters, 20*(4), 2432–2442. doi:10.1021/acs.nanolett.9b05159 PMID:32097014

Xiong, W., Ndokoye, P., & Leung, M. K. (2020). Noble Metal–Based Nanosensors for Environmental Detection. Advanced Nanomaterials for Pollutant Sensing and Environmental Catalysis. Elsevier.

Xu, L., Shoaie, N., Jahanpeyma, F., Zhao, J., Azimzadeh, M., & Al, K. T. (2020). Optical, electrochemical, and electrical (nano) biosensors for detection of exosomes: A comprehensive overview. *Biosensors & Bioelectronics, 161*, 112222. doi:10.1016/j.bios.2020.112222 PMID:32365010

Yadegari, A., Khezri, J., Esfandiari, S., Mahdavi, H., Karkhane, A. A., Rahighi, R., Heidarimoghadam, R., Tayebi, L., Hashemi, E., & Farmany, A. (2019). Bottom-up synthesis of nitrogen and oxygen co-decorated carbon quantum dots with enhanced DNA plasmid expression. *Colloids and Surfaces. B, Biointerfaces, 184*, 110543. doi:10.1016/j.colsurfb.2019.110543 PMID:31627102

Yan, Z., Yuan, H., Zhao, Q., Xing, L., Zheng, X., Wang, W., Zhao, Y., Yu, Y., Hu, L., & Yao, W. (2020). Recent developments of nanoenzyme-based colorimetric sensors for heavy metal detection and the interaction mechanism. *Analyst (London), 145*(9), 3173–3187. doi:10.1039/D0AN00339E PMID:32222739

Yang, T., & Duncan, T. V. (2021). Challenges and potential solutions for nanosensors intended for use with foods. *Nature Nanotechnology, 16*(3), 251–265. doi:10.103841565-021-00867-7 PMID:33712739

Zahra, Z., Habib, Z., Chung, S., & Badshah, M. A. (2020). Exposure route of TiO2 NPs from industrial applications to wastewater treatment and their impacts on the agro-environment. *Nanomaterials (Basel, Switzerland), 10*(8), 1469. doi:10.3390/nano10081469 PMID:32727126

Zhou, D., Huang, H., Wang, Y., Wang, Y., Hu, Z., & Li, X. (2019). A yellow-emissive carbon nanodot-based ratiometric fluorescent nanosensor for visualization of exogenous and endogenous hydroxyl radicals in the mitochondria of live cells. *Journal of Materials Chemistry. B, Materials for Biology and Medicine, 7*(23), 3737–3744. doi:10.1039/C9TB00289H

Zhu, H.-W., Ge, J., Zhao, H.-Y., Shi, L.-A., Huang, J., Xu, L., & Yu, S.-H. (2020). Sponge-templating synthesis of sandwich-like reduced graphene oxide nanoplates with confined gold nanoparticles and their enhanced stability for solar evaporation. *Science China Materials, 63*(10), 1957–1965. doi:10.100740843-020-1446-5

Zopf, D., Pittner, A., Dathe, A., Grosse, N., Csáki, A., Arstila, K., & Uhlrich, G. (2019). Plasmonic nanosensor array for multiplexed DNA-based pathogen detection. *ACS Sensors, 4*(2), 335–343. doi:10.1021/acssensors.8b01073 PMID:30657315

Chapter 3
Nano-Structured Spinel Ferrites and Their Applications as Antimicrobial Agents, Gas Sensors, and Dye Adsorbents

Laxmi J. Hathiya
https://orcid.org/0000-0001-5466-8768
Saurashtra University, India

Hiren H. Joshi
Saurashtra University, India

ABSTRACT

Ferrites–the electro-ceramic materials have been considered highly important magnetic materials for more than half a century. The crystal structure of ferrimagnetic oxides falls into one of the following categories: spinel, magnetoplumbite(hexagonal), garnet, and orthoferrite. The development of spinel ferrite nanoparticles (SFNPs) has become more popular owing to their fascinating properties, such as their potential adsorption properties, superparamagnetic behavior, high stability, high value of resistivity, saturation magnetization, coercivity, low power losses, and ease of functionalization. These advantageous features make them suitable for use in various applications. This chapter aims to discuss the structure, general properties, and some recent applications of spinel ferrites. Commonly used synthesis techniques for nanocrystalline ferrites are discussed. The applications of nano-structured ferrites like antimicrobial activity, gas sensors, and dye degradation are discussed.

DOI: 10.4018/978-1-6684-7358-0.ch003

INTRODUCTION

Nanomaterials are particles with one or more external and internal structure dimensions or surface dimensions falling in the range of 1 nm–100 nm (Rana et al., 2021). When the particle size of materials is reduced to the nanoscale, they show very different properties compared to what they show on the macroscale (Jacob & Khadar, 2010; McHenry & Laughlin, 2000). The nano-sized particles possess a high surface area hence, the effectiveness also gets high. There are two basic types of fabrication methods for nanomaterials: top down and bottom up. Nanomaterials can be constructed by a top-down approach, producing very small structures from larger pieces of the material. For example, laser ablation, ball-milling, arc-discharge methods, etc. They can also be fabricated by bottom-up techniques, producing from atom by atom or molecule by molecule. i.e., chemical vapor deposition, sol-gel method, co-precipitation method, etc. Nanomaterials are important in various areas from basic research to various applications in electronics, medical, catalysis, cosmetics, and energy.

The history of ferrites began with the discovery of stones that would attract iron. A large amount of these stones was found in the district of Magnesia in Asia Minor and hence this mineral is named magnetite (Fe_3O_4) (Shaikh et al., 2016). The first application of magnetite was the "Lodestones", which were used by early navigators to detect magnetic poles (Mills, 2004). Lodestone is a naturally available mineral that is rich in iron oxide and hence behaves like a permanent magnet. The etymology of ferrites is from the Latin word 'Ferrum' which meaning is iron (Garcia-Munoz et al., 2020). Ferrites are a large class of oxides containing Fe^{3+} and other metal cations that have been investigated and applied in many fields for more than 50 years. They are classified according to their crystal structure and the arrangement of oxygen anions around the metal cations, such as spinel, garnet, magnetoplumbite, and orthoferrite. Among the various types of ferrites, spinel ferrites are widely used in electrical, magnetic, and medical applications. The systematic study on the relationship between chemical composition and magnetic properties of various ferrites was firstly reported by Hilpert in 1909 (Hilpert, 1909). At that time, Hilpert had successfully prepared spinel ferrites, such as magnesium, cobalt, manganese, copper, and zinc. The commendable aspects of his work were that he already had attempted to prepare barium, lead, and calcium ferrites and for that, he had applied for a German patent (Hilpert, 1909) on his fore-handed idea of the ferrite core.

An important ferrite material is magnetite, $Fe^{2+}Fe^{3+}{}_2O_4$ (typically known as Fe_3O_4), a natural spinel oxide and probably the oldest magnetic material which is a very active research field at present, with the practical applications due to its interesting properties associated with the coexistence of ferrous and ferric cations. Magnetite is naturally available in large quantities, chemically stable, and environmentally robust. It is a weak hard ferrite material. Another important material by its structure and applications is maghemite or γ-Fe_2O_3, which can be considered a defective spinel (Valenzuela, 2012).

BACKGROUND

The spinel structure of ferrites was first determined by Bragg and Nishikawa in 1915 (Murdock & Simmons, 1992; Valenzuela, 1994). About 2 decades after Hilpert's work, Forestier started the chemical study on the preparation of various ferrites as well as measurements of their saturation magnetization and Curie temperature (Forestier, 1928). However, he could not get success in improving the magnetic properties of the ferrites to that of a commercial level.

The actual use of ferrite materials was begun in 1930 (Gore & Jadhav, 2020) and its structural, electrical, and magnetic properties are considerably studied by many researchers including Snoek (Snoek, 1947). Modern telecommunication technology would not be possible without ferrites. The large-scale production of television in the 1950s was the main reason for the expansion of the ferrite industry (Jaswal & Singh, 2014). The first commercial soft ferrite product in Korea was the radio antenna core in 1970. In 1976, Sam wha electronics Co. Ltd prepared high-quality soft ferrite for use in new technology.

ICF (International Conference on Ferrites) has provided a forum for exchanging information on the science and technology of ferrites and related materials. Various ICFs have taken place from 1970 to 2000. During this period many researchers have reported results on the ferrites.

Nano-structured ferrites have extraordinary electric and magnetic properties that are comparatively different from micro sized, tailoring them to provide various applications such as high-density information storage (Kasapoglu et al., 2007), photocatalysis (Cao et al., 2009; Amir et al., 2015), ferrofluids (Harris et al., 2009), magnetic drug delivery (Qu et al., 2006), gas sensors (Liu et al., 2005), etc.

MAIN FOCUS OF THE CHAPTER

Classification of Ferrites

According To Magnetic Properties

Ferrites can be divided into two families based on their magnetic coercivity (their resistance to being demagnetized) (Okamoto, 2009).

Hard Ferrites

Hard ferrites have high coercivity which means they are difficult to demagnetize. These ferrites are also known as permanent magnetic materials as they can hold their magnetism after being magnetized. They form a large class of ceramic materials. Hard ferrites are compounds of iron and barium or strontium oxides. The color of hard ferrites varies from dark grey to black. They are very hard and brittle. They have high saturation flux density, high eddy current losses, low susceptibility and permeability, high retentivity and coercivity (Vedratnam et al., 2020). Therefore, they are used to make magnets for devices such as refrigerator, loudspeakers, and small electric motors.

Soft Ferrites

Soft ferrites are ferrimagnetic materials that do not hold their magnetism after being magnetized. They give a steeply rising magnetization curve, narrow hysteresis loop, and small energy losses during magnetization. Soft ferrites have low coercivity hence, they can easily be magnetized and demagnetized. Soft ferrites are ceramic insulators with cubic structure MFe_2O_4, where M is the transition metal ions. They are used to make ferrite cores for inductors, transformers, switched-mode power supplies, and microwave components because of their low coercivity, high permeability, high Curie temperature, and low losses at high frequencies (Bhandare et al., 2017; Anupama et al., 2018; Assar et al., 2014).

In 1930, Japan and the Netherlands started research on soft ferrites. However, in 1945, J. L. Snoek from the Philips Research Laboratories, Netherlands successfully synthesized soft ferrites for commercial applications (Shaikh et al., 2020). The term soft ferrites have no concern with their physical properties but refer to their magnetic characteristics. Soft ferrites are ceramic electromagnetic materials that are dark grey or black in appearance with very hard and brittle properties. The magnetic properties of soft ferrites arise from interactions between metallic ions occupying positions relative to the oxygen ions in their spinel crystalline structure (Naseri et al., 2014).

According to the Crystal Structure

Based on the crystal structure ferrites can be classified into four types i.e., spinel, garnet, ortho, and hexagonal ferrites (Table 1).

Table 1. classification of ferrites according to crystal structure

Sr. No.	Types	Molar ratio	Representation
1	Spinel	Fe_2O_3–1MO	MO – Transition metal oxide
2	Garnet	$5Fe_2O_3$–$3M_2O_3$	M_2O_3 – Rare earth metal oxide
3	Ortho	$AFeO_3$	A – Rare earth elements (Yb, Ho, Dy, Y, Er, etc.)
4	Hexagonal	$6Fe_2O_3$–1MO	MO – Divalent metal oxide from the IIA group (CaO, BaO, SrO, etc.)

1. Spinel Ferrites

Spinel ferrites (SFs) are magnetic iron oxide containing a cubic spinel structure with a general formula AB_2O_4 where A is the divalent metal ion like Mn, Ni, Zn, Co, Fe, Mg, etc., B is the trivalent metal ion such as Fe, Al or Cr, and O is Oxygen. Spinel ferrites have the crystal structure of the natural spinel $MgAl_2O_4$ which was firstly determined by Bragg (Bragg et al., 1915).

Figure 1. Crystal structure of spinel ferrite

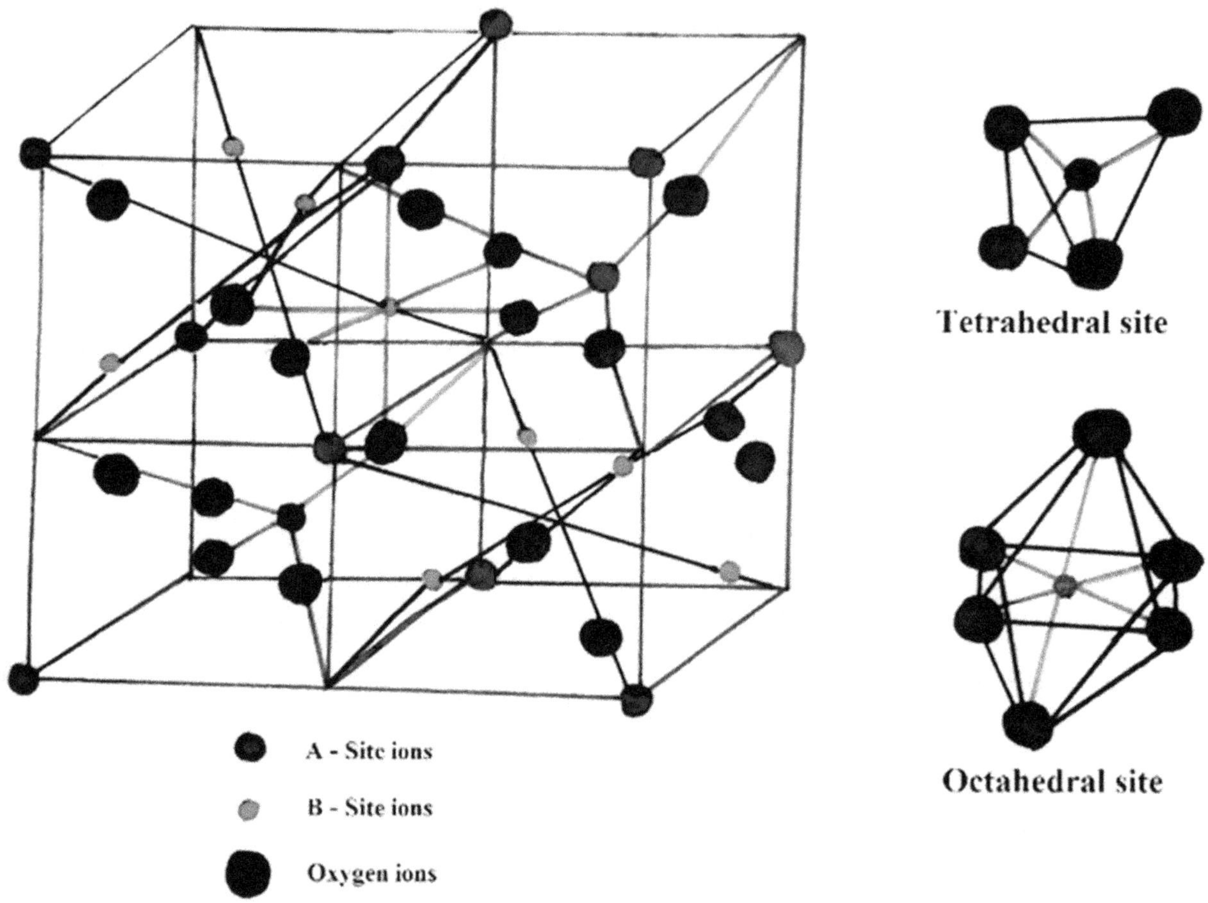

As shown in figure 1 in the spinel structure the oxygen ions are closely packed into *an fcc* (face-centered cubic) lattice having two kinds of interstitial sub-lattice for the metal ions which are named as tetrahedral (A) site and octahedral [B] site (Han et al., 2004). Spinel ferrites generally crystalize in the cubic form with the Fd-3m space group. The unit cell of spinel ferrite crystal structure consists of a cubic closed-pack arrangement of oxygen ions with 64 tetrahedral (A-site) and 32 octahedral interstitial sites [B-site]. Out of this, 8 tetrahedral (A-site) and 16 octahedral [B-site] sites are occupied by the metal cations (Vasic et al., 2013). Spinel ferrites have remarkable characteristics of magnetic and insulating properties. They are soft magnetic materials with higher electrical resistivity, and lower eddy current losses (Singh et al., 2012; Sivakumar et al., 2010), high specific heat, low melting point, chemical stability, mechanical hardness (Moran et al., 2002; Melagiriyappa & Jayanna, 2009), etc.

A notable characteristic of spinel structure is that it is able to form an exceedingly large variety of total solid solutions. That means the composition of ferrite can be strongly modified, while the basic crystalline structure remains the same. For example, the Zn-Ni system, with the general formula $Zn_xNi_{1-x}Fe_2O_4$, where $0 \leq x \leq 1$. In this system, the parent ferrite materials are $NiFe_2O_4$ (for $x = 0$) and $ZnFe_2O_4$ ($x = 1$). $ZnFe_2O_4$ is a normal spinel whereas $NiFe_2O_4$ is an inverse spinel and the properties of the compositions of these ferrites are different. $NiFe_2O_4$ is ferrimagnetic with a Curie temperature »858 K, while $ZnFe_2O_4$

is antiferromagnetic, with a Neel temperature of about 9 K. This means that the general properties of the ferrite can be easily tailored just by varying the composition (Valenzuela, 2012).

Spinel ferrites are widely used in many applications due to their availability, effectiveness, and flexibility. Spinel ferrites have been extensively used in several applications including recording heads, antenna rods, loading coils, microwave devices, and core material for power transformers in electronics and telecommunication applications (Adam et al., 2002; Harris et al., 2009; Kulikowski, 1984). SFs are used to prepare the magnetic fluids (Lazarevic et al., 2012; Gozuak et al., 2009), in magnetic recording (Singh et al., 2012; Saez-Puche et al., 2008), drug delivery (Yadav et al., 2015; Lazarevic et al., 2012; Gozuak et al., 2009), microwave (Moran et al., 2002; Nejati K & Zabihi, 2012), radio frequency application (Souriou et al., 2010), digital tapes (Yadav et al., 2015), catalyst (Paramesh et al., 2016; Nejati K & Zabihi R, 2012), switching circuits in computers (Paramesh et al., 2016), magnetic refrigeration system (Saez-Puche et al., 2008), sensors (Lazarevic et al., 2012; Sutka & Gross, 2016; Rathore et al., 2015), magnetic resonance imaging (Gozuak et al., 2009), etc.

2. Garnet Ferrites

The garnet ferrites accommodate a large trivalent rare earth ion of large magnetic moments (Moretti & Ottonello, 1998). This type of ferrites reveals the structure of the silicate mineral garnet. They crystallize in the 12-sided structure or dodecahedral. The general formula of garnet ferrites is $Me_3Fe_5O_{12}$. In these magnetic garnets, Me is usually yttrium (Y) or one of the rare earth ions (Nur-E-Alam et al., 2018). The crystal structure and the unit cell dimensions of the rare-earth ion garnets were first reported by Bertaut and Forrat (Cunningham & Anderson, 1960) and then by Geller and Gilleo (Geller & Gilleo, 1957; Gilleo & Geller ,1958). The garnet structure is very stable as all the sites are occupied by cations, giving high stability to the compound (Song et al., 2019).

3. Ortho Ferrites

The ortho ferrites have the generic formula $RFeO_3$ where R is yttrium or a rare earth ion (Bamzai & Bhat, 2014). These ferrites are cubic ferrites with slightly distorted perovskite structures. This structure is often acquired by a material, which has a complicated molecular arrangement consistent with cubic symmetry and the structure gets orthorhombic rather than cubic (Shaikh et al., 2020). The rare earth ion perovskites are also known as ortho ferrites (Haye et al., 2015). They were first studied by Forestier and Guillot Guillain (Shaikh et al., 2020).

4. Hexagonal Ferrites

The hexagonal ferrites (HFs) are a group of ferrites possessing a hexagonal crystal structure with a unique c axis. This class of ferrites suggests magnetoplumbite structure hence, the symmetry of magnetoplumbite structure is hexagonal. Moreover, it has a major preferred axis called the c-axis and a minor axis called a-axis (Delacotte et al., 2018). HFs are ferrimagnetic and their magnetic properties are associated with their crystalline structures. The common formula of hexagonal ferrite is $MFe_{12}O_{19}$, where M is usually Ba, Pb, Sr, etc. (Pullar, 2012; Belrhazi et al., 2019; Went et al., 1952). Hexaferrites can be divided into two main groups: uniaxial hexaferrite and hexaplana ferrites. The lattice and structure of HFs are alike, with closely packed oxygen ions. The lattice has three different sites which are octahedral, trigonal bipyramid, and tetrahedral occupied by metals. There are six types of hexaferrites indicated by M, W, Y, X, Z, and U letters (Dairy et al., 2019; Jotania, 2014; Izadkhah et al., 2017; Mahmood et al., 2018; Stergiou & Litsardakis, 2016; Kitagawa et al., 2010). M, W, Y, X, Z, and U consist (MO + MeO) and Fe_2O_3 in the ratios of 1:6, 3:8, 4:6, 4:14, 5:12, and 6:18, respectively where M can be Ca, Ba, La, Pb, Sr ions, Me is a transition metal ion like Mn, Mg, Co, Zn or a combination of cations similar to spinel. Moreover, the replacement of Fe^{3+} ions with other trivalent cations such as Ga^{3+}, Al^{3+}, and In^{3+} is also possible. In the crystalline structure of HFs, the divalent and trivalent metallic cations are located in interstitial sites of the structure, while heavy ions like Ba or Sr enter the oxygen layers. Many Hexaferrites contain two molecular units for M, W, and Z-type ferrite, three molecular units for X and Y-type ferrite, but only one molecular unit for U-type ferrite (Vedrtnam et al., 2020).

Most of these materials are ferrimagnetic, but some are also antiferromagnetic (Hibst, 1982). Hexaferrites are used in permanent magnets, microwave devices, recording heads, home appliances, electronic communications, radar absorbing, inductor coils, etc (Jotania, 2014; Stergiou & Litsardakis, 2016).

SOLUTIONS AND RECOMMENDATIONS

Classification of Spinel Ferrites

Based on the favorable occupancy of metal ions in the sites, the spinel ferrites can be divided into three types: (i) normal ferrites (where all divalent cations occupy A-site), (ii) inverse ferrites (where all divalent cations occupy B-site) and (iii) mixed ferrites (where divalent cations are distributed in A and B-sites) (Ramakrishna et al., 2018; Hajalilou, 2017).

1. Normal Spinel

In the normal spinel structure, divalent metallic cations (Me^{2+}) occupy the tetrahedral site whereas, trivalent Fe^{3+} cations accommodate the octahedral site. The general formula for normal spinel ferrites is $(Me^{2+})\ [Fe^{3+}]_2\ O_4$, the square bracket indicates B-site occupancy. Example: $(Zn)^A\ [Fe_2]^B\ O_4$

2. Inverse Spinel

In this type of spinel, divalent cations are on B-site, but trivalent cations are equally distributed among the A and B-sites. The general formula for inverse spinel is $(Fe^{3+})\ [Me^{2+}\ Fe^{3+}]\ O_4$. Example: $(Fe)^A\ [Ni\ Fe]^B\ O_4$

3. Mixed Spinel

In some cases, the divalent cation has not any specific site preference among A and B-site, and thus it circulates in both sites leading to a formation of mixed spinel. The general formula for mixed spinel ferrites can be given as $(Me^{2+}_{1-\delta} Fe^{3+}_{\delta})^A\ [Me^{2+}_{\delta} Fe^{3+}_{2-\delta}]^B\ O_4$, where 'δ' is the degree of inversion. The degree of inversion depends on the synthesis method (Kambale et al., 2010). Example: $(Mg_{0.120}\ Fe_{0.880})^A\ [Mg_{0.880}\ Fe_{1.120}]^B\ O_4$ (Hathiya et al., 2022).

Synthesis Methods

The synthesis methods play a significant role to produce high-quality ferrite materials, in terms of purity, morphology, and stability. Various synthesis methods are being used for ferrite material preparation and each synthesis technique has its advantages and disadvantages. The basic categories of synthesis methods are:

(i) Top-down
(ii) Bottom-up

In the top-down approach, materials that might be in micron size are pulverized to form nano-sized particles whereas, in the bottom-up, the ions are chemically combined to form the fine particles. The techniques for preparing spinel ferrite nanoparticles are divided into two categories: (i) chemical and (ii) conventional solid-state routes. Each of them has its benefits and drawbacks. For example, by applying the chemical technique, it is possible to produce finer crystallite-sized particles at lower temperatures as compared to the conventional solid-state routes. However, the chemical homogeneity of the solution is often disturbed by the differences in the chemistries of the cations. To avoid this problem, the starting precursor solution containing various forms of metal complexes must be chemically modified (Kakihana, 1996). On the other hand, the conventional solid-state routes are relatively simple techniques but sometimes it is difficult to control on powder quality and particle size during the process. Furthermore, this type of synthesis process needs a higher temperature, and as a result, an expensive furnace is required.

Figure 2. Nanomaterial synthesis routes

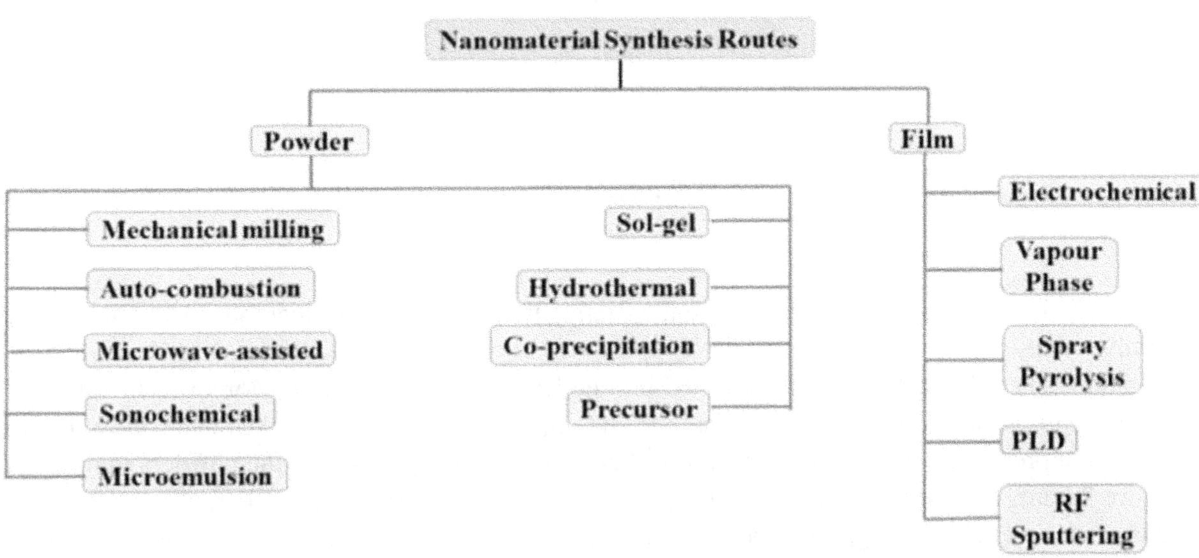

Figure 2 shows commonly used synthesis routes to prepare nanomaterial. The flow chart contains both top-down and bottom-up synthesis routes. The materials are being prepared in the form of powder and film at the nanoscale by using these synthesis techniques.

Some of the commonly used synthesis techniques for spinel ferrite powders are described as follows:

1. Solid-State

The term solid-state reaction suggests that there are not any interactions with a controlled vapor phase or a solvent medium. The solid-state method is known as the simplest route to synthesizing various materials. This method produces polycrystalline ferrite nanomaterials from solid reagents at relatively high temperatures (Alarifi et al., 2009). This technique is mostly applied in the fabrication of bulk crystals, glasses, ceramics, and piezo electrics. Many researchers have performed this method to prepare nanocrystalline spinel ferrites (Qin et al., 2009; Jeseentharani et al., 2013).

Figure 3. Synthesis process of solid-state method

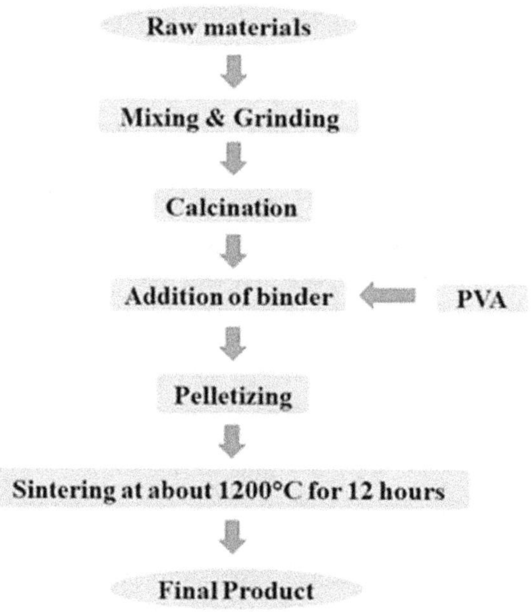

Figure 3 shows the general steps for sample preparation by solid-state method. In this technique generally oxides are used as raw materials. They are mixed and ground and then calcined at preferred temperature. Then the material is transformed into pellets and being sintered. Moreover, this method includes the grinding and mixing of metal nitrates or sulphates with NaOH or NaCl in an agate mortar or mills for a preferred time. After the removal of NaCl or NaOH by washing, the powders may dry at 80 °C for 2 hours and then annealed at 700 °C for 2 hours (Jeseentharani et al., 2013).

It is a low-cost, time-saving, convenient, environment-friendly, and low-energy consumption process (Varma et al., 2008). Furthermore, in comparison to other synthesis techniques, the solid-state process does not require a complex apparatus, solvent, or solution. Some drawbacks of this technique are long milling time, sintering at higher temperatures for long times, and for that, it needs an expensive furnace, and sometimes it is difficult to control the formation of unwanted phases (Ghazanfar et al., 2005; Zahi et al., 2007). Similar to ceramic process, in the solid-state reaction technique, the precursor metal oxides are mixed and ground in a pestle mortar and annealed at high temperatures followed by grinding in pestle mortar to ensure homogenization of the precursors. Afterward, second annealing is to be performed to obtain ferrite powders. Some authors have synthesized cobalt ferrite nanoparticles employing a solid-state reaction technique and observed the saturation magnetization of 57.0 emu/g and coercivity of 590 Oe (Sheenu et al., 2016; Pandit et al., 2014).

2. Mechanical Milling

Mechanical milling is an example of a ''top-down'' approach to the synthesis of ferrite nanoparticles. Mechanical milling is a typical technique for the preparation of NPs using high-energy planetary balls, shakers, or tumbler mills (Marinca et al., 2016). There are different types of mills available when working with the ball milling synthesis method namely planetary mill, vibrating mill, oscillating mill, etc. Amongst all, the planetary mill is efficient in the production of iron oxide nanoparticles (Antony et al., 2021). The name of this milling type is called, as the milling container rotates in two separate parallel axes in a manner just like the revolution of a planet around the sun (Arbain et al., 2011). This method is widely used for metals, cement, glass, ceramics, silicate products, fireproof materials, chemical fertilizers, etc. (Srivastava & Yadav, 2012).

The major advantage of mechanical milling is that it offers the possibility of scaling up to a large number of materials while its most serious negative point is the contamination of NPs during longtime milling (Manova et al., 2004) by using milling tools during milling, which may change the stoichiometry of the as-obtained SFs (Marinca et al., 2016). The production of nanomaterials using planetary ball-milling also has been developed on an industrial scale (Fokina et al., 2004). The particle size and properties of the ferrite nanoparticles are highly dependent on the rotational speed and rotational time of the ball milling. Furthermore, the material-to-ball ratio (in gram) is also an important factor in ball-milling, and generally, it is taken as 1:10. Dry milling and wet milling are both techniques that can be done for material fabrication.

Figure 4. Synthesis chart of ball-milling method

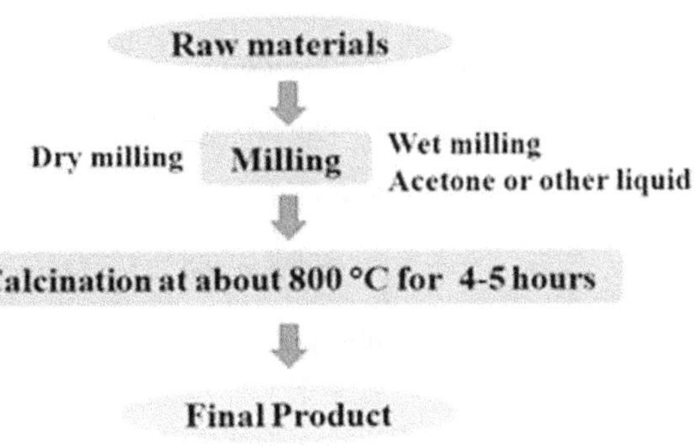

The synthesis of ferrite nanoparticles (FNPs) by mechanical milling method is relatively simple and cost-effective. A number of FNPs have been synthesized using the mechanical milling technique. Figure 4 shows the ball-milling method for sample preparation. This method includes less steps as compared to other synthesis methods. In this technique, the raw materials and the balls are taken as per the preferred ratio. Then milling is done for several hours at fixed rpm. In the end, the sample is calcined to get the final product. Some Spinel ferrites synthesized by this method are $CuFe_2O_4$ (Marinca et al., 2013), $CoFe_2O_4$ (Manova et al, 2004), $NiFe_2O_4$ (Zhang et al., 2015; Yan et al., 2015), and $Ni_{1-x}Mn_xFe_2O_4$ (Marinca et al., 2016).

3. Auto-Combustion

Auto-combustion synthesis method (also called auto-ignition, low-temperature self-combustion, or self-propagation) which includes the combination of formation of chemical sol-gel and combustion process, which has shown great potential in the fabrication of spinel-type ferrite nanoparticles (Sutka & Mezinskis, 2012). This method is also used to synthesize advanced spinel ferrites, the one-dimensional (1D) and two-dimensional (2D) nanostructures (Guo et al., 2010; Xiang et al., 2009; Zhang et al., 2007; Gupta et al., 2005). This technique has been employed for the synthesis of many different spinel ferrites MFe_2O_4, where M could be Mn, Ni, Cu, Cd, Mg, Li, Zn, and Co ion or their composition (Sutka et al., 2011; Azadmanjiri et al., 2007; Randhawa et al., 2007; Sutka et al., 2010; Thant et al., 2010; Yue et al., 2000; Yue et al., 2001; Nayak P K, 2008; Selvan et al., 2003; Lee et al., 2007). The auto-combustion method involves exothermic and self-propagating redox auto-combustion reaction and does not require a high-temperature sintering process (Hathiya et al., 2022). This technique offers a simple preparation process, low processing time, good yield of nano ferrite products with the expected stoichiometry, high degree of homogeneity, and controlled particle size. (Patil et al., 2002; Dalt et al., 2009; Mukasyan et al., 2007; Costa et al., 2007). Generally, in this method, metal nitrates are used as precursors. In the auto-combustion technique, some organic fuels are used which trigger the combustion process. Most popular fuels are urea ($CO(NH_2)_2$), citric acid ($C_6H_8O_7$), and glycine (NH_2CH_2COOH) (Sutka & Mezinskis, 2012). Additionally, NH_4OH or NH_4NO_3 are used to modify the nature of the combustion reaction.

Figure 5. Synthesis process of auto-combustion technique

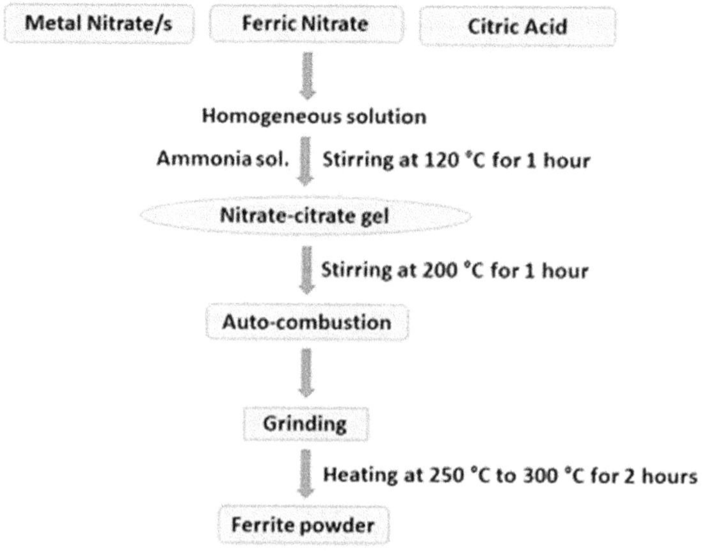

Figure 5 displays the synthesis process of ferrite nanoparticles by auto-combustion technique. As shown in the figure all the metal nitrates salts would dissolve in double distilled water separately and the required amount of aqueous citric acid solution is added as a coordinating agent(fuel). An appropriate amount of ammonia solution can be added to this nitrate-citrate solution to adjust the pH value. For the combustion initiation, different heating devices such as muffle furnace (Xue et al., 2007) or hot plate (Hathiya et al., 2020) may be used. The mixed solution is then stirred and heated at about 120°C to 200°C for 2-3 hours to obtain a highly viscous gel. After that, a fast flameless auto-combustion reaction starts which produces the burnt ferrite powder. The combustion starts at the hottest zones of the beaker and forms fluffy ash from the bottom to the top of the beaker like the eruption of a volcano. This burnt ferrite product appears like a branched tree (Hathiya et al., 2022). The reaction continues till the whole citrate complex is consumed (Bhukal & Singhal, 2014). Finally, the burnt ferrite is ground and baked into a muffle furnace from 250 °C to 300 °C for 2 hours.

4. Sol-gel

The sol-gel method is a widely employed chemical route to synthesize the spinel ferrite in the form of ultrafine powder. In this method, the metal alkoxide solution undergoes hydrolysis and condensation polymerization reactions to produce a gel. This technique offers the possibility of controlling not only the size and distribution of particles but also their shape. e.g., powders, fibers, and films.

Figure 6. Flow chart of sol-gel synthesis method

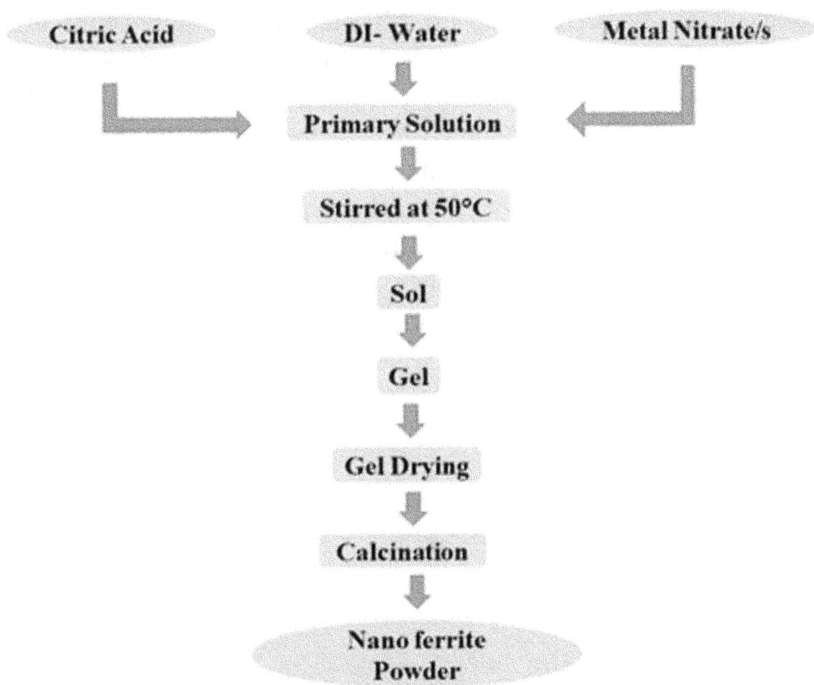

The process involves the preparation of a sol, which is a dispersion of solid particles of metal salts (Figure 6). Then adding water which modifies the pH of the sol and reduces the repulsion between particles. This results in a large increase in the viscosity of the system which forms the gel (Valenzuela, 1994). In the end, the water molecules are condensed and the gel phase changes into the powder form (Kolahalam et al., 2019). The size of the particles depends on the type of organic solvent, the temperature, and the rate of removal of the liquid. Some additional heat can be given to get the fine crystalline powder of spinel ferrites (Sriondee et al., 2018; Heju et al., 2019; Yu et al., 2019).

The advantages of this method include simple, low cost, can be carried out at a relatively low temperature, and achievable of narrow particle size distribution (Mukhtar et al., 2015; Jesudoss et al., 2016; Sharma et al., 2015). However, some essential care should be taken for better control of the reaction, the factors are stirring rate, the concentration of sol, and annealing temperature (Carta et al., 2009). The major drawback of this method is the lack of purity of the final product, and thus thermal treatment is essential after the synthesis to get the required purity of the sample. However, composition control, homogeneity, particle size, and particle distribution can be well achieved by this method. Some SFs synthesized by this method are $CoFe_2O_4$ (Olsson et al., 2005; Venturini et al., 2016), $ZnFe_2O_4$ (Mukhtar et al., 2015), $NiFe_2O_4$ (Sharma et al., 2015), $MnFe_2O_4$ (Carta et al., 2009), $CuFe_2O_4$ (Sharma et al., 2015), and Pr-substituted Mg-Zn ferrites (Mukhtar et al., 2015).

5. Co-Precipitation

Co-precipitation synthesis is one of the most popular techniques which is widely used to prepare ferrite nanoparticles. This method involves the formation of aqueous solutions of chlorides, sulphates, or nitrates of Fe^{3+} and divalent metal ions such as Ni, Mg, Co, Mn, etc., in the required concentrations for the formation of ferrite or its composition. The name of this synthesis process suggests there are simultaneous precipitation forms with the help of NaOH (Co-precipitating agent). In this method, the homogeneity of the ferrite product is associated with pH and changes in the concentration of the suspension during the precipitation. The different metal hydroxides can have different solubility depending on pH, and the cation ratios in the precipitations can be different from those in the solution (Valenzuela, 1994). To solve this problem, the solution of metal ions is usually added to the alkaline solution dropwise and slowly (McColm & Clark, 1988). And if the alkaline solution is added to the solution of metal ions, the pH of the suspension must be brought to a value above 10 to avoid differences in precipitate composition (Takada & Kiyama, 1971).

Figure 7. synthesis process by co-precipitation technique

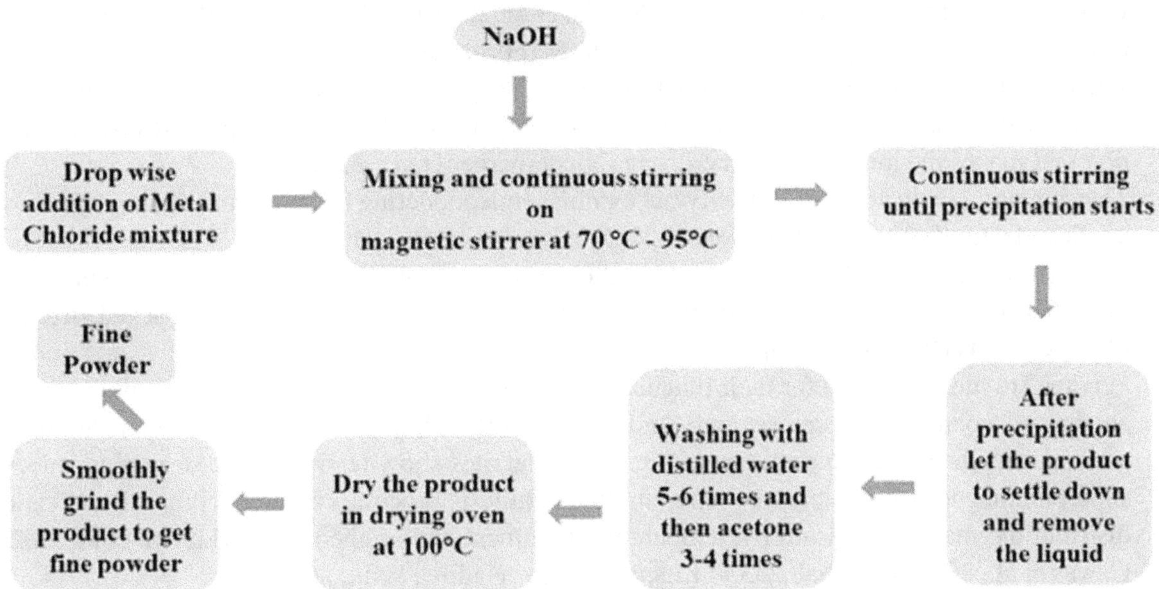

As shown in figure 7 in the process of this synthesis method, the aqueous solutions of metal salts in the preferred ratio are prepared with distilled water and then mixed and continuously stirred till the clear solution is made and then after kept at 60 °C. Then the mixture of these solutions is added abruptly to the boiling solution of NaOH (Amiri & Shokrollahi, 2013). This solution is continuously stirred at constant rpm from 70°C to 95°C with the help of a magnetic stirrer for 45 minutes to 1 hour (Rahman et al., 2013; Vaidyanathan et al., 2007; Zhang et al., 2010). During this process, the nanocrystalline ferrites are formed by the transformation of metal salts into hydroxides which takes place immediately, and the conversion of these hydroxides in the form of ferrites (Arulmurugana et al., 2005). This happens because of the dehydration of the final solution and atomic rearrangement which results in the conversion of the intermediate hydroxide phase into ferrite (Jeyadevan et al., 2003). The time duration of stirring (1 hour) is sufficient for the transformation of hydroxides into spinel ferrite (Amiri & Shokrollahi, 2013). The pH value of the solution is maintained at 12-13. These particles should be washed several times with distilled water followed by acetone and dried at room temperature (Vaidyanathan, et al., 2007). Then after the product is dried in an electric oven at a temperature of 100 °C to remove the water contents (Gul et al., 2007; Iqbal & Ashiq, 2008; Rahman et al., 2013). In the end, the dried powder is ground homogeneously with a cleaned agate mortar and pestle to get the fine powder of ferrite nanoparticles.

General Properties of Spinel Ferrites

The properties of ferrites depend on the chemical composition as well as the process of synthesis method. Some general properties of spinel ferrites are as follows:

1. Spinel ferrites are grey or black in color which is due to the absorption that occurs only in the visible range for ferrite crystal structure (Shaikh et al., 2020). This can also be ascribed to the approximately equal energies of the 3d and 4s states (Fernandes et al., 2019).
2. Their melting points are difficult to measure as the oxygen loss occurs at high temperatures.
3. They are very hard and brittle which is because the oxygen ions have a close-packed spatial formation that makes the ionic bonds very strong (Niederberger et al., 2004).
4. They are non-conductors of electricity but exhibit semiconducting behavior under the influence of applied electric field (Khairy & Gouda, 2015).
5. Nano-structured spinel ferrites have a high dielectric constant (Hathiya, et al., 2022) which is of the order of thousands at lower frequencies. This happens because of the close-packed structure of oxygen ions (Prakash & Baijal,1985).
6. Ferrites are mostly known for their magnetic properties with important parameters like magnetic anisotropy, coercivity, hysteresis loop, and susceptibility. They are highly magnetic materials. In general, they have high curie temperature. Spinel ferrites show ferrimagnetic, ferromagnetic, or superparamagnetic behavior depending on their synthesis process, chemical composition, cation distribution, magnetic anisotropy, temperature treatment and applied field (Hathiya et al., 2022; Gawas et al., 2017; Jani et al., 2008; Ligemza, 1981; Hathiya et al., 2020).

Applications of Nano-structured Spinel Ferrites

Nanocrystalline spinel ferrites have achieved considerable attention because of their wide applications in various fields due to their marked properties.

Figure 8. Applications of nano ferrites

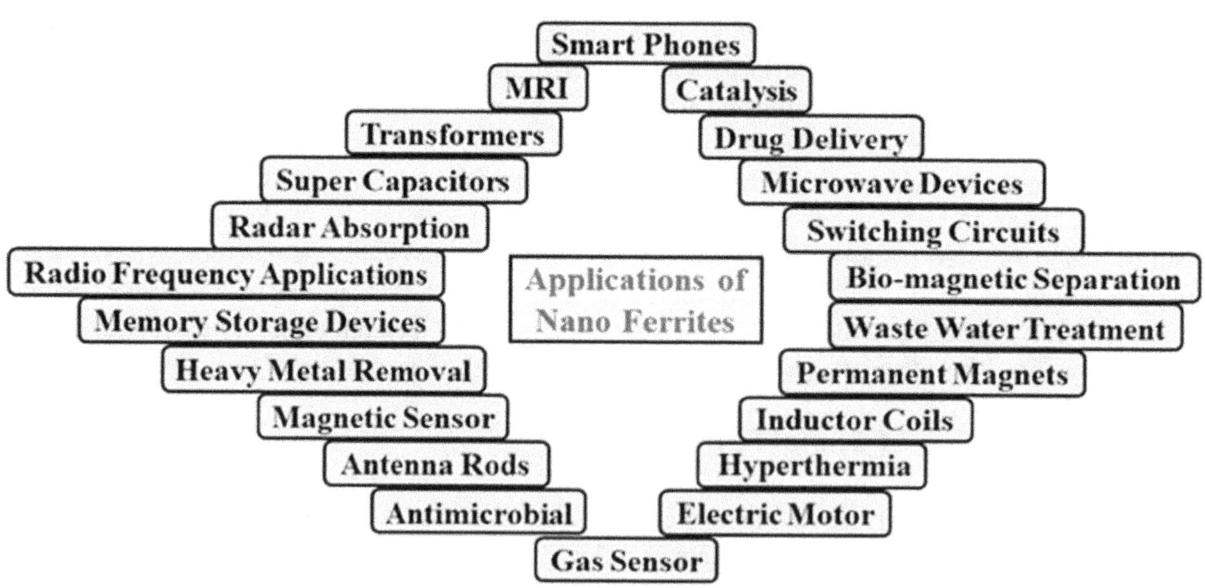

Figure 8 shows various applications of nano-structured ferrite materials. It shows that the nano ferrites are widely applied in many fields from smart phones to medical applications. In industry, Spinel ferrite nanoparticles are used in the manufacturing of electronic devices (Zhang et al., 2015; Mohamed et al., 2010; Hasanzadeh et al., 2015). They are also used in wastewater treatment (Kefeni et al., 2017; Brar et al., 2010). In biomedical applications, they are used in magnetic resonance imaging (MRI) (Zhu et al., 2012; Sohn et al., 2015; Lee et al., 2015), hyperthermia (Spirou et al., 2018; Obaidat et al., 2015; Spirou et al., 2018), drug delivery (Wu et al., 2015; Phumying et al., 2013; Reddy et al., 2012), and bio-magnetic separation (Plouffe et al., 2014). They are useful for the preparation of humidity sensors (Arshaka et al., 2002), magnetoelastic sensor (Atif et al., 2018), and biosensors (Rocha-Santos, 2014; Carregal-Romero et al., 2013; Lee et al., 2014). In addition, spinel ferrite-based NPs have strong antimicrobial activity against some pathogenic microorganisms (Naik et al., 2019; Maksoud et al., 2018). They are also used as adsorbents and catalysts (Senapati et al., 2011; Faria et al., 2014; Ibrahim et al., 2016; Mahfouz, et al., 2015; Zhao et al., 2014; Flores et al., 2012; Hema et al., 2016; El-Fouani et al., 2019), for detection of glucose in urine (Su et al., 2012), used as diagnostic tools for the detection of H_2O_2 (González et al., 2016; Shi et al., 2011), for biodiesel production (Bharti et al., 2021), high-frequency electromechanical transducers (Yamaguchi et al., 1974), as photosensor and pressure sensor (Seki et al., 1994) power system-on-chip for smartphone applications (Lee et al., 2011), microwave integrated circuits (MIC) subassemblies (Newton et al., 1975), patch antennas (Yang et al., 2010), gun hybrid oscillators (Gondek et al., 1984), magnetic recording (Morisako et al., 2002; Iwasaki, 1984), in microwave absorption (Mathews et al., 2020), ferrite-core memories, radar absorption, and microwave antenna (Bradley, 1980). They are highly used in permanent magnets (Mathur et al., 2010; L´opez-Ortega et al., 2015), low voltage and large current DC-DC converter (Kowase et al., 2005), in smartphones (Lee et al., 2011), and high-density recording (Matsushita et al., 1998).

Among these some recent and important applications of nano-structured spinel ferrites are discussed as follows:

Antimicrobial Agent

Nano-structured spinel ferrites are widely used as an antimicrobial agent for a few years. They are potential to inhibit some microorganisms and behave as an antibacterial and antifungal agent.

Figure 9. Schematic diagram of antimicrobial activity process

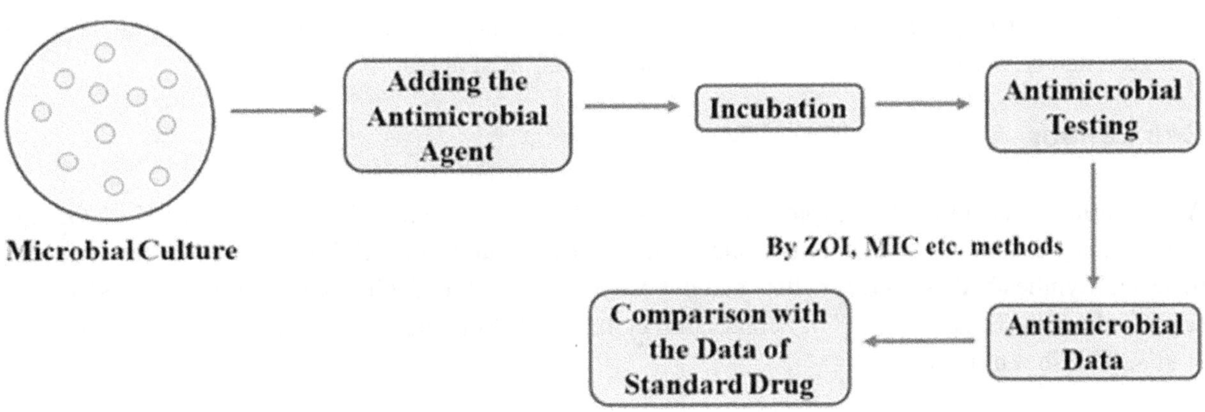

The schematic presentation of the procedure of antimicrobial activity is shown in Figure 9. Firstly, the required diluted concentrations of the antimicrobial agent are prepared. Thereafter this material is added to the microbial culture and incubated at 37°C for overnight. By this method, the growth of the microbial culture is inhibited. The inhibition of microbial organisms is tested by various antimicrobial activity methods like ZOI, MIC, etc. At last, the antimicrobial data are compared with the data of standard drugs to study how much the tested antimicrobial agent is capable.

Mostly the bacteria like E. Coli, P. Aeruginosa, S. Aureus, and S. Pyogenus and fungi like C. Albicans, N. Niger, and A. Clavatus are being tested by nanocrystalline spinel ferrites. Several spinel nano ferrites have been studied by some researchers. It has been found that the antimicrobial effectiveness of nano-structured $CoFe_2O_4$ was high towards the fungi at the zone of inhibition of 15 mm (Kombaiah et al., 2018). Zn^{+2} doped $CoFe_2O_4$ showed a high inhibition affinity for Gram-positive E. coli (Sanpo et al., 2012; Ashour et al., 2018) and the similar results were also obtained by Ag doped $CoFe_2O_4$ (Gingasu et al., 2016). The composition like $CoCuFe_2O_4$ gave good results for some pathogens (Naik et al., 2018). $Mn_{0.5}Mg_{0.5}Fe_2O_4$ NPs showed impressive antimicrobial behavior at low concentration (20.0 µg/ml) against S. Aureus (25.0 mm ZOI), C. Albicans (21.0 mm ZOI), E. Feacium (20.4 mm ZOI), and E. Coli (18.8 mm ZOI) (Maksoud, et al., 2020). $Ag_{0.5}Cr_{2.5}O_4$ nanoparticles gave strong antibacterial activity against Gram-positive microorganisms (El-Bassuony, 2020). Chitosan-encapsulated zinc ferrite nanoparticles showed excellent antibacterial activity against both Gram-positive and Gram-negative bacteria (Sharma et al., 2020). The citric acid-modified manganese ferrite nanoparticles were able to inhibit C. Albicans with a minimal inhibitory concentration of 250 µg/mL (Lopez et al., 2016). Some authors studied the antimicrobial testing on $MnFe_2O_4$, $ZnFe_2O_4$, and the compositions of Mn and Zn ferrite for Staphylococcus Aureus, Lactobacillus Amylovorus, E. Coli, Pseudomonas, and C. Albicans. Among them, $Mn_{0.25}Zn_{0.75}Fe_2O_4$ and $ZnFe_2O_4$ gave high antimicrobial activity (Bharamagoudar et al., 2018). Antimicrobial results of citric acid functionalized magnesium ferrite (CA-$MgFe_2O_4$) nanoparticles showed maximum activities against S. aureus, E. Coli, and C. Albicans with ZOI of 20.0, 16.0, and 14.0 mm, respectively (El-Khawaga et al., 2021). The P. Aeruginosa was tested by $CoFe_2O_4$ and $NiFe_2O_4$ nanoparticles in vitro by using the MIC method and the wound infection was induced in the rats. The effect of nanoparticles was tested in vivo which showed that P. Aeruginosa was the predominant bacterial type that caused wound infection. The minimum inhibitory concentration of $NiFe_2O_4$ and $CoFe_2O_4$ nanoparticles was 32 µg /ml and 16 µg /ml respectively (Majed et al., 2021). Some researchers studied the antimicrobial activity of metals-substituted spinel cobalt ferrite nanoparticles with compositions $M_xCo_{1-x}Fe_2O_4$ (where M = Zn, Cu, Mn; x = 0.0, 0.25, 0.5, and 0.75) and they observed that most effective composition was $Zn_{0.75}Co_{0.25}Fe_2O_4$ (20.0 ppm) showing activity against S. Aureus, E. Columbae, and A. Viridians by 15.0, 13.0, and 12.0 mm ZOI, respectively (Maksoud et al., 2018). Thus, we can say that the nano-structured spinel ferrites (pure ferrite and their composition) are excellent antimicrobial agent.

Gas Sensor

Air pollution caused by toxic, explosive, and flammable gases, such as CO, H_2S, NH_3, CH_4, NO_2, and LPG is one of the critical factors that affect human health and other living things. For that, various materials synthesized as porous pellets and deposited in the form of thick or thin films are used as gas-sensing devices (Baik et al., 2000; Schierbaum et al., 1992; Semancik & Cavicchi, 1991; Ivanovskaya et al., 2000; Ihokura et al., 1994).

Figure 10. Schematic presentation of gas sensing experiment

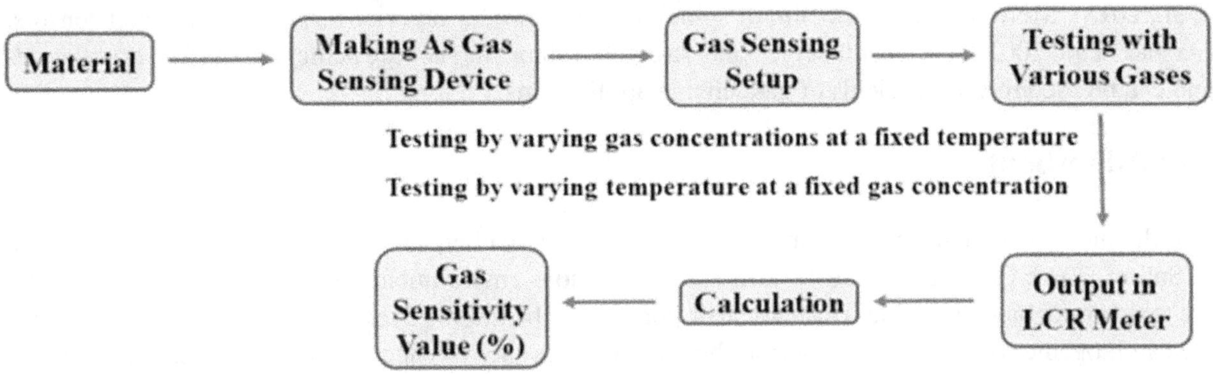

Figure 10 shows the schematic presentation of the gas sensing experiment. As shown in the figure, at the very first step, the material is fabricated by means of the gas sensing device in the form of a pellet, thin film or thick film, and the electrodes are mounted on the surface of this device. Then, this device is attached to the gas sensing setup which contains a gas cylinder/injection, thermocouple, variac, temperature indicator, and precision LCR meter. The required amount of gas at a predefined temperature is injected which would be detected by the device. The output is measured in the form of resistance by using the LCR meter. The gas sensitivity (%) of the tested material is evaluated by using the value of resistance in gas (output value of the experiment) with reference to the resistance in the air.

There are two types of semiconductor-based gas sensor devices: p-type and n-type (Korotcenkov, 2007; Korotcenkov, 2013). A large number of n- and p-type nano oxides have been studied for their gas sensing properties (Arshak et al., 2002; Xiangfeng et al., 2003; Bodade et al., 2012; Arshak et al., 2006). Several materials have been reported for use in semiconductor gas sensors, such as TiO_2 (n-type) (Devi et al., 2002), carbon nanotube (p-type) (Barnard, 2012), ZnO (n-type) (Liao et al., 2007), α-Fe_2O_3 (n-type) (Reddy et al., 2000; Malyshev et al., 1994), V_2O_5 (p-type) (Yu et al., 2012), Co_3O_4 (p-type) (Li et al., 2005), CuO (p-type) (Yang et al., 2013), etc. Nano-structured spinel ferrites have attained excellent results in gas sensor applications because of their simple electronic structure, high surface-to-volume ratio and stability, fast response and recovery time, and low cost (Satyanarayana et al., 2003; Wu et al., 2019; Caltun et al., 2007). Various spinel ferrites MFe_2O_4 (M = Mg, Cd, Cu, Ni, Zn) have been studied for LPG, CO, N_2O, C_2H_5OH, CH_4, and CH_3COCH_3 (Rezlescu et al., 2009; Mukherjee & Majumder, 2010; Rezlescu et al., 2007). The thick film of Ni-Ferrite showed good sensitivity and response to H_2 gas at an operating temperature of 175 °C (Kamble & Mathe, 2008). Some researchers have reported that Ni-O and Fe_2O_3 thick film gas sensors showed the highest sensitivity to toluene. Tungsten doped $Ni_{0.5}Zn_{0.5}W_xFe_{2-x}O_4$ (x = 0.0, 0.2, 0.4, 0.6, 0.8, 1.0) was tested for H_2 gas which gave a fast response to sensitivity of gas (Pathania et al., 2019). Spinel nanoparticles of $MgFe_2O_4$ showed the highest response (>80%) for LPG gas with response and recovery times of 15 s and 23 s, respectively (Nadargi et al., 2022). The gas sensing properties of various sizes of $CoFe_2O_4$ nanoparticles were studied for LPG, ethanol, and ammonia gas which exhibited that the sensitivity increases very fast by reducing particle size. The sensitivity of the smallest particle size (5.8 nm) for LPG, ethanol and ammonia was found at 72%, 44%, and 40% respectively (Rathore et al., 2015) which suggests that cobalt ferrite behaves as a good LPG gas sensor. $Ni_{0.25}Zn_{0.75}Fe_2O_4$ thin film gas sensor displayed good selectivity to chlorine gas at room temperature with the highest sensitivity (~50%) at room temperature (Pawar et al., 2015). Hydrogen gas sensing test by lithium substituted nickel ferrite thin films exhibited the optimal sensitivity (Manikandan et al., 2018). Moreover, the thick film of spinel ferrite showed a superior response than pellet sensors. (Arshak et al., 2002). It is clear that the nano-structured spinel ferrites are acute sensors for several gases and can be used in various fields of gas sensing applications.

Dye Adsorbent

The chemical dyes are widely used in textile, plastic, paper, leather, and cosmetic industries (Hashemian & Salimi, 2010; Cottet et al., 2014). This causes a serious environmental problem as most of the dyes are toxic or not biocompatible (Hashemian & Foroghi, 2014). Some dyes are harmful to humans as they are carcinogenic which may cause serious health problems such as cancer or skin diseases (Eren, 2009). The discharge of these kinds of dyes into the water may create environmental problems and so it is a very challenging task. Many adsorbents have been studied by several researchers (Zhang et al., 2007; Stafiej & Pyrzynska, 2008; Mihoc & Ianos, 2014; Fan et al., 2012; Shariati-Rad et al., 2014; Pastrana-Martínez et al., 2012; Satheesh et al., 2016; Xie et al., 2012; Muthukumaran et al., 2016; Mak & Chen, 2004). Nano-structured spinel ferrites are very popular for dye adsorption as they are chemically stable, porous, possess interesting magnetic properties, cost-effective, and can be re-utilized.

Figure 11. Schematic diagram of dye adsorption experiment

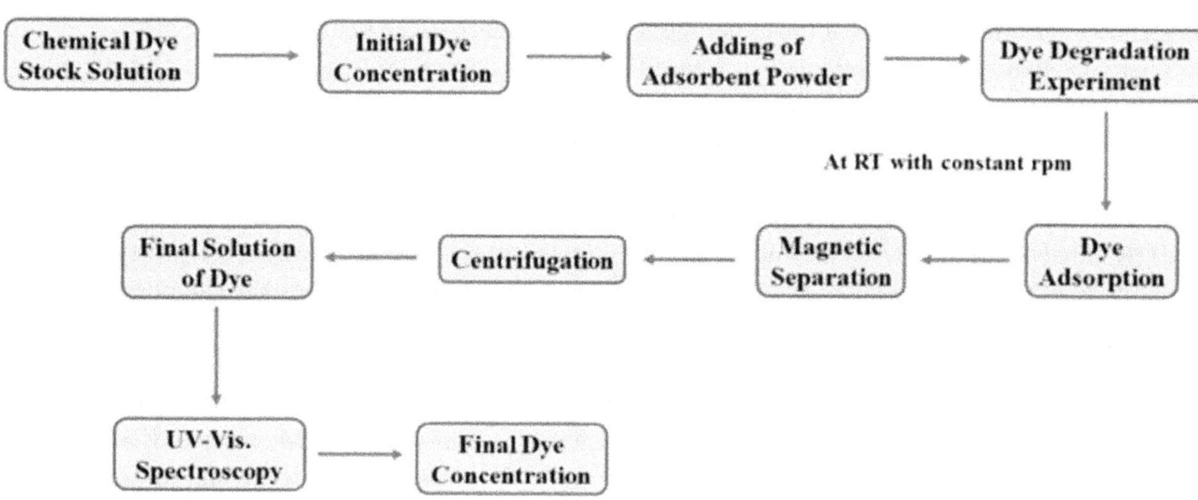

The schematic presentation of the dye adsorption experiment is shown in Figure 11. To perform this experiment, the stock solution of chemical dye (adsorbate) is prepared. The required concentrations of dye (Initial concentrations) are prepared using this stock solution. The adsorbent (by which material we want to study the adsorption capacity) powder in fixed quantity is added to the dye solution. Then this solution is kept at a fixed rpm for a few hours to study the adsorption of dye. After performing the dye adsorption experiment, the dye solution is taken out by magnetic separation and treated with centrifugation to get the final concentration of the dye solution. The adsorption capacity (%) is calculated by taking the absorbance value of the final dye solution by using the UV-Vis. spectroscopy.

The adsorption capacity of porous nano-crystalline $CoFe_2O_4$ composite was studied for brilliant blue dye removal and that gave impressive results (Khan et al., 2015). The photocatalytic activity of Al-substituted nickel ferrite $NiAl_xFe_{2-x}O_4$, where x = (0.00, 0.05, 0.10, and 0.15) was observed for the degradation of methyl orange dye showed good adsorption of dye (Lassoued et al., 2018). The adsorption capacity of nanocrystalline $Mn_{0.6}Co_{0.2}Ni_{0.2}Fe_2O_4$ was observed at 59.33% and 61.00% for 0.1 mM and 0.2 mM concentrations of Methylene Blue (MB) dye (Hathiya et al., 2020). The dye removal efficiency of magnetic ferrite nanoparticle (Nickle–zinc ferrite) (MFN)–alginate composite was tested and the maximum dye adsorption capacity was observed at 106 mg/g, 25 mg/g, and 56 mg/g for BB9, BB41, and BR18 dye, respectively (Mahmoodi, 2013). The results obtained from the adsorption by nano-structured metals-substituted spinel cobalt ferrite nanoparticles with nominal composition [$M_xCo_{1-x}Fe_2O_4$; (M = Zn, Cu, Mn; x = 0.0, 0.25, 0.5 and 0.75)] were promising photocatalyst achieving 96.0% removal of MB after 100 min (Maksoud et al., 2018). The role of $MgFe_2O_4$ nanoparticles in the adsorption of MB dye (89.73% in 240 minutes), suggested the possible application of the ferrite for wastewater treatment (Nguyen et al., 2019). The MB dye adsorption by spinel $Ni_xZn_{1-x}Fe_2O4$ (x = 0.0 to 1.0) nanoparticles gave the superior results (Padmapriya et al., 2016). The adsorption capacity of polyaniline (PANI)–nickel ferrite ($NiFe_2O_4$) nanocomposite was observed at 87.8% for a 40 mg/L concentration of MB dye (Patil et al., 2015). In this way, the spinel ferrite nanoparticles are much more efficient for chemical dye adsorption.

FUTURE RESEARCH DIRECTIONS

As described in this chapter, the research and application of nano-structured spinel ferrites will continue to increase each year, even in the future, with advanced applications. The future of ferrites will be more progressive in science and technology. The production of spinel ferrite nanoparticles with extremely high-performance values will be mass-produced soon. Although nano-ferrites already have been used in the electrical components of conventional automobiles, one of the new impressive applications of ferrite is the magnetic parts of the direct-current driving motor in electrically powered automobiles (Sugimoto, 1999). The transformation of solar energy to hydrogen energy using ferrites as catalysts has opened a new direction in applications of ferrites (Tamaura et al., 1997). The research on biomagnetism will soon improve the living world (Matsunaga et al., 1998). Moreover, due to their eye-catching magnetic properties they can be recovered and reused in various applications.

CONCLUSION

Nanosized spinel ferrites have received splendid attention in the last decade due to their special properties, such as chemical, mechanical, and thermal stability, large coercivity, moderate saturation magnetization, high Curie temperature, low eddy current loss, and high electrical resistance. Every type of ferrites has its unique properties and can be used in several applications, accordingly. Amongst all the reviewed synthesis methods, the solid-state and auto-combustion techniques stand as superlative routes for synthesizing fine and homogenous nanostructured ferrites. They are used in many fields, including material science, physics, chemistry, biology, engineering, and medical science. Notably, these magnetic ferrite nanoparticles have exhibited outstanding results in the field of medicine because of their biocompatibility, low toxicity, and magnetic properties. The nano-structured spinel ferrites are excellent gas sensors. They have achieved remarkable chemical dye adsorption applications due to their porosity, high chemical stability, good surface morphology, unique magnetic properties, and reusability.

REFERENCES

Abarrategui, C. L., Espi, V. F., Alvarez, M. B. L., Pereira, C. D., & Garay, H. (2016). The intrinsic antimicrobial activity of citric acid-coated manganese ferrite nanoparticles is enhanced after conjugation with the antifungal peptide Cm-p5. *International Journal of Nanomedicine*, *11*, 3849–3857. doi:10.2147/IJN.S107561 PMID:27563243

Adam, J. D., Davis, L. E., Dionne, G. F., Schloemann, E. F., & Stitzer, S. N. (2002). Ferrite devices and materials. *IEEE Transactions on Microwave Theory and Techniques*, *50*(3), 721–737. doi:10.1109/22.989957

Alarifi, A., Deraz, N. M., & Shaban, S. (2009). Structural, morphological and magnetic properties of $NiFe_2O_4$ nano-particles. *Journal of Alloys and Compounds*, *486*(1-2), 501–506. doi:10.1016/j.jallcom.2009.06.192

Amir, M., Baykal, A., Guner, S., Sertkol, M., Sozeri, H., & Toprak, M. (2015). Synthesis and Characterization of CoxZn1−xAlFeO4 Nanoparticles. *Journal of Inorganic and Organometallic Polymers and Materials*, *25*(4), 747–754. doi:10.100710904-014-0153-6

Amiri, S., & Shokrollahi, H. (2013). Magnetic and structural properties of RE doped Co-ferrite (RE = Nd, Eu, Gd) nano-particles synthesized by co-precipitation. *Journal of Magnetism and Magnetic Materials*, *345*, 18–23. doi:10.1016/j.jmmm.2013.05.030

Antony, V. (2021). A review on synthesis, characterization and potential biological applications of superparamagnetic iron oxide nanoparticles. *Current Research in Green and Sustainable Chemistry*, *4*, 100042. doi:10.1016/j.crgsc.2020.100042

Anupama, M. K., Srinatha, N., Matteppanavar, S., Angadi, B., Sahoo, B., & Rudraswamy, B. (2018). Effect of Zn substitution on the structural and magnetic properties of nanocrystalline $NiFe_2O_4$ ferrites. *Ceramics International*, *44*(5), 4946–4954. doi:10.1016/j.ceramint.2017.12.087

Arbain, R., Othman, M., & Palaniandy, S. (2011). Preparation of iron oxide nanoparticles by mechanical milling. *Minerals Engineering*, *24*(1), 1–9. doi:10.1016/j.mineng.2010.08.025

Arshak, K., Nicholson, M., Arshak, A., Moore, E. G., Gaidan, I., & Cunniffe, C. (2006). Examining the effects of polymer binder in Fe2O3/ZnO thick film sensors on the response to propanol vapour. In the *29th International Spring Seminar on Electronics Technology,* (pp. 489-494). UPM.

Arshak, K., Twomey, K., & Egan, D. (2002). A Ceramic Thick Film Humidity Sensor Based on MnZn Ferrite. *Sensors (Basel)*, *2*(2), 50–61. doi:10.339020200050

Arulmurugana, R., Jeyadevan, B., Vaidyanathan, G., & Sendhilnathan, S. (2005). Effect of zinc substitution on Co-Zn and Mn-Zn ferrite nanoparticles prepared by co-precipitation. *Journal of Magnetism and Magnetic Materials*, *288*, 470–477. doi:10.1016/j.jmmm.2004.09.138

Ashour, A. H., El-Batal, A. I., Maksoud, M. I. A. A., El-Sayyad, G. S., Labib, S., Abdeltwab, E., & El-Okr, M. M. (2018). Antimicrobial activity of metal-substituted cobalt ferrite nanoparticles synthesized by sol-gel technique. *Particuology, 40*, 141–151. doi:10.1016/j.partic.2017.12.001

Assar, S. T., Abosheiasha, H. F., & El Nimr, M. K. (2014). Comparison study 1 of the magnetic permeability and dc conductivity of Co-Ni-Li ferrite nanoparticles and their bulk counterparts. *Journal of Magnetism and Magnetic Materials, 354*, 1–6. doi:10.1016/j.jmmm.2013.10.029

Atif, M., Asghar, M. W., Nadeem, M., Khalid, W., Ali, Z., & Badshah, S. (2018). Synthesis and investigation of structural, magnetic and dielectric properties of zinc substituted cobalt ferrites. *Journal of Physics and Chemistry of Solids, 123*, 36–42. doi:10.1016/j.jpcs.2018.07.010

Azadmanjiri, J., Salehani, H. K., Barati, M. R., & Farzan, F. (2007). Preparation and electromagnetic properties of $Ni_{1-x}Cu_xFe_2O_4$ nanoparticle ferrites by sol-gel auto-combustion method. *Materials Letters, 61*(1), 84–87. doi:10.1016/j.matlet.2006.04.011

Baik, N. S., Sakai, G., Miura, N., & Yamazoe, N. (2000). Hydrothermally treated sol solution of tin oxide for thin-film gas sensor. *Sensors and Actuators. B, Chemical, 63*(1-2), 74–79. doi:10.1016/S0925-4005(99)00513-4

Bamzai, K., & Bhat, M. (2014). Electrical and magnetic properties of some rare earth orthoferrites ($RFeO_3$ where R = Y, Ho, Er) systems. *Integrated Ferroelectrics, 158*(1), 108–122. doi:10.1080/1058 4587.2014.957591

Barnard, A. S. (2012). Modelling of the reactivity and stability of carbon nanotubes under environmentally relevant conditions. *Physical Chemistry Chemical Physics, 14*(29), 10080–10093. doi:10.1039/c2cp40862g PMID:22576146

Belrhazi, H., Hafidi, M. Y. E., & Hafidi, M. E. (2019). Permanent magnets elaboration from $BaFe_{12}O_{19}$ hexaferrite material: Simulation and prototype. *Research and Developments in Material Science, 11*, 1–5.

Bhandare, S. V., Kumar, R., Anupama, A. V., Choudhary, H. K., Jali, V. M., & Sahoo, B. (2017). Annealing temperature-dependent structural and magnetic properties of $MnFe_2O_4$ nanoparticles grown by sol-gel auto-combustion method. *Journal of Magnetism and Magnetic Materials, 433*, 29–34. doi:10.1016/j.jmmm.2017.02.040

Bharamagoudar, R. C., Patil, A. S., Mathad, S. N., Kumbar, V. M., & Kankanawadi, L. B. (2018). Magnetic and antibacterial studies of Nanoferrites prepared by self propagating high-temperature synthesis route. *Acta Chemica IASI, 26*(2), 249–262. doi:10.2478/achi-2018-0016

Bharti, M. K., Chalia, S., Thakur, P., Sridhara, S. N., Thakur, A., & Sharma, P. B. (2021). Nanoferrites heterogeneous catalysts for biodiesel production from soybean and canola oil: A review. *Environmental Chemistry Letters, 19*(5), 3727–3746. doi:10.100710311-021-01247-2 PMID:33967660

Bhukal, S., Shivali, & Singhal, S. (2014). Magnetically separable copper substituted cobalt-zinc nanoferrite photocatalyst with enhanced photocatalytic activity. *Materials Science in Semiconductor Processing, 26*, 467–476. doi:10.1016/j.mssp.2014.05.023

Bodade, A. B., Bodade, A. B., Wankhade, H. G., Chaudhari, D. G., & Kothari, D. C. (2012). Conduction mechanism and gas sensing properties of $CoFe_2O_4$ nanocomposite thick films for H_2S gas. *Talenta, 89*, 183–188. doi:10.1016/j.talanta.2011.12.013 PMID:22284478

Bradley, F. N. (1980). *Materials for magnetic functions*, 247-290. Hayden book company, Inc.

Bragg, W. H. (1915). The structure of the spinel group of crystals. *Philosophical Magazine, 30*(176), 305–315.

Brar, S. K., Verma, M., Tyagi, R. D., & Surampalli, R. Y. (2010). Engineered nanoparticles in wastewater and wastewater sludge-Evidence and impacts. *Waste Management (New York, N.Y.), 30*(3), 504–520. doi:10.1016/j.wasman.2009.10.012 PMID:19926463

Caltun, O. F., Rao, G. S. N., Rao, K. H. B., Rao, P., Kim, C. G., Kim, C. O., Dumitru, I., Lupu, N., & Chiriac, H. (2007). High magnetostrictive cobalt ferrite for sensor applications. *Sensor Letters, 5*(1), 45–47. doi:10.11661.2007.027

Cao, S. W., Zhu, Y. J., Cheng, G. F., & Huang, Y. H. (2009). $ZnFe_2O_4$ nanoparticles: Microwave-hydrothermal ionic liquid synthesis and photocatalytic property over phenol. *Journal of Hazardous Materials, 171*(1–3), 431–435. doi:10.1016/j.jhazmat.2009.06.019 PMID:19564076

Carregal-Romero, S., Caballero-Diaz, E., Beqa, L., Abdelmonem, A. M., Ochs, M., Huhn, D., Suau, B. S., Valcarcel, M., & Parak, W. J. (2013). Muliplexed sensing and imaging with colloidal nano- and microparticles. *Annual Review of Analytical Chemistry (Palo Alto, Calif.), 6*(1), 53–81. doi:10.1146/annurev-anchem-062012-092621 PMID:23451718

Carta, D., Casula, M. F., Falqui, A., Loche, D., Mountjoy, G., Sangregorio, C., & Corrias, A. (2009). A structural and magnetic investigation of the inversion degree in ferrite nanocrystals MFe_2O_4 (M = Mn, Co, Ni). *The Journal of Physical Chemistry C, 113*(20), 8606–8615. doi:10.1021/jp901077c

Costa, A. C. F. M., Morelli, M. R., & Kiminami, R. H. G. A. (2007). Microstructure and magnetic properties of $Ni_{1-x}Zn_xFe_2O_4$ synthesized by combustion reaction. *Journal of Materials Science, 42*(3), 779–783. doi:10.100710853-006-1440-6

Cottet, L., Almeida, C. A., Naidek, N., Viantea, M. F., Lopesa, M. C., & Debacher, N. A. (2014). Adsorption characteristics of montmorillonite clay modified with iron oxide with respect to methylene blue in aqueous media. *Applied Clay Science, 95*, 25–31. doi:10.1016/j.clay.2014.03.023

Cunningham, J. R. Jr, & Anderson, E. E. (1960). Samarium substitution in yttrium iron garnet. *Journal of Applied Physics, 31*(5), S45–S46. doi:10.1063/1.1984598

Dairy, A. R. A., Al-Hmoud, L. A., & Khatatbet, H. A. (2019). Magnetic and structural properties of barium hexaferrite nanoparticles doped with titanium. *Symmetry, 11*(6), 732. doi:10.3390ym11060732

Dalt, S. D., Takimi, A. S., Sousa, V. C., & Bergmann, C. P. (2009). Magnetic and structural characterization of nanostructured $MgFe_2O_4$ synthesized by combustion reaction. *Particulate Science and Technology, 27*(6), 519–527. doi:10.1080/02726350903328662

Delacotte, C., Whitehead, G. F. S., Pitcher, M. J., Robertson, C. M., Sharp, P. M., Dyer, M. S., Alaria, J., Claridge, J. B., Darling, G. R., Allan, D. R., Winter, G., & Rosseinsky, M. J. (2018). Structure determination and crystal chemistry of large repeat mixed-layer hexaferrites. *International Union of Crystallography Journal*, *5*(6), 681–698. doi:10.1107/S2052252518011351 PMID:30443353

Devi, G. S., Hyodo, T., Shimizu, Y., & Egashira, M. (2002). Synthesis of mesoporous TiO_2-based powders and their gas-sensing properties. *Sensors and Actuators. B, Chemical*, *87*(1), 122–129. doi:10.1016/S0925-4005(02)00228-9

El-Bassuony, A. A. H. (2020). Influence of High Annealing Temperature on Structural, Magnetic and Antimicrobial Activity of Silver Chromite Nanoparticles for Biomedical Applications. *Journal of Inorganic and Organometallic Polymers and Materials*, *30*(5), 1821–1828. doi:10.100710904-019-01306-w

El-Fouani, A. H., Aamouche, A., Mohseni, F., Amaral, J. S., Tobaldi, D. M., & Pullar, R. C. (2019). Effect of surfactants on the optical and magnetic properties of cobalt zinc ferrite $Co_{0.5}Zn_{0.5}Fe_2O_4$. *Journal of Alloys and Compounds*, *774*, 1250–1259. doi:10.1016/j.jallcom.2018.09.393

El-Khawaga, A. M., Farrag, A. A., Elsayed, M. A., El-Sayyad, G. S., & El-Batal, A. I. (2021). Promising Antimicrobial and Azo Dye Removal Activities of Citric Acid-Functionalized Magnesium Ferrite Nanoparticles. *Journal of Cluster Science*, *33*(1), 197–213. doi:10.100710876-020-01944-y

Eren, E. (2009). Investigation of a basic dye removal from aqueous solution onto chemically modified Unye bentonite. *Journal of Hazardous Materials*, *166*(1), 88–93. doi:10.1016/j.jhazmat.2008.11.011 PMID:19111978

Fan, H., Zhao, X., Yang, J., Shan, X., Yang, L., Zhang, Y., Li, X., & Gao, M. (2012). ZnO-graphene composite for photocatalytic degradation of methylene blue dye. *Catalysis Communications*, *29*, 29–34. doi:10.1016/j.catcom.2012.09.013

Faria, M. C. S., Rosemberg, R. S., Bomfeti, C. A., Monteiro, D. S., Barbosa, F., Oliveira, L. C. A., Rodriguez, M., Pereira, M. C., & Rodrigues, J. L. (2014). Arsenic removal from contaminated water by ultrafine d-FeOOH adsorbents. *Chemical Engineering Journal*, *237*, 47–54. doi:10.1016/j.cej.2013.10.006

Fernandes, R. J. C., Maghalhaes, C. A. B., Amorim, C. O., Amaral, V. S., Almeida, B. G., Castanheira, E. M. S., & Coutinho, P. J. G. (2019). Magnetic nanoparticles of zinc/calcium ferrite decorated with silver for photodegradation of dyes. *Materials (Basel)*, *12*(21), 3582. doi:10.3390/ma12213582 PMID:31683585

Flores, R. G., Andersen, S. L. F., Maia, L. K. K., Jose, H. J., & Moreira, R. (2012). Recovery of iron oxides from acid mine drainage and their application as adsorbent or catalyst. *Journal of Environmental Management*, *111*, 53–60. doi:10.1016/j.jenvman.2012.06.017 PMID:22820746

Fokina, E. L., Budim, N. I., Kochnev, V. G., & Chernik, G. G. (2004). Planetary mills of periodic and continuous action. *Journal of Materials Science*, *39*(16-17), 5217–5221. doi:10.1023/B:JMSC.0000039213.44891.7d

Forestier, H. (1928). Magnetic Transformations of the Sesquioxide of Iron, of Its Solid Solution and of Its Ferromagnetic Combinations (In Fr.). *Annales de Chimie. (Paris)*, *9*, 316–401.

Garcia-Muñoz, P., Fresno, F., de la Peña O'Shea, V. A., & Keller, N. (2020). Ferrite Materials for Photoassisted Environmental and Solar Fuels Applications. In M. Muñoz-Batista, A. Navarrete Muñoz, & R. Luque (Eds.), *Heterogeneous Photocatalysis. Topics in Current Chemistry Collections* (pp. 107–162). Springer. doi:10.1007/978-3-030-49492-6_4

Gawas, U. B., Verenkar, V. M. S., Meena, S. S., & Bhatt, P. J. (2017). Influence of Mn Substitution on Mossbauer and Magnetic Properties of Ni-Zn Ferrite Nanoparticles. *Journal of Superconductivity and Novel Magnetism*, *30*(11), 3241–3246. doi:10.100710948-017-4149-7

Geller, S., & Gilleo, M. A. (1957). The crystal structure and ferrimagnetism of yttrium-iron garnet, $Y_3Fe_2(FeO_4)_3$. *Journal of Physics and Chemistry of Solids*, *3*(1-2), 30–36. doi:10.1016/0022-3697(57)90044-6

Ghazanfar, U., Siddiqi, S. A., & Abbas, G. (2005). Structural analysis of the Mn-Zn ferrites using XRD technique. *Journal of Material Science and Engineering: B*, *118*(1), 84-86.

Gilleo, M. A., & Geller, S. (1958). Magnetic and crystallographic properties of substituted yttrium-iron garnet, $3Y_2O_3$ xM_2O_3 $(5\text{-}x)Fe_2O_3$. *Physical Review*, *110*(1), 73–78. doi:10.1103/PhysRev.110.73

Gingasu, D., Mindru, I., Patron, L., Moreno, J. M. C., Mocioiu, O. C., Preda, S., Stanica, N., Nita, S., Dobre, N., Popa, M., Gradisteanu, G., & Chifiriuc, M. C. (2016). Green synthesis methods of $CoFe_2O_4$ and Ag-$CoFe_2O_4$ nanoparticles using Hibiscus extracts and their antimicrobial potential. *Journal of Nanomaterials*, *2016*, 2106756. doi:10.1155/2016/2106756

Gondek, J. J., Koprowski, J., & Wojcicki, M. A. (1984).. . *Active and Passive Electronic Components*, *11*, 147.

Gonzalez, J. J., Mazario, E., Menendez, N., Marcos, J. S., Bonilla, A. M., & Herrasti, P. (2016). Comparison of ferrite nanoparticles obtained electrochemically for catalytical reduction of hydrogen peroxide. *Journal of Solid State Electrochemistry*, *20*(4), 1191–1198. doi:10.100710008-015-2938-0

Gore, S. K., & Jadhav, S. S. (2020). Basics of ferrites. In R. S. Mane & V. V. Jadhav (Eds.), *Spinel Ferrite Nanostructures for Energy Storage Devices* (pp. 1–11). Elsevier. doi:10.1016/B978-0-12-819237-5.00001-8

Gozuak, F., Koseoglu, Y., Baykal, A., & Kavas, H. (2009). Synthesis and characterization of $Co_xZn_{1-x}Fe_2O_4$ magnetic nanoparticles via a PEG-assisted route. *Journal of Magnetism and Magnetic Materials*, *321*(14), 2170–2177. doi:10.1016/j.jmmm.2009.01.008

Gul, I. H., Abbasi, A. Z., Amin, F., Anis-ur-Rehman, M., & Maqsood, A. (2007). Structural, magnetic and electrical properties of $Co_{1-x}Zn_xFe_2O_4$ synthesized by co-precipitation method. *Journal of Magnetism and Magnetic Materials*, *311*(2), 494–499. doi:10.1016/j.jmmm.2006.08.005

Guo, L., Shen, X., Meng, X., & Feng, Y. (2010). Effect of Sm^{3+} ions doping on structure and magnetic properties of nanocrystalline $NiFe_2O_4$ fibers. *Journal of Alloys and Compounds*, *490*(1-2), 301–306. doi:10.1016/j.jallcom.2009.09.182

Gupta, N., Verma, A., Kashyap, S. C., & Dube, D. C. (2005). Dielectric behavior of spin-deposited nanocrystalline nickel-zinc ferrite thin films processed by citrate-route. *Solid State Communications*, *134*(10), 689–694. doi:10.1016/j.ssc.2005.02.037

Hajalilou, A., Kamari, H. M., & Shameli, K. (2017). Dielectric and electrical characteristics of mechanically synthesized Ni-Zn ferrite nanoparticles. *Journal of Alloys and Compounds, 708*, 813–826. doi:10.1016/j.jallcom.2017.03.030

Han, M., Vestal, C. R., & Zhang, Z. J. (2004). Quantum couplings and magnetic properties of $CoCr_xFe_{2-x}O_4$ (0<x<1) spinel ferrite nanoparticles synthesized with reverse micelle method. *The Journal of Physical Chemistry B, 108*(2), 583–587. doi:10.1021/jp035966m

Harris, V. G., Geiler, A., Chen, Y., Yoon, S. D., Wu, M., Yang, A., Chen, Z., He, P., Parim, P. V., Zuo, X., Patton, C. E., Abe, M., Acher, O., & Vittoria, C. (2009). Recent advances in processing and applications of microwave ferrites. *Journal of Magnetism and Magnetic Materials, 321*(14), 2035–2047. doi:10.1016/j.jmmm.2009.01.004

Hasanzadeh, M., Shadjou, N., & de la Guardia, M. (2015). Iron, and iron-oxide magnetic nanoparticles as signal-amplification elements in electrochemical biosensing. *Trends in Analytical Chemistry, 72*, 1–9. doi:10.1016/j.trac.2015.03.016

Hashemian, S., & Foroghimoqhadam, A. (2014). Effect of copper doping on $CoTiO_3$ ilmenite type nanoparticles for removal of congo red from aqueous solution. *Chemical Engineering Journal, 235*, 299–306. doi:10.1016/j.cej.2013.08.089

Hashemian, S., & Salimi, M. (2010). Nano composite a potential low cost adsorbent for removal of cyanine acid. *Chemical Engineering Journal, 188*, 57–63. doi:10.1016/j.cej.2012.02.008

Hathiya, L. J., Baraliya, J. D., Das, A., Sen, D., Gismelseed, A. M., Yousif, A. A., & Joshi, H. H. (2022). Anomalous magnetic behaviour at nano-scale of Mn^{2+}-substituted magnesio-ferrite synthesized by auto-combustion technique. *Indian Journal of Physics, 96*(8), 2323–2335. doi:10.100712648-021-02169-z

Hathiya, L. J., Baraliya, J. D., Jethva, H. O., & Joshi, H. H. (2022). Structural, dielectric and ac electrical properties of nano-structured $Mn_{0.6}Co_{0.2}Ni_{0.2}Fe_2O_4$ ternary ferrite synthesized by citrate-gel auto-combustion technique. *Indian Journal of Pure and Applied Physics, 60*, 59–71.

Hathiya, L. J., Tanna, A. R., Baraliya, J. D., & Joshi, H. H. (2020). Study of ternary nano-structured ferrite Mn0.6Co0.2Ni0.2Fe2O4 for its magnetic and chemical dye degradation properties. In *Proceedings of the 3rd International Conference on Condensed Matter and Applied Physics (ICC-2019), AIP Conference Proceedings, 2220*, 020155. AIP.

Haye, E., Capon, F., Barrat, S., Boulet, P., Andre, R., Carteret, C., & Bruyere, S. (2015). Properties of rare-earth orthoferrites perovskite driven by steric hindrance. *Journal of Alloys and Compounds, 657*, 631–638. doi:10.1016/j.jallcom.2015.10.135

Heju, X., Xin, L., Liu, L., Pang, D., Jiao, Y., Cong, R., & Yu, W. (2019). Large area MoS_2/Si heterojunction-based solar cell through sol-gel method. *Materials Letters, 238*, 13–16. doi:10.1016/j.matlet.2018.11.051

Hema, E., Manikandan, A., Gayathri, M., Durka, M., Antony, A. S., & Venkatraman, B. R. (2016). Role of Mn^{2+} - doping on structural, morphological, optical, magnetic and catalytic properties spinel $ZnFe_2O_4$ nanoparticle. *Journal of Nanoscience and Nanotechnology, 16*(6), 5929–5943. doi:10.1166/jnn.2016.11037 PMID:27427654

Hibst, H. (1982). Hexagonal ferrites from melts and aqueous solutions, magnetic recording materials. *Angewandte Chemie International Edition*, *21*(4), 270–282. doi:10.1002/anie.198202701

Hilpert, S. (1909). Correspondence as to Structure and Origin in Magnetic Properties of Ferrite and Iron Oxide. (In Ger.). *Berichte der Deutschen Chemischen Gesellschaft*, *42*, 2248–2261. doi:10.1002/cber.190904202121

Hilpert, S. (1909). Manufacturing Method of Magnetic Materials Having Small Electric Conductivity for Electric and Magnetic Apparatus. *German Pat., 226, 347, 227, 787.*

Ibrahim, I., Ali, I. O., Salama, T. M., Bahgat, A. A., & Mohamed, M. M. (2016). Synthesis of magnetically recyclable spinel ferrite (MFe_2O_4, M = Zn, Co, Mn) nanocrystals engineered by sol gel-hydrothermal technology: High catalytic performances for nitroarenes reduction. *Applied Catalysis B: Environmental*, *181*, 389–402. doi:10.1016/j.apcatb.2015.08.005

Ihokura, K., & Watson, I. (1994). The Stannic Oxide Gas Sensor-Principles and Applications. *CRC Press, Boa Raton.*

Iqbal, M. J., & Ashiq, M. N. (2008). Physical and electrical properties of Zr-Cu substituted strontium hexaferrite nanoparticles synthesized by co-precipitation method. *Chemical Engineering Journal*, *136*(2-3), 383–389. doi:10.1016/j.cej.2007.05.046

Ivanovskaya, M., Bogdanov, P., Faglia, G., & Sberveglieri, G. (2000). The features of thin film and ceramic sensors at the detection of CO and NO_2. *Sensors and Actuators. B, Chemical*, *68*(1-3), 344–350. doi:10.1016/S0925-4005(00)00455-X

Iwasaki, S. (1984). Perpendicular magnetic recording-evolution and future. *IEEE Transactions on Magnetics*, *20*(5), 657–662. doi:10.1109/TMAG.1984.1063563

Izadkhah, H., Zare, S., Somu, S., Lombardi, F., & Vittoria, C. (2017). Utilizing alternate target deposition to increase the magnetoelectric effect at room temperature in a single phase M-type hexaferrite. *MRS Communications*, *7*(2), 97–101. doi:10.1557/mrc.2017.36

Jacob, J., & Khadar, M. A. (2010). VSM and Mossbauer study of nanostructured hematite. *Journal of Magnetism and Magnetic Materials*, *322*(6), 614–621. doi:10.1016/j.jmmm.2009.10.025

Jani, K. H., Chhantbar, M. C., & Joshi, H. H. (2008). Study of magnetic ordering in $MnAl_xCr_xFe_{2-2x}O_4$. *Journal of Magnetism and Magnetic Materials*, *320*(18), 2208–2214. doi:10.1016/j.jmmm.2008.03.061

Jaswal, L., & Singh, B., Ferrite materials: A chronological review, *Journal of Integrated Science and Technology, 2*(2), 69-71.

Jauhar, S., Kaur, J., Goyal, A., & Singhal, S. (2016). Tuning the properties of cobalt Ferrite: A road towards diverse Applications. *RSC Advances*, *6*(100), 97694–97719. doi:10.1039/C6RA21224G

Jeseentharani, V., George, M., Jeyaraj, B., Dayalan, A., & Nagaraj, K. S. (2013). Synthesis of metal ferrite (MFe_2O_4, M = Co, Cu, Mg, Ni, Zn) nanoparticles as humidity sensor materials. *Journal of Experimental Nanoscience*, *8*(3), 358–370. doi:10.1080/17458080.2012.690893

Jesudoss, S. K., Vijaya, J. J., Kennedy, L. J., Rajan, P. I., Al-Lohedan, H. A., Ramalingam, R. J., Kaviyarasu, K., & Bououdina, M. (2016). Studies on the efficient dual performance of $Mn_{1-x}Ni_xFe_2O_4$ spinel nanoparticles in photodegradation and antibacterial activity. *Journal of Photochemistry and Photobiology. B, Biology, 165*, 121–132. doi:10.1016/j.jphotobiol.2016.10.004 PMID:27776260

Jeyadevan, B., Chinnasamy, C. N., Shinoda, K., Tohji, K., & Oka, H. (2003). Mn-Zn ferrite with higher magnetization for temperature sensitive magnetic fluid. *Journal of Applied Physics, 93*(10), 8450–8452. doi:10.1063/1.1543135

Jotania, R. (2014). Crystal structure, magnetic properties and advances in hexaferrites: A brief review. *AIP Conference Proceedings, 1621*, 596–599. doi:10.1063/1.4898528

Kakihana, M. (1996). Invited review "sol–gel" preparation of high temperature superconducting oxides. *Journal of Sol-Gel Science and Technology, 6*(1), 7–55. doi:10.1007/BF00402588

Kambale, R. C., Adhate, N. R., Chougule, B. K., & Kolekar, Y. D. (2010). Magnetic and dielectric properties of mixed spinel Ni-Zn ferrite synthesized by citratenitrate combustion method. *Journal of Alloys and Compounds, 491*(1-2), 372–377. doi:10.1016/j.jallcom.2009.10.187

Kamble, R., & Mathe, V. (2008). Nanocrystalline nickel ferrite thick film as an efficient gas sensor at room temperature. *Sensors and Actuators. B, Chemical, 131*(1), 205–209. doi:10.1016/j.snb.2007.11.003

Kasapoglu, N., Birsoz, B., Baykal, A., Koseoglu, Y., & Toprak, M. S. (2007). Synthesis and magnetic properties of octahedral ferrite $Ni_xCo_{1-x}Fe_2O_4$ nanocrystals. *Central European Journal of Chemistry, 5*(2), 570–580.

Kefeni, K. K., Mamba, B. B., & Msagati, T. A. M. (2017). Application of spinel ferrite nanoparticles in water and wastewater treatment: A review. *Separation and Purification Technology, 188*, 399–422. doi:10.1016/j.seppur.2017.07.015

Khairy, M., & Gouda, M. E. (2015). Electrical and optical properties of nickel ferrite/polyaniline nanocomposite. *Journal of Advanced Research, 6*(4), 555–562. doi:10.1016/j.jare.2014.01.009 PMID:26199745

Khan, M. A., Alam, M. M., Naushad, M., Alothman, Z. A., Kumar, M., & Ahamad, T. (2015). Sol-gel assisted synthesis of porous nano-crystalline $CoFe_2O_4$ composite and its application in the removal of brilliant blue-R from aqueous phase: An ecofriendly and economical approach. *Chemical Engineering Journal, 279*, 416–424. doi:10.1016/j.cej.2015.05.042

Kitagawa, Y., Hiraoka, Y., Honda, T., Ishikura, T., Nakamura, H., & Kimura, T. (2010). Low-field magnetoelectric effect at room temperature. *Nature Materials, 9*(10), 797–802. doi:10.1038/nmat2826 PMID:20693994

Kolahalam, L. A., Viswanath, I. V. K., Diwakar, B. S., Govindh, B., Reddy, V., & Murthy, Y. L. N. (2019). Review on nanomaterials: Synthesis and applications. *Materials Today: Proceedings, 18*, 2182–2190. doi:10.1016/j.matpr.2019.07.371

Kombaiah, K., Vijaya, J. J., Kennedy, L. J., Bououdina, M., Ramalingam, R. J., & Al-Lohedan, H. A. (2018). Okra extract-assisted green synthesis of $CoFe_2O_4$ nanoparticles and their optical, magnetic, and antimicrobial properties. *Materials Chemistry and Physics*, *204*, 410–419. doi:10.1016/j.matchemphys.2017.10.077

Korotcenkov, G. (2007). Metal oxides for solid-state gas sensors: What determines our choice? *Materials Science and Engineering B*, *139*(1), 1–23. doi:10.1016/j.mseb.2007.01.044

Korotcenkov, G. (2013). Handbook of Gas Sensor Materials: Properties, Advantages and Shortcomings for Applications. New Trends and Technologies (Integrated Analytical Systems) (Vol. 2). Springer. doi:10.1007/978-1-4614-7165-3

Kowase, I., Sato, T., Yamasawa, K., & Miura, Y. (2005). A planar inductor using Mn-Zn ferrite/polyimide composite thick film for low-Voltage and large-current DC-DC converter. *IEEE Transactions on Magnetics*, *41*(10), 223–224. doi:10.1109/TMAG.2005.855164

Kulikowski, J. (1984). Soft magnetic ferrites-development or stagnation? *Journal of Magnetism and Magnetic Materials*, *41*(1-3), 56–62. doi:10.1016/0304-8853(84)90136-7

Lassoued, A., Lassoued, M. S., Dkhil, B., Ammar, S., & Gadri, A. (2018). Substituted effect of Al^{3+} on structural, optical, magnetic and photocatalytic activity of Ni ferrites. *Journal of Magnetism and Magnetic Materials*, *476*, 124–133. doi:10.1016/j.jmmm.2018.12.062

Lazarevic, Z. Z., Jovalekic, C., Sekulic, D., Slankamenac, M., Romcevic, M., Milutinovic, A., & Romcevic, N. Z. (2012). Characterization of nanostructured spinel $NiFe_2O_4$ obtained by soft mechanochemical synthesis. *Science of Sintering*, *44*(3), 331–339. doi:10.2298/SOS1203331L

Lee, H. T., Shin, H., Cheon, J., & Weissleder, R. (2015). Recent developments in magnetic diagnostic systems. *Chemical Reviews*, *115*(19), 10690–10724. doi:10.1021/cr500698d PMID:26258867

Lee, J., Hong, Y. K., Bae, S., Jalli, J., Park, J., Abo, G., Donohoe, G., & Choi, B. C. (2011). Integrated Ferrite Film Inductor for Power System-on-Chip (PowerSoC) Smart Phone Applications. *IEEE Transactions on Magnetics*, *47*(2), 304–307. doi:10.1109/TMAG.2010.2092420

Lee, J. H., Kim, J. W., Levy, M., Kao, A., Noh, S. H., Bozovic, D., & Cheon, J. (2014). Magnetic nanoparticles for ultrafast mechanical control of inner ear hair cells. *ACS Nano*, *8*(7), 6590–6598. doi:10.1021/nn5020616 PMID:25004005

Lee, S. P., Chen, Y. J., Ho, C. M., Chang, C. P., & Hong, Y. S. (2007). A study on synthesis and characterization of the core-shell materials of $Mn_{1-x}Zn_xFe_2O_4$-polyaniline. *Materials Science and Engineering B*, *143*(1-3), 1–6. doi:10.1016/j.mseb.2007.06.029

Li, W. Y., Xu, L. N., & Chen, J. (2005). Co_3O_4 Nanomaterials in Lithium-Ion Batteries and Gas Sensors. *Advanced Functional Materials*, *15*(5), 851–857. doi:10.1002/adfm.200400429

Liao, L., Lu, H. B., Li, J. C., He, H., Wang, D. F., Fu, D. J., Liu, C., & Zhang, W. F. (2007). Size Dependence of Gas Sensitivity of ZnO Nanorods. *The Journal of Physical Chemistry C*, *111*(5), 1900–1903. doi:10.1021/jp065963k

Ligemza, S. (1981). Mossbauer Study of the $3d^5$ Shell Deformation of Fe^{3+} Ions in $MnFe_2O_4$. *Physica Status Solidi. B, Basic Research*, *105*(1), 353–358. doi:10.1002/pssb.2221050141

Liu, Y. L., Liu, Z. M., Yang, Y., Yang, H. F., Shen, G. L., & Yu, R. Q. (2005). Simple synthesis of Mg-Fe_2O_4 nanoparticles as gas sensing materials. *Sensors and Actuators. B, Chemical*, *107*(2), 600–604. doi:10.1016/j.snb.2004.11.026

Lo'pez-Ortega, A. L., Lottini, E., Fernandez, C., & Sangregorio, C. (2015). Exploring the magnetic properties of cobalt-ferrite nanoparticles for the development of a rare-earth-free permanent magnet. *Chemistry of Materials*, *27*(11), 4048–4056. doi:10.1021/acs.chemmater.5b01034

Mahfouz, M. G., Galhoum, A. A., Gomaa, N. A., Abdel-Rehem, S. S., Atia, A. A., Vincent, T., & Guibal, E. (2015). Uranium extraction using magnetic nano-based particles of diethylenetriamine-functionalized chitosan: Equilibrium and kinetic studies. *Chemical Engineering Journal*, *262*, 198–209. doi:10.1016/j.cej.2014.09.061

Mahmood, S. H., Al-Shiab, Q., Bsoul, I., Maswadeh, Y., & Awadallah, A. (2018). Structural and magnetic properties of $(Mg,Co)_2W$ hexaferrites. *Current Applied Physics*, *18*(5), 590–598. doi:10.1016/j.cap.2018.02.013

Mahmoodi, N. M. (2013). Magnetic ferrite nanoparticle-alginate composite: Synthesis, characterization and binary system dye removal. *Journal of the Taiwan Institute of Chemical Engineers*, *44*(2), 322–330. doi:10.1016/j.jtice.2012.11.014

Majed, H. M., Khalef, H. Y., Awadh, H. A., Noomi, B. S., Jafar, N. A., & Hadi, K. A. (2021). Evaluation the safety and synergistic effect of $NiFe_2O_4$ nanoparticles with antibiotic against Pseudomonas aeruginosa. *Iraqi Journal of Veterinary Sciences*, *35*(1), 71–77. doi:10.33899/ijvs.2020.126298.1294

Mak, S. Y., & Chen, D. H. (2004). Fast adsorption of methylene blue on polyacrylic acid-bound iron oxide magnetic nanoparticles. *Dyes and Pigments*, *61*(1), 93–98. doi:10.1016/j.dyepig.2003.10.008

Maksoud, M. I. A., El-Sayyad, S. G., Ashour, A. H., El-Batal, A. I., Abd-Elmonem, M. S., Hendawy, H. A. M., Abdel-Khalek, E. K., Labib, S., Abdeltwab, E., & El-Okr, M. M. (2018). Synthesis and characterization of metals-substituted cobalt ferrite $(Co_{(1-x)})M_xFe_2O_4$; ($M = Zn, Cu, Mn; x = 0.05$) nanoparticles as antimicrobial agents and sensors for Anagrelide determination in biological samples. *Materials Science and Engineering C*, *92*, 644–656. doi:10.1016/j.msec.2018.07.007 PMID:30184791

Maksoud, M. I. A. A., El-Sayyad, G. S., Ashour, A. H., El-Batal, A. I., Elsayed, M. A., Gobara, M., El-Khawaga, A. M., Abdel-Khalek, E. K., & El-Okr, M. M. (2018). Antibacterial, antibiofilm, and photocatalytic activities of metals- substituted spinel cobalt ferrite nanoparticles. *Microbial Pathogenesis*, *127*, 144–158. doi:10.1016/j.micpath.2018.11.045 PMID:30502518

Maksoud, M. I. A. A., El-Sayyad, G. S., Soliman, A. A. L. I., El-Bahnasawy, H. H., & Ashour, A. H. (2020). Influence of Mg^{2+} substitution on structural, optical, magnetic, and antimicrobial properties of Mn-Zn ferrite nanoparticles. *Journal of Materials Science Materials in Electronics*, *31*(3), 2598–2616. doi:10.100710854-019-02799-4

Malyshev, V. V., Eryshkin, A. V., Koltypin, E. A., Varfolomeev, A. E., & Vasiliev, A. A. (1994). Gas sensitivity of semiconductor Fe_2O_3-based thick-film sensors to CH_4, H_2, NH_3. *Sensors and Actuators. B, Chemical, 19*(1-3), 434–436. doi:10.1016/0925-4005(93)01031-X

Manikandan, V., Kim, J. H., Mirzaei, A., Kim, S. S., Vigneselvan, S., Singh, M., & Chandrasekaran, J. (2018). Effect of temperature on gas sensing properties of lithium (Li) substituted ($NiFe_2O_4$) nickel ferrite thin film. *Journal of Molecular Structure, 1177*, 485–490. doi:10.1016/j.molstruc.2018.09.085

Manlika, S., Dungsuwan, W., & Thountom, S. (2018). Synthesis and characterization of $Bi_{0.5}(Na_{1-x}K_x)_{0.5}TiO_3$ powders by sol-gel combustion method with glycine fuel. *Ceramics International, 44*, S168–S171. doi:10.1016/j.ceramint.2018.08.122

Manova, E., Kunev, B., Paneva, D., Mitov, I., Petrov, L., Estournes, C., D'Orlean, C., Rehspringer, J. L., & Kurmoo, M. (2004). Mechano-Synthesis, Characterization, and Magnetic Properties of Nanoparticles of Cobalt Ferrite, $CoFe_2O_4$. *Chemistry of Materials, 16*(26), 5689–5696. doi:10.1021/cm049189u

Marinca, T. F., Chicina, I., Isnard, O., & Neamtu, B. V. (2016). Nanocrystalline/nanosized manganese substituted nickel ferrites - $Ni_{1-x}Mn_xFe_2O_4$ obtained by ceramic-mechanical milling route. *Ceramics International, 42*(4), 4754–4763. doi:10.1016/j.ceramint.2015.11.155

Martinez, L. M. P., Torres, S. M., Likodimos, V., Figueiredo, J. L., Faria, J. L., Falaras, P., & Silva, A. M. (2012). Advanced nanostructured photocatalysts based on reduced graphene oxide–TiO_2 composites for degradation of diphenhydramine pharmaceutical and methyl orange dye. *Applied Catalysis B: Environmental, 123*, 241–256. doi:10.1016/j.apcatb.2012.04.045

Mathews, S. A., Babu, D. R., Saravanan, P., & Hayakawa, Y. (2020). Microwave absorption studies of $(Ba_{0.5}Sr_{0.5}Fe_{12}O_{19})_{1-x}/(NiFe_2O_4)_x$ hard/soft ferrite nanocomposites. *Materials Chemistry and Physics, 252*, 123063. doi:10.1016/j.matchemphys.2020.123063

Mathur, P., Thakur, A., Lee, J. H., & Singh, M. (2010). Sustained electromagnetic properties of Ni-Zn-Co nanoferrites for the high-frequency applications. *Materials Letters, 64*(24), 2738–2741. doi:10.1016/j.matlet.2010.08.056

Matsunaga, T., Takeyama, H., Tsujimura, N., & Kamiya, S. (1998). Magnetic Bacteria and Biomineralization. *Journal of the Magnetics Society of Japan, 22*(S_1_ISFA_97), 434–436. doi:10.3379/jmsjmag.22.S1_434

Matsushita, N., Ichinose, M., Nakagawa, S., & Naoe, M. (1998). Preparation and characteristics of Co-Zn ferrite rigid disks without protective layers for high density recording. *IEEE Transactions on Magnetics, 34*(4), 1639–1641. doi:10.1109/20.706641

McColm, I. J., & Clark, N. J. (1988). *Forming, Shaping and Working of High Performance Ceramics*. Blackie.

McHenry, M. E., & Laughlin, D. E. (2000). Nano-scale materials development for future magnetic applications. *Acta Materialia, 48*(1), 223–238. doi:10.1016/S1359-6454(99)00296-7

Melagiriyappa, E., & Jayanna, H. S. (2009). Structural and magnetic susceptibility studies on samarium substituted magnesium-zinc ferrites. *Journal of Alloys and Compounds*, *482*(1-2), 147–150. doi:10.1016/j.jallcom.2009.03.112

Mihoc, G., Ianos, R., & Pacurariu, C. (2014). Adsorption of phenol and p-chlorophenol from aqueous solutions by magnetic nanopowder. *Water Science and Technology*, *69*(2), 385–391. doi:10.2166/wst.2013.727 PMID:24473310

Mills, A. A. (2004). The lodestone: History, Physics, and Formation. *Annals of Science*, *61*(3), 273–319. doi:10.1080/00033790310001642812

Mohamed, R. M., Rashad, M. M., Haraz, F. A., & Sigmund, W. (2010). Structure and magnetic properties of nanocrystalline cobalt ferrite powders synthesized using organic acid precursor method. *Journal of Magnetism and Magnetic Materials*, *322*(14), 2058–2064. doi:10.1016/j.jmmm.2010.01.034

Moran, E., Blesa, M. C., Medina, M. E., Tornero, J. D., Menendez, N., & Amador, U. (2002). Nonstoichiometric spinel ferrites obtained from α-$NaFeO_2$ via molten media reactions. *Inorganic Chemistry*, *41*(23), 5961–5967. doi:10.1021/ic011302k PMID:12425621

Moretti, R., & Ottonello, G. (1998). An appraisal of endmember energy and mixing properties of rare earth garnets. *Geochimica et Cosmochimica Acta*, *62*(7), 1147–1173. doi:10.1016/S0016-7037(98)00033-7

Morisako, A., Naka, T., Ito, K., Takizawa, A., Matsumoto, M., & Hong, Y. K. (2002). Properties of Ba-ferrite/AlN double layered films for perpendicular magnetic recording media. *Journal of Magnetism and Magnetic Materials*, *242*, 304–310. doi:10.1016/S0304-8853(01)01230-6

Mukasyan, A. S., Epstein, P., & Dinka, P. (2007). Solution combustion synthesis of nanomaterials. *Proceedings of the Combustion Institute*, *31*(2), 1789–1795. doi:10.1016/j.proci.2006.07.052

Mukherjee, K., & Majumder, S. B. (2010). Reducing gas sensing behaviour of nanocrystalline magnesium–zinc ferrite powders. *Talanta*, *81*(4-5), 1826–1832. doi:10.1016/j.talanta.2010.03.042 PMID:20441981

Mukhtar, M. W., Irfan, M., Ahmad, I., Ali, I., Akhtar, M. N., Khan, M. A., Abbas, G., Rana, M. U., Ali, A., & Ahmad, M. (2015). Synthesis and properties of Pr-substituted MgZnferrites for core materials and high-frequency applications. *Journal of Magnetism and Magnetic Materials*, *381*, 173–178. doi:10.1016/j.jmmm.2014.12.072

Murdock, E. S., Simmons, R. F., & Davidson, R. (1992). Roadmap for 10 Gbit/in/sup 2/ media: Challenges. *IEEE Transactions on Magnetics*, *28*(5), 3078–3083. doi:10.1109/20.179719

Muthukumaran, C., Sivakumar, V. M., & Thirumarimurugan, M. (2016). Adsorption isotherms and kinetic studies of crystal violet dye removal from aqueous solution using surfactant modified magnetic nanoadsorbent. *Journal of the Taiwan Institute of Chemical Engineers*, *63*, 354–362. doi:10.1016/j.jtice.2016.03.034

Nadargi, D., Umar, A., Nadargi, J., Patil, J., Mulla, I., Akbar, S., & Suryavanshi, S. (2022). Spinel Magnesium Ferrite ($MgFe_2O_4$): A Glycine-Assisted Colloidal Combustion and Its Potentiality in Gas-Sensing Application. *Chemosensors (Basel, Switzerland)*, *10*(9), 361. doi:10.3390/chemosensors10090361

Naik, C. C., Gaonkar, S. K., Furtado, I., & Salker, A. V. (2018). Effect of Cu^{2+} substitution on structural, magnetic and dielectric properties of cobalt ferrite with its enhanced antimicrobial property. *Journal of Materials Science Materials in Electronics*, *29*(17), 14746–14761. doi:10.100710854-018-9611-1

Naik, M. M., Naik, H. S. B., Nagaraju, G., Vinuth, M., Vinu, K., & Viswanath, R. (2019). Green synthesis of zinc doped cobalt ferrite nanoparticles: Structural, optical photocatalytic and antibacterial studies. *Nano Structures & Nano Objects*, *19*, 100322. doi:10.1016/j.nanoso.2019.100322

Naseri, M. G., Halimah, M. K., Dehzangi, A., Kamalianfar, A., Saion, E. B., & Majlis, B. Y. (2014). A Comprehensive overview on the structure and comparison of magnetic properties of nanocrystalline synthesized by a thermal treatment method. *Journal of Physics and Chemistry of Solids*, *75*(3), 315–327. doi:10.1016/j.jpcs.2013.11.004

Nayak, P. K. (2008). Synthesis and characterization of cadmium ferrite. *Materials Chemistry and Physics*, *112*(1), 24–26. doi:10.1016/j.matchemphys.2008.05.018

Nejati, K., & Zabihi, R. (2012). Preparation and magnetic properties of nano size nickel ferrite particles using hydrothermal method. *Chemistry Central Journal*, *6*(1), 23–28. doi:10.1186/1752-153X-6-23 PMID:22462726

Newton, B., Pearson, R., & Williams, J. (1975). The design and realisation of M.I.C. subassemblies on ferrite substrates. *IEEE Transactions on Magnetics*, *11*(5), 1270–1272. doi:10.1109/TMAG.1975.1058899

Nguyen, L. T. T., Nguyen, L. T. H., Manh, N. C., Quoc, D. N., Quang, H. N., Nguyen, H. T. T., Nguyen, D. C., & Bach, L. G. (2019). A Facile Synthesis, Characterization, and Photocatalytic Activity of Magnesium Ferrite Nanoparticles via the Solution Combustion Method. *Hindawi Journal of Chemistry*, *2019*(8), 3428681. doi:10.1155/2019/3428681

Niederberger, M., Pinna, N., Polleux, J., & Antonietti, M. (2004). A general soft-chemistry route to perovskites and related materials: Synthesis of $BaTiO_3$, $BaZrO_3$, and $LiNbO_3$ nanoparticles. *Angewandte Chemie*, *43*(17), 2270–2273. doi:10.1002/anie.200353300 PMID:15108141

Nur-E-Alam, M., Vasiliev, M., Belotelov, V., & Alameh, K. (2018). Properties of ferrite garnet (Bi, Lu, $Y)_3(Fe, Ga)_5O_{12}$ thin film materials prepared by RF magnetron sputtering. *Nanomaterials (Basel, Switzerland)*, *8*(5), 355. doi:10.3390/nano8050355 PMID:29789463

Obaidat, I. M., Issa, B., & Haik, Y. (2015). Magnetic properties of magnetic nanoparticles for efficient hyperthermia. *Nanomaterials (Basel, Switzerland)*, *5*(1), 63–89. doi:10.3390/nano5010063 PMID:28347000

Okamoto, A. (2009). The Invention of Ferrites and Their Contribution to the Miniaturization of Radios. *IEEE Globecom Workshops*, 1–42. doi:10.1109/GLOCOMW.2009.5360693

Olsson, R. T., Salazar, A. G., Hedenqvist, S. M., Gedde, U. W., Lindberg, F., & Savage, S. J. (2005). Controlled synthesis of near stoichiometric cobalt ferrite nanoparticles. *Chemistry of Materials*, *17*(20), 5109–5118. doi:10.1021/cm0501665

Padmapriya, G., Manikandan, A., Krishnasamy, V., Jaganathan, S. K., & Antony, S. A. (2016). Spinel $Ni_xZn_{1-x}Fe_2O_4$ (0.0 £ x £ 1.0) nano-photocatalysts: Synthesis, characterization and photocatalytic degradation of methylene blue dye. *Journal of Molecular Structure*, *1119*, 39–47. doi:10.1016/j.molstruc.2016.04.049

Pandit, R., Sharma, K. K., Kaur, P., & Kumar, R. (2014). Cation distribution controlled dielectric, electrical and magnetic behavior of In^{3+} substituted cobalt ferrites synthesized via solid-state reaction technique. *Materials Chemistry and Physics*, *148*(3), 988–999. doi:10.1016/j.matchemphys.2014.09.009

Paramesh, D., Kumar, K. V., & Reddy, P. V. (2016). Influence of nickel addition on structural and magnetic properties of aluminum substituted Ni-Zn ferrite nanoparticles. *Processing and Application of Ceramics*, *10*(3), 161–167. doi:10.2298/PAC1603161P

Pathania, A., Thakur, P., Trukhanov, A. V., Trukhanov, S. V., Panina, L. V., Luders, U., & Thakur, A. (2019). Development of tungsten doped Ni-Zn nano-ferrites with fast response and recovery time for hydrogen gas sensing application. *Results in Physics*, *15*, 102531. doi:10.1016/j.rinp.2019.102531

Patil, K. C., Aruna, S. T., & Mimani, T. (2002). Combustion synthesis: An update. *Current Opinion in Solid State and Materials Science*, *6*(6), 507–512. doi:10.1016/S1359-0286(02)00123-7

Patil, M. R., & Shrivastava, V. S. (2015). Adsorptive removal of methylene blue from aqueous solution by polyaniline-nickel ferrite nanocomposite: A kinetic approach. *Desalination and Water Treatment*, *57*(13), 5879–5887. doi:10.1080/19443994.2015.1004594

Pawar, C. S., Gujar, M. P., & Mathe, V. L. (2015). Synthesis and Characterization of Nanocrystalline Nickel Zinc Ferrite for Chlorine gas sensor at room temperature. In *59th DAE solid state physics symposium 2014, AIP Conference Proceedings, 1665(1),* 050060. AIP. 10.1063/1.4917701

Phumying, S., Labuayai, S., Swatsitang, E., Amornkitbamrung, V., & Maensiri, S. (2013). Nanocrystalline spinel ferrite (MFe_2O_4, M = Ni, Co, Mn, Mg, Zn) powders prepared by a simple aloe vera plant-extracted solution hydrothermal route. *Materials Research Bulletin*, *48*(6), 2060–2065. doi:10.1016/j.materresbull.2013.02.042

Plouffe, B. D., Murthy, S. K., & Lewis, L. H. (2014). Fundamentals and application of magnetic particles in cell isolation and enrichment. *Reports on Progress in Physics*, *78*(1), 016601. doi:10.1088/0034-4885/78/1/016601 PMID:25471081

Prakash, C., & Baijal, J. S. (1985). Dielectric behaviour of tetravalent titanium-substituted Ni-Zn ferrites. *Journal of the Less Common Metals*, *107*(1), 51–57. doi:10.1016/0022-5088(85)90240-1

Pullar, R. C. (2012). Hexagonal ferrites: A review of the synthesis, properties, and applications of hexaferrite ceramics. *Progress in Materials Science*, *57*(7), 1191–1334. doi:10.1016/j.pmatsci.2012.04.001

Qin, R., Li, F., Jiang, W., & Liu, L. (2009). Salt-assisted low temperature solid state synthesis of high surface area $CoFe_2O_4$ nanoparticles. *Journal of Materials Science and Technology*, *25*, 69–72.

Qu, Y., Yang, H., Yang, N., Fan, Y., Zhu, H., & Zou, G. (2006). The effect of reaction temperature on the particle size, structure and magnetic properties of coprecipitated $CoFe_2O_4$ nanoparticles. *Materials Letters*, *60*(29-30), 3548–3552. doi:10.1016/j.matlet.2006.03.055

Rahman, S., Nadeem, K., Anis-ur-Rehman, M., Mumtaz, M., Naeem, S., & Letofsky-Papst, I. (2013). Structural and magnetic properties of ZnMg-ferrite nanoparticles prepared using the co-precipitation method. *Ceramics International*, *39*(5), 5235–5239. doi:10.1016/j.ceramint.2012.12.023

Ramakrishna, K. S., Srinivas, Ch., Prajapat, C. L., Meena, S. S., Mehar, M. V. K., Potukuchi, D. M., & Sastry, D. L. (2018). Structural and magnetic investigations: Study of magnetocrystalline anisotropy and magnetic behavior of 0.1% Cu^{2+} substituted Ni-Zn ferrite nanoparticles. *Ceramics International*, *44*(1), 1193–1200. doi:10.1016/j.ceramint.2017.10.011

Rana, G., Dhiman, P., Kumar, A., Vo, D. N., Sharma, G., Sharma, S., & Nausad, M. (2021). Recent advances on nickel nano-ferrite: A review on processing techniques, properties and diverse applications. *Chemical Engineering Research & Design*, *175*, 182–208. doi:10.1016/j.cherd.2021.08.040

Randhawa, B. S., Dosanjh, H. S., & Kumar, N. (2007). Synthesis of lithium ferrite by precursor and combustion methods: A comparative study. *Journal of Radioanalytical and Nuclear Chemistry*, *274*(3), 581–591. doi:10.100710967-006-6924-y

Rathore, D., Kurchania, R., & Pandey, R. K. (2015). Gas Sensing Properties of Size Varying $CoFe_2O_4$ Nanoparticles. *IEEE Sensors Journal*, *15*(9), 4961–4966. doi:10.1109/JSEN.2015.2432035

Reddy, G. C. V., Seela, K. K., & Manorama, S. V. (2000). Preparation of Fe_2O_3 by the hydrazine method: Application as an alcohol sensor. *International Journal of Inorganic Materials*, *2*, 301–305. doi:10.1016/S1466-6049(00)00044-1

Reddy, L. H., Arias, J. L., Nicolas, J., & Couvreur, P. (2012). Magnetic nanoparticles: Design and characterization, toxicity and biocompatibility, pharmaceutical and bio-medical applications. *Chemical Reviews*, *112*(11), 5818–5878. doi:10.1021/cr300068p PMID:23043508

Rezlescu, N., Rezlescu, E., Tudorache, F., & Popa, P. D. (2007). Some spinel oxide compounds as reducing gas sensors. *Sensors & Transducers*, *78*, 1134–1142.

Rezlescu, N., Rezlescu, E., Tudorache, F., & Popa, P. D. (2009). Gas sensing properties of porous Cu-, Cd- and Zn-ferrites. *Romanian Reports in Physics*, *61*(2), 223–234.

Rocha-Santos, T. A. P. (2014). Sensors and biosensors based on magnetic nanoparticles. *Trends in Analytical Chemistry*, *62*, 28–36. doi:10.1016/j.trac.2014.06.016

Saez, P. R., Fernandez, M. J., Gutierrez, V. B., Gomez, R., & Marquina, V. (2008). Ferrites nanoparticles MFe_2O_4 (M = Ni and Zn): Hydrothermal synthesis and magnetic properties. *Bulletin of the Spanish Society of Ceramics and Glass*, *47*, 133–137.

Sanpo, N., Berndt, C. C., & Wang, J. (2012). Microstructural and antibacterial properties of zinc-substituted cobalt ferrite nanopowders synthesized by sol-gel methods. *Journal of Applied Physics*, *112*(8), 1–7. doi:10.1063/1.4761987

Satheesh, R., Vignesh, K., Rajarajan, M., Suganthi, A., Sreekantan, S., Kang, M., & Kwak, B. S. (2016). Removal of congo red from water using quercetin modified α-Fe_2O_3 nanoparticles as effective nanoadsorbent. *Materials Chemistry and Physics*, *180*, 53–65. doi:10.1016/j.matchemphys.2016.05.029

Satyanarayana, L., Reddy, K. M., & Manorama, S. V. (2003). Synthesis of nanocrystalline $Ni_{1-x}Co_xMn_xFe_{2-x}O_4$: A material for liquefied petroleum gas sensing. *Sensors and Actuators. B, Chemical*, *89*(1-2), 62–67. doi:10.1016/S0925-4005(02)00429-X

Schierbaum, K. D., Weimar, U., & Gopel, W. (1992). Comparison of ceramic, thick-film and thin-film chemical sensors based upon SnO_2. *Sensors and Actuators. B, Chemical*, *7*(1-3), 709–716. doi:10.1016/0925-4005(92)80390-J

Seki, K., Nakayama, Y., Chiba, S., & Murakami, K. (1994). Construction of Pressure and Photo Sensors Using Ferrite Thick Films. *IEEE Translation Journal on Magnetics in Japan*, *9*(2), 92–96. doi:10.1109/TJMJ.1994.4565832

Selvan, R. K., Augustin, C. O., Berchmans, L. J., & Saraswathi, R. (2003). Combustion synthesis of $CuFe_2O_4$. *Materials Research Bulletin*, *38*(1), 41–54. doi:10.1016/S0025-5408(02)01004-8

Semancik, S., & Cavicchi, R. E. (1991). The growth of thin, epitaxial SnO2 films for gas sensing applications. *Thin Solid Films*, *206*(1-2), 81–87. doi:10.1016/0040-6090(91)90397-G

Senapati, K. K., Borgohain, C., & Phukan, P. (2011). Synthesis of highly stable $CoFe_2O_4$ nanoparticles and their use as magnetically separable catalyst for Knoevenagel reaction in aqueous medium. *Journal of Molecular Catalysis A Chemical*, *339*(1-2), 24–31. doi:10.1016/j.molcata.2011.02.007

Shaikh, S. F., Ubaidullah, M., Rajaram, S. M., & Al-Enizi, A. M. (2020). Types, Synthesis methods and applications of ferrites. In R. Mane & V. Jadhav (Eds.), *Spinel Ferrite Nanostructures for Energy Storage Devices* (pp. 51–82). Elsevier. doi:10.1016/B978-0-12-819237-5.00004-3

Shariati-Rad, M. S., Irandoust, M., Amri, S., Feyzi, M., & Jafari, F. (2014). Magnetic solid phase adsorption, preconcentration and determination of methyl orange in water samples using silica coated magnetic nanoparticles and central composite design. *International Nano Letters*, *4*(4), 91–101. doi:10.100740089-014-0124-5

Sharma, R., Bansal, S., & Singhal, S. (2015). Tailoring the photo-Fenton activity of spinel ferrites (MFe_2O_4) by incorporating different cations (M = Cu, Zn, Ni, and Co) in the structure. *RSC Advances*, *5*(8), 6006–6018. doi:10.1039/C4RA13692F

Sharma, R. P., Raut, S. D., Kadam, A. S., Mulani, R. M., & Mane, R. S. (2020). In-vitro antibacterial and anti-biofilm efficiencies of chitosan-encapsulated zinc ferrite nanoparticles. *Applied Physics. A, Materials Science & Processing*, *126*(10), 824. doi:10.100700339-020-04007-1

Shi, W., Zhang, X., He, S., & Huang, Y. (2011). $CoFe_2O_4$ magnetic nanoparticles as a peroxidase mimic mediated chemiluminescence for hydrogen peroxide and glucose. *Chemical Communications*, *47*(38), 10785–10787. doi:10.1039/c1cc14300j PMID:21892508

Singh, S., Ralhan, N. K., Kotnala, R. K., & Verma, K. C. (2012). Nanosize dependent electrical and magnetic properties of $NiFe_2O_4$ ferrite. *Indian Journal of Pure and Applied Physics*, *50*, 739–743.

Sivakumar, N., Narayanasamy, A., Greneche, J. M., Murugaraj, R., & Lee, Y. S. (2010). Electrical and magnetic behaviour of nanostructured $MgFe_2O_4$ spinel ferrite. *Journal of Alloys and Compounds*, *504*(2), 395–402. doi:10.1016/j.jallcom.2010.05.125

Snoek, J. L. (1947). Gyromagnetic resonance in ferrites. *Nature*, *160*(4055), 90. doi:10.1038/160090a0 PMID:20256149

Sohn, C. H., Park, S. P., Choi, S. H., Park, S. H., Kim, S., Xu, L., Kim, S. H., Hur, J. A., Choi, J., & Choi, T. H. (2015). MRI molecular imaging using GLUT1 antibody-Fe_3O_4 nanoparticles in the hemangioma animal model for differentiating infantile hemangioma from vascular malformation. *Nanomedicine; Nanotechnology, Biology, and Medicine*, *11*(1), 127–135. doi:10.1016/j.nano.2014.08.003 PMID:25168935

Song, Z., Zhou, D., & Liu, Q. (2019). Tolerance factor and phase stability of the garnet structure. *Acta Crystallographica. Section C, Crystal Structure Communications*, *75*(10), 1353–1358. doi:10.1107/S2053229619011975 PMID:31589151

Souriou, D., Mattei, J. L., Chevalier, A., & Queffelec, P. (2010). Influential parameters on electromagnetic properties of nickel-zinc ferrites for antenna miniaturization. *Journal of Applied Physics, 107,* 09A518.

Spirou, S. V., Basini, M., Lascialfari, A., Sangregorio, C., & Innocenti, C. (2018). Magnetic hyperthermia and radiation therapy: Radiobiological principles and current practice. *Nanomaterials (Basel, Switzerland)*, *8*(6), 1–22. doi:10.3390/nano8060401 PMID:29865277

Srivastava, R., & Yadav, B. C. (2012). Ferrite Materials: Introduction, Synthesis Techniques, and Applications as Sensors. *International Journal of Green Nanotechnology*, *4*(2), 141–154. doi:10.1080/19430892.2012.676918

Stafiej, A., & Pyrzynska, K. (2008). Solid phase extraction of metal ions using carbon nanotubes. *Microchemical Journal*, *89*(1), 29–33. doi:10.1016/j.microc.2007.11.001

Stergiou, C. A., & Litsardakis, G. (2016). Y-type hexagonal ferrites for microwave absorber and antenna applications. *Journal of Magnetism and Magnetic Materials*, *405*, 54–61. doi:10.1016/j.jmmm.2015.12.027

Su, L., Feng, J., Zhou, X., Ren, C., Li, H., & Chen, X. (2012). Colorimetric detection of urine glucose based $ZnFe_2O_4$ magnetic nanoparticles. *Analytical Chemistry*, *84*(13), 5753–5758. doi:10.1021/ac300939z PMID:22702236

Sugimoto, M. (1999). The Past, Present, and Future of Ferrites. *Journal of the American Ceramic Society*, *82*(2), 269–280. doi:10.1111/j.1551-2916.1999.tb20058.x

Sutka, A., & Gross, K. A. (2016). Spinel ferrite oxide semiconductor gas sensors. *Sensors and Actuators. B, Chemical*, *222*, 95–105. doi:10.1016/j.snb.2015.08.027

Sutka, A., Gross, K. A., Mezinskis, G., & Knite, M. (2011). The effect of heating conditions on the properties of nano- and microstructured Ni-Zn ferrite. *Physica Scripta*, *83*(2), 025601–025606. doi:10.1088/0031-8949/83/02/025601

Sutka, A., & Mezinskis, G. (2012). Sol-gel auto-combustion synthesis of spinel-type ferrite nanomaterials. *Frontiers of Materials Science*, *6*(2), 128–141. doi:10.100711706-012-0167-3

Sutka, A., Mezinskis, G., Pludons, A., & Lagzdina, S. (2010). Characterization of sol-gel auto-combustion derived spinel ferrite nano-materials. *Energetika*, *56*(3-4), 254–259.

Takada, T., & Kiyama, M. (1971). *Preparation of ferrites by wet method.* In Ferrites: Proceedings of the International Conference, Y. Hoshino, S. Iida and M. Sugimoto (Eds.), University of Tokyo Press, Tokyo, pp. 69-71.

Tamaura, Y., Kojima, M., Hasegawa, N., Tsuji, M., Ehrensberger, K., & Steinfeld, A. (1997). Solar Energy Conversion into H2 Energy Using Ferrites. Journal of Physics, IV, 673-674.

Thant, A. A., Srimala, S., Kaung, P., Itoh, M., Radzali, O., & Fauzi, M. N. A. (2010). Low temperature synthesis of $MgFe_2O_4$ soft ferrite nanocrystallites. *Journal of the Australian. Ceramic Society*, *46*(1), 11–14.

Vaidyanathan, G., Sendhilnathan, S., & Arulmurugan, R. (2007). Structural and magnetic properties of $Co_{1-x}Zn_xFe_2O_4$ nanoparticles by co-precipitation method. *Journal of Magnetism and Magnetic Materials*, *313*(2), 293–299. doi:10.1016/j.jmmm.2007.01.010

Valenzuela, R. (1994). *Magnetic ceramics*. Cambridge Press. doi:10.1017/CBO9780511600296

Valenzuela, R. (2012). Novel Applications of Ferrites. *Physics Research International*, 591839.

Varma, P. C. R., Manna, R. S., Benerjee, D., Varma, M. R., Suresh, K. G., & Nigam, A. K. (2008). Magnetic properties of $CoFe_2O_4$ synthesized by solid state, citrate precursor and polymerized complex methods: A comparative study. *Journal of Alloys and Compounds*, *453*(1-2), 298–303. doi:10.1016/j.jallcom.2006.11.058

Vasic, M. V., Bozin, E. S., Bessais, L., Stojanovic, G., Luburic, U. K., Abeykoon, M., Jancar, B., Meden, A., Kremenovic, A., & Antic, B. (2013). Thermal Evolution of Cation Distribution/Crystallite Size and Their Correlation with the Magnetic State of Yb-Substituted Zinc Ferrite Nanoparticles. *The Journal of Physical Chemistry C*, *117*(23), 12358–12365. doi:10.1021/jp403459t

Vedrtnam, A., Lalauni, K., Dubey, S., & Kumar, A. (2020). A comprehensive study on structure, properties, synthesis and characterization of ferrites. *AIMS Materials Science*, *7*(6), 800–835. doi:10.3934/matersci.2020.6.800

Venturini, J., Piva, D. H., da Cunha, J. B. M., & Bergmann, C. P. (2016). Effect of the thermal treatment on the magnetic and structural properties of cobalt ferrite particles. *Ceramics International*, *42*(14), 15183–15188. doi:10.1016/j.ceramint.2016.06.120

Went, J. J., Rathenau, G. W., Gorter, E. W., & Oosterhout, G. W. (1952). Hexagonal iron-oxide compounds as permanent-magnet materials. *Physical Review*, *86*(3), 424–425. doi:10.1103/PhysRev.86.424.2

Wu, K., Li, J., & Zhang, C. (2019). Zinc ferrite based gas sensors: A review. *Ceramics International*, *45*(9), 11143–11157. doi:10.1016/j.ceramint.2019.03.086

Wu, W., Wu, Z., Yu, T., Changzhong, J., & Kim, W. S. (2015). Recent progress on magnetic iron oxide nanoparticles: Synthesis, surface functional strategies, and biomedical applications. *Science and Technology of Advanced Materials*, *16*(2), 23501. doi:10.1088/1468-6996/16/2/023501 PMID:27877761

Xiang, J., Shen, X., & Meng, X. (2009). Preparation of Co-substituted MnZn ferrite fibers and their magnetic properties. *Materials Chemistry and Physics*, *114*(1), 362–366. doi:10.1016/j.matchemphys.2008.09.037

Xiangfeng, C., & Chenmou, Z. (2003). Sulfide-sensing characteristics of MFe_2O_4 (M = Zn, Cd, Mg and Cu) thick film prepared by co-precipitation method. *Sensors and Actuators. B, Chemical*, *96*(3), 504–508. doi:10.1016/S0925-4005(03)00626-9

Xie, G., Xi, P., Liu, H., Chen, F., Huang, L., Shi, Y., Hou, F., Zeng, Z., Shao, C., & Wang, J. (2012). A facile chemical method to produce superparamagnetic graphene oxide–Fe_3O_4 hybrid composite and its application in the removal of dyes from aqueous solution. *Journal of Materials Chemistry*, *22*(3), 1033–1039. doi:10.1039/C1JM13433G

Xue, H., Li, Z., Wang, X., & Fu, X. (2007). Facile synthesis of nanocrystalline zinc ferrite via self-propagating combustion method. *Materials Letters*, *61*(2), 347–350. doi:10.1016/j.matlet.2006.04.061

Yadav, R. S., Havlica, J., Hnatko, M., Sajgalik, P., Alexander, C., Palou, M., Bartonickova, E., Bohac, M., Frajkorova, F., Masilko, J., Zmrzly, M., Kalina, L., Hajduchova, M., & Enev, V. (2015). Magnetic properties of $Co_{1-x}Zn_xFe_2O_4$ spinel ferrite nanoparticles synthesizedby starch-assisted sol-gel autocombustion method and its ballmilling. *Journal of Magnetism and Magnetic Materials*, *378*, 190–199. doi:10.1016/j.jmmm.2014.11.027

Yamaguchi, M., Naoe, M., Yasuda, T., & Yamanaka, S. (1974). Fabrication and Characteristics of Ferrite Films for High-Frequency Electroacoustic Transducers. *Electrical Engineering in Japan*, *94*(7), 295–302.

Yan, Z., Gao, J., Li, Y., Zhang, M., & Guo, M. (2015). Hydrothermal synthesis and structure evolution of metal-doped magnesium ferrite from saprolite laterite. *RSC Advances*, *5*(112), 92778–92787. doi:10.1039/C5RA17145H

Yang, C., Su, X., Wang, J., Cao, X., Wang, S., & Zhang, L. (2013). Facile microwave assisted hydrothermal synthesis of varied-shaped CuO nanoparticles and their gas sensing properties. *Sensors and Actuators. B, Chemical*, *185*, 159–165. doi:10.1016/j.snb.2013.04.100

Yang, G. M., Xing, X., Obi, O., Daigle, A., Liu, M., Stoute, S., Naishadham, K., & Sun, N. (2010). Loading effects of self-biased magnetic films on patch antennas with substrate/superstrate sandwich structure. *IET Microwaves, Antennas & Propagation*, *4*(9), 1172–1181. doi:10.1049/iet-map.2009.0429

Yu, M., Liu, X., Wang, Y., Zheng, Y., Zhang, J., Li, M., Lan, W., & Su, Q. (2012). Gas sensing properties of p-type semiconducting vanadium oxide nanotubes. *Applied Surface Science*, *258*(24), 9554–9558. doi:10.1016/j.apsusc.2012.05.120

Yu, S., Jing, W., Tang, M., Xu, T., Yin, W., & Kang, B. (2019). Fabrication of Nd:YAG transparent ceramics using powders synthesized by citrate sol-gel method. *Journal of Alloys and Compounds*, *772*, 751–759. doi:10.1016/j.jallcom.2018.09.184

Yue, Z., Zhou, J., Li, L., & Gui, Z. (2000). Synthesis of nanocrystalline NiCuZn ferrite powders by sol-gel auto-combustion method. *Journal of Magnetism and Magnetic Materials*, *208*(1-2), 55–60. doi:10.1016/S0304-8853(99)00566-1

Yue, Z., Zhou, J., Li, L., & Gui, Z. (2001). Effect of copper on the electromagnetic properties of Mg-Zn-Cu ferrites prepared by sol-gel auto-combustion method. *Materials Science and Engineering B*, *86*(1), 64–69. doi:10.1016/S0921-5107(01)00660-2

Zahi, S., Daud, A. R., & Hashim, M. (2007). A comparative study of nickel-zinc ferrites by sol-gel route and solid-state reaction. *Materials Chemistry and Physics*, *106*(2), 452–456. doi:10.1016/j.matchemphys.2007.06.031

Zhang, C. Y., Shen, X. Q., Zhou, J. X., Jing, M. X., & Cao, K. (2007). Preparation of spinel ferrite Ni-Fe_2O_4 fibres by organic gel-thermal decomposition process. *Journal of Sol-Gel Science and Technology*, *42*(1), 95–100. doi:10.100710971-006-1515-5

Zhang, H., Zhang, X., Lin, H., Wang, K., Sun, X., Xu, N., Li, C., & Ma, Y. (2015). Graphene and maghemite composites based supercapacitors delivering high volumetric capacitance and extraordinary cycling stability. *Electrochimica Acta*, *156*, 70–76. doi:10.1016/j.electacta.2015.01.041

Zhang, Y., Yang, Z., Yin, D., Liu, Y., Fei, C. L., Xiong, R., Shi, J., & Yan, G. L. (2010). Composition and magnetic properties of cobalt ferrite nano-particles prepared by the co-precipitation method. *Journal of Magnetism and Magnetic Materials*, *322*(21), 3470–3475. doi:10.1016/j.jmmm.2010.06.047

Zhang, Z., Shan, Y., Wang, J., Ling, H., Zang, S., Gao, W., Zhao, Z., & Zhang, H. (2007). Investigation on the rapid degradation of congo red catalyzed by activated carbon powder under microwave irradiation. *Journal of Hazardous Materials*, *147*(1-2), 325–333. doi:10.1016/j.jhazmat.2006.12.083 PMID:17293037

Zhang, Z., Yao, G., Zhang, X., Ma, J., & Lin, H. (2015). Synthesis and characterization of nickel ferrite nanoparticles via planetary ball milling assisted solid-state reaction. *Ceramics International*, *41*(3), 4523–4530. doi:10.1016/j.ceramint.2014.11.147

Zhao, X., Wang, W., Zhang, Y., Wu, S., Li, F., & Liu, J. P. (2014). Synthesis and characterization of gadolinium doped cobalt ferrite nanoparticles with enhanced adsorption capability for Congo Red. *Chemical Engineering Journal*, *250*, 164–174. doi:10.1016/j.cej.2014.03.113

Zhu, X., Zhou, J., Chen, M., Shi, M., Feng, W., & Li, F. (2012). Core-shell Fe_3O_4@$NaLuF_4$:Yb, Er/Tm nanostructure for MRI, CT and upconversion luminescence tri-modality imaging. *Biomaterials*, *33*(18), 4618–4627. doi:10.1016/j.biomaterials.2012.03.007 PMID:22444645

Chapter 4
Plant Extract-Based Silver Nanoparticles and Their Bioactiviy Investigations

Dharshini Perumal
Universiti Putra Malaysia, Malaysia

Nur Farahah Mohd Khairuddin
Universiti Putra Malaysia, Malaysia

Jie Hui Wong
Universiti Putra Malaysia, Malaysia

Che Azurahanim Che Abdullah
Universiti Putra Malaysia, Malaysia

ABSTRACT

Green synthesised nanoparticles are a new source of inspiration for clinical research, as this method uses natural materials, such as plant-derived materials, for synthesis. Green nanoparticle synthesis using plant extracts is simple, inexpensive, chemicals, resulting in biologically active shape- and size-dependent products with no contaminants or byproducts. Silver nanoparticles (AgNPs) have unique physiochemical properties and promising biological applications. To assess AgNPs' functionality, they must be characterized. Natural compounds found in plant extracts reduce and stabilise AgNPs. Green synthesis is more environmentally friendly than conventional methods, but many questions remain. To understand the plant-mediated process, many studies are needed. This proposed book chapter examines the synthesis, characterization, and bio activities of green-synthesised AgNPs. The goal is to provide an overview of green AgNP synthesis and their bioactivity assessment.

DOI: 10.4018/978-1-6684-7358-0.ch004

1. INTRODUCTION

Nanotechnology is emerging as a new field of study in the modern era. It is concerned with the production of nanomaterials and nanoparticles. Nanotechnology-related multidisciplinary research has revolutionized numerous fields, including biotechnology, material science, biochemistry, botany, and polymer science (Gupta et al., 2020). Nanoparticles are nanometre-sized particles (1 to 100 nm) that dominate global research (Joseph et al.,2021). Nanoparticles exhibit extraordinary physical strength, chemical reactivity, optical effects, electrical conductivity, and magnetism owing to their small sizing. Nanoparticles exhibit unique properties such as a large surface-to-volume ratio and a strong quantum confinement effect, in addition to other biological, physical, and chemical characteristics that make them superior to bulk materials (Adebayo-Tayo et al., 2019). Figure 1, displays nanoparticles are used in a variety of fields (Almasoud et al., 2021; Ijaz et al., 2020).

Figure 1. Applications of nanoparticles in various fields

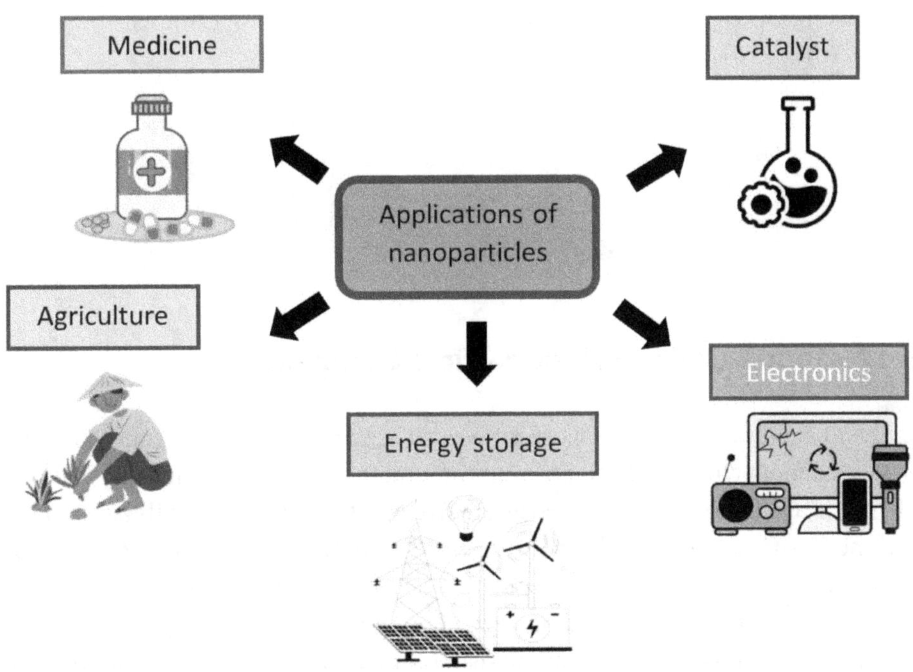

1.1 Approaches for Nanoparticle Fabrication

Traditionally, nanoparticle fabrication has been accomplished through either a top-down or bottom-up approach, as illustrated in Figure 2. While, both techniques, employ distinct synthesis principles, they both result in nanoparticles with the desired properties. Bulk materials were fragmented into bit-to-bit particles using a top-down approach, resulting in the fine production of nanoparticles. These nanoparticles were created using photolithographic techniques, grinding, sputtering, and milling. The top-down approach is a viable technique capable of producing nanoparticles on a large scale. However, the disadvantages of top-down processing include the fact that it is expensive, time consuming, results in surface structural defects, and, in some cases, damages nanoparticles (Patil & Chandrasekaran, 2020). The bottom-up approach entails the assembly of nanoparticles from individual atoms, molecules, or clusters. Chemical, electrochemical, sol-gel, chemical vapor deposition, pyrolysis, and bio-assisted synthesis techniques are used to synthesize nanoparticles (Habibullah et al., 2021; Ijaz et al., 2020). Bottom-up approaches are becoming increasingly advantageous and are now frequently used for nanoparticle synthesis.

Figure 2. Top-down and bottom-up nanoparticle synthesis

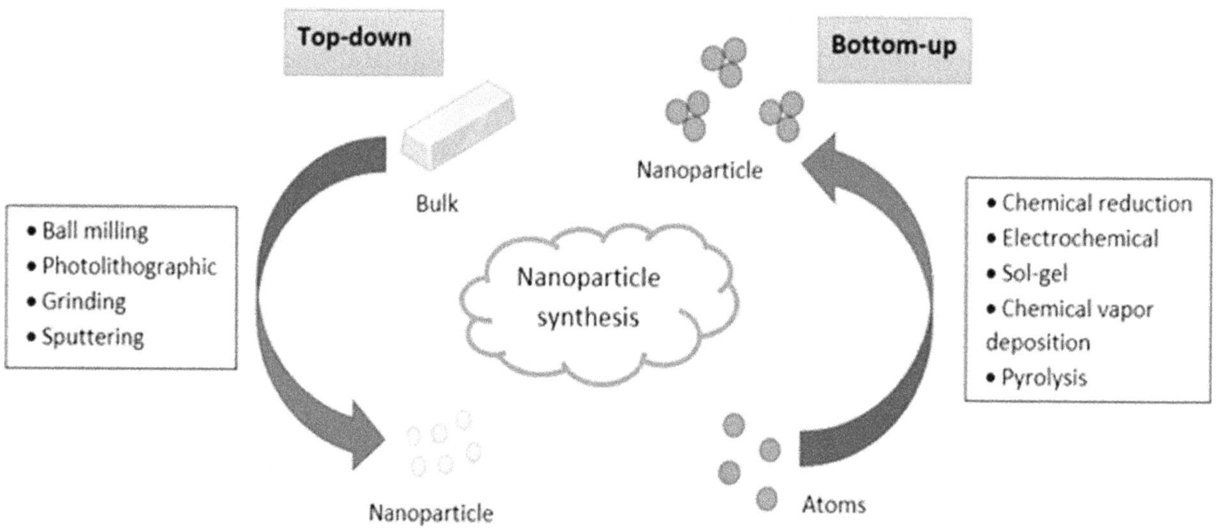

1.2 Different Methods Used in Nanoparticle Synthesis

Several physical and chemical strategies were employed at the turn of the twentieth century to enhance the effectiveness of nanoparticles. These conventional methods, on the other hand, involve exorbitant and harmful chemical and cannot be described as environmentally friendly, as illustrated in Figure 3. The physical process necessitates the use of expensive equipment, elevated temperatures, and pressures, as well as ample space for machine setup. Chemical processes involve the use of hazardous chemicals, which can be harmful to both the environment and biological organism. Spray pyrolysis, electrodeposition, wet chemical, chemical microemulsion, microwave assisted combustion, and chemical and direct precipitation are all examples of chemical techniques. Thermal evaporation, molecular beam epitaxy, and pulsed laser deposition are all physical approaches(Agarwal et al., 2017; Ijaz et al., 2020). Both physical and chemical processes require additional capping and stabilizing agents. According to the literature, some of the toxic substances used in physical and chemical processed may end up with nanoparticles, posing a risk in the field of medical applications. As a result, we required a method for nanoparticle generation that was both environmentally friendly and cost effective. Biological approaches to nanoparticle synthesize are environmentally sustainable; this is referred to as "green synthesis" because the reagents used are non-toxic and safe, posing no danger or risk to living organisms. These approaches are not prohibitively expensive, and they do not necessitate a high cost of power. These methods are non-polluting towards the environment (Y. Khan et al., 2018).

Figure 3. The advantages and disadvantages of biological, physical, and chemical methods (Habibullah et al., 2021; Mukherjee & Patra, 2017)

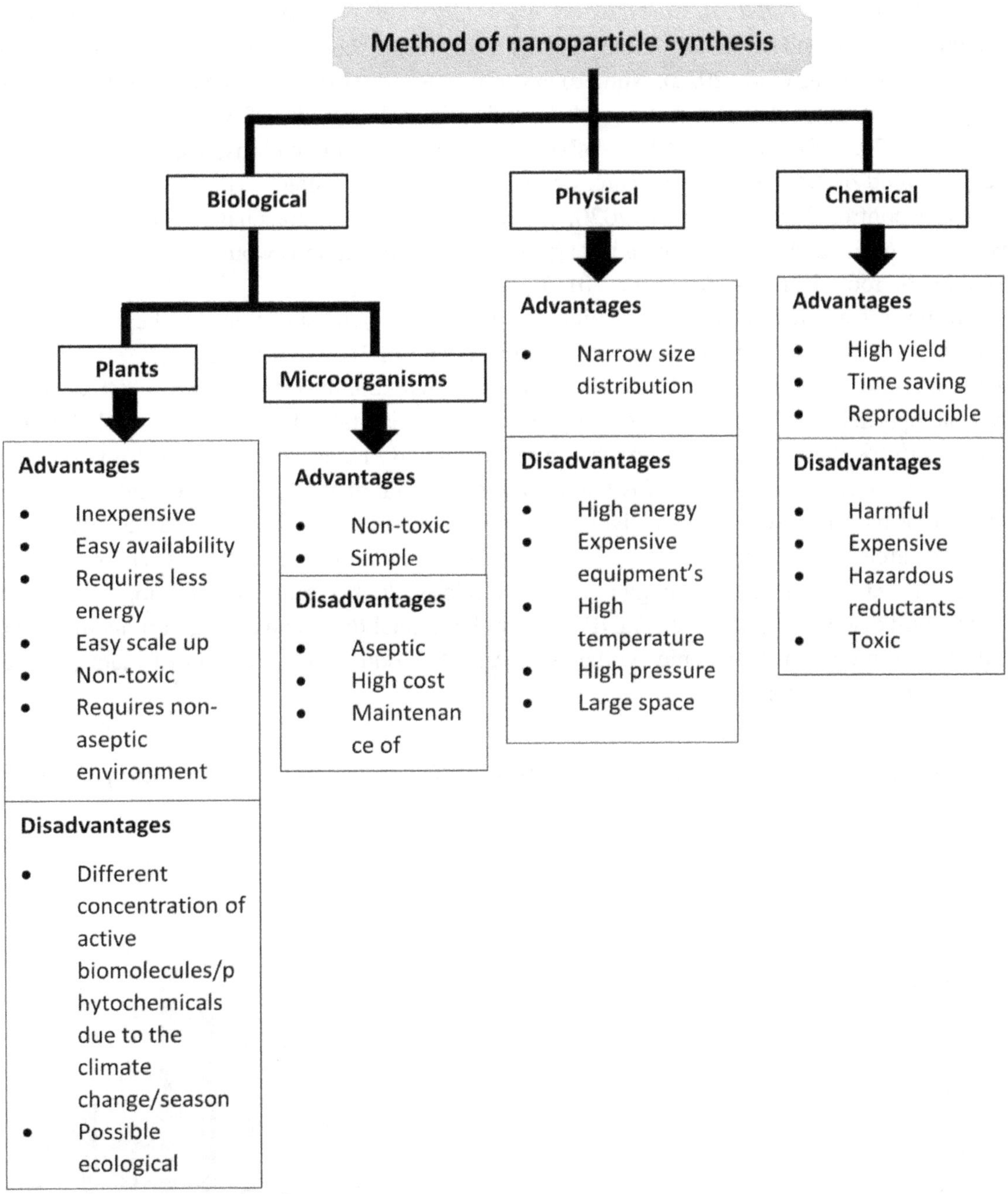

1.3 Silver Nanoparticles

Recent breakthroughs and accumulating data indicate that nanotechnology-driven revolutions based on the use of metal-based nanoparticles may be able to eradicate bacterial infection, combat the growth of resistant germs, and serve as a potential alternative to conventional antibiotics. The synthesis and properties of silver (Ag), gold (Au), platinum (Pt), copper (Cu), Zinc (Zn), and palladium (Pd) nanoparticles have piqued the interest of nanomedicine researchers (Pugazhendhi et al., 2018).

Silver nanoparticles (AgNPs) are the most commercially viable noble metal nanoparticles, particularly in consumer products and the medical field. AgNPs have recently been recognized as a significant agent in medical science (Das et al., 2022). Numerous studies have demonstrated the antiviral, antibacterial, antifungal, antioxidant, anticancer, antimicrobial, and antiplatelet activity of AgNPs (Gupta et al., 2020; Neupane et al., 2022; Shahi et al., 2018). AgNPs have numerous applications, such as medicine, water purification, food packaging, personal care, dentistry, diagnostic biological probes, sensing, catalysis, and pollution control (Mat Yusuf et al., 2020; Subha et al., 2018). Ag ions have been recognized as effective antimicrobial and tissue regenerating agents for decades. Ag ions wound dressings may be used to treat diabetic foot ulcers (Bakker et al., 2012).

According to statistical data analysis, Figure 4 depicts an increasing trend in the number of published research papers in the field of biogenic synthesis of AgNPs. We tracked, analysed, categorised, and gathered scholarly content on the Lens.org platform that contained the keyword "Green synthesis of silver nanoparticles". The Lens, a free and open platform maintained by Cambia, a global non-profit organization (https://www.lens.org/), serves as a comprehensive resource connecting numerous sources of information. Between 2001 and 2021, Lens published a total of 8,636 scholarly works. The number of publications has increased dramatically, from three in 2001 to 661 in 2021. Thus, this review aims to stimulate researchers to investigate natural resources capable of producing AgNPs via diverse plant and their organs to integrate nanotechnology and biotechnology, dubbed nanobiotechnology. Additionally, this review will discuss how to create AgNPs that are beneficial to humans using various methods. We have discussed in detail the biogenic synthesis of AgNPs produced by a variety of plants, as well as their bioactivity applications.

Figure 4. Annual publications for green synthesis of AgNPs from 2001 to 2021 (Data collected from Lens.org)

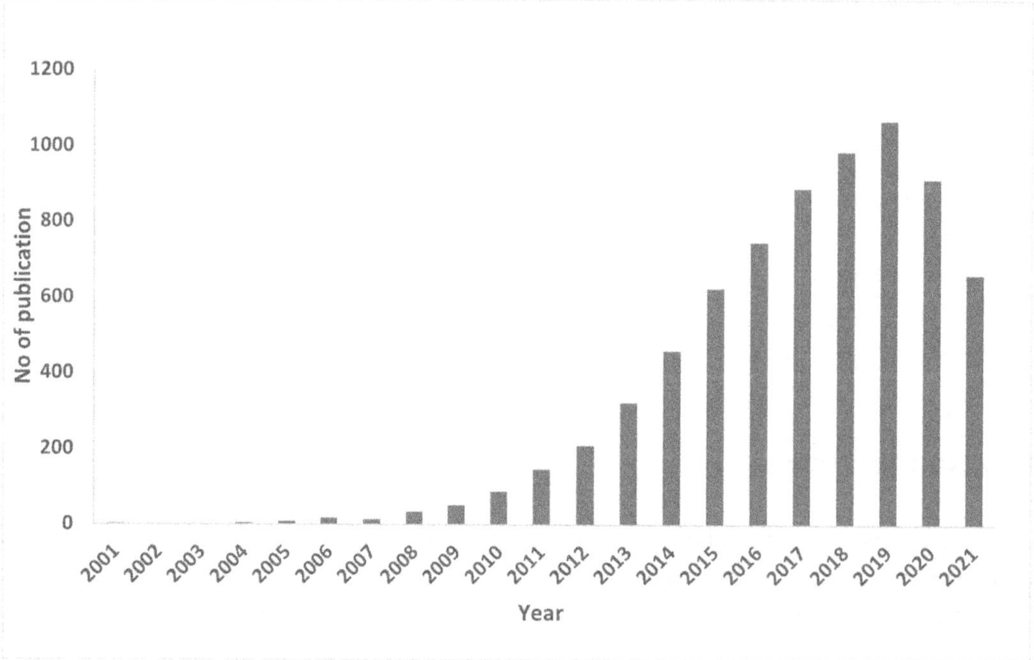

1.4 Green Synthesis of Silver Nanoparticles

The most recent environmentally friendly approach is green chemistry technology. The green synthesis approach outperforms chemical and physical methods in terms of speed, cost, environmental friendliness, and ease of scaling up for large-scale production without the use of high pressure, high temperature, high energy, or harmful chemicals. Green synthesis frequently uses microbes or plant components as a reducing agent in the synthesis of nanoparticles as shown in Figure 5 (H. Singh et al., 2018). Unlike microorganisms, plants have been extensively used as a reductant in the synthesis of AgNPs. Our institution emphasizes plant biomass due to its widespread renewability, availability, and environmental friendliness, as well as its significant application in the synthesis of nanoparticles. Due to the unique physiochemical properties of plants and significant advancements in green chemistry and plant-based research, every part of the plant is used to synthesize metal nanostructures (Vanlalveni et al., 2021).

Figure 5. Representatives image of plant and microorganism of green synthesis

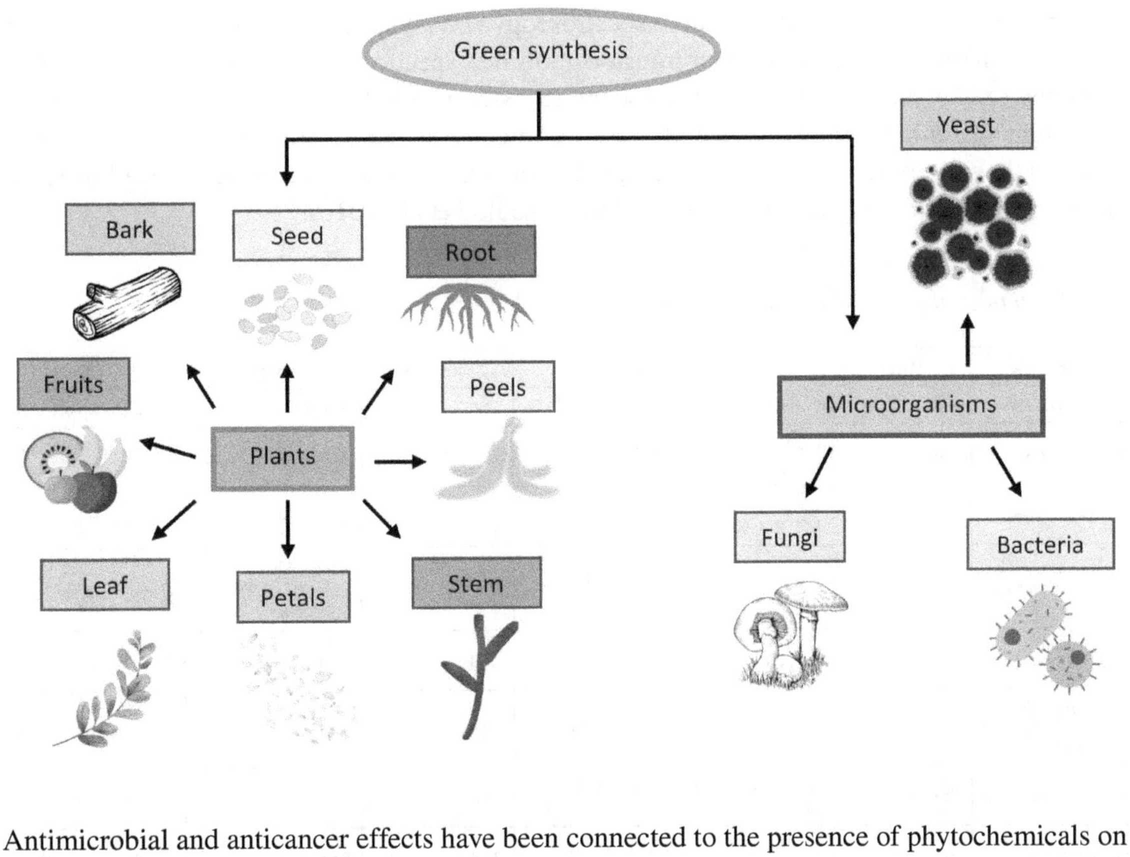

Antimicrobial and anticancer effects have been connected to the presence of phytochemicals on the surface of green produced AgNPs. The phytochemical constituents of plants, which include alkaloids, ketones, terpenoids, polysaccharides, phenols, vitamins, amines, amides, aromatic dicarboxylic acids, enzymes, and tannins, aren't always essential for plant growth routine activities, but it also serve as a rich source of chemical compounds used to reduce metals in bulk to metal ions (Jabeen et al., 2021; Subramaniam et al., 2022). Many secondary metabolites in plant extracts are sugars, polyphenols, flavonoids, and terpenoids. Plant extracts contain phytochemicals and secondary metabolites that have the ability to cap and reduce Ag^+ ions to Ag^0 (Sellami et al., 2021). Typically, a reducing agent derived from a plant extract doubles as a capping and stabilizing agent (Salayová et al., 2021).

Numerous plants have been linked to the production of AgNPs (Table 1), and the diagram below summarises their presence (Figure 6). The following procedures are used to create nanoparticles: The collected parts are thoroughly cleaned with tap water to remove dirt. Prior to crushing, the plant components are cleaned and dried in the shade for a minimum of ten to fifteen days. To obtain the plant extract, the powdered plant is soaked and boiled in deionized distilled water. The infusion is then filtered to remove any remaining insoluble material from the extract. When a plant extract is added to a silver nitrate solution, it initiates the reduction of pure Ag^+ ions to Ag^0, which can be quantified using a variety of techniques (Raj et al., 2021). Numerous parameters affect the production of AgNPs, including the concentration of plant extract, the concentration of salts, the reaction time, the temperature, and the solution pH (Perumal et al., 2022).

The shape, surface area, size, and dispersity of nanoparticles define them. The two approaches used to investigate the properties of nanoparticles are quantitative as well as qualitative. A range of complex approaches are included in these techniques such as surface enhanced Raman spectroscopy (SERS), atomic force microscopy (AFM), ultraviolet-visible spectroscopy (UV-Vis), transmission electron microscopy (TEM), x-ray diffraction (XRD), Fourier transform infrared spectroscopy (FTIR), x-ray photoelectron spectroscopy (XPS), scanning electron microscopy (SEM), energy dispersive spectroscopy (EDS) and dynamic light scattering (DLS). Several of these techniques can be used to investigate the properties of AgNPs, allowing for the resolution of parameters such as particle size, shape, crystallinity, fractal dimension, pores diameter, and surface area (Mohammadlou et al., 2016).

Table 1. Selection of parts of plants used to synthesize AgNPs

Scientific name	Parts	Characterization	Size (nm)	Shape	Ref.
Annona senegalensis	Bark	UV-Vis, SEM, TEM, EDX	11- 24.76	Spherical	(Olabemiwo et al., 2021)
Hyptis suaveolens	Callus	UV-Vis, SEM, TEM, EDX, XRD, DLS, FTIR	12 to 25	Spherical	(Botcha & Prattipati, 2020)
Syzygium malaccense	Fruit	UV-Vis, FTIR, XRD, SEM, EDX, HRTEM, XPS	35	Spherical	(Herbin et al., 2022)
Cucumis prophetarum	Leaf	UV-Vis, DLS, FTIR, XRD, SEM, EDAX	30 – 50	Irregularly granulated, ellipsoidal	(Hemlata et al., 2020)
Acacia senegal (L.)	Gum	UV-Vis, FTIR, XRD, SEM, TEM, EDAX	<10	Oval, spherical,	(Al-Ansari et al., 2021)
Zingiber officinale *Curcuma longa*	Rhizome	UV-Vis, FTIR, XRD, TEM, SEM	20-51	Spherical	(Venkatadri et al., 2020)
Mangifera indica	Seed	UV-Vis, DLS, FTIR, XRD, EDX, TEM	26.85	Spherical	(Donga & Chanda, 2021)
Duchesnea indica	Root	UV-Vis, TEM, SEM, XRD, EDX, FTIR, DLS	20.49	Spherical	(Ilahi et al., 2021)
Rosa santana	Petal	UV-Vis, FTIR, XRD, TEM DLS	14.48	Spherical	(Jahan et al., 2019)
Cassia fistula	Pod	UV-Vis, DLS, TEM, FTIR	20.6	Spherical	(I. Singh et al., 2021)
Citrus sinensis	Peel	UV-Vis, TEM, SEM, DLS, FTIR, XRD	16	Spherical	(Nahar et al., 2021)
Garcinia mangostana	Stem	UV-Vis, SEM, XRD, EDX, FTIR	30	Spherical	(Karthiga, 2018)

Figure 6. Schematic diagram of plant-mediated synthesis of silver nanoparticles and their applications

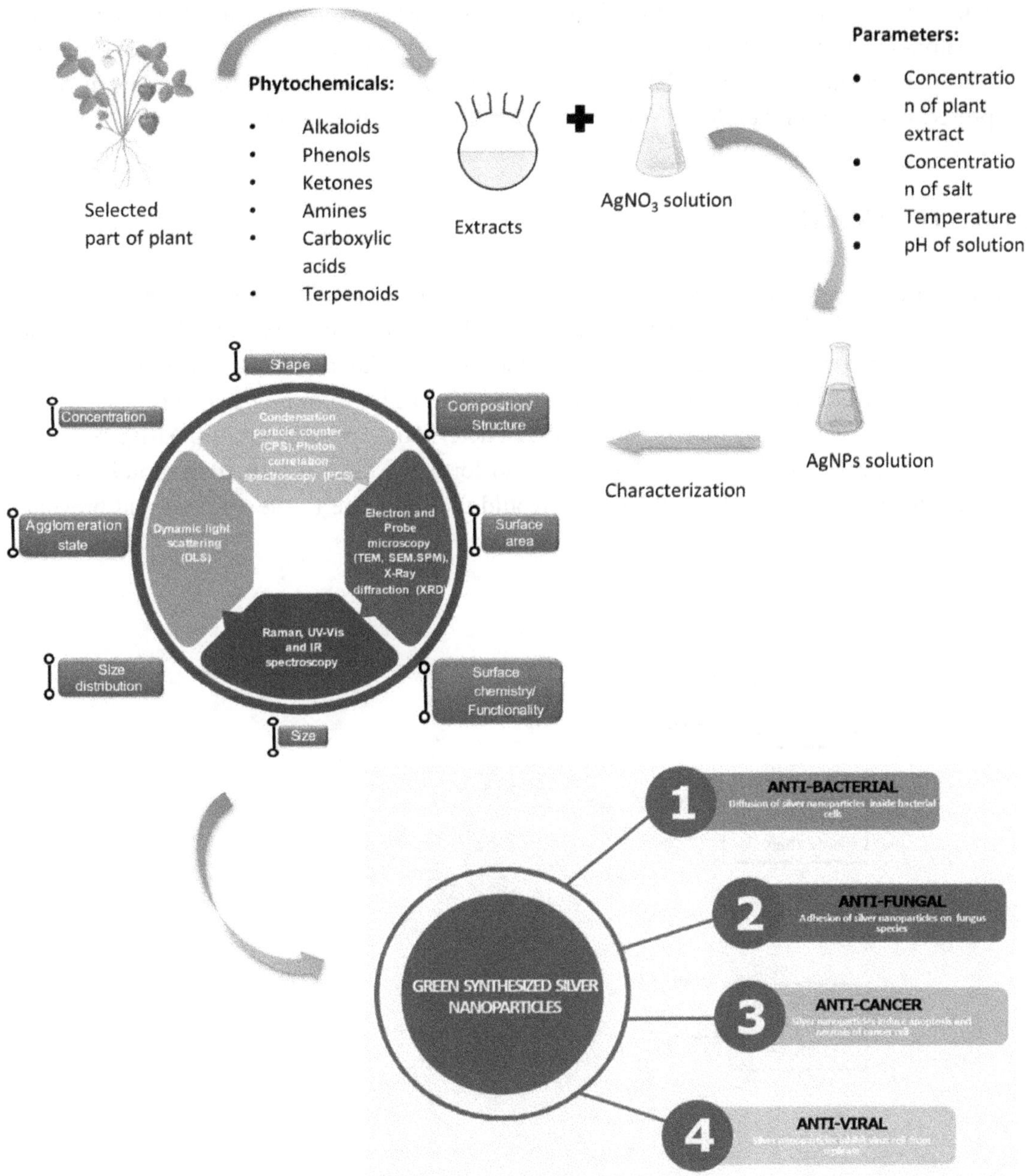

BIOLOGICAL APPLICATIONS OF GREEN SYNTHESIS AGNPS

Global expansion of new nanomaterial production has been facilitated by the exponential growth of nanoscience and nanotechnology research. AgNPs have found application in a wide variety of fields. Numerous properties are not present in macroparticles. Since the surface area of nanoscale AgNPs increases several folds, they are advantageous for surface coating. The bioactive of AgNPs are discussed in greater detail below (Rosman et al., 2021).

Although AgNPs are used in a variety of applications, they have garnered the most attention in the medical field due to the growing number of life-threatening diseases worldwide and the issue of multidrug resistance in non-specific drug delivery. However, AgNPs adaptability to a wide variety of disease is well established (Shanmuganathan et al., 2019). More precisely, this review may assist readers in developing research projects and comprehending the utility of AgNPs through extensive studies.

ANTI-BACTERIAL ACTIVITY

For centuries, antibiotics have been used to treat bacterial infections in humans. However, vancomycin, erythromycin, and penicillin G are ineffective against *Staphylococcus aureus*, *Escherichia coli*, and *Pseudomonas aeruginosa* (Huq, 2020). In bacteria, the antibacterial mechanism is associated with ribosome and mitochondrial membrane and wall damage, as well as inhibition of thiol-groups. AgNPs have a greater antibacterial activity against gram-negative bacteria than against gram-positive bacteria since the cell wall of gram-negative bacteria is thinner and forms a stable electrostatic bond with AgNPs, allowing AgNPs to diffuse into the cell. AgNPs would then migrate to the nucleus and prevent DNA replication, resulting in cell lysis (Figure 7) (Behzad et al., 2021).

Figure 7. Diagram of AgNPs uptake in E-coli bacteria

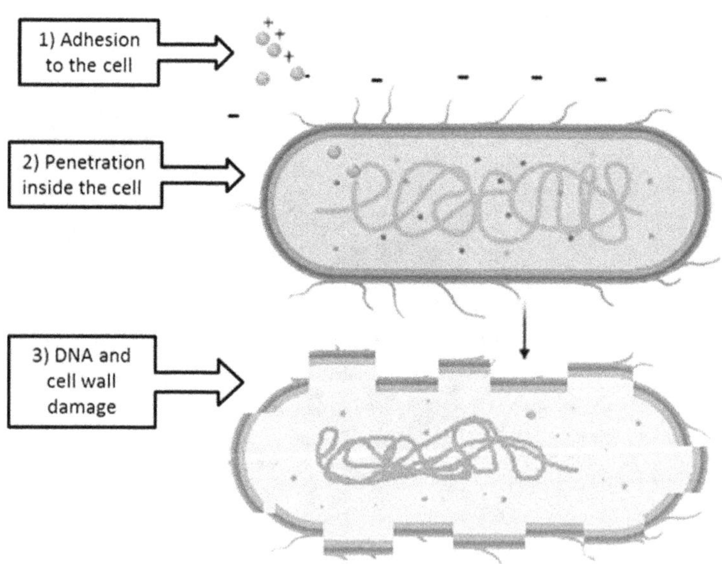

The size, shape, and surface charge of the synthesized AgNPs are important to consider (Table 2). Smaller sized AgNPs has better antibacterial properties as it provides high surface area for interaction with microbial (Manosalva et al., 2019). Triangular shaped AgNPs is also better than spherical and rod shaped (A. Singh et al., 2020). Positively charge (Ag^{+}) have greater antibacterial activity than neutral charge (Ag^{0}) and greatest than negative charge (Ag^{-}); $Ag^{+} > Ag^{0} > Ag^{-}$ (Abbas et al., 2015). The reason is that microbes or bacteria that is normally in negatively charge tend to adhere to AgNPs that has opposite surface charge from them (Salleh et al., 2020).

Table 2. Anti-bacterial activity of green synthesized AgNPs

Scientific name	Parts	Size (Shape)	Bacterial strain	Minimum concentration (Inhibition zone)	Ref.
Symplocos racemosa	Leaves	20 – 50 nm (worm-like) 50 –100 nm (irregular flower) 70 –100 nm (spherical particles and dendrite-like)	*Pseudomonas aeruginosa*	**50 µg/ml** (4 mm)	(Kumar Panda et al., 2021)
Bryophyllum pinnatum	Leaves	38 nm (spherical)	*Escherichia coli*	**N/A** (11.90 mm)	(Aryan, Ruby, & Mehata, 2021)
Crataegus ambigua	Leaves	32 nm (spherical)	*Listeria ivanovii* *Staphylococcus aureus* *Mycobacterium smegmatis*	**15 mg/ml** (7.5–10 mm)	(Ojemaye et al., 2020)
Achillea millefolium	Leaves	20.77 nm (spherical) 18.53 nm (rectangular) 14.27 nm (cubical shapes)	*Staphylococcus aureus*	**N/A** (14.33 ± 0.33 mm)	(Yousaf et al., 2020)
			Escherichia coli	(10.33 ± 0.33 mm)	
			Pseudomonas aeruginosa	(13.67 ± 0.33 mm)	
Rosemary	Leaves	14 nm (Spherical)	*Escherichia coli*	**10 mg/ml** (12 mm)	(Awwad et al., 2020)
			Staphylococcus aureus	(13 mm)	
			Shigella	(9 mm)	
			Listeria	(13 mm)	
Lysiloma acapulcensis	Aerial	5 nm (Spherical)	*Escherichia coli*	**0.25 µg/ml** (18.0±1.3 mm)	(Garibo et al., 2020)
			Pseudomonas Aeruginosa	**0.25 µg/ml** (15.0±0.5 mm)	
			Staphylococcus aureus	**0.25 µg/ml** (16.0±1.0 mm)	
Moringa oleifera	Flower	8 - 13 mm (Spherical)	*Klebsiella pneumoniae*	**N/A** (17 mm)	(Bindhu et al., 2020)
			Staphylococcus aureus	(29 mm)	

By synthesizing AgNPs from high concentration of plant extract, their antibacterial activity is enhanced, and their particle size is decreased. As the concentration of flavonoids and phenolic compounds extracted from the plant increases, the nanoparticle size decreases (Bindhu et al., 2020). Flavonoids (–C–O–C, CHO, –OH, COOH, –C=O, and CH_3) (Hussain, Raja, Iqbal, & Aslam, 2019), as well as polyphenols (C=C, -C-O- and –C-H), are responsible for the reduction in the size of AgNP (Salayová et al., 2021). Polyphenols are secondary metabolites found in plants that act as an antioxidant and protect the plant from ultraviolet radiation and pathogen attack. Additionally, there are three solvents that are effective in the synthesis of AgNPs: aqueous, ethanol, and methanol. Ethanol and methanol are more effective at dissolving polyphenols in plants into AgNPs (Yousaf et al., 2020).

There is concern that AgNPs synthesized by plants could be detrimental to the environment and agriculture due to their ability to inhibit a wide variety of microbes. However, the effect is negligible. AgNPs can absorb pesticides, and photocatalytic degradation of nanoparticles can be used to detoxify wastewater (Castillo-Henríquez et al., 2020). Additionally, there is concern about the cytotoxicity of AgNPs in human cells. *Bacillus sp.* MB353- synthesized AgNPs concentration greater than 5 μg/ml are toxic towards human cells. Above this concentration, it appears to be more toxic to fibroblast cells than to Rhabdomyosarcoma (RMS) cells above this concentration (T. Khan et al., 2020). Certain green synthesized AgNPs require higher concentration to be toxic, for example *Averrhoa bilimbi* synthesized AgNPs are cytotoxic to Vero cells and human epithelial (HEp-2) cells at the concentration of 76.9 μg/ml and 49.49 μg/ml, respectively. This nanoparticle has been determined to be safe and non-toxic to human cells (Sagadevan et al., 2019).

ANTI-FUNGAL ACTIVITY

Antifungal medications are typically used to treat fungal infections. However, some fungi-related diseases are difficult to treat, particularly those caused by eukaryotic fungi, because treating them may have unintended consequences for the host (Waghule et al., 2020). Fungi cells, like mammalian cells, contain nucleus, cytoplasm, and mitochondria. However, they differ from mammalian cells in that their membrane cell is composed of ergosterol. This ergosterol is the primary target for drug development (Aboody & Mickymaray, 2020). The walls of fungal cells are literally rigid. Glucans are the primary component of fungal cell walls. Glucan synthase synthesize glucose monomers from pre-existing glucans, thereby assisting in the maintenance of fungal cell wall integrity. Numerous fungi species, particularly *Candida, Malassezia, Saccharomyces, Histoplasm, and Trichosporon,* produce biofilms. This biofilm protects them from adverse conditions and increases their resistance to antifungal agents (Garg et al., 2020).

AgNPs synthesized from plants have tremendous antifungal potential. This is because the flavonoids found in plants act as effective antifungal agents against a broad spectrum of pathogenic organisms (Aboody & Mickymaray, 2020). However, the complete antifungal mechanism of green synthesis AgNPs is still unknown. Recent research on AgNPs demonstrates that they can adhere to and killing fungal cells (Table 3).

Table 3. Anti-fungal activity of green synthesized AgNPs

Scientific name	Parts	Size	Fungal strain	Concentration	Ref.
Alhagi graecorum	Leaves	22 – 36 nm	*Candida sp.*	0.01 mmol/ml	(Hawar et al., 2022)
Corncob xylan	Fruit	102 nm	*Candida albicans*	7.5 µg/ml	(Viana et al., 2020)
			Candida parapsilosis	7.5 µg/ml	
			Cryptococcus neoformans	7.5 µg/ml	
Melia azedarach	Leave	18 – 30 nm	*Verticillium dahlia*	60 ppm (*in vitro*)	(Jebril et al., 2020)
				20 ppm (*in vivo*)	
Citrus limetta	Peel	18 nm	*Candida sp.*	4.75 - 21.4 µg/ml	(Dutta et al., 2020)
Teucrium polium L	Stems and flowers	10 – 100 nm	*Fusarium oxysporum*	50 ppm	(Ghojavand et al., 2020)
Ferulago macrocarpa	Flower	14 – 25 nm	*Candida albicans*	500 µg/ml	(Azarbani & Shiravand, 2020)

ANTI-CANCER ACTIVITY

AgNPs exert anticancer activity via a variety of mechanisms. Through modulation of vascular endothelial growth factor (VEGF) receptor-2 phosphorylation, AgNPs can inhibit VEGF-induced retinal neovascularization and inhibit extracellular signal-related kinase (ERK½) activation (Ratan et al., 2020). Additionally, AgNPs can induce apoptosis (cell death) in breast cancer cells by increasing matrix metalloproteinases (MMP) levels, a process associated with tumor surface disintegration and cell death (Figure 8) (Gomathi et al., 2019). AgNPs can also cause necrosis, a type of cell injury caused by insufficient blood flow to the tissue. Whether cancer cells die via necrosis, apoptosis, or another mechanism is dependent on the AgNPs concentration. At low concentrations, necrosis and apoptosis frequently coexist, whereas at high concentrations, necrosis occurs exclusively (Ahn et al., 2019). Necrosis and apoptosis occur because of the generation of reactive oxygen species within the cells.

Figure 8. Schematic mechanism of green synthesis AgNPs against breast cancer cell

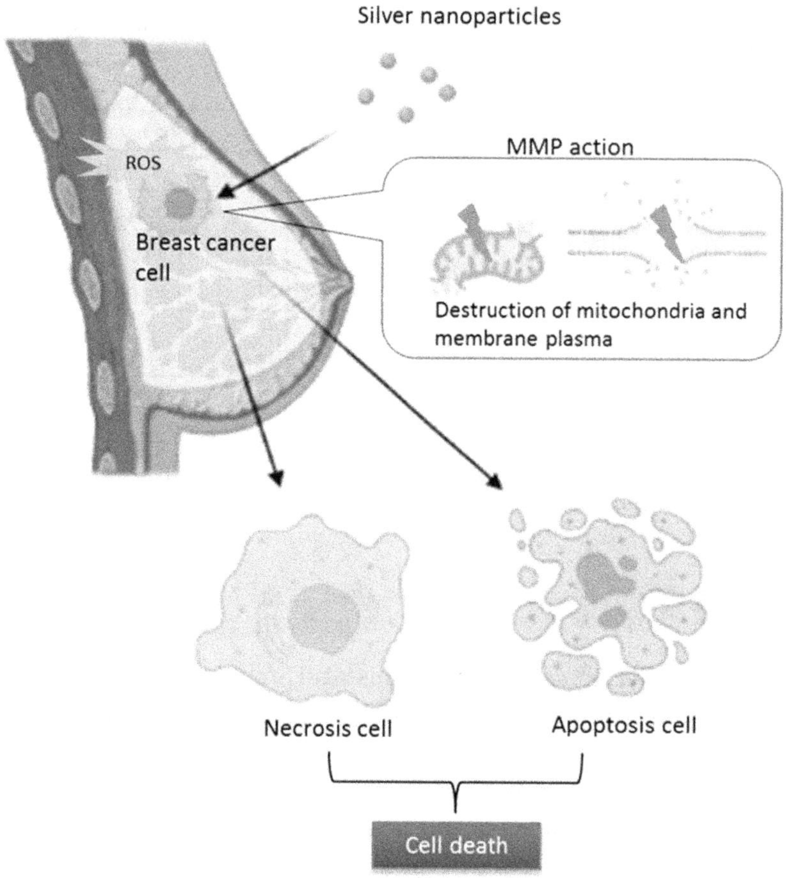

AgNPs' anticancer activity is quantified by their inhibitory concentration at 50% (IC_{50}) value (Table 4). In the laboratory the difference between living and dead cells can be observed using a cell imaging station and liquid staining. To differentiate between living and dead cells, acridine orange and ethidium bromide can be used a liquid staining. Live cells are colored green, whereas dead/ apoptotic cells are colored red (Gomathi et al., 2019). Another sign of infection is a change in the shape of the infected cells (normally from elongated to circular), a breakdown of membrane integrity, cytoplasmic condensation, and cell clumping. Under an inverted microscope, the shape changes are also visible (Soliman et al., 2020) (Bharadwaj et al., 2021).

Table 4. Anti-cancer activity of green synthesized AgNPs

Scientific name	Parts	Size	Cell lines	IC_{50} value	Ref.
Diospyros malabarica	Fruit	17.4 nm	Human primary glioblastoma (U87-MG)	58.63 ± 5.74 μg/ml	(Bharadwaj et al., 2021)
Hypericum Perforatum L.	Aerial	20 – 50 nm	Human cervical cancer (Hela) Human liver cancer (HepG2) Human lung cancer A549	35.88 μg/ml	(Alahmad et al., 2021)
Ferula persica	Aerial	15 nm	Human breast cancer (MCF-7) Human adenocarcinoma gastric (AGS)	11.07 μg/ml (AGS) 21.28 μg/ml (MCF-7)	(Hashemi et al., 2021)
Poa annua	Leaves	36.66 ± 7.85 nm	Cellosaurus (SCC7)	431.2 μg/mL	(Gul et al., 2021)
Naringi crenulata	Leaves	20 – 40 nm	Breast cancer (HER2)	72.88 μg/mL	(Vallinayagam et al., 2021)
Gloriosa superba	Tuber	7 –14 nm	A549	46.54 μg/mL	(Murugesan et al., 2021)
Asafoetida	Roots	90 – 95 nm	HepG2	0.01 ppm	(Satsangi, 2020)
Benincasa hispida	Peel	26 ± 2 nm	HeLa	0.066 μg/ml	(Soliman et al., 2020)

The cytotoxicity of synthesized AgNPs associated with bioactive is attributed to phytochemicals. The functional groups of bioactive coated on AgNPs contribute to cancer cell cytotoxicity and death (Ahmed et al., 2020; Behzad et al., 2021). They would interact with protein functional groups prevalent in cancer cells, as well as with nitrogen bases and R groups found in nuclear material (Vallinayagam et al., 2021). Another reason is that the atypical metabolism and accelerated cell proliferation of cancerous cells results in increased AgNPs uptake (Soliman et al., 2020). Finally, cancer cells have a slightly acidic pH, which results in a greater release of Ag^+ ions than normal cells (Jorge de Souza et al., 2019).

Antiviral Activity

Recent outbreaks of infection disease caused by novel virus-like pathogens have developed resistance to antiviral drugs currently available. Virus-related infections can be difficult to treat due to the virus's rapid replication and spread. AgNPs are highly effective antiviral agents. This has been demonstrated to be effective against a wide range of viruses, including Influenza A, noroviruses, poliovirus, SARS-CoV-2 and Chikungunya (Ratan et al., 2021). The antiviral concept of AgNPs is based on the nanoparticle's ability to inhibit viral replication and intracellular spread through interaction with envelops glycoprotein (Jain et al., 2021). AgNPs first bind to the virus's outer coat, preventing it from binding to cell receptors, and then to the virus's DNA or RNA, inhibiting replication or propagation (Salleh et al., 2020). The antiviral mechanism of green synthesis AgNPs has not been fully established (Haggag et al., 2019; Salleh et al., 2020). However, as shown in Table 5, the antiviral properties of green synthesized AgNPs are most likely due to alkaloids, phenols, and flavonoids extracted from the bioactive.

Molecular docking is a technique for discovering antiviral compounds. It is a technique for determining the ligand-binding sites on a protein by analysing the interaction energies of a methyl probe with the protein. Higher energy binding sites indicate greater interaction (Vincent et al., 2020). Such as *Psidium guajava*, which contains Quercetin, Beta-Caryophyllene, Longifollen, Resveratrol, Guaijaverin, Limonene and Beta-bisabolene among other phytochemicals. Longifollen has the lowest minimum binding energy with the virus's cysteine protease, which is – 8.26. For Quercetin, Beta-Caryophyllene, Resveratrol, Guaijaverin, Limonene and Beta-bisabolene, the values are −6.66, −6.27, −6.18, −5.89, −5.82 and −5.44, respectively. As a result, Longifollen is the primary compound isolated from *Psidium guajava* plants that inhibits the Chikungunya virus (Sharma et al., 2021).

Table 5. Antiviral activity of green synthesized AgNPs

Scientific name	Part	Size (nm)	Viral strain	Concentration	Ref.
Psidium guajava	Leaves	75 - 99	Chikungunya	15 μg/ml	(Sharma et al., 2021)
Zingiber officinale	Fruit	N A	Chikungunya	62.5 μg/ml 100% inhibition	(Kaushik et al., 2020)
Spirulina platensis and *Oscillatoria sp*	Leaves	14.4 - 48.9	Herps Simplex (HSV-1)	31.25 μl 49.23% inhibition	(El-Sheekh et al., 2020)
Lampranthus coccineus	Aerial	10.12 - 27.89	Hepatovirus A-10 (HAV-10)	11.71 μg/ml	(Haggag et al., 2019)
			HSV-1	36.36 μg/ml	
			Coxsackie B virus type 4 (CoxB4)	12.74 μg/ml	

COVID-19 is currently wreaking havoc throughout the world. The novel virus's genomic sequence is 78.5% identical to the previously discovered SARS-CoV-2. AgNPs have been shown to suppress extracellular SARS-CoV-2 infection at concentration as low as 1ppm. As a result, AgNPs may be useful in combating the COVID-19 pandemic. However, no significant study, particularly one involving green synthesized AgNPs, has been published to date (Allawadhi et al., 2021). Continuous research into surface coating AgNPs for antiviral applications is undoubtedly in need of rapid development through multidisciplinary collaboration between chemistry and biomedical sciences to combat emerging viruses.

CONCLUSION

Different processes, such as chemical, physical, and biological, are used to generate different nanoparticles. Among these, green synthesis of AgNPs is more appealing since the procedure is straightforward, cost-effective, easy to operate, and timesaving. Natural compounds contained in plant extracts have been shown to work as both reducing and stabilizing agents in the AgNPs synthesis. The use of plant extracts in the green synthesis of AgNPs offers several advantages, including environment friendliness, biocompatibility, and cost-effectiveness. AgNPs are anticipated to play a significant role in a range of nanotechnology-based systems due to their unique features.

Although green chemistry-based biological synthesis is more environmentally friendly than traditional methods, some difficulties remain unanswered. The obstacles consist of inconsistent formation and particle size, reproducibility, and understanding of the pathway involved in producing nanoparticles via biological entities. The knowledge on the mechanism of nanoparticle production differs amongst plant species as well as its parts. As a result, numerous research is required to analyze and comprehend the real plant mediated process. This is an unprecedented that will necessitate significantly more venture capital to fully utilize living entities in the green synthesis of metal nanoparticles.

The review's discussion of the reported biological potentials of AgNPs encourages scientists to use disparate plants as a precursor of reducing agents to acquire the desired particle size and shape for various medicinal purposes. Furthermore, there is still a need to determine the optimal way for increasing product yield efficiency so that the biological technique can be used on a large scale.

REFERENCES

Abbas, A., Yasamin, G., Ahmad, G., Bahram, H., Samira, D., Mohammadreza, N., & Hashem, S. (2015). The Effect of Charge at the Surface of Silver Nanoparticles on Antimicrobial Activity against Gram-Positive and Gram-Negative Bacteria: A Preliminary Study. *Journal of Nanomaterials*, 1–8.

Adebayo-Tayo, B., Salaam, A., & Ajibade, A. (2019). Green synthesis of silver nanoparticle using Oscillatoria sp. extract, its antibacterial, antibiofilm potential and cytotoxicity activity. *Heliyon*, *5*(10), e02502. doi:10.1016/j.heliyon.2019.e02502 PMID:31667375

Agarwal, H., Venkat Kumar, S., & Rajeshkumar, S. (2017). A review on green synthesis of zinc oxide nanoparticles – An eco-friendly approach. *Resource-Efficient Technologies*, *3*(4), 406–413. doi:10.1016/j.reffit.2017.03.002

Ahmed, T., Shahid, M., Noman, M., Bilal Khan Niazi, M., Zubair, M., Almatroudi, A., Khurshid, M., Tariq, F., Mumtaz, R., & Li, B. (2020). Bioprospecting a native silver-resistant Bacillus safensis strain for green synthesis and subsequent antibacterial and anticancer activities of silver nanoparticles. *Journal of Advanced Research*, *24*, 475–483. doi:10.1016/j.jare.2020.05.011 PMID:32566283

Ahn, E. Y., Jin, H., & Park, Y. (2019). Assessing the antioxidant, cytotoxic, apoptotic and wound healing properties of silver nanoparticles green-synthesized by plant extracts. *Materials Science and Engineering C*, *101*, 204–216. doi:10.1016/j.msec.2019.03.095 PMID:31029313

Al Aboody, M. S., & Mickymaray, S. (2020). Anti-fungal efficacy and mechanisms of flavonoids. *Antibiotics (Basel, Switzerland)*, *9*(45), 1–43. PMID:31991883

Al-Ansari, M. M., Al-Dahmash, N. D., & Ranjitsingh, A. J. A. (2021). Synthesis of silver nanoparticles using gum Arabic: Evaluation of its inhibitory action on Streptococcus mutans causing dental caries and endocarditis. *Journal of Infection and Public Health*, *14*(3), 324–330. doi:10.1016/j.jiph.2020.12.016 PMID:33618277

Alahmad, A., Feldhoff, A., Bigall, N. C., Rusch, P., Scheper, T., & Walter, J. G. (2021). Hypericum perforatum L.-mediated green synthesis of silver nanoparticles exhibiting antioxidant and anticancer activities. *Nanomaterials (Basel, Switzerland)*, *11*(487), 1–25. doi:10.3390/nano11020487 PMID:33673018

Allawadhi, P., Singh, V., Khurana, A., Khurana, I., Allwadhi, S., Kumar, P., Banothu, A. K., Thalugula, S., Barani, P. J., Naik, R. R., & Bharani, K. K. (2021). Silver nanoparticle based multifunctional approach for combating COVID-19. *Sensors International*, *2*(May), 100101. doi:10.1016/j.sintl.2021.100101 PMID:34766057

Almasoud, N., Alhaik, H., Almutairi, M., Houjak, A., Hazazi, K., Alhayek, F., Aljanoubi, S., Alkhaibari, A., Alghamdi, A., Soliman, D. A., Alomar, T. S., & Awad, M. A. (2021). Green nanotechnology synthesized silver nanoparticles: Characterization and testing its antibacterial activity. *Green Processing and Synthesis*, *10*(1), 510–528. doi:10.1515/gps-2021-0048

Aryan, R., Ruby, & Mehata, M. S. (2021). Green synthesis of silver nanoparticles using Kalanchoe pinnata leaves (life plant) and their antibacterial and photocatalytic activities. *Chemical Physics Letters*, *778*, 138760–138770. doi:10.1016/j.cplett.2021.138760

Awwad, A. M., & Salem, M., N., M.Aqarbeh, M., & M.Abdulaziz, F. (2020). Green synthesis, characterization of silver sulfide nanoparticles and antibacterial activity evaluation. *Chemistry International*, *6*(1), 42–48.

Azarbani, F., & Shiravand, S. (2020). Green synthesis of silver nanoparticles by Ferulago macrocarpa flowers extract and their antibacterial, antifungal and toxic effects. *Green Chemistry Letters and Reviews*, *13*(1), 41–49. doi:10.1080/17518253.2020.1726504

Bakker, K., Apelqvist, J., & Schaper, N. C. (2012). Practial guidelines on the management and prevention of the diabetic foot 2011. *Diabetes/Metabolism Research and Reviews*, *28*(30), 225–231. doi:10.1002/dmrr.2253 PMID:22271742

Behzad, F., Naghib, S. M., kouhbanani, M. A. J., Tabatabaei, S. N., Zare, Y., & Rhee, K. Y. (2021). An overview of the plant-mediated green synthesis of noble metal nanoparticles for antibacterial applications. *Journal of Industrial and Engineering Chemistry*, *94*, 92–104. doi:10.1016/j.jiec.2020.12.005

Bharadwaj, K. K., Rabha, B., Pati, S., Choudhury, B. K., Sarkar, T., Gogoi, S. K., Kakati, N., Baishya, D., Kari, Z. A., & Edinur, H. A. (2021). Green synthesis of silver nanoparticles using diospyros malabarica fruit extract and assessments of their antimicrobial, anticancer and catalytic reduction of 4-nitrophenol (4-np). *Nanomaterials (Basel, Switzerland)*, *11*(8), 1999–2043. doi:10.3390/nano11081999 PMID:34443829

Bindhu, M. R., Umadevi, M., Esmail, G. A., Al-Dhabi, N. A., & Arasu, M. V. (2020). Green synthesis and characterization of silver nanoparticles from Moringa oleifera flower and assessment of antimicrobial and sensing properties. *Journal of Photochemistry and Photobiology. B, Biology*, *205*, 111836. doi:10.1016/j.jphotobiol.2020.111836 PMID:32172135

Botcha, S., & Prattipati, S. D. (2020). Callus Extract Mediated Green Synthesis of Silver Nanoparticles, Their Characterization and Cytotoxicity Evaluation Against MDA-MB-231 and PC-3 Cells. *BioNanoScience*, *10*(1), 11–22. doi:10.100712668-019-00683-3

Castillo-Henríquez, L., Alfaro-Aguilar, K., Ugalde-álvarez, J., Vega-Fernández, L., de Oca-Vásquez, G. M., & Vega-Baudrit, J. R. (2020). Green synthesis of gold and silver nanoparticles from plant extracts and their possible applications as antimicrobial agents in the agricultural area. *Nanomaterials, 10*(1763), 1763–1797.

Das, P., Dutta, T., Manna, S., Loganathan, S., & Basak, P. (2022). Facile green synthesis of non-genotoxic, non-hemolytic organometallic silver nanoparticles using extract of crushed, wasted, and spent Humulus lupulus (hops): Characterization, anti-bacterial, and anti-cancer studies. *Environmental Research, 204*(PA), 111962.

Donga, S., & Chanda, S. (2021). Facile green synthesis of silver nanoparticles using Mangifera indica seed aqueous extract and its antimicrobial, antioxidant and cytotoxic potential (3-in-1 system). *Artificial Cells, Nanomedicine, and Biotechnology*, *49*(1), 292–302. doi:10.1080/21691401.2021.1899193 PMID:33733973

Dutta, T., Ghosh, N. N., Das, M., Adhikary, R., Mandal, V., & Chattopadhyay, A. P. (2020). Green synthesis of antibacterial and antifungal silver nanoparticles using Citrus limetta peel extract: Experimental and theoretical studies. *Journal of Environmental Chemical Engineering*, *8*(4), 104019. doi:10.1016/j.jece.2020.104019

El-Sheekh, M. M., Shabaan, M. T., Hassan, L., & Morsi, H. H. (2020). Antiviral activity of algae biosynthesized silver and gold nanoparticles against Herps Simplex (HSV-1) virus in vitro using cell-line culture technique. *International Journal of Environmental Health Research*, 1–12. PMID:32627584

Garg, A., Sharma, G. S., Goyal, A. K., Ghosh, G., Si, S. C., & Rath, G. (2020). Recent advances in topical carriers of anti-fungal agents. *Heliyon*, *6*(8, e04663), 1–12. doi:10.1016/j.heliyon.2020.e04663 PMID:32904164

Garibo, D., Borbón-Nuñez, H. A., de León, J. N. D., García Mendoza, E., Estrada, I., Toledano-Magaña, Y., Tiznado, H., Ovalle-Marroquin, M., Soto-Ramos, A. G., Blanco, A., Rodríguez, J. A., Romo, O. A., Chávez-Almazán, L. A., & Susarrey-Arce, A. (2020). Green synthesis of silver nanoparticles using Lysiloma acapulcensis exhibit high-antimicrobial activity. *Scientific Reports*, *10*(1), 1–11. doi:10.103841598-020-69606-7 PMID:32732959

Ghojavand, S., Madani, M., & Karimi, J. (2020). Green synthesis, characterization and antifungal activity of silver nanoparticles using stems and flowers of Felty Germander. *Journal of Inorganic and Organometallic Polymers and Materials*, *30*(8), 1–11. doi:10.100710904-020-01449-1

Gomathi, A. C., Rajarathinam, S. R. X., Sadiq, A. M., Rajeshkumar, S., Rajarathinam, S. R. X., & Sadiq, A. M. (2019). Anticancer activity of silver nanoparticles synthesized using aqueous fruit shell extract of Tamarindus indica on MCF-7 human breast cancer cell line. *Journal of Drug Delivery Science and Technology*, 1–28.

Gul, A. R., Shaheen, F., Rafique, R., Bal, J., Waseem, S., & Park, T. J. (2021). Grass-mediated biogenic synthesis of silver nanoparticles and their drug delivery evaluation: A biocompatible anti-cancer therapy. *Chemical Engineering Journal*, *407*, 1–54. doi:10.1016/j.cej.2020.127202

Gupta, N., Goyal, S., & Pawar, S. (2020). Green synthesis of silver nanoparticles based on moringa oleifera leaf extract and its application in antimicrobial activity. *International Journal of Advanced Science and Technology, 29*(8 Special Issue), 1045–1052.

Habibullah, G., Viktorova, J., & Ruml, T. (2021). Current Strategies for Noble Metal Nanoparticle Synthesis. *Nanoscale Research Letters*, *16*(1), 47. doi:10.118611671-021-03480-8 PMID:33721118

Haggag, E. G., Elshamy, A. M., Rabeh, M. A., Gabr, N. M., Salem, M., Youssif, K. A., Samir, A., Bin Muhsinah, A., Alsayari, A., & Abdelmohsen, U. R. (2019). Antiviral potential of green synthesized silver nanoparticles of lampranthus coccineus and malephora lutea. *International Journal of Nanomedicine*, *14*, 6217–6229. doi:10.2147/IJN.S214171 PMID:31496682

Hashemi, Z., Mohammadyan, M., Naderi, S., Fakhar, M., Biparva, P., Akhtari, J., & Ebrahimzadeh, M. A. (2021). Green synthesis of silver nanoparticles using Ferula persica extract (Fp-NPs): Characterization, antibacterial, antileishmanial, and in vitro anticancer activities. *Materials Today Communications*, *27*(102264), 1–10. doi:10.1016/j.mtcomm.2021.102264

Hawar, S. N., Al-Shmgani, H. S., Al-Kubaisi, Z. A., Sulaiman, G. M., Dewir, Y. H., & Rikisahedew, J. J. (2022). Green Synthesis of Silver Nanoparticles from Alhagi graecorum Leaf Extract and Evaluation of Their Cytotoxicity and Antifungal Activity. *Journal of Nanomaterials*, *2022*, 1–8. doi:10.1155/2022/1058119

Hemlata, M., Meena, P. R., Singh, A. P., & Tejavath, K. K. (2020). Biosynthesis of Silver Nanoparticles Using Cucumis prophetarum Aqueous Leaf Extract and Their Antibacterial and Antiproliferative Activity against Cancer Cell Lines. *ACS Omega*, *5*(10), 5520–5528. doi:10.1021/acsomega.0c00155 PMID:32201844

Herbin, H. B., Aravind, M., Amalanathan, M., Mary, M. S. M., Lenin, M. M., Parvathiraja, C., Siddiqui, M. R., Wabaidur, S. M., & Islam, M. A. (2022). Synthesis of Silver Nanoparticles Using Syzygium malaccense Fruit Extract and Evaluation of Their Catalytic Activity and Antibacterial Properties. *Journal of Inorganic and Organometallic Polymers and Materials*, *32*(3), 1–13. doi:10.100710904-021-02210-y

Huq, M. A. (2020). Green synthesis of silver nanoparticles using pseudoduganella eburnea MAHUQ-39 and their antimicrobial mechanisms investigation against drug resistant human pathogens. *International Journal of Molecular Sciences*, *21*(4), 1–14. doi:10.3390/ijms21041510 PMID:32098417

Hussain, M., Raja, N. I., Iqbal, M., & Aslam, S. (2019). Applications of Plant Flavonoids in the Green Synthesis of Colloidal Silver Nanoparticles and Impacts on Human Health. *Iranian Journal of Science and Technology. Transaction A, Science*, *43*(3), 1381–1392. doi:10.100740995-017-0431-6

Ijaz, I., Gilani, E., Nazir, A., & Bukhari, A. (2020). Detail review on chemical, physical and green synthesis, classification, characterizations and applications of nanoparticles. *Green Chemistry Letters and Reviews*, *13*(3), 59–81. doi:10.1080/17518253.2020.1802517

Ilahi, I., Khuda, F., Umar Khayam Sahibzada, M., Alghamdi, S., Ullah, R., Zakiullah, Dablool, A. S., Alam, M., Khan, A., & Ali Khan Khalil, A. (2021). Synthesis of silver nanoparticles using root extract of Duchesnea indica and assessment of its biological activities. *Arabian Journal of Chemistry*, *14*(5), 103110. doi:10.1016/j.arabjc.2021.103110

Jabeen, S., Qureshi, R., Munazir, M., Maqsood, M., Munir, M., Shah, S. S. H., & Rahim, B. Z. (2021). Application of green synthesized silver nanoparticles in cancer treatment - A critical review. *Materials Research Express*, *8*(9), 092001. doi:10.1088/2053-1591/ac1de3

Jahan, I., Erci, F., & Isildak, I. (2019). Microwave-Assisted Green Synthesis of Non-Cytotoxic Silver Nanoparticles Using the Aqueous Extract of Rosa santana (rose) Petals and Their Antimicrobial Activity. *Analytical Letters*, *52*(12), 1860–1873. doi:10.1080/00032719.2019.1572179

Jain, N., Jain, P., Rajput, D., & Patil, U. K. (2021). Green synthesized plant-based silver nanoparticles: therapeutic prospective for anticancer and antiviral activity. *Micro and Nano Systems Letters, 9*(1).

Jebril, S., Khanfir Ben Jenana, R., & Dridi, C. (2020). Green synthesis of silver nanoparticles using Melia azedarach leaf extract and their antifungal activities: In vitro and in vivo. *Materials Chemistry and Physics*, *248*, 122898. doi:10.1016/j.matchemphys.2020.122898

Jorge de Souza, T. A., Rosa Souza, L. R., & Franchi, L. P. (2019). Silver nanoparticles: An integrated view of green synthesis methods, transformation in the environment, and toxicity. *Ecotoxicology and Environmental Safety, 171*(December 2018), 691–700.

Joseph, J., Keren Deborah, S., Raghavi, R., Mary Shamya, A., & Aruni, W. (2021). Green synthesis of silver nanoparticles using Phyllanthus amarus Seeds and their antibacterial activity assessment. *Biomedical and Biotechnology Research Journal*, *5*(1), 35–38.

Karthiga, P. (2018). Preparation of silver nanoparticles by Garcinia mangostana stem extract and investigation of the antimicrobial properties. *Biotechnology Research and Innovation*, *2*(1), 30–36. doi:10.1016/j.biori.2017.11.001

Kaushik, S., Jangra, G., Kundu, V., Yadav, J. P., & Kaushik, S. (2020). Anti-viral activity of Zingiber officinale (Ginger) ingredients against the Chikungunya virus. *Virusdisease*, *31*(3), 270–276. doi:10.100713337-020-00584-0 PMID:32420412

Khan, T., Yasmin, A., & Townley, H. E. (2020). An evaluation of the activity of biologically synthesized silver nanoparticles against bacteria, fungi and mammalian cell lines. *Colloids and Surfaces. B, Biointerfaces*, *194*, 111167. doi:10.1016/j.colsurfb.2020.111156 PMID:32512312

Khan, Y., Qasim Nasar, M., & Numan, M. (2018). Biomimetic Synthesis of Silver Nanoparticles for Breast Cancer Therapeutics and Its Mechanism Bio safety assessments of genetically modified maize View project Green synthesis of Silver Nano particles View project. *International Journal of Nanotechnology and Nanomedicine*, *3*(1), 1–9.

Kumar Panda, M., Kumar Dhal, N., Kumar, M., Manjari Mishra, P., & Kumar Behera, R. (2021). Green synthesis of silver nanoparticles and its potential effect on phytopathogens. *Materials Today: Proceedings*, *35*, 233–238. doi:10.1016/j.matpr.2020.05.188

Manosalva, N., Tortella, G., Cristina Diez, M., Schalchli, H., Seabra, A. B., Durán, N., & Rubilar, O. (2019). Green synthesis of silver nanoparticles: Effect of synthesis reaction parameters on antimicrobial activity. *World Journal of Microbiology & Biotechnology*, *35*(88), 1–9. doi:10.100711274-019-2664-3 PMID:31134435

Mat Yusuf, S. N. A., Che Mood, C. N. A., Ahmad, N. H., Sandai, D., Lee, C. K., & Lim, V. (2020). Optimization of biogenic synthesis of silver nanoparticles from flavonoid-rich Clinacanthus nutans leaf and stem aqueous extracts: Biogenic Synthesis of C. nutans AgNPs. *Royal Society Open Science*, *7*(7), 200065. doi:10.1098/rsos.200065 PMID:32874618

Mohammadlou, M., Maghsoudi, H., & Jafarizadeh-Malmiri, H. (2016). A review on green silver nanoparticles based on plants: Synthesis, potential applications and eco-friendly approach. *International Food Research Journal*, *23*(2), 446–463.

Mukherjee, S., & Patra, C. R. (2017). Biologically synthesized metal nanoparticles: Recent advancement and future perspectives in cancer theranostics. *Future Science OA*, *3*(3), FSO203. doi:10.4155/fsoa-2017-0035 PMID:28884002

Murugesan, A. K., Pannerselvam, B., Javee, A., Rajenderan, M., & Thiyagarajan, D. (2021). Facile green synthesis and characterization of Gloriosa superba L. tuber extract-capped silver nanoparticles (GST-AgNPs) and its potential antibacterial and anticancer activities against A549 human cancer cells. *Environmental Nanotechnology, Monitoring & Management*, *15*, 100460–100471. doi:10.1016/j.enmm.2021.100460

Nahar, K., Hafezur Rahaman, M., Arifuzzaman Khan, G., Khairul Islam, M., & Md Al-Reza, S. (2021). Green synthesis of silver nanoparticles from Citrus sinensis peel extract and its antibacterial potential. *Asian Journal of Green Chemistry*, *5*, 135–150.

Neupane, N. P., Kushwaha, A. K., Karn, A. K., Khalilullah, H., Uzzaman Khan, M. M., Kaushik, A., & Verma, A. (2022). Anti-bacterial efficacy of bio-fabricated silver nanoparticles of aerial part of Moringa oleifera lam: Rapid green synthesis, In-Vitro and In-Silico screening. *Biocatalysis and Agricultural Biotechnology, 39*, 102229.

Ojemaye, M. O., Okoh, S. O., & Okoh, A. I. (2020). Silver nanoparticles (AgNPs) facilitated by plant parts of Crataegus ambigua Becker AK extracts and their antibacterial, antioxidant and antimalarial activities. *Green Chemistry Letters and Reviews*, *14*(1), 49–59.

Olabemiwo, O. M., Akintelu, S. A., Waheed, A. A., Okunlola, D. S., Akinwale, D. R., Adeyinka, G. C., Adeniji, T. W., & Adebisi, S. A. (2021). Green synthesis of silver nanoparticles using stem bark extract of Annona senegalensis: Characterization and its antibacterial potency. *Current Research in Green and Sustainable Chemistry*, *4*, 100219. doi:10.1016/j.crgsc.2021.100219

Patil, S., & Chandrasekaran, R. (2020). Biogenic nanoparticles: A comprehensive perspective in synthesis, characterization, application and its challenges. *Journal of Genetic Engineering and Biotechnology*, *18*(1), 67. doi:10.118643141-020-00081-3 PMID:33104931

Perumal, D., Zulkifli, S. N., Een, L. G., Albert, E. L., Muhamad Yusop, M. A., & Che Abdullah, C. A. (2022). Green biosynthesis of metallic nanoparticles and their future biomedical applications. In Green Nanomaterials for Industrial Applications (pp. 41–70). doi:10.1016/B978-0-12-823296-5.00006-X

Pugazhendhi, A., Edison, T. N. J. I., Karuppusamy, I., & Kathirvel, B. (2018). Inorganic nanoparticles: A potential cancer therapy for human welfare. *International Journal of Pharmaceutics*, *539*(1–2), 104–111. doi:10.1016/j.ijpharm.2018.01.034 PMID:29366941

Raj, S., Trivedi, R., & Soni, V. (2021). Biogenic Synthesis of Silver Nanoparticles, Characterization and Their Applications—A Review. *Surfaces*, *5*(1), 67–90. doi:10.3390urfaces5010003

Ratan, Z. A., Haidere, M. F., Nurunnabi, M., Shahriar, S. M., Ahammad, A. J. S., Shim, Y. Y., Reaney, M. J. T., & Cho, J. Y. (2020). Green chemistry synthesis of silver nanoparticles and their potential anticancer effects. *Cancers (Basel)*, *12*(4), 1–26. doi:10.3390/cancers12040855 PMID:32244822

Ratan, Z. A., Mashrur, F. R., Chhoan, A. P., Shahriar, S. M., Haidere, M. F., Runa, N. J., Kim, S., Kweon, D.-H., Hosseinzadeh, H., & Cho, J. Y. (2021). Silver nanoparticles as potential antiviral agents. *Pharmaceutics, 13*(12), 1–25. doi:10.3390/pharmaceutics13122034 PMID:34959320

Rosman, N. S. R., Harun, N. A., Idris, I., & Wan Ismail, W. I. (2021). Nanobiotechnology: Nature-inspired silver nanoparticles towards green synthesis. *Energy & Environment, 32*(7), 1183–1206. doi:10.1177/0958305X21989883

Sagadevan, S., Vennila, S., Singh, P., Lett, J. A., Johan, M. R., Marlinda, A. R., Muthiah, B., & Lakshmipathy, M. (2019). Facile synthesis of silver nanoparticles using Averrhoa bilimbi L and Plum extracts and investigation on the synergistic bioactivity using in vitro models. *Green Processing and Synthesis, 8*(1), 873–884. doi:10.1515/gps-2019-0058

Salayová, A., Bedlovičová, Z., Daneu, N., Baláž, M., Lukáčová Bujňáková, Z., Balážová, L., & Tkáčiková, L. (2021). Green synthesis of silver nanoparticles with antibacterial activity using various medicinal plant extracts: Morphology and antibacterial efficacy. *Nanomaterials (Basel, Switzerland), 11*(4), 1005. doi:10.3390/nano11041005 PMID:33919801

Salleh, A., Naomi, R., Utami, N. D., Mohammad, A. W., Mahmoudi, E., Mustafa, N., & Fauzi, M. B. (2020). The potential of silver nanoparticles for antiviral and antibacterial applications: A mechanism of action. *Nanomaterials (Basel, Switzerland), 10*(8), 1–20. doi:10.3390/nano10081566 PMID:32784939

Satsangi, N. (2020). Synthesis and Characterization of Biocompatible Silver Nanoparticles for Anticancer Application. *Journal of Inorganic and Organometallic Polymers and Materials, 30*(6), 1907–1914. doi:10.100710904-019-01372-0

Sellami, H., Khan, S. A., Ahmad, I., Alarfaj, A. A., Hirad, A. H., & Al-Sabri, A. E. (2021). Green synthesis of silver nanoparticles using olea europaea leaf extract for their enhanced antibacterial, antioxidant, cytotoxic and biocompatibility applications. *International Journal of Molecular Sciences, 22*(22), 12562. doi:10.3390/ijms222212562 PMID:34830442

Shahi, D., Bhattarai, E., Poudel, M., Pradhan, P., Pradhananga, R. R., & Awal, S. C. (2018). Green Synthesis of Silver Nanoparticles Using Different Plant Materials and Their Antibacterial Activity. *International Journal of Applied Sciences and Biotechnology, 6*(4), 294–301. doi:10.3126/ijasbt.v6i4.22112

Shanmuganathan, R., Karuppusamy, I., Saravanan, M., Muthukumar, H., Ponnuchamy, K., Ramkumar, V. S., & Pugazhendhi, A. (2019). Synthesis of Silver Nanoparticles and their Biomedical Applications - A Comprehensive Review. *Current Pharmaceutical Design, 25*(24), 2650–2660. doi:10.2174/1381612825666190708185506 PMID:31298154

Sharma, Y., Kawatra, A., Sharma, V., Dhull, D., Kaushik, S., Yadav, J. P., & Kaushik, S. (2021). In-vitro and in-silico evaluation of the anti-chikungunya potential of Psidium guajava leaf extract and their synthesized silver nanoparticles. *Virusdisease, 32*(2), 260–265. doi:10.100713337-021-00685-4 PMID:33869673

Singh, A., Gautam, P. K., Verma, A., Singh, V., Shivapriya, P. M., Shivalkar, S., Sahoo, A. K., & Samanta, S. K. (2020). Green synthesis of metallic nanoparticles as effective alternatives to treat antibiotics resistant bacterial infections: A review. *Biotechnology Reports (Amsterdam, Netherlands)*, *25*, 1–11. doi:10.1016/j.btre.2020.e00427 PMID:32055457

Singh, H., Du, J., Singh, P., & Yi, T. H. (2018). Role of green silver nanoparticles synthesized from Symphytum officinale leaf extract in protection against UVB-induced photoaging. *Journal of Nanostructure in Chemistry*, *8*(3), 359–368. doi:10.100740097-018-0281-6

Singh, I., Gupta, S., Gautam, H. K., Dhawan, G., & Kumar, P. (2021). Antimicrobial, radical scavenging, and dye degradation potential of nontoxic biogenic silver nanoparticles using Cassia fistula pods. *Chemical Papers*, *75*(3), 979–991. doi:10.100711696-020-01355-3

Soliman, W. E., Khan, S., Rizvi, S. M. D., Moin, A., Elsewedy, H. S., Abulila, A. S., & Shehata, T. M. (2020). Therapeutic applications of biostable silver nanoparticles synthesized using peel extract of benincasa hispida: Antibacterial and anticancer activities. *Nanomaterials (Basel, Switzerland)*, *10*(10), 1–13. doi:10.3390/nano10101954 PMID:33008104

Subha, V., Ernest Ravindran, R., Preethi, R., Cyrril, J., Kirubanandan, S., & Renganathan, S. (2018). Silver Nanoparticles - Green Synthesis with Aq. Extract of Stems Ipomoea Pes-Caprae, Characterization, Antimicrobial and Anti-Cancer Potential. *International Journal of Medical Nano Research*, *5*(1), 1–9. doi:10.23937/2378-3664.1410024

Subramaniam, S., Kumarasamy, S., Narayanan, M., Ranganathan, M., Rathinavel, T., Chinnathambi, A., & Whangchai, K. (2022). Spectral and structure characterization of Ferula assafoetida fabricated silver nanoparticles and evaluation of its cytotoxic, and photocatalytic competence. *Environmental Research*, *204*(PA), 111987.

Vallinayagam, S., Rajendran, K., & Sekar, V. (2021). Green synthesis and characterization of silver nanoparticles using Naringi crenulate leaf extract: Key challenges for anticancer activities. *Journal of Molecular Structure*, *1243*, 130829–130831. doi:10.1016/j.molstruc.2021.130829

Vanlalveni, C., Lallianrawna, S., Biswas, A., Selvaraj, M., Changmai, B., & Rokhum, S. L. (2021). Green synthesis of silver nanoparticles using plant extracts and their antimicrobial activities: A review of recent literature. *RSC Advances*, *11*(5), 2804–2837. doi:10.1039/D0RA09941D PMID:35424248

Venkatadri, B., Shanparvish, E., Rameshkumar, M. R., Arasu, M. V., Al-Dhabi, N. A., Ponnusamy, V. K., & Agastian, P. (2020). Green synthesis of silver nanoparticles using aqueous rhizome extract of Zingiber officinale and Curcuma longa: In-vitro anti-cancer potential on human colon carcinoma HT-29 cells. *Saudi Journal of Biological Sciences*, *27*(11), 2980–2986. doi:10.1016/j.sjbs.2020.09.021 PMID:33100856

Viana, R. L. S., Fidelis, G. P., Medeiros, M. J. C., Morgano, M. A., Alves, M. G. C. F., Passero, L. F. D., & Rocha, H. A. O. (2020). Green synthesis of antileishmanial and antifungal silver nanoparticles using corn cob xylan as a reducing and stabilizing agent. *Biomolecules*, *10*(9), 1–21. PMID:32854282

Vincent, S., Arokiyaraj, S., Saravanan, M., & Dhanraj, M. (2020). Molecular Docking Studies on the Anti-viral Effects of Compounds From Kabasura Kudineer on SARS-CoV-2 3CLpro. *Frontiers in Molecular Biosciences*, *7*, 1–12. doi:10.3389/fmolb.2020.613401 PMID:33425994

Waghule, T., Sankar, S., Rapalli, V. K., Gorantla, S., Dubey, S. K., Chellappan, D. K., Dua, K., & Singhvi, G. (2020). Emerging role of nanocarriers based topical delivery of anti-fungal agents in combating growing fungal infections. *Dermatologic Therapy*, *33*(6), 1–34. doi:10.1111/dth.13905 PMID:32588940

Yousaf, H., Mehmood, A., Ahmad, K. S., & Raffi, M. (2020). Green synthesis of silver nanoparticles and their applications as an alternative antibacterial and antioxidant agents. *Materials Science and Engineering C*, *112*, 110901–110908. doi:10.1016/j.msec.2020.110901 PMID:32409057

Chapter 5
Smart Nanomaterials for Photo-Catalytic Applications

Enas Amdeha
Egyptian Petroleum Research Institute, Egypt

ABSTRACT

Smart nanomaterials are grabbing more significant attention. Photocatalysts are catalysts that can be stimuli by light with an appropriate wavelength. Photocatalysis is deemed to be a promising approach for the utilization of photo energy and has been extensively studied for many processes. Therefore, the synthesis of nanomaterials with tailored-made catalytic characteristics, are of great importance. For applications using visible light and hence solar energy, the modification of photocatalysts can occur via the formation of heterojunction nanocomposites by inhibiting the recombination of charge carriers. These applications are driven by solar energy, which is conducive to the sustainable development of energy resources with no impact on the environment. This chapter will include recent enhancements in the smart nanomaterials for photocatalytic applications, especially in the fields of removing environmental pollutants, self-cleaning surfaces, water splitting for hydrogen production to provide clean fuel resources, selective alcohol oxidation and CO2 reduction in the environment.

INTRODUCTION

Smart materials are a type of material that may have one or more properties changed simply by changing the external environment. Such changes in material characteristics enable the creation of so-called responsive materials, which have a significant impact on the rapid advancement of materials science, water treatment, and energy generation applications. The ability of smart materials to adjust to external environmental changes such as temperature, pH, light, moisture, and fluid composition in order to activate their unique functionalities is one of their most remarkable qualities (Wieszczycka et al., 2021; Yoshida & Lahann, 2008). Since nanotechnology's rapid development a few decades ago, new unknown possibilities for smart materials applications have emerged. Professor Richard Feynman coined the term "nanotechnology" in 1959 during a conference on the future of technology held at the California Institute of Technology (Caltech). The conference's provocative title, "There is plenty of room at the bottom,"

DOI: 10.4018/978-1-6684-7358-0.ch005

was only the beginning of what we now know as nanotechnology. Nanotechnology is an interdisciplinary field of study that combines engineering with biology, chemistry, and physics. Nanotechnology is typically seen as a scientific instrument that "steps across the limit" of downsizing, resulting in materials that behave differently compared to macroscopic scale.

More particular, as a solid substance's dimensions approach the nanoscale, its physical and chemical properties might diverge dramatically from those of the identical material at the macroscale (Gopakumar et al., 2018). Nanomaterials have evolved as a fascinating class of materials that includes a wide range of samples with at least one dimension in the 1–100 nm range. The sensible design of nanoparticles can result in extremely high surface areas. Nanomaterials can be made with magnetic, electrical, optical, mechanical, and catalytic capabilities that are vastly superior to those of their bulk counterparts. The size, shape, synthesis conditions, and appropriate functionalization of nanomaterials may all be precisely controlled to provide the desired qualities (Ahlawat & Narayan, 2020; Wellia et al., 2017). Organic (primarily polymer-based nanoparticles), inorganic (metal, metal oxide, metalloid nanoparticles), and hybrid nanomaterials are all examples of smart nanomaterials (Wieszczycka et al., 2021). Smart nano photocatalysts are light-responsive nanomaterials that trigger reactions when exposed to light. Photocatalysis is a multifunctional process that has applications in science, physics, chemical engineering, and chemistry. Photocatalysis has a wide range of benefits and applications. It is a cost-effective and simple method of producing energy, removing pollution, and reducing CO_2. The absorption of light by a solid substance initiates a photocatalytic reaction. Semiconductor photocatalysis is commonly thought to be a reaction that takes place on the semiconductor's solid surface. Because not all catalysts have photocatalytic capabilities, catalysis is classed differently.

Photocatalysis, on the other hand, considers processes that lead to energy storage and increased efficiency in pollution removal and water splitting. Non-toxic, stable, highly photoactive, and low-cost photocatalysts are suitable. Templates with zeolites were extended to micro structured silica, alumina, and other oxides by heterogeneous photocatalysis, allowing sub nanoclusters to be stabilized down to the nanoscale. The supports not only stabilize the nanoparticles, but they also act synergistically with the nanoparticle surfaces to activate substrates in a manner similar to the positive synergy reported in alloys between two transition metal atoms or between a transition metal and a main group atom (L. Liu & Corma, 2018). Nanocomposites refer to complex nanoparticle support ensembles. For a competent and well-established photocatalytic scheme, the surface properties, material selection, crystallinity, structure, electronic band, and design of a heterogeneous photocatalyst must all be carefully considered. For photocatalysis, a large surface area is advantageous because more light will absorb on it, resulting in a large number of electrons and holes for pollutant destruction. Higher crystallinity with fewer defects can reduce charge carrier recombination, which improves the material's photocatalytic efficiency during catalytic processes. Edge recombination and Fermi level trapping are reduced when the eminence boundary is large. For heterogeneous photocatalysis, material selection is critical.

In addition to the best bandgap and band-edge association, the effective energy level at the heterojunction border can determine carrier transportation competency and direction. In general, photocatalysis turns photon energy into chemical energy, which necessitates the use of a material that develops a transitory state upon light absorption, forming an electron–hole pair, in order to form chemical species. The photocatalytic redox processes will then flow from a higher energy state to a lower energy one. Stability and durability under sustained irradiation are two requirements for a photocatalyst to maintain its reusability over multiple cycles. UV photocatalysis, on the other hand, has a narrow solar spectrum of only 4%, compared to 46% for visible light (Haw et al., 2016; Luo et al., 2014; Mostafa & Amdeha, 2022).

Nano-photocatalysts can be made in two ways: top-down and bottom-up. Top-down methods include mechanical milling, laser ablation, etching, sputtering, and electro-explosion. Another way is the bottom-up method, which is used in reverse to produce and build NPs from comparatively simpler substances (atoms), such as CVD, solvothermal, hydrothermal, sol–gel, soft and hard templating, and reverse micelle processes, Figure 1 (Baig et al., 2021; Khanna et al., 2019; Rane et al., 2018).

Figure 1. The synthesis of nanomaterials via top-down and bottom-up approaches.
Adapted with permission from (Khanna et al., 2019) (Copyright 2019, Elsevier)

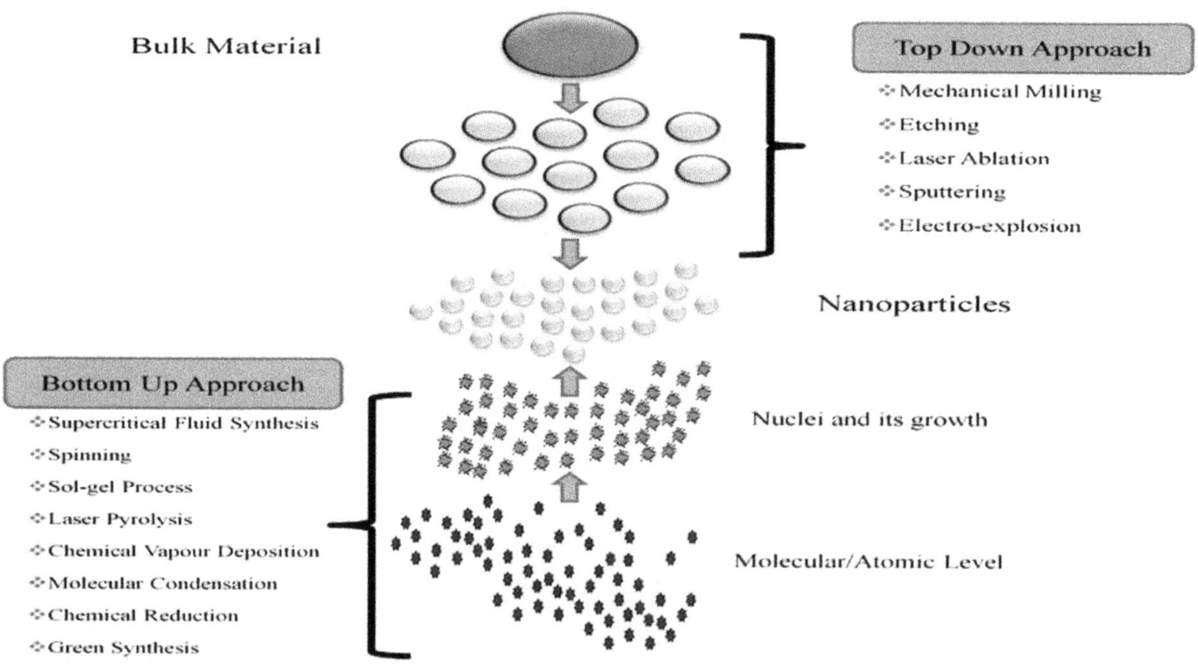

For the investigation of different physico-chemical properties of photocatalysts, many characterization techniques have been used. These techniques include X-ray diffraction (XRD), X-ray photoelectron spectroscopy (XPS), Raman, Fourier transform infrared (FTIR), polarized optical microscopy (POM), scanning and transmission electron microscopy (SEM & TEM), atomic force microscope (AFM), energy dispersive X-ray (EDX), Brunauer–Emmett–Teller (BET), and dynamic light scattering (DLS) (Khan et al., 2019). For morphological properties: microscopic techniques e.g. POM, SEM and TEM are used. For structural properties to study the composition and nature of bonding materials for more information about the bulk properties of the subject materials, XRD, EDX, XPS, FTIR, Raman, BET, and Zeta size analyzer are used. For surface area properties and particle size, SEM, TEM, XRD, AFM, and DLS are applied.

Optical characteristics are very important in photocatalytic applications, thus researchers learned how to use this technique to figure out how their photochemical processes work. These definitions are based on the well-known Beer-Lambert law as well as basic photo principles. The absorbance, reflectance, luminescence, and phosphorescence properties of NPs can be determined using these techniques. As a result, knowing the absorption and reflectance values of these materials is always interesting in order to understand the underlying mechanism for each application. The well-known optical tools UV–visible (UV–Vis) and photoluminescence (PL) can be used to explore the optical properties of NPs materials.

The diffuse reflectance spectrometer (DRS) is a fully equipped instrument for measuring optical absorption, transmittance, and reflectance in the UV/vis range. PL is a good approach for studying the optical characteristics of photoactive NPs and other nanomaterials, in addition to UV. This method provides extra information regarding the materials' absorption or emission capacity, as well as their impact on the overall photo-excitation time. As a result, it gives valuable information regarding charge recombination and half-life of excited materials in their conductance band, which is useful for all photo applications.

The limitations of photocatalysis are the recombination of charge carriers, unsuitable band edge potentials, and low absorption of light (Tahir et al., 2020).Considering the unique properties for smart NPs, specially, photocatalysts, and the growing energy demands and environmental remediation as the top priority in the research field worldwide (Demir et al., 2019; H. Wang et al., 2014; Y. Wang et al., 2016), this chapter will shed light on the different photocatalytic applications of the smart nanomaterials in the fields of removing environmental pollutants, self-cleaning surfaces, water splitting to provide clean fuel resources, selective alcohol oxidation and CO_2 reduction in the environment.

1. Photocatalytic for Wastewater Treatment Applications

With the global shortage of clean water, it's more crucial than ever to create efficient wastewater treatment methods. Organic pollutants have been found in both ground and surface water, as well as in drinking water. These pollutants are hazardous to human, animal, and environmental health. Treatments for eliminating organic pollutants from wastewater that are both low-cost and highly efficient are urgently needed. Current water treatment technologies can absorb or concentrate contaminants from one phase to the next, but they can't entirely destroy or remove them. Other water treatment technologies, such as filtration, sedimentation, and chemical oxidation, have a number of drawbacks, including the inability to entirely remove contaminants, high costs, and long processing times. New methods are necessary to improve water quality (Abou-Hadid et al., 2022; Amdeha, 2021; Iqbal et al., 2019; Tahir et al., 2018). Photocatalysis is one such technology that may be used to remove pollutants in the air and water.

Solution combustion synthesis (SCS) route provides a rapid and energy saving single step pathway for the large-scale synthesis of porous nanoparticles with high surface area (Ravishankar et al., 2015; Z. Tian et al., 2009). (Negi et al., 2018) has successfully prepared well crystalline Cerium Oxide (CeO_2) nanoparticles through solution combustion synthesis (SCS) using $(NH_4)_2[Ce(NO_3)_6]$ and $C_4H_6O_6$ as oxidizer and fuel, Figure 2. The synthesized (CeO_2) nanoparticles were applied for sensing and photocatalytic degradation of picric acid (PA) in its aqueous solution. The properties of the synthesized (CeO_2) were examined in terms of FESEM, HRTEM, EDS, XRD, FTIR, BET and UV–Vis spectroscopy techniques. XRD results revealed a well-crystallized, pure-flake-like mesoporous material to be formed having crystallite size of 18.86 nm with very small lattice strain of 0.0078. FESEM and attached EDS confirmed the purity and the sample composition as well as demonstrated that the (CeO_2) sample is well dispersed and formed in high yield. The BET results depicted a Type-IV isotherm with hysteresis loop for the mesoporous materials according to the IUPAC classification with 36.85m^2/g surface area which is higher that of the commercial ceria (2m^2/g) (J. Yang et al., 2014). By using the synthesized (CeO_2) nanoparticles (0.5 g/l) as a photocatalyst for PA degradation (10 ppm) under UV light irradiation, almost complete degradation can be obtained within 150 min. These findings point out that the synthesized CeO_2 nanomaterial can be efficiently used as an effective photocatalyst.

Figure 2. (a) Schematic diagram of solution combustion synthesis leading to the formation of porous nanostructures. (b) Photocatalytic degradation mechanism of PA in the presence of CeO_2 as photocatalyst.
Adapted with permission from Ref. (Negi et al., 2018) (Copyright 2018, Elsevier)

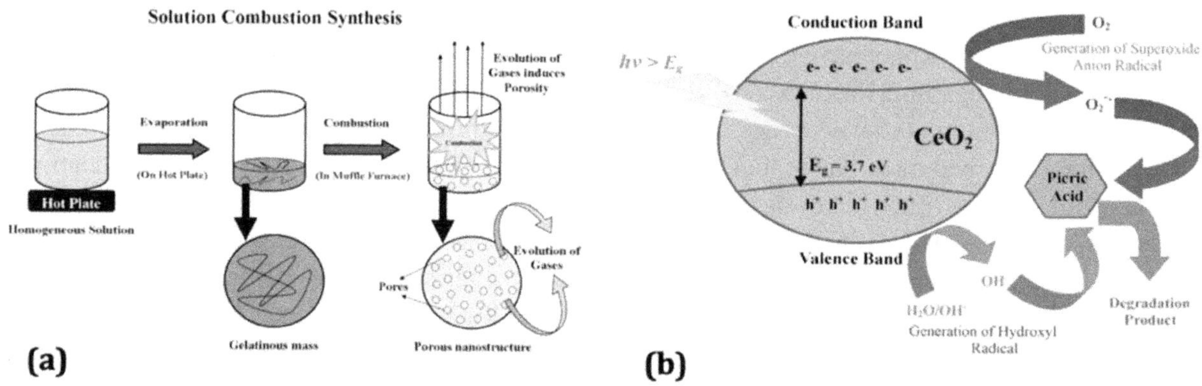

(Haw et al., 2016) developed a highly $CoFe_2O_4$ decorated 3D urchin-like TiO_2 nanocomposite ($CoFe_2O_4$–TiO_2) for the usage as smart magnetically-recyclable photocatalyst for the photodegradation of methylene blue dye (MB). The 3D urchin-like TiO_2 microparticles are hydrothermally prepared then decorated with $CoFe_2O_4$ magnetic nanoparticles (NPs) through a co-precipitation method, Figure 3. The crystallinity, morphologies and the optical properties of the synthesized samples were elucidated. Under UV irradiation for six hours, the removal percentages of MB and the kinetic rate constants were (45.54%, 0.1073 h^{-1}), (79.91%, 0.2605 h^{-1}) and (98.89%, 0.7432 h^{-1}) in the presence of the commercial rutile TiO_2, pure 3D urchin-like TiO_2 and ($CoFe_2O_4$–TiO_2), respectively. The enhancement efficiency of 3D urchin-like TiO_2 over the commercial rutile TiO_2 may be due to the high surface area to volume ratio endowed by the hierarchical structure of the urchin-like TiO_2, which provides excellent accessibility for the water molecules to diffuse into the space between the nanorods of the urchin-like structure. As a consequence, a large number of hydroxyl radicals are produced and this has greatly assisted in triggering the degradation process (J. Tian et al., 2015). However, the surface area of 3D urchin-like TiO_2 (68.52 m^2/g) is higher than that of $CoFe_2O_4$–TiO_2 (42.38 m^2/g), the ($CoFe_2O_4$–TiO_2) still exhibits the boosted photocatalytic behaviour due to the enhancement in the charge separation efficiency for the nanocomposite as it possesses the lowest recombination rate of the photoexcited charge carriers, as evidenced by the photoluminescence measurement (PL). Moreover, the $CoFe_2O_4$–TiO_2 nanocomposite is magnetically-retrievable for sequential recycling, and the results indicate that the nanocomposite shows a relatively consistent photocatalytic performance with negligible loss of magnetic properties. Hence, it is perceived that this study could potentially serve as a potential route for the advancement of photocatalyst technology, especially for developing recyclable magnetic-nano photocatalyts for wastewater treatment.

Figure 3. Schematic diagram showing the preparation steps (a) the stabilization procedure in an ice bath with the addition of the as-prepared 50 wt% titanium tetrachloride solution, tetra butyl titanate (TBT) and toluene serve as the solvent. The mixture is magnetically stirred for one hour before being transferred into a Teflon-lined autoclave; (b) the mixture is poured into the autoclave and heated in an oven at 150 °C for 24 h to synthesize 3-dimensional urchin-like TiO_2; (c) the preparation of $CoFe_2O_4$ nanoparticle decorated 3D urchin-like TiO_2 via the co-precipitation of Co^{2+} and Fe^{3+} ions under alkaline conditions (labelled as $CoFe_2O_4$–3D urchin-like TiO_2 nanocomposites).
Adapted with permission from Ref. (Haw et al., 2016) (Copyright 2016, RSC)

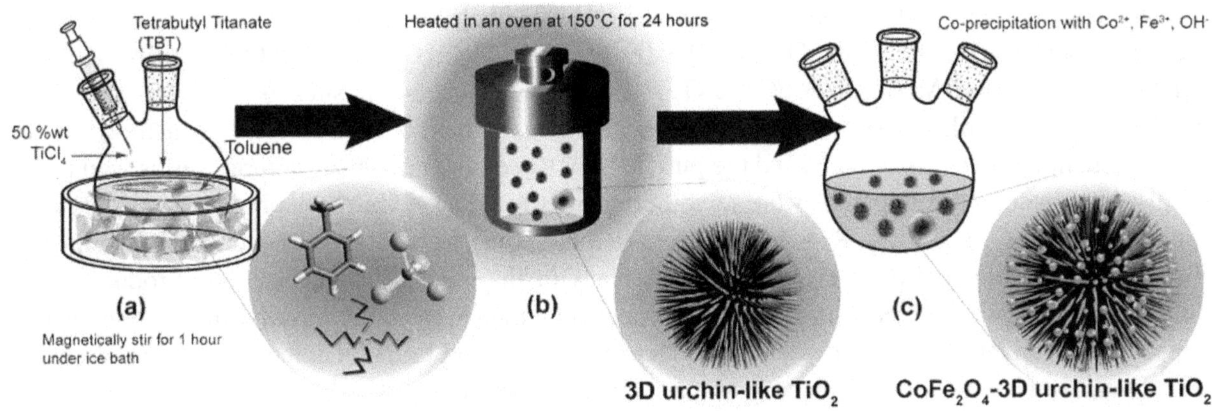

(G. Wang et al., 2022) designed Z-scheme CdS/WS_2 heterojunction consisting of CdS nanoparticles and WS_2 nanosheets via a facile and green two-step hydrothermal method. The prepared CdS/WS_2 has high efficiency towards both photodegradation of organic dyes and photoreduction of Cr (VI). To adjust the energy band structure in CdS/WS_2 composites with the enhanced separation efficiency for photo-generated electron-hole to decrease the recombination rate for photogenerated carrier, the molar ratio of CdS and WS_2 is varied to improve the photocatalytic activities. The 30% CdS/WS_2 sample exhibited the highest photocatalytic activity with apparent rate constant of 0.06153 min^{-1}, 0.02378 min^{-1}, and 0.0846 min^{-1} for RhB, MB and Cr (VI), respectively, Figure 4. The CdS/WS_2 composite has good stability for more than four cycles with 88.1% degradation efficiency for the fourth cycle of the RhB. The scavengers experiment revealed that h^+ and $^{\bullet}O_2^-$ are the major reactive species for the photodegradation of dyes while e^- and $^{\bullet}O_2^-$ are for the photoreduction of Cr(VI).

Figure 4. Apparent rate constants of (a) RhB, (b) MB and (c) Cr (IV) for different catalysts.
Adapted with permission from Ref. (G. Wang et al., 2022) (Copyright 2022, Elsevier)

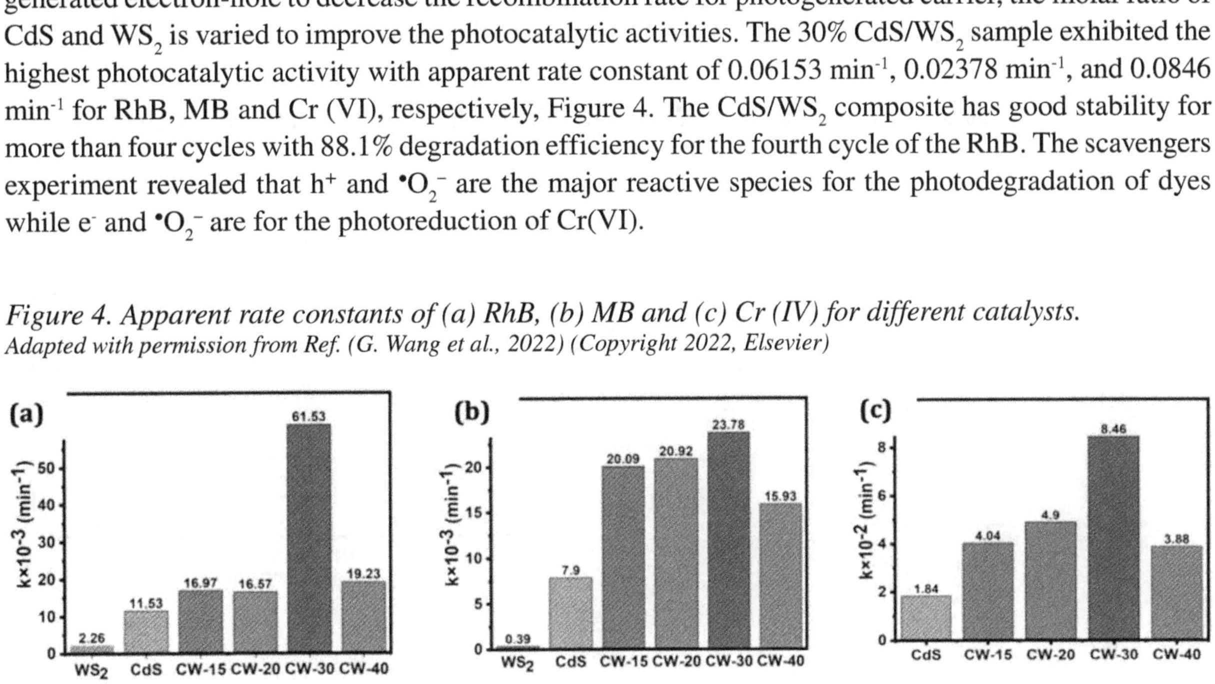

(Rauf et al., 2022) prepared cerium oxide (CeO_2), cerium sulfide (Ce_2S_3) and cerium oxide-cerium sulfide composite (CeO_2/Ce_2S_3) by co-precipitation method. The XRD and SEM were employed to characterize the prepared samples. RhB dye is used to evaluate the photocatalytic performance of the prepared samples under UV light irradiation. The photolysis of RhB after two hours was only 15%, while in the presence of the prepared samples were 71%, 80% and 94% for Ce_2S_3, CeO_2 and CeO_2/Ce_2S_3 composite, respectively. Moreover, the efficiency of the composite towards the degradation of other three dyes were evaluated and the RhB is the highest dye to be degraded as the order of degradation were Rhodamine B < Methylene blue < Congo red < Bromothymol blue.

(Bagherzadeh et al., 2021) synthesized MIL-101(Fe), cobalt ferrite ($CoFe_2O_4$) magnetic nanoparticles, binary (MIL-101(Fe)/$CoFe_2O_4$, MIL-101(Fe)/GO and $CoFe_2O_4$/GO), and ternary MIL-101(Fe)/$CoFe_2O_4$/3% GO and MIL-101(Fe)/$CoFe_2O_4$/7% GO) magnetic composites based upon the MIL-101(Fe). In this regard, GO and $CoFe_2O_4$ were prepared using the modified Hummer's method and the hydrothermal method, respectively, and the pure MIL-101(Fe) and the binary composites as well as the ternary composites were prepared by the solvothermal method. The visible light photocatalytic degradation of Direct Red 23 (DR23), Reactive Red 198 (RR198) dyes as well as Tetracycline Hydrochloride (TC) antibiotic were studied to evaluate the efficiency of the prepared samples. The introduce of the $CoFe_2O_4$ and GO was to create magnetic properties and enhance the optical properties for the ternary samples, respectively. GO not only a strong electron acceptor led to the efficient separation of charge carriers and thus reduction of their recombination but also increased the absorption of visible light of MIL/$CoFe_2O_4$/3% GO and hence, it has the lowest band gap and the lowest recombination rate, Figure 5. The enhanced optical properties of MIL/$CoFe_2O_4$/3% GO are proved by the DRS, PL, EIS and electrochemical analyses which explain the best performance towards dyes degradation (99% after 70 min under visible light) compared to the other samples. Also, the BET-BJH analysis proved that the MIL/Co/3% GO had the highest specific surface area, total pore volume and mean pore diameter compared to the pure MIL-101(Fe) as well as the sample with higher content of GO. The scavenger test exhibits that the h^+ and $^\bullet O_2^-$ are the main active species for the degradation. Moreover, the composite has good stability and can be recycled up to five times with small loss of activity (only 8% loss).

Figure 5. (a) UV–vis/DRS analysis and (b) PL spectra of the MIL-101(Fe), MIL/Co/ 3% GO and MIL/Co/ 7% GO.

Adapted with permission from Ref. (Bagherzadeh et al., 2021) (Copyright 2021, Elsevier)

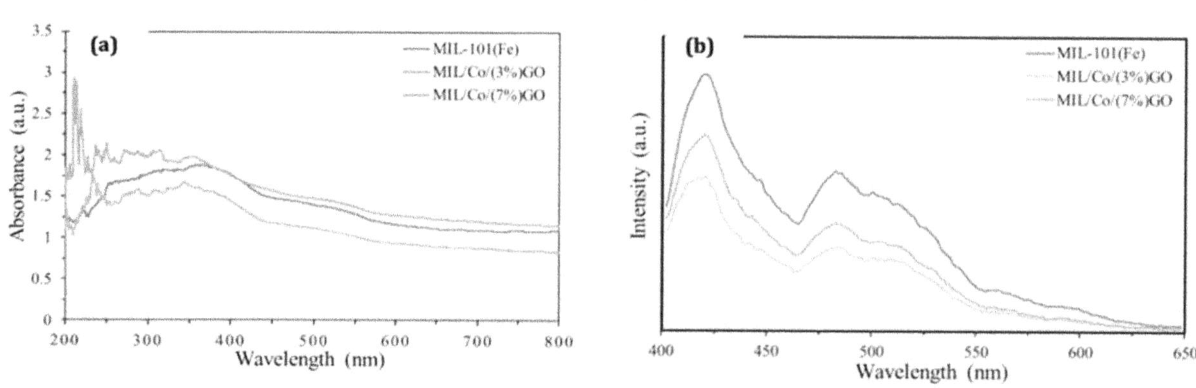

Visible-light-driven photocatalysis is a green and efficient strategy for wastewater treatment, where carbon materials-based semiconductors showed excellent performance in this regard.

(Hashem et al., 2021) reported a facile and green one-pot room-temperature ultrasonic route for the preparation of novel ternary nanocomposite (CdS@ZnO/g-C_3N_4) of cadmium sulfide quantum dots (CdS QDs), zinc oxide nanoparticles (ZnO NPs), and graphitic carbon nitride nanosheets (g-C_3N_4 NSs) as well as the corresponding binary nanocomposites, Figure 6. The photodegradation of Rhodamine B dye (RhB) under visible light was applied to assess the photocatalytic efficiency of the prepared photocatalysts. The formation of the cubic and hexagonal CdS QDs, hexagonal Wurtzite ZnO NPs, and hexagonal g-C_3N_4 NSs is confirmed by XRD. The oxidation states and chemical composition of CdS@ ZnO/g-C_3N_4 is revealed by XPS. Also, FE-SEM and HR-TEM graphs exhibited the fabrication of homogenous, well-dispersed CdS QDs over the ZnO/g-C_3N_4 matrix, revealing the success of the proposed synthesis route. The band gap of the CdS@ZnO/g-C_3N_4 is estimated by DRS using Tauc plot and it was 2.85 eV confirming the good visible-light efficiency. Moreover, the CdS@ZnO/g-C_3N_4 showed the lowest recombination rate over the other composites via the PL measurements. These enhanced optical properties were the main reason of the excellent photocatalytic activity of the CdS@ZnO/g-C_3N_4 over binary and single samples. The CdS@ZnO/g-C_3N_4 can completely degrade RhB dye solution after two or three hours under UV or visible light, respectively, due to the synergetic effect between CdS, ZnO, and g-C_3N_4 in the composite. This synergetic performance could be ascribed to the perfect electronic band configuration of the three components, leading to the establishment of several combined synergetic Z-Scheme/Type-II photocatalytic heterojunctions. The scavengers test investigated that the superoxide anion radicals $^{\bullet}O_2^-$ were the main active species, revealing that the predominant degradation pathway is reductive. Finally, the photocatalytic activity of the CdS@ZnO/g-C_3N_4 composite presented competing results and is expected to show promising performance if employed in the photodegradation of water pollutants or when employed in other photocatalytic applications.

Figure 6. (a) Schematic diagram of the aqueous-based synthesis of soluble CdS QDs, (b) CTAB template-assisted sol-gel synthesis of ZnO NPs, (c) thermal polymerization of urea to g-C_3N_4 NSs, and (d) the ultrasonic-assisted synthesis of ternary CdS@ZnO/g-C_3N_4 composites.
Adapted with permission from Ref. (Hashem et al., 2021) (Copyright 2021, Elsevier)

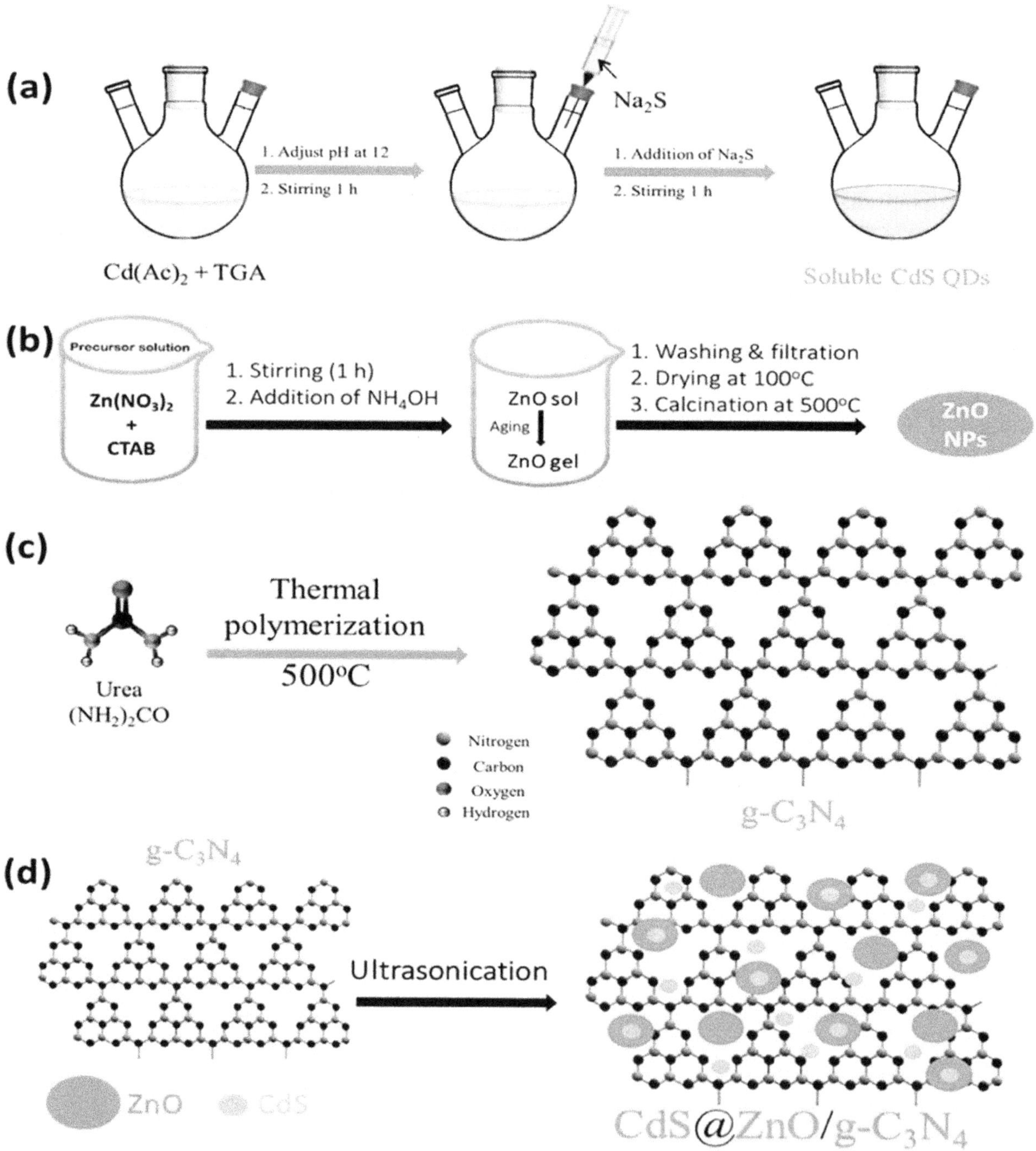

Phosphorus doping could form a new Fermi level and reduce the intermediate band gap, thereby prolonging the lifetime of photo-generated electron carriers. The well-designed core–shell nanostructure is conducive to the rapid transfer of carriers. The formation of Z-scheme heterojunction can provide a higher redox potential and reduce the photo-corrosion phenomenon of CdS. (Fang et al., 2021) synthesized P-doping CdS nanorods@NiFe layered double hydroxide (LDH) Z-scheme heterostructure photocatalyst (P-CdS@NiFe- LDH) via hydrothermal and calcination techniques, Figure 7. The prepared composite was used as a visible light responsive photocatalyst for hydrogen generation and photocatalytic degradation of bisphenol A (BPA); BPA is a plastic pollutant that threatens children's health and is widely used in mineral water bottles and food packaging boxes. The prepared P-CdS@NiFe-LDH sample exhibits the highest degradation per cent of BPA (98%) as well as the highest hydrogen generation rate, about 10.05 mmol h^{-1} g^{-1}, which is about 39 times higher than that of the pristine CdS sample (0.27 mmol h^{-1} g^{-1}). This amazing catalytic performance can be attributed to the synergistic effect of P-doping and Z-scheme heterojunction. The P doping produces a new Fermi level, and consequently reduces the band gap which improved the utilization of visible light energy. Also, the construction of the Z-scheme heterojunction increased the redox potential of the photocatalyst and inhibited the charge recombination of CdS. With the addition of H_2O_2, the photocatalytic degradation performance can be improved significantly due to the photo-Fenton reaction of iron-based substances which generate more free radicals and increased the degradation rate.

Figure 7. Schematic illustration for the formation of P-CdS@NiFe-LDH Z-scheme heterojunctions.
Adapted with permission from Ref. (Fang et al., 2021) (Copyright 2021, Elsevier)

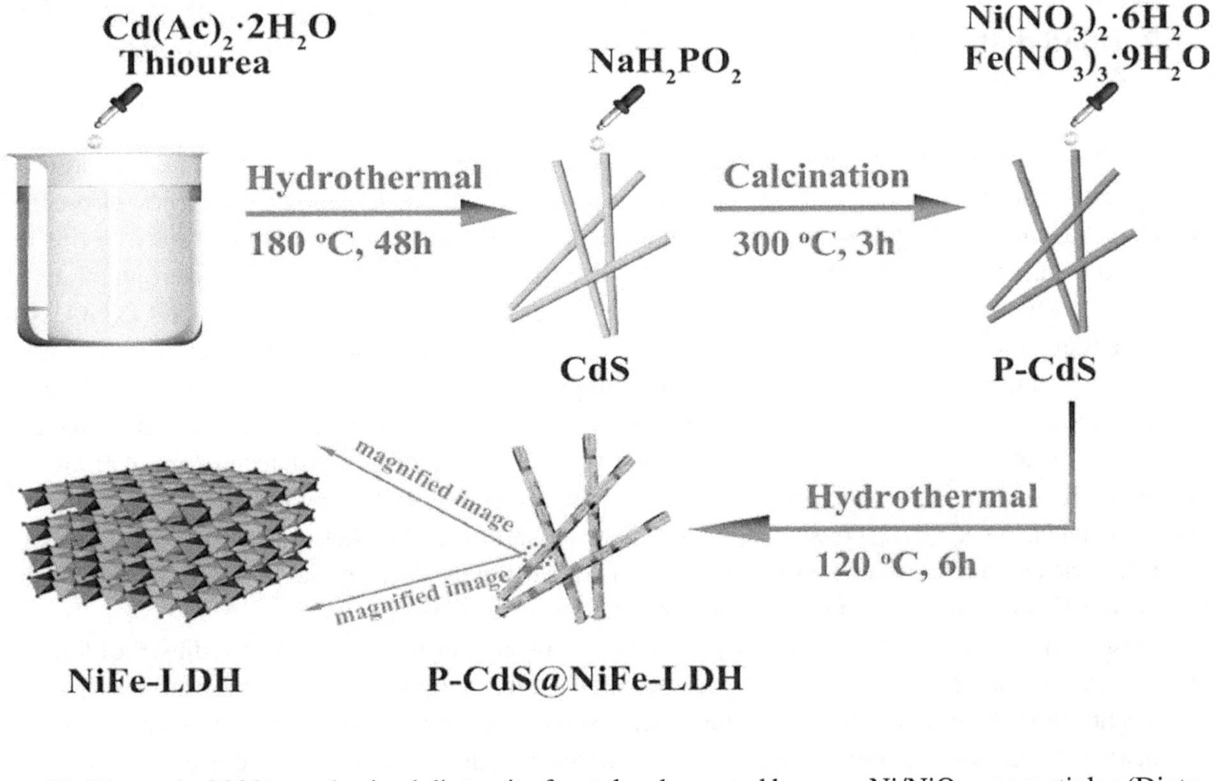

(G. Liu et al., 2020) synthesized diatomite frustules decorated by nano Ni/NiO nanoparticles (Diatomite@Ni/NiO), Figure 8, as a novel visible light responsive photocatalyst for effective degradation of malachite green dye (MG) and photocatalytic-reduction of Cr (VI) ions. The composite exhibits good texture properties with high surface area (400 m^2/g) and 5.8 nm as average pore diameter. Also, the optical properties are enhanced and the band gap is reduced up to 1.71 eV. These characteristics explained the high efficiency of the composite towards the MG photodegradation and the Cr(VI) photoreduction. The 50 mg of (Diatomite@Ni/NiO) composite can attain complete removal of 25 ppm MG or Cr(VI) after 30 min. while, the complete removal of 50 ppm can be attained after one hour. The MG photocatalytic degradation controlled mainly by the generated electron-hole pairs and the $^{\bullet}O_2^-$ species while the photocatalytic-reduction of Cr (VI) controlled directly by the excited electrons of Ni/NiO and partially by the formed $^{\bullet}O_2^-$ radicals.

Figure 8. Schematic diagram for the decoration of diatomite by Ni/NiO in diatomite@Ni/NiO nanocomposite.
Adapted with permission from Ref. (G. Liu et al., 2020) (Copyright 2020, Elsevier)

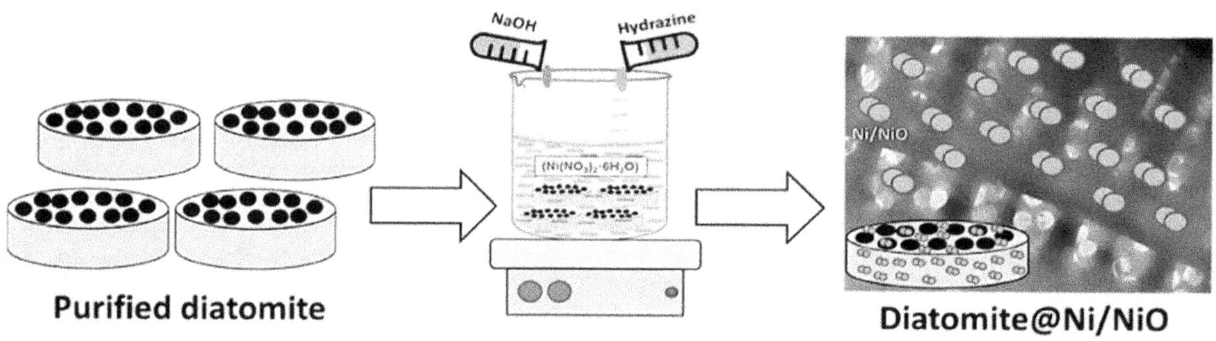

BiOCl as a representative layered bismuth-based photocatalyst with Sillén-structure has aroused wide public concern on photocatalytic degradation. However, the photocatalytic efficiency of pristine BiOCl is currently restricted by its low optical absorption and charge separation efficiency. In this regard, (Xu et al., 2022) constructed a binary (In_2O_3-BiOCl) step-scheme (S-scheme) heterojunctions of In_2O_3 nanoparticle and BiOCl micron-sheet through the molten salt method using a $LiNO_3$-KNO_3 system. The prepared composite is studied for the photocatalytic degradation of ciprofloxacin (CIP) under Xenon lamp illumination. The 20%In_2O_3-BiOCl sample exhibited 91% of CIP removal after 35 min, which is accounting for 39.6 and 3.2 times greater than In_2O_3 and BiOCl, respectively. This enhanced and highest efficiency related to the gained higher optical absorption performance of In_2O_3-BiOCl in the range (380–700 nm) than the pristine BiOCl. Additionally, this efficiency is attributed to the novel S-scheme mechanism, which boosted highly efficient separation of the photo-induced carriers. The PL spectrometric test and transient photocurrent response results demonstrated that In_2O_3-BiOCl composite exhibited efficient separation of photo-generated charge carriers.

Constructing heterojunction photocatalysts has been proven as an ingenious tactic to adjust the lifetimes of the higher redox-active charge-carriers toward highly-efficient photocatalytic applications. Numerous recent efforts focused on the short-range spatial separation of charge-carriers in the heterojunction photocatalysts. However, engineering a long-range transfer channel in the well-designed heterojunction photocatalyst to further spatially separate the charge-carriers is still a huge challenge. In this regard, (Lu et al., 2022) synthesized 2D-layered heterojunction photocatalysts of SnS_2/r-GO/g-C_3N_4 nanosheets (NSs) via a simple photo-assisted self-assembly technique, Figure 9. In this process, the photoactive semiconductors of 2D SnS_2 and g-C_3N_4 NSs convert GO into r-GO and simultaneously were anchored onto the surface of the r-GO NSs. The prepared composite is examined toward the photocatalytic redox reactions in which photocatalytic oxidation degradation of Rhodamine B (RhB) and photocatalytic reduction of protons for H_2 generation are tested under visible light. The presence of r-GO in the composite, make the later possess high surface area with good conductivity which induces a long-range spatial separation of the photoinduced charge-carriers. Upon visible light irradiation, the photoinduced electrons on the conduction band (CB) of SnS_2 NSs could transfer to the valence band (VB) of g-C_3N_4 NSs, thereby extending the lifetimes of the photoinduced electron of g-C_3N_4 NSs and the photoinduced hole of SnS_2 but without reducing their intrinsic redox ability. These enhanced characteristics give the SnS_2/r-GO/g-C_3N_4 composite the highest activity for the photo redox reaction. Thus, the rate constant of the SnS_2/r-GO/g-C_3N_4 NSs for the photocatalytic oxidation degradation of RhB was ~9.2 and ~4.6 times higher than that of the SnS_2 and g-C_3N_4 NSs, respectively. Meanwhile, the photocatalytic protons-reduction ability of this sample was ~ 68 and ~ 3.4 times beyond that of the SnS_2 and g-C_3N_4 NSs, respectively.

Figure 9. Schematic diagram showing the photo-assisted self-assembly synthesis strategy of the all 2D-layered heterojunction photocatalyst of SnS_2/r-GO/g-C_3N_4 NSs.
Adapted with permission from Ref. (Lu et al., 2022) (Copyright 2022, Elsevier)

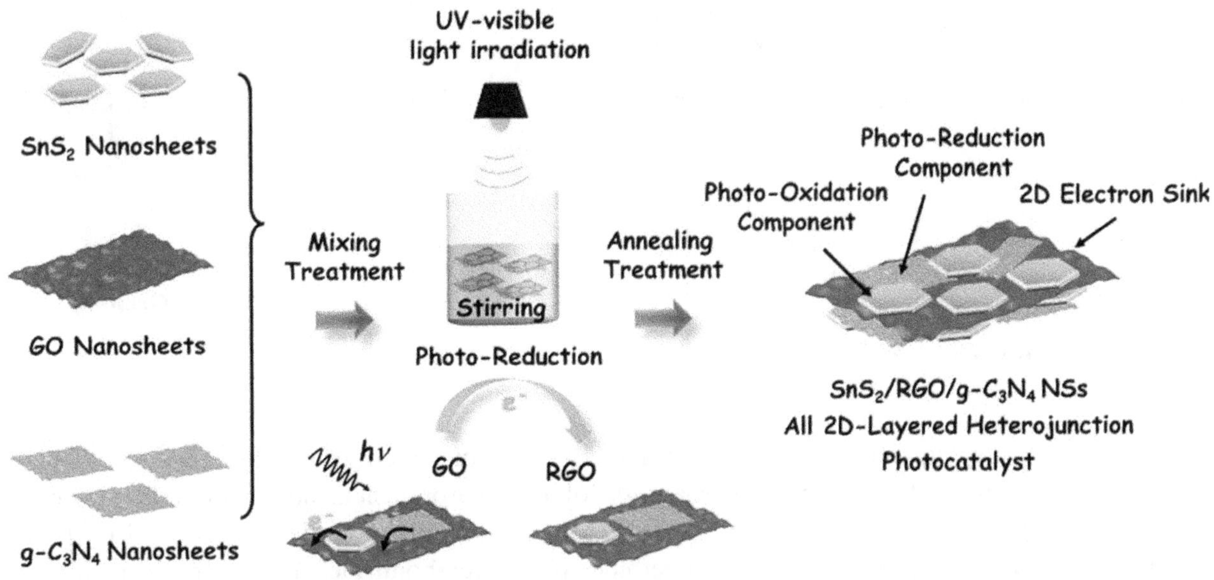

(Chu et al., 2022) reported the preparation of sodium and boron co-doped graphitic carbon nitride (NaB-CN) by thermal polymerization, Figure 10, for H_2O_2 production via photocatalytic oxygen reduction and 2,4-dichlorophenol (2,4-DCP) degradation via photocatalytic oxidation coupled with Fenton oxidation. The co-doping of the Na and B obviously enhanced the photocatalytic performance of CN for H_2O_2 production and 2,4-DCP degradation as a result of the effective enhancement in light absorption, electron-hole (e^--h^+) pairs separation and charge carrier mobility. The H_2O_2 production rate of the composite sample under visible light in 1% ethanol aqueous solution was 22.3 $mgL^{-1}h^{-1}$ which is 15 times higher than that of the un-doped CN. Interestingly, during the photocatalytic degradation of 2,4-DCP, the 2,4-DCP and some degradation intermediates could act as electron donors to promote H_2O_2 production, which allowed the application of Fenton oxidation to further enhance degradation performance after photocatalytic oxidation without adding H_2O_2 externally. This developed two-step oxidation process combined photocatalytic oxidation with Fenton oxidation, has presented the outstanding performance of 2,4-DCP degradation. The radical trapping analysis manifested that h^+ and $\bullet O_2^-$ were the two important oxidative species for the photocatalytic 2,4- DCP degradation, and $\bullet O_2^-$ played the dominant role. Also, the contribution of •OH can be excluded. Moreover, the stability of the (2 g/l) composite is tested for the (100 ppm) 2,4-DCP and the results revealed that the composite can degrade up to 80% of the dye after the sixth cycle. The results of this work indicate the recommended photocatalyst possesses relatively stable and appreciable photocatalytic activity, exhibiting its great potential in industrial application for H_2O_2 production and industrial organic wastewater treatment.

Figure 10. Schematic illustration for the preparation of NaB-CN photocatalysts.
Adapted with permission from Ref. (Chu et al., 2022) (Copyright 2022, Elsevier)

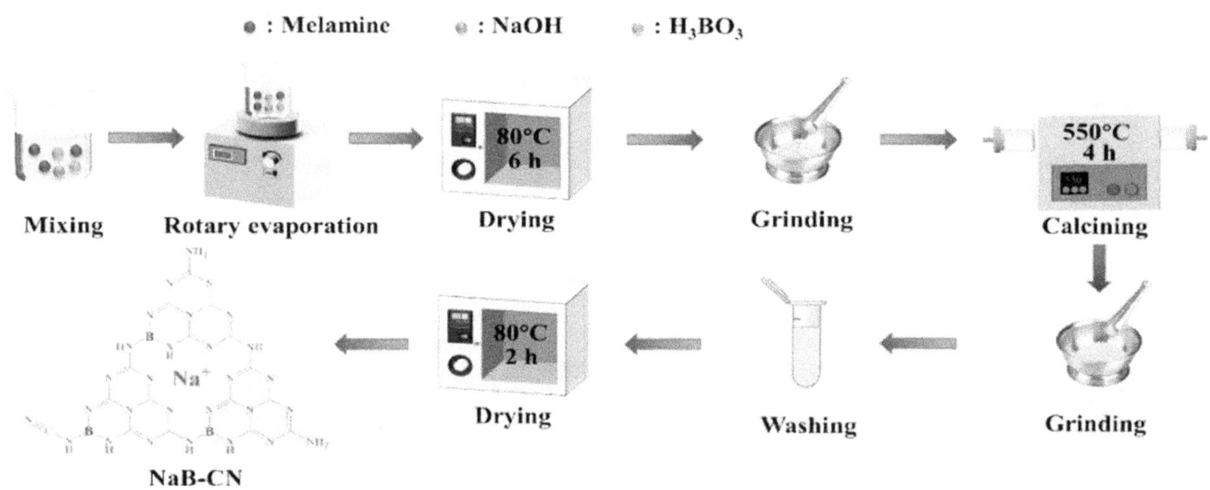

Photocatalytic degradation and hydrogen production using solar energy through semiconductor photocatalysts are deemed to be a powerful approach for solving environmental and energy crisis. However, the biggest challenge in photocatalysis is the efficient separation of photo-induced carriers. MXenes, a novel two-dimensional (2D) transition metal carbides/nitrides/carbonitrides, have received increasing attention since Ti_3C_2 was discovered (Naguib et al., 2011). In general, MXenes are obtained by selective etching the element A layers of the bulk MAX, where M, A, and X represent early transition metals, IIIA/IVA group elements, and a carbon and/or nitrogen, respectively (Naguib et al., 2014). Hitherto, MXenes have been widely studied and applied in the fields of energy storage and conversion (Zhao et al., 2018), catalysis (Kuang et al., 2019), sensing (Kim et al., 2018), heavy metals and dye adsorption (C. Cai et al., 2020; Feng et al., 2020). Additionally, based on its excellent electrical conductivity caused by metallic nature, MXenes have been clearly certified as highly efficient co-catalysts in boosting the photocatalytic activity, which can effectively promote the separation and transfer of photo-induced electron-hole pairs (T. Cai et al., 2018; Kuang et al., 2020; K. Li et al., 2020). So far, TiO_2/MXene photocatalysts with excellent photocatalytic activity have been widely studied and reported. (H. Li et al., 2022) prepared mesoporous TiO_2 nanoparticles using surfactant as structure-directing agent and then used the electrostatic self-assembly technique to anchor the prepared TiO_2 on different amount of the highly conductive Ti_3C_2 MXene co-catalyst, (TiO_2/Ti_3C_2). The mesoporous TiO_2 are homogenously dispersed on the surface of the layer structured Ti_3C_2 nanosheets. Moreover, the prepared composites reveal the apparently enhanced light absorption performance, photo-induced carriers' separation and electron transfer ability from TiO_2 to Ti_3C_2, thus boosting the photocatalytic activity toward the methyl orange (MO) degradation and H_2 production compared to the virgin TiO_2. The 3% TiO_2/Ti_3C_2 as optimized composite can degrade MO up to 99.6% after 40 min with $•O_2^-$ and •OH as the main active species for the degradation process. Additionally, the photocatalytic H_2 production rate was 218.85 μmol $g^{-1}h^{-1}$ for the optimized composite which was 5.6 times as compared to the pristine meso-TiO_2. Meanwhile, the unique mesoporous structure in TiO_2 and layer structure in Ti_3C_2 donate abundant surface-active sites for adsorption and surface reaction of reactants, which further promote photocatalytic activity. The discussed work demonstrates the potential of MXene family materials as noble-metal-free co-catalysts in the field of photocatalysis to solve the environmental and energy crisis.

The removal of water pollution of persistent pharmaceutical products and heavy metals is highly demanded. It is urgently needed to develop highly efficient and eco-friendly photocatalysts to meet the requirements of current green environmental development. Eco-friendly enzymatic photocatalyst with potential solar light utilization is an effective alternative to environmental remediation. (Arif et al., 2022) demonstrate the combined effect of artificial enzyme hemin immobilization on the surface of semiconductor photocatalyst $Bi_4Ti_3O_{12}$ (BTO) to fabricate a novel biomimetic oxygen vacancy-rich 3D photocatalyst hemin-$Bi_4Ti_3O_{12}$ nanocomposite (HBTO) by using a facile one-step solvothermal method. The HBTO nanocomposite photocatalysts showed high photocatalytic efficiency for the photodegradation of tetracycline hydrochloride (TCHCL) and photo-reduction of Cr(VI) to Cr(III) under visible irradiation. The 1.0-HBTO as an optimum ratio exhibited more than 99.9% for both TCHCL photodegradation and Cr(VI) photoreduction which is more than that of BTO (33% for TCHCL and 58.7% for Cr(VI)). The great photocatalytic performance of HBTO samples was attributed to the high surface area, improved pore volume and enhanced photo-absorption and the photoelectrochemical which were highly improved with ~3.8 times higher photocurrent density than BTO. Moreover, the boosted charge transfer, and efficient separation of photogenerated charge species which arise from the immobilization of hemin act as an electron-conducting bridge and a surface oxygen absorption site. Furthermore, the reduction with methanol using hemin-methanol during the synthesis process has boosted the formation of oxygen vacancies in HBTO nanocomposite. This work sheds light on an efficient approach of a synergistic effect of artificial enzyme and semiconductor photocatalyst to fabricate highly efficient biomimetic photocatalyst HBTO for environmental remediation and other photocatalytic application.

(C. Wu et al., 2022) loaded $Bi_2Sn_2O_7$ nanoparticles on the surface of mesoporous MIL-100(Fe) to fabricate a novel $Bi_2Sn_2O_7$@MIL-100(Fe) heterojunction composite photocatalyst via a hydrothermal method. The $Bi_2Sn_2O_7$@MIL-100(Fe) composite exhibited enhanced photocatalytic degradation of ciprofloxacin (CIP) under visible light. The rate constant of the composite (0.0143 min^{-1}) is 3.86 and 1.96 times higher than that of $Bi_2Sn_2O_7$ (0.0037 min^{-1}) and MIL-100(Fe) (0.0073 min^{-1}), respectively. This superior activity is due to the composite heterojunction which promote the separation of photo-generated carriers, leading to the fortification of photo-generated electrons on MIL-100(Fe). Also, MIL-100(Fe) with astonishing surface area can provide more active sites for photocatalytic reaction, where $•O_2^-$ and h^+ radicals are the main active species in the CIP degradation process. The $Bi_2Sn_2O_7$@MIL-100(Fe) composite supplies a promising method for synthesizing MOF-based photocatalyst with high surface area in the field of environmental protection and renewable resources.

Iron/steel making industry is a weed that produces large quantities of slag and dust. Using these wastes to synthesize photocatalysts for wastewater treatment is of great importance. In this regard, (Amdeha et al., 2021) extracted sodium silicate (Na_2SiO_3) and zinc from Blast Furnace Slag (BFS) and Electric Arc Furnace Dust (EAFD), respectively, to synthesize MCM-41, ZnO and ZnS. Then, ZnS and ZnO were sono-chemically loaded on the MCM-41 to prepare ZnS-ZnO/MCM-41 nanocomposite for the Cr (IV) photoreduction. Small-angle XRD, FTIR, HRTEM and N_2 adsorption–desorption techniques have proved the formation of waste-based mesoporous MCM-41 and ZnO with high purity, high surface area, good homogeneity and uniform distribution. The ZnS-ZnO/MCM-41 composite exhibited high Cr(VI) photoreduction efficiency (94%) after three hours under UV irradiation which is higher than those of ZnO/MCM-41 (58%) and ZnO (40.6%). The scavengers test reveals that the •OH radicals are the main active species for the reduction. Moreover, the composite showed high stability and activity even after the fourth cycle (84.7%) of photoreduction. Based on these results, ZnS–ZnO/MCM-41 composite material seems to be high efficiency, green, stable, environment, and economical alternative to be used as a photocatalyst for the reduction of Cr (VI).

(El-Salamony et al., 2021) prepared silica (SiO_2)-based nanocomposites via impregnation method, namely, 20 TiO_2-SiO_2, 20 CeO_2-SiO_2, 20 TiO_2-CeO_2-SiO_2, and 40 TiO_2-CeO_2-SiO_2. These composites used for (25 ppm) Fuchsine acid (FA) dye degradation for three hours under both UV and visible light irradiation after one hour of adsorption in dark to attain equilibrium. The composites exhibited higher photo-efficiency under visible light than UV light except for the TiO_2-SiO_2 sample due to its high bandgap (3.2 eV). The CeO_2-SiO_2 exhibited an excellent photocatalytic activity (84.8%, 1.1×10^{-2} min^{-1}) under visible light over the other composites which is in line with the adsorption results. Moreover, the h^+ and $•O_2^-$ are the predominant species for the degradation over CeO_2-SiO_2 composite.

2. Self-Cleaning

Wool fabrics (the less studied textile obtained from home or industry) and model glass plate samples deposited at low temperature with different amounts of ZnO or TiO_2 nanoparticles by (Frunza et al., 2018). Sputtering (SP) and sol-gel (SG) were used for TiO_2 deposition while sputtering (SP) or electroless (E) for ZnO. The coated samples were characterized structurally and morphologically, by XRD, SEM and XPS and were checked for photocatalytic degradation of MB in the PCC2 checker apparatus under both UV and vis light. Sputtering leads to wurtzite type ZnO crystallites but deposited TiO_2 does not appear crystalline. Reflectance data in the visible spectral range of highest absorption band served for direct analysis of MB degradation. The coated fabrics exhibited an increased MB photodegradation over the uncoated materials, and followed the first order kinetic equation.

(Nozari et al., 2021) prepared silica-supported zinc oxide (ZnO) catalyst (ZnO@SiO_2) on polyester fabric (PET) in alkali conditions at 90 ºC via a chemical in-situ method to impart the stable photo-active ZnO catalyst on the textiles with multi-functional properties under lights and acidic media for anti-bacterial, self-cleaning and flame-retardant applications. Zinc acetate and sodium silicate as two precursors were used in ammonia aqueous solution to form ZnO@SiO_2 nanoparticles on PET fabric. The ZnO@SiO_2 catalyst on the PET fabric was characterized via XRD, FE-SEM, EDX, and FTIR. The ZnO@SiO_2 controlled and reduced the deficiencies of ZnO NPs on the PET fabric, the nanoparticles size was about 28.27 nm and the core shape (ZnO NPs) was changed from nano-pin to the diverse geometrical shapes.

Interestingly, ZnO@SiO_2 enhanced the photocatalytic activities of the fabric through discoloration of MB dye in short time under both ambient and acidic conditions as a result of the covering of silica with the lower band gap on the ZnO NPs. In addition, in-situ synthesis of ZnO@SiO_2 created linkages between amine groups of the aminolyzed fabric and the formed ions improved the tensile strength in comparison with the ZnO NPs fabric. Also, ZnO@SiO_2 acted as thermo-catalyst on the fabric due to inherent thermal resistance of Si, considerably increased the anti-dripping properties of PET during burning. In addition, ZnO@SiO_2 (PET) exhibited about 100% of bacterial reduction of *Escherichia coli*. The ZnO@SiO_2 enhanced the photocatalytic and thermo properties of PET fabric.

(Abualnaja et al., 2021) prepared titanium dioxide (TiO_2) and metallic silver/titanium dioxide (Ag°/TiO_2) nanocomposite to be coated on the polyester fabric surface via the straightforward pad-dry-curing technique for photocatalytic, self-cleaning and antimicrobial applications. Ag°/TiO_2 nanocomposite was fabricated in a two-step method including a low temperature TiO_2 sol-gel process followed by deposition of metallic Ag on TiO_2 via ultraviolet irradiation. Ag°/TiO_2 nanocomposite was deposited onto oxygen plasma activated polyester fabric using a simple route. The prepared Ag°/TiO_2 showed highly improved absorbance in the visible region and high photo-catalytic and antibacterial activities toward the MB degradation and *Escherichia coli*, respectively, compared to TiO_2 nanoparticles immobilized onto oxygen plasma-activated polyester. This significant enhancement in self-cleaning and antibacterial performance can be ascribed to silver nanoparticles loaded on the surface of TiO_2.

Membrane fouling is an ongoing challenge in the membrane filtration process. Developing the high-efficient and robust antifouling membranes for oily emulsions to achieve the purpose of prolonging the service life of the membrane shows great demand because of the restriction of hydration anti-fouling mechanism strategy to against pollution. (Zhang et al., 2022) designed and fabricated a hydrophilic photo-Fenton self-cleaning catalyst – GO/Fe-MOF assembled stabilized polyacrylonitrile (PAN) nanofibrous membrane for oily wastewater separation by introducing inter-penetrating "intercept plate" – GO/M88A heterostructure via ultrasonic assistance, mussel-inspired fixation and hydrothermal synthesis methods, Figure 11. Benefiting from the regulable amounts of GO nanosheets on the SPAN@GO/M88A membrane surface, the obtained membrane possessed sub-micrometer skin layer (0.17–1.02 μm), excellent water wettability (~0 °) and underwater oleophobic (~160.6°), which guarantee the balance of fast permeation flux (7083 ± 47 L m^{-2} h^{-1}) and separation efficiency (>99%) for diesel/water emulsion with small emulsified oils. Meanwhile, faced the surfactant-stabilized emulsified oil droplets with smaller size, the charged intercept plate not only realized "size sieving" results, but produced strong electrostatic force to achieve demulsification separation. Moreover, introduced hydrophilic (MIL-88A (Fe)) into modified membrane endowed underwater anti-oil-fouling and photo-Fenton catalytic degradation performance for organic contaminants. Meanwhile, the charged GO/M88A heterostructure can assist to achieve demulsification for separation of surfactant-stabilized emulsions by strong electrostatic force. More importantly, attached and irremovable contaminants on the membrane can be quickly removed by photo-Fenton self-cleaning method, and achieved over 96% of recovery flux recovery ratio, which effectively prolonged the service life of the membrane. The recently designed SPAN@GO/M88A membrane with hydration antifouling and photo- Fenton self-cleaning properties can stimulate more studies on revivable separation membranes for multifunctional applications.

Figure 11. Design and fabricate processing of SPAN@GO/M88A nanofibrous membrane.
Adapted with permission from Ref. (Zhang et al., 2022) (Copyright 2022, Elsevier)

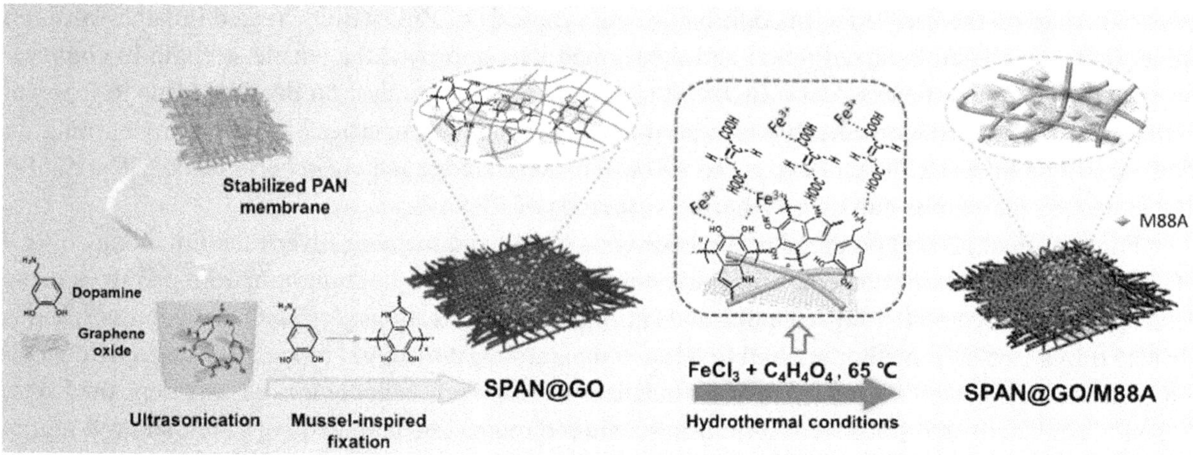

(Gao et al., 2021) fabricated a novel photo-Fenton self-cleaning PAA@NM88B/GO membrane with polyacrylic acid (PAA), NH_2-MIL-88B(Fe) (NM88B), and graphene oxide (GO) via a simple vacuum filter method equipped with the nylon membrane substrate for the separation and degradation of dyes, Figure 12. The membrane (PAA@NM88B/GO composite) retained the laminated structure of GO, and PAA@NM88B was used as the photo-Fenton catalytic nano-filler to intercalate GO. The results showed that by introducing porous NM88B and PAA cross-linking modification, PAA@NM88B/GO membrane had excellent hydrophilicity, water flux and dye rejection performance. The flux of the PAA@NM88B/GO membrane was 68.21 L m^{-2} h^{-1} bar^{-1}, which is about 15 times higher than that of GO membrane. In addition, the composite membrane had a high rejection (> 97%) for methylene blue (MB), crystal violet (CV) and Congo red (CR) and can also be used to degrade textile wastewater in real life with good stability and antifouling performance due to the synergistic effect of membrane adsorption and photocatalysis.

Figure 12. Schematic illustration of the fabrication of PAA@NM88B/GO membrane.
Adapted with permission from Ref. (Gao et al., 2021) (Copyright 2021, Elsevier)

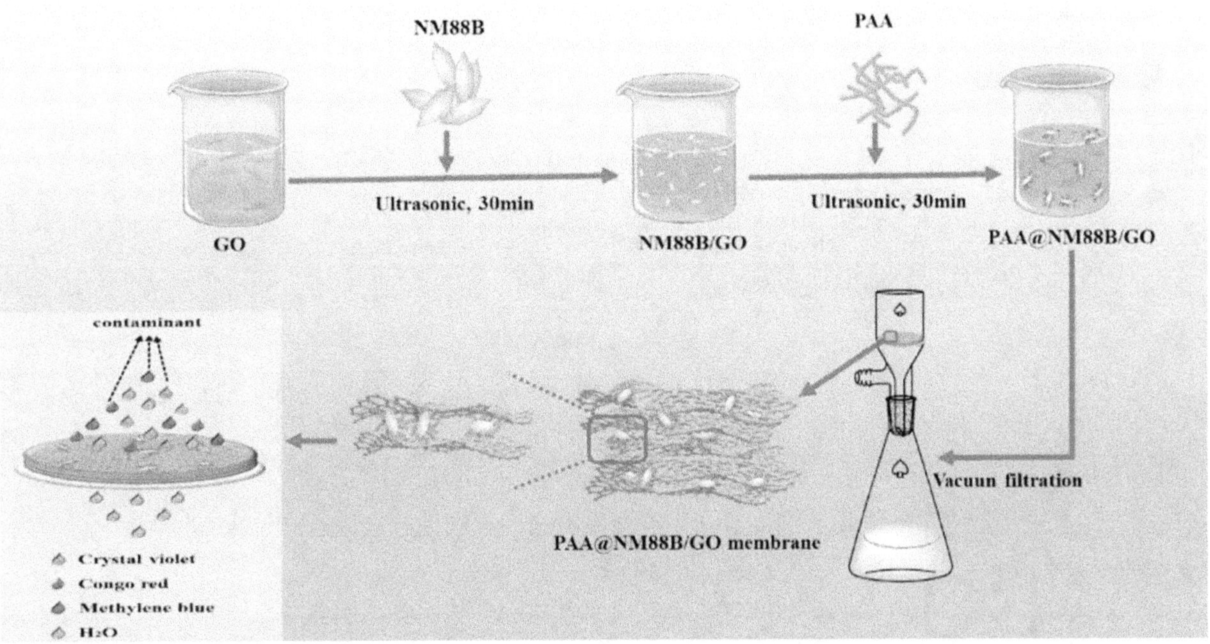

For the first time, (Yue et al., 2021) fabricated a self-cleaning photo-Fenton membrane containing a reactive layer by engineering partially reduced graphene oxide/Ag nanoparticles/MIL-88A (pr-GO/Ag/M88A, pGAM) photocatalysts on the PVDF substrate membranes via a one-step vacuum filtration method, Figure 13. Benefiting from the high conductivity of pr-GO and the surface plasmon resonance (SPR) effect of Ag nanoparticles (Ag NPs), the photo-sensitivity of the pGAM is significantly enhanced for the dye removal with self-cleaning properties. Based on the synergistic filtration/photo-Fenton processes, the pGAM membrane exhibited 99.7% of dye removal compared to the dark condition and the permeability was 189 L m^{-2} h^{-1} bar $^{-1}$. Notably, the nano-channel structure of the membrane did not collapse under high pressure (0.1–0.4 MPa) and the permeability was 183 L m^{-2} h^{-1} bar $^{-1}$ with high dye removal of about 98.1% even after the tenth cycle. The membrane also exhibits antifouling properties with a high-water flux recovery (>90%). When the membrane is used directly as a photo-Fenton catalyst, it exhibited 90% of MB degradation after one hour. According to the scavenger tests, •OH and $•O_2^-$ are the reactive species responsible for the membrane anti-fouling behavior.

Figure 13. Flowchart of the fabrication process of the pGAM membranes.
Adapted with permission from Ref. (Yue et al., 2021) (Copyright 2021, Elsevier)

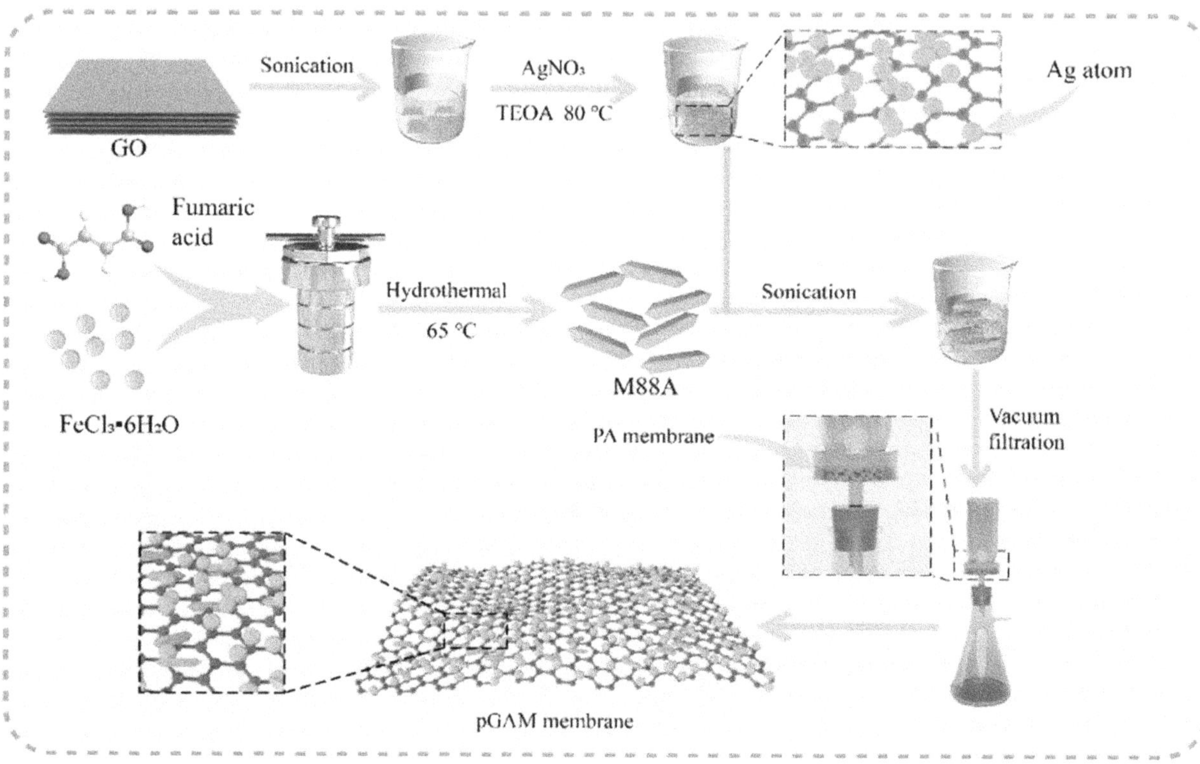

3. Water Splitting for Hydrogen Production

Solar energy as abundant, inexhaustible, and broadly spread is the most appealing renewable alternative to fossil fuels. However, because solar irradiation is diffuse and intermittent, solar energy must be efficiently transformed into chemical fuels that are storable, transportable, and used on demand (Green & Bremner, 2016; Tachibana et al., 2012). This need has prompted the development of sustainable artificial photosynthetic systems with the goal of simulating natural photosynthesis in order to manufacture fuels from H_2O and CO_2 using solar energy, Figure 14, (Kudo & Miseki, 2009; Zhu & Wang, 2017). Solar-to-chemical energy conversion is a cost-effective and environmentally friendly way of storing solar energy as long-term energy carriers.

Hydrogen is one of the most appealing possible solar fuels produced by artificial photosynthesis since it has a high energy density and does not emit pollutants during combustion (Jafari et al., 2016; Mazloomi & Gomes, 2012). Solar water splitting, which combines solar energy with water (a plentiful resource), provides a platform for hydrogen synthesis; a schematic of a prospective sustainable hydrogen generation and usage plant is shown in Figure 15. Water is fed into a solar water splitting module and transformed to hydrogen, which is then used in high-efficiency power systems such as internal combustion engines and fuel cells. As a result, H_2O and sunlight are used to generate energy.

Furthermore, hydrogen can be mixed with CO_2 extracted from the atmosphere or emissions from coal or natural gas-burning facilities to produce methanol, synthetic natural gas, or polymers and fertilizer precursors (She et al., 2017). These kinds of sustainable processes would have various benefits in terms of climate change mitigation and compatibility with present infrastructure. The efficient, cost-effective breakdown of water on the scale necessary to meet global energy demands (tens of terawatts) is one of the most critical obstacles connected with achieving such technologies (Lewis, 2016). Exploring an innovative and efficient photocatalyst to reduce energy and environmental challenges is a crucial research goal in this area.

Figure 14. Schematic illustration of natural and artificial photosynthetic systems.
Adapted with permission from Ref. (Q. Wang & Domen, 2020) (Copyright 2020, ACS)

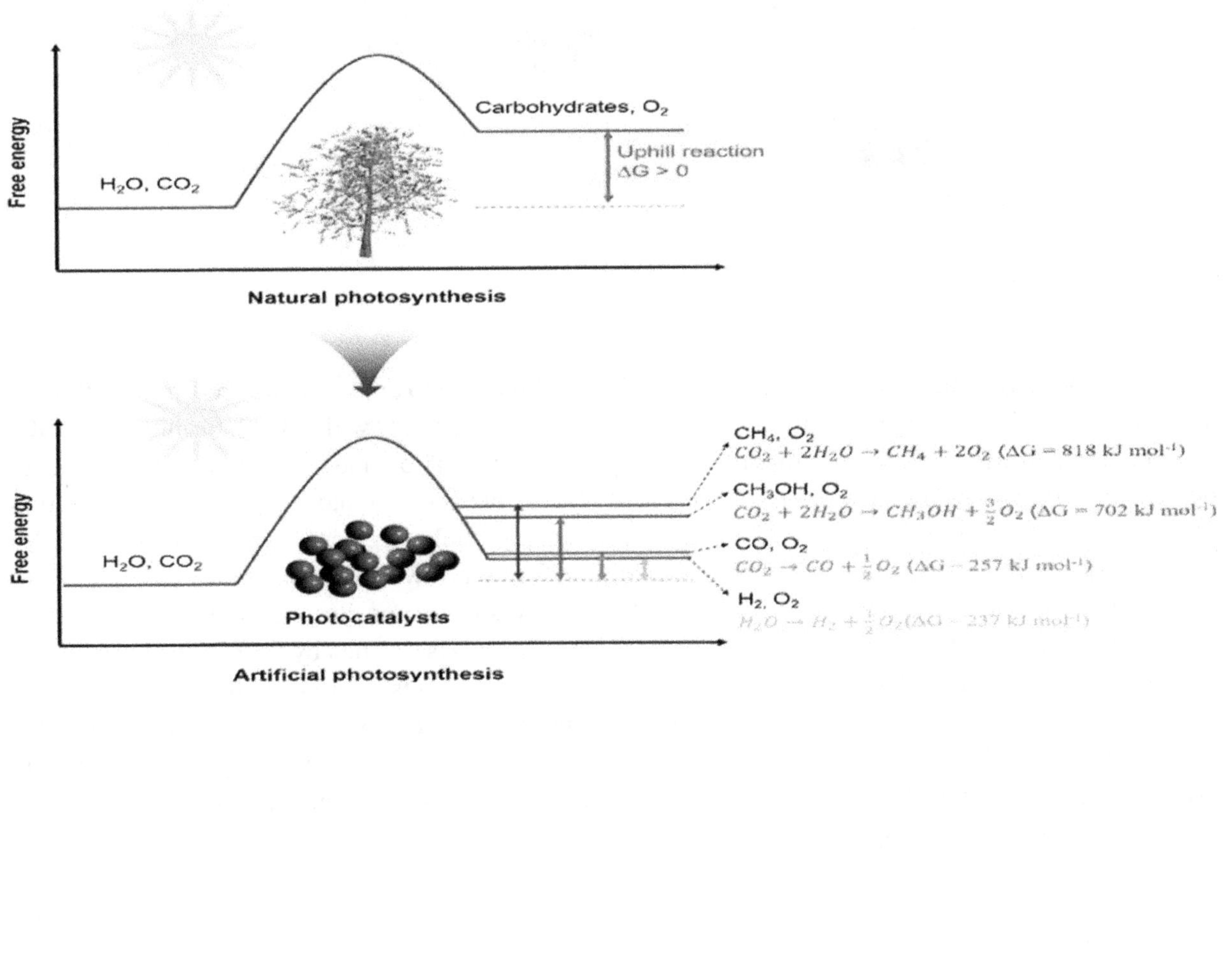

Figure 15. Potential sustainable hydrogen fuel cycle based on photocatalytic water splitting.
Adapted with permission from Ref. (Q. Wang & Domen, 2020) (Copyright 2020, ACS)

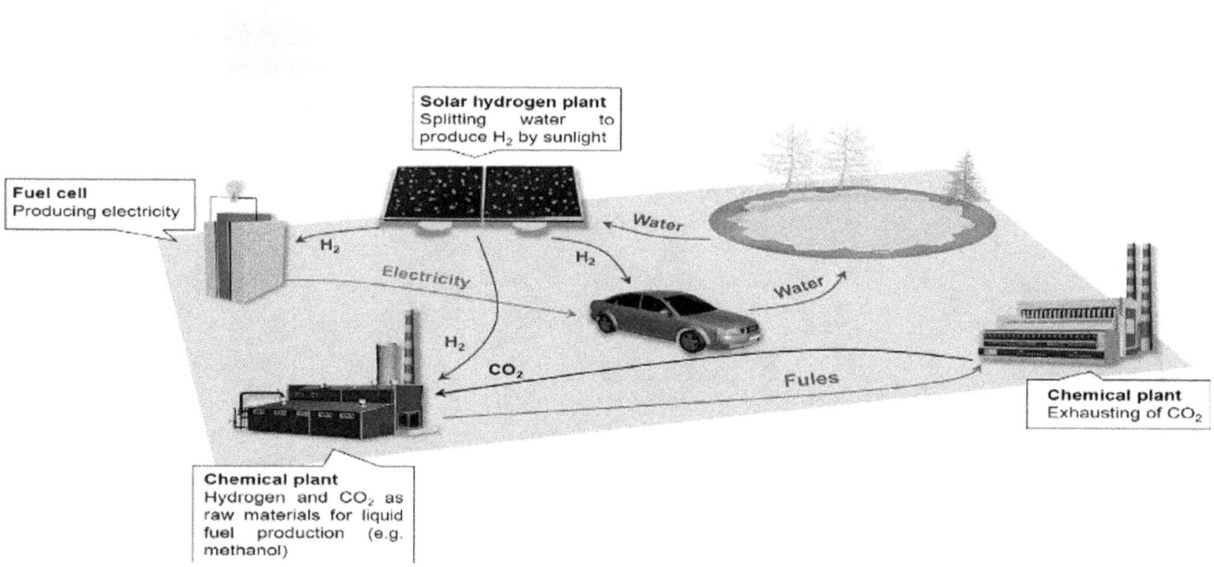

(Rakibuddin & Kim, 2020) used a simple wet chemical method to dope Samarium (Sm^{+3}) with ZnO and one step solid-state transformation method to couple Sm–doped ZnO with g-C_3N_4 forming (SAZOC) composite, Figure 16, at different proportions for hydrogen production under visible light. The fruitful doping of Sm^{+3} into ZnO which enhances the ZnO to be visible active photocatalyst and the formation of heterojunction between g-C_3N_4 and Sm–doped ZnO nanoparticles are confirmed by XPS and TEM, respectively. Additionally, the higher charge separation and the lower electrons-holes recombination through the SAZOC composites are implied by PL. All SAZOC composites exhibited outstanding H_2 production and the highest H_2 production rate (10.25 mmol g^{-1} h^{-1}) is attained by SAZOC10 sample. The pH2; acidic medium; is preferred for H_2 production by these composites. Moreover, these composites are stable up to ten hours with 30.3–90.02 mmol g-1 of H_2 production. The boosted photo activity was due to the synergistic effects of ZnO and g-C_3N_4, and Sm^{+3} which extended the visible light absorption and effectively enhanced the charge separation by acting as co-catalyst in the reaction making it promising for H_2 generation via water splitting.

Figure 16. Schematic representation for the preparation of SAZOC composites.
Adapted with permission from Ref. (Rakibuddin & Kim, 2020) (Copyright 2020, Elsevier)

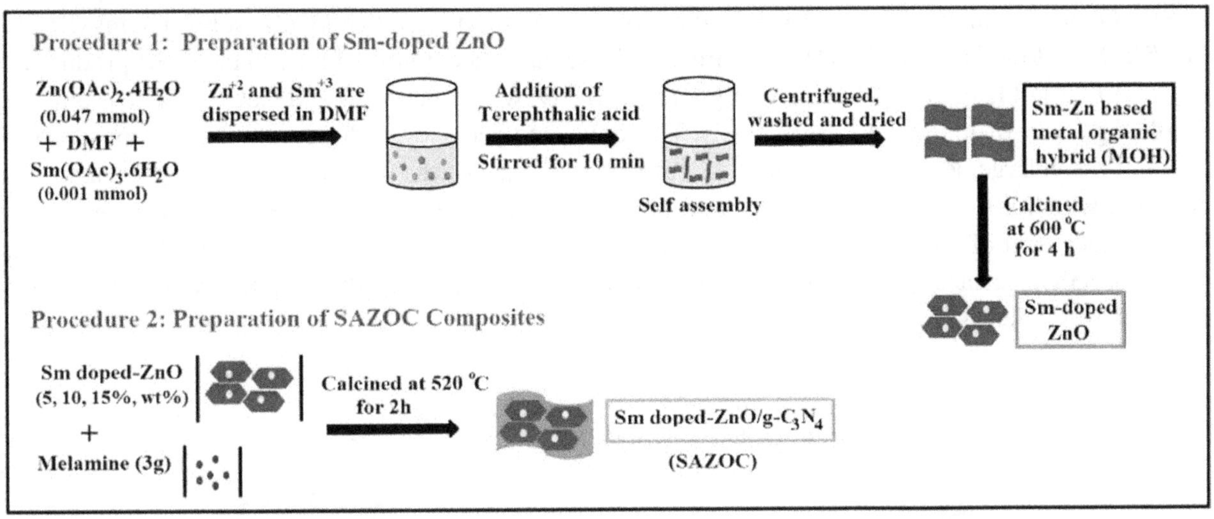

(Gu et al., 2019) loaded g-C_3N_4 nanosheets (CN) on the surface of $Mg_{1.2}Ti_{1.8}O_5$ nanoflakes (MT) at different ratios via facile sol-gel and calcination methods to fabricate Z-scheme (2D/2D MT/CN) heterojunction composites for H_2 production and photocatalytic inactivation of *Escherichia coli* (*E. coli*). The MT/CN-10 exhibited the highest H_2 production rate (3128.3 μmolh^{-1}g^{-1}) which was 3.57 and 7.24 times higher than those of MT and CN, respectively. Also, it has good stability and reusability as there is no significant in the H_2 production even after the fourth cycle. Furthermore, the inactivating performance of Escherichia coli (E. coli) over the same sample was 1.95 and 2.06 times compared to MT and CN, respectively. This improved photocatalytic performance is attributed to the establishment of heterojunction between MT and CN as a result of the extremely negative conduction band (CB) of CN and the extremely positive valence band (VB) of MT, which can lead a superior interfacial charge transfer under solar light irradiation generating more electrons and active radicals for H_2 production and bacteria inactivation, respectively. In addition, the efficient harvesting of light and the large specific surface area of 2D MT/CN nanosheets are of great importance of the enhancement of the photocatalytic efficiency.

Taking into consideration the high demand for globally clean energy, chemists and engineers searched for methods for the simultaneous degradation of actual industrial effluent and hydrogen production. (Meshram et al., 2022) synthesized a new MIL-53@TiO_2 composite via solution combustion for water splitting applications in the absence of any agent. Under three hours of visible light, MIL-53 did not exhibit any activity toward H_2 production, while the H_2 evolution were 0.26 mmol and 7.9 mmol for TiO_2 and MIL-53@TiO_2, respectively. Revealing that the H_2 generation rate of MIL-53@TiO_2 (26.33 mmol h^{-1}g^{-1}) is 30 times higher than TiO_2 (0.86 mmol h^{-1}g^{-1}). This significant performance is related to the enhanced optical properties; observed by PL; as the MIL-53@TiO_2 composite has the superior charge separation property and less recombination of the charge carrier pair. The MIL-53@TiO_2 results exhibited economic feasibility to be extended for practical large-scale applications as it is stable and can be recycle up to four cycles with high H_2 production rate of 23.5 mmol h^{-1}g^{-1}.

Nevertheless, CeO_2 has been hardly mentioned as a photocatalyst for H_2 evolution due to the high photoelectron-hole pair recombination and limited absorption of visible light. The morphology controlling strategy to prepare CeO_2 as visible active photocatalyst with regular shape and good performance toward H_2 production is presented by (Xia et al., 2021). They used vulcanized and modified MOFs to construct ZnS/CeO_2 photocatalysts with totally different morphologies structure (ZnS/CeO_2 solid spheres (S) and ZnS/CeO_2 hollow dodecahedra (D)), Figure 17. These catalysts exhibited an enhanced H_2 production rate of 397 μmol h^{-1} g^{-1} and 417 μmol h^{-1} g^{-1}, respectively, while pure ZnS and CeO_2 exhibited only 82 μmol h^{-1} g^{-1} and 32 μmol h^{-1} g^{-1}, respectively. This enhanced behavior of these modified composites is a result of the enlargement of the visible light absorption range and the increase of surface area and nano porosity, which afford more active sites and promote the transfer of photogenerated electrons-holes. The enhancement of the (D) sample over the (S) sample is due to the presence of more charge transfer channels at the interface of the (D) sample. Additionally, the prepared ZnS/CeO_2 photocatalysts has good stability even after the fifth cycle (25 h).

Figure 17. The preparation processes for ZnS/CeO_2 (D) and ZnS/CeO_2 (S).
Adapted with permission from Ref. (Xia et al., 2021) (Copyright 2021, Elsevier)

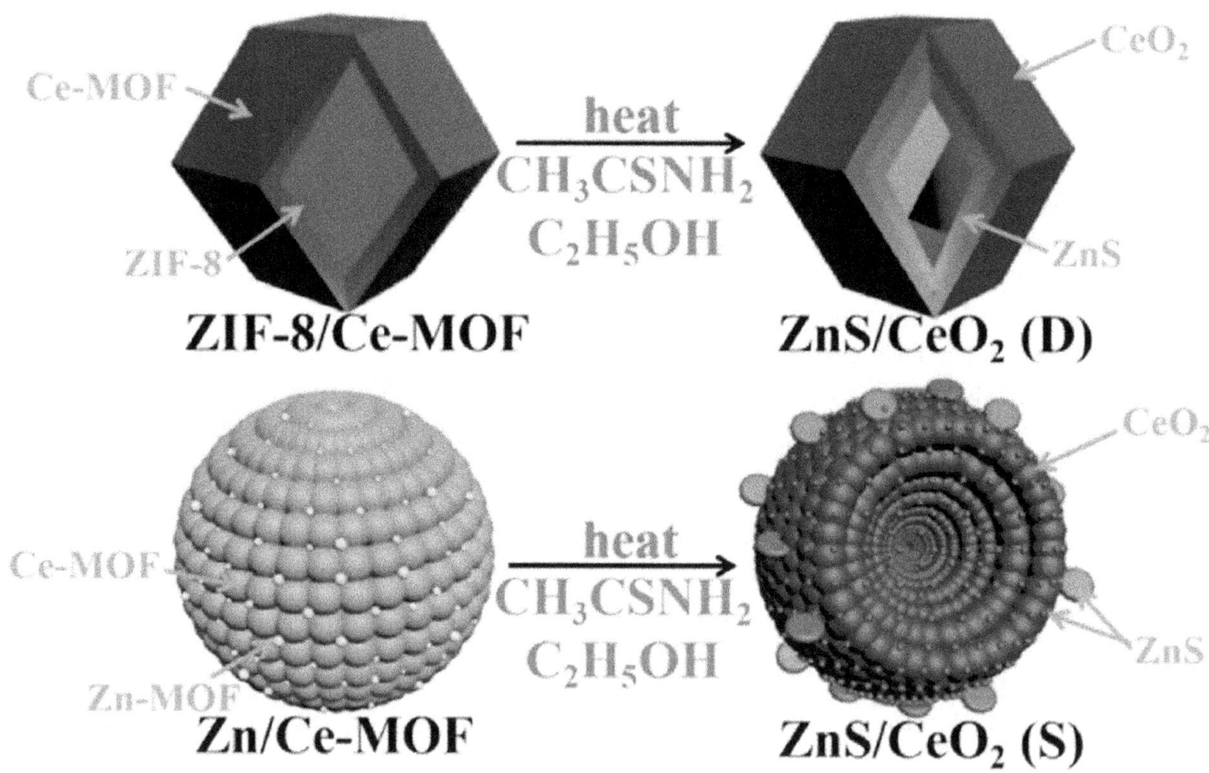

(Moscow et al., 2022) used a facile microwave irradiation method, Figure 18, to synthesize Erbium (Er) and Yttrium (Y) doped monoclinic-tetragonal $BiVO_4$ nanocatalysts; (Er–$BiVO_4$ and Y–$BiVO_4$) for photoelectrochemical water splitting and photocatalytic degradation of dyes and acetaldehyde. The introduction of rare earth metal (Er / Y) affected and stabilized the monoclinic-tetragonal crystalline structure of $BiVO_4$ and lead to the formation of a mixed crystal phase structure as indicated by XRD. After three hours under visible light, the Y–$BiVO_4$ degraded about 85%, 91% and 93% for MO, RhB, and

MB respectively. Also, acetaldehyde photodecomposition was studied. The enhanced activity is due to the small bandgap, leading to extending the photoabsorption capability and reducing the recombination rate of photoexcited electron-hole pairs through the formation inner energy state of the rare earth metals.

Figure 18. Schematic illustration of Er & Y doped $BiVO_4$ synthesis.
Adapted with permission from Ref. (Moscow et al., 2022) (Copyright 2022, Elsevier)

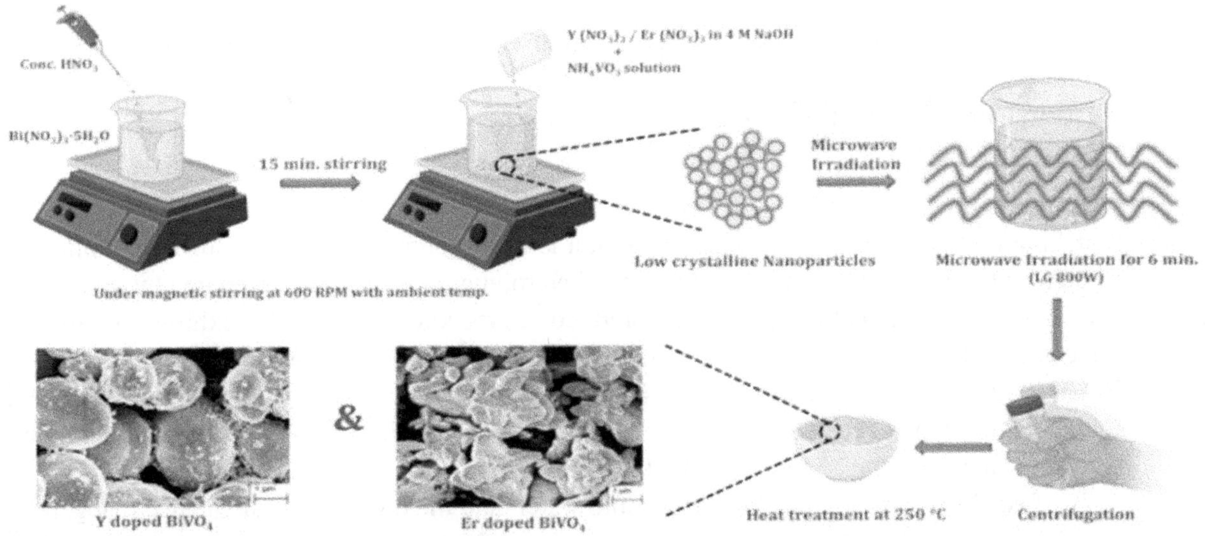

To produce H_2 from water splitting, coating a protective agent or promoter on the surface of the photocatalyst is a significant challenge. In this regard, (Qiu et al., 2022) used hydrothermal and precipitation methods, Figure 19, to prepare CdS@$Mg(OH)_2$ core/shell composite nanorods photocatalyst by coating $Mg(OH)_2$ on CdS surface. Also, Pt/CdS@$Mg(OH)_2$ composite is prepared by the photo deposited method. Under visible light, the H_2 production rate of CdS@$Mg(OH)_2$ (283.4 μmol g-^{1h}-$^{1)}$ was 3.3 times that of CdS nanorods (87 μmol g-1^{h}-1$^{)}$. Moreover, this composite can exhibit such greater H2 $_{p}$roduction rate (239.6 μmol g-1h^{-1}) ^{u}p to the fourth cycle (20 h) which about > 8 times compared to the pure CdS (29.2 μmol g-1h-$^{1)}$. ^{A}dditionally, the H2 pr$_{o}$duction rate and stability of Pt/CdS@Mg(OH)2 we$_{r}$e 3.3 and 2.4 times that of the Pt/CdS. The enhanced H2 pr$_{o}$duction performance can be attributed to the coated Mg(OH)2 in$_{t}$erfacial nanolayer, which not only passivates the surface of CdS semiconductor, reducing defect density and improving stability, but also provides a hydrophilic surface for CdS to better contact with water.

Figure 19. Schematic illustrations of the synthesis processes of CdS@Mg(OH)$_2$ photocatalyst.
Adapted with permission from Ref. (Qiu et al., 2022) (Copyright 2022, Elsevier)

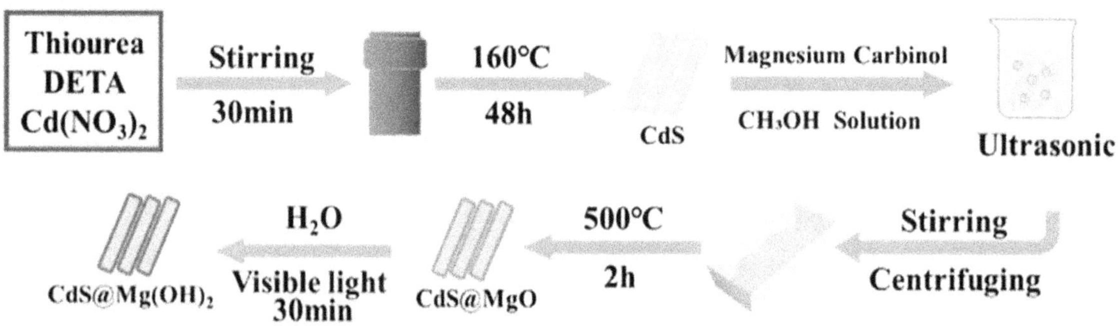

The matching of HER/OER process is a significant issue for photocatalytic overall water splitting. The prepared MnOx/$_g$-C3N$_4$/$_C$dS/Pt hollow core–shell heterojunction composite; prepared by (Pan et al., 2021) via a continuous chemical-annealing-photoreduction procedure (Figure 20); exhibits an obvious HER enhancement of about ~ 80 folds than single CdS. A significant overall water splitting activity of about 1303.39(H2)$_/$ 641.60(O2) μmolg-1^{h-}1 ^{w}ith suitable photocatalytic stability can be mainly ascribed to the remarkable matching of HER/OER process via Pt/MnOx, formation of g-C3N$_4$/$_C$dS heterojunction and hollow core–shell structure. Moreover, the Pt nanoparticles with quickly charge carriers transport can accept/ diffuse the photo-generated electrons into water, and the high chemical valance manganese ions can excite the hole related species and accelerate the H2O$_2$ $_d$ecomposition. The formation of g-C3N$_4$/$_C$dS heterojunction can promote photo-generated carriers separation/transport via the appropriate potential, and the hollow structure can increase the HER/OER active sites, solar efficiency. Additionally, the spherical structure can improve the physical stability and the thin heterojunction can restrain the photo-corrosion to improve the stability.

Figure 20. The schematic diagram of MnO$_x$/g-C$_3$N$_4$/CdS/Pt hollow core–shell heterojunction.
Adapted with permission from Ref. (Pan et al., 2021) (Copyright 2021, Elsevier)

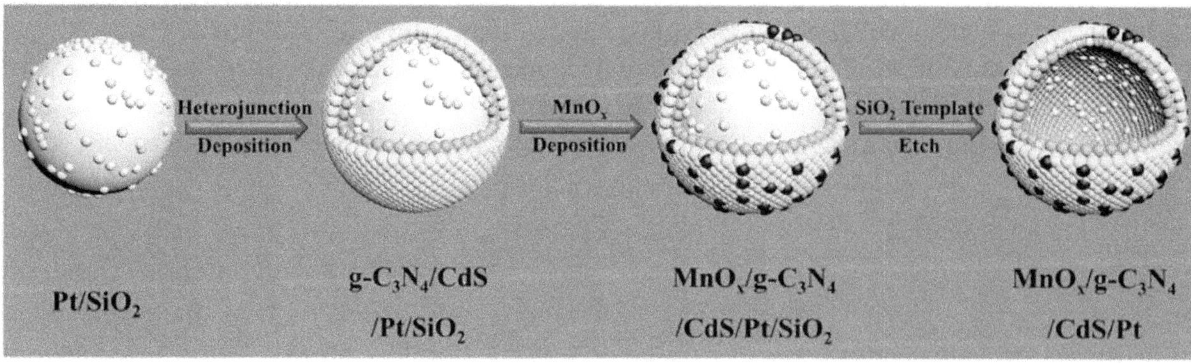

The incorporation of a possible electron acceptor and reservoir r-GO is an excellent choice of material to maintain e - h pairs separated for a long period by defeat the charge carrier recombination. In this regard, (Sharmin et al., 2021) used a simple hydrothermal route to synthesis r-GO incorporated Fe/MgO nanocomposites (Fe/ MgO-r-GO) followed by thermal treatment (400 ºC or 500 ºC) under Argon (Ar) as an inert atmosphere (Figure 21) to evaluate their performance towards photodegradation of RhB dye and photocatalytic H_2 production via water splitting. The Fe/MgO nanoparticles are well dispersed throughout the r-GO as elucidated by TEM. Under solar irradiation, the Fe/ MgO-r-GO (500 ºC) exhibited the highest RhB photodegradation (86%) which is higher than that of commercially Degussa P25 TiO_2 and also is more than 3.4 times that of MgO (25%). Also, the Fe/ MgO-r-GO (500 ºC) sample can recycled up to four cycles with trivial loss in activity which may be according to the loss in catalyst through the recovery process. Additionally, the H_2 production of Fe/MgO-r-GO (500 ºC) after four hours was 1741 $\mu molg^{-1}$ which is about four times that of P25 (445 $\mu molg^{-1}$). This favorable photocatalytic performance of the Fe/MgO-r-GO (500 ºC) can be attributed to the improved morphological and surface features due to heat treatment under inert atmosphere as well as deteriorated charge carrier separation with increased light absorption capacity imputed to r-GO incorporation.

Figure 21. A schematic diagram of the synthesis scheme for r-GO supported Fe/MgO nanocomposites.
Adapted with permission from Ref. (Sharmin et al., 2021) (Copyright 2021, Elsevier)

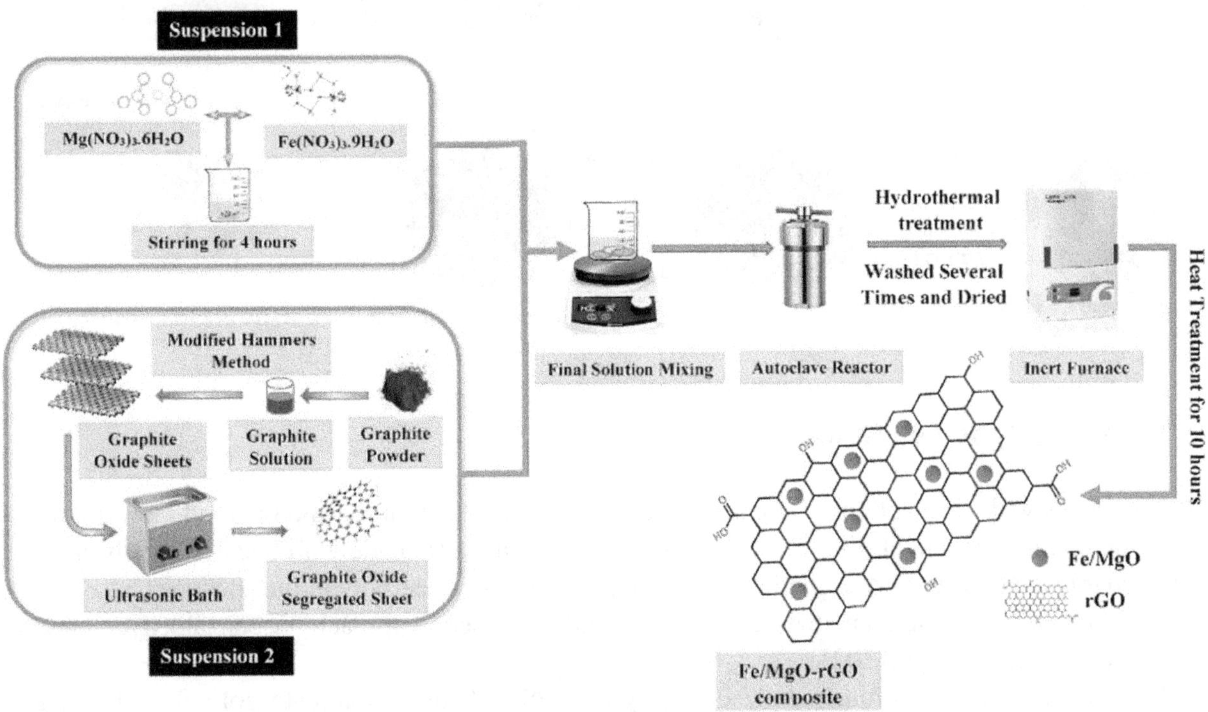

Photocatalytic water splitting is one of the prospective green energy technologies, particularly, for hydrogen production from wastewater under natural sunlight irradiation. (Sekar et al., 2022) prepared biomass activated carbon (b-AC)-anchored anatase titanium oxide (a-TiO_2) nanocomposites (a-TiO_2/b-AC) ultrasonically using the sol-gel-a-TiO_2 nanoparticles and the KOH-activated citron-derived b-AC nanoflakes, Figure 22. The a-TiO_2/b-AC has a high surface area (495 m^2/g) with mesoporous characteristics (pore diameter = 4.59 nm, pore surface area = 153m^2/g, pore volume = 0.5636 cm^3/g) and an aggregated morphology with the interconnected spherical nanoparticles. As a result of these characteristics, under solar irradiation (740 W/m^2), the a-TiO_2/b-AC hybridized nanocomposite exhibited high photocatalytic H_2 production rate (400 mL h^{-1}) from sulphide (0.15 M) wastewater splitting.

Figure 22. Synthesis procedures for the fabrication of the a-TiO_2/b-AC nanocomposites.
Adapted with permission from Ref. (Sekar et al., 2022) (Copyright 2022, Elsevier)

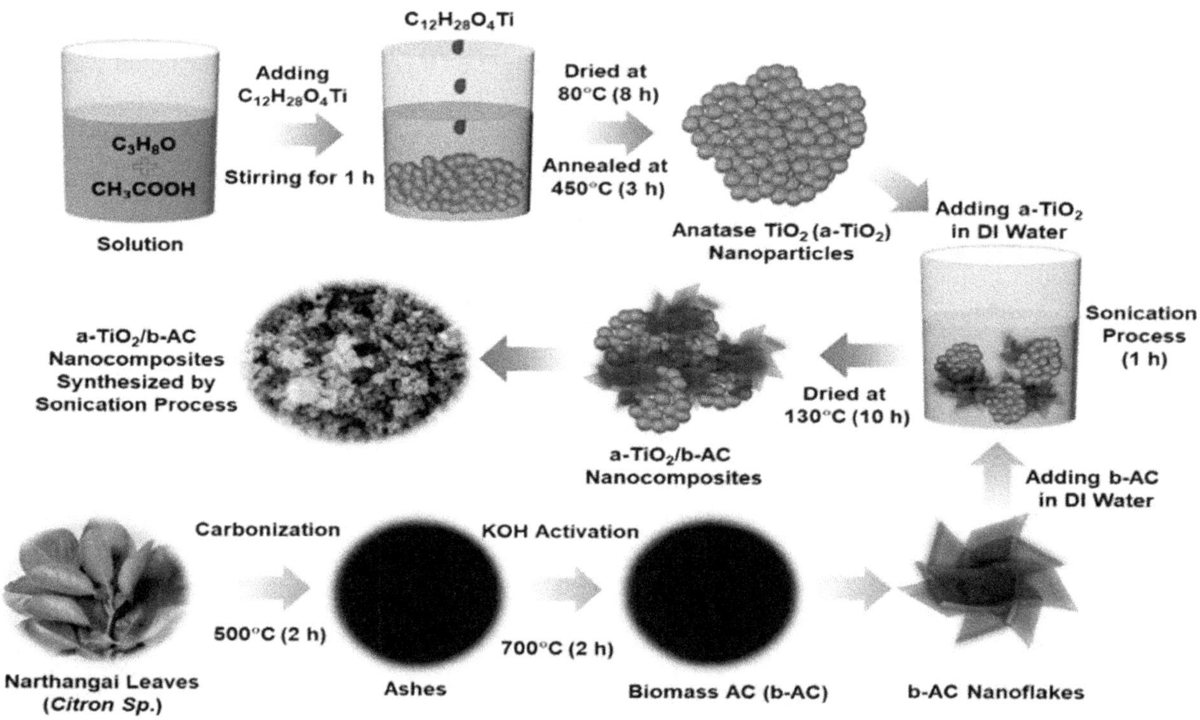

4. Photocatalytic Oxidation

Because of its wide range of applications in medicine, chemical synthesis, and the food industry, benzaldehyde is a highly sought-after chemical. However, it is usually made with hazardous solvents like trifluoro toluene or acetonitrile, and its conversion, particularly selectivity in the aqueous phase, is still lacking. As a result, finding an environmentally friendly synthetic approach for benzaldehyde production is critical (L. Wu et al., 2021).

(S. Li et al., 2018) used functionalized polystyrene (PS) spheres as template, sol-gel, hydrothermal reaction, and calcination to prepare double-shelled and sandwiched TiO_2@Pt@CeO_2 hollow spheres, Figure 23, for photo-reduction of Cr(VI) and selective photo-oxidation of benzyl alcohol into benzaldehyde in the presence of water under solar light at ambient conditions. The Pt nanoparticles (NPs) were controllably loaded between the TiO_2 and CeO_2 shells and the morphology of the hollow TiO_2@Pt@CeO_2 was uniform. Under solar-light irradiation, the rate of Cr(VI) photoreduction was 0.992, 1.040, 1.424 and 1.901 μmol h-^{1} for CeO2, TiO2, TiO2@CeO2 and TiO2@Pt@CeO2, respectively. Also, TiO2@Pt@CeO2 exhibited the highest conversion rate of benzyl alcohol. This amazing behaviour could attribute to the combination of TiO2 and CeO2 as photocatalyst and oxygen buffer, the double-shelled and sandwiched nanostructure, and the addition of Pt NPs as cocatalyst and electron trap site, which could store and shuttle photo-generated electrons, reduce the recombination of the electron-hole, and then enhance photo-generation of active radicals. This conclusion was verified by the electron paramagnetic resonance (EPR) spectroscopy.

Figure 23. A general "template + sol-gel + calcination" method for preparing the double-shelled TiO_2@CeO_2 and TiO_2@Pt@CeO_2 nanocomposites.
Adapted with permission from Ref. (S. Li et al., 2018) (Copyright 2018, Elsevier)

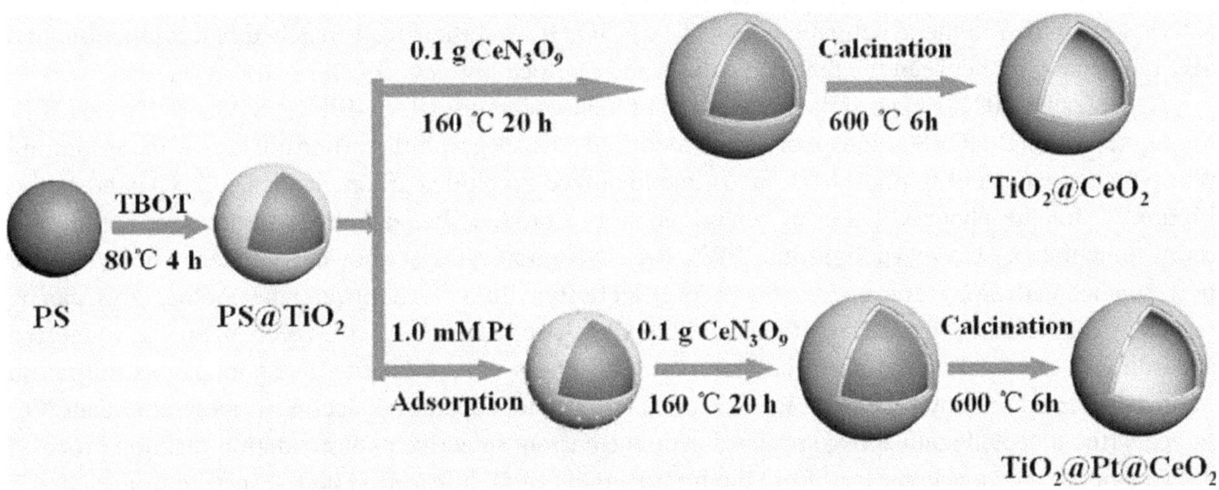

(Cerdan et al., 2019) used a facile onepot mechanochemical method to synthesize graphitic carbon nitride (CN) enriched with ZnO or Fe_2O_3 photocatalysts without the production of any potentially hazardous waste for selective photooxidation of benzyl alcohol to benzaldehyde. Both composite materials possess an enhanced activity and benzaldehyde selectivity compared to pure CN as a result of the inclusion of oxides phases. The most active composite was Fe_2O_3/g-C_3N_4 (Fe/CN) with 20% conversion and 70% benzaldehyde selectivity which is considerably high compared to the pure CN and TiO_2 P25 commercial reference. This enhanced activity is due to the outstanding characteristics. In which, the amorphous structure could be modifying the charge handling of CN, limiting the CO_2 photo-production due to the complete mineralization of the alcohol. On the other hand, SEM and XPS clearly indicated that the enhancement of the activity and selectivity to benzaldehyde was associated with the concentration of the metal component on the surface, as well as with the homogeneity in the structure. Analysis of charge carrier transfer pathways indicate that performance of samples is intimately related to a dominant Z-scheme route. This route increases the available active holes of the oxide component and electrons of the CN mainly for the Fe/CN composite sample.

(Hosseini & Amoozadeh, 2018) synthesized TiO_2-P25-SO_3H as a covalently grafted solid acid nano photocatalyst for the selectively oxidation of different aromatic alcohols to the corresponding carbonyl compounds (aldehydes) in the presence of nitrobenzene as available industrial oxidant with excellent yields under visible light irradiation (blue LED, λ>420 nm). The successful immobilization of sulfonic acid groups was evaluated by FT-IR and TGA. The XRD presented crystallinity retention of nano-TiO_2-P25 after treatment by acidic reagent. SEM results confirmed nanometersized particles of the photo-catalyst and the presence of Ti, S and O atoms was demonstrated in the EDX spectrum. This method has some notable advantages such as operational simplicity, excellent photo-catalytic performance, mild reaction condition, simple work-up without column chromatographic purification, environmental friendly and reusability of the photo-catalyst for five consecutive cycles with consistent activity.

TiO_2(B) is typical crystal structure for TiO_2, besides anatase, rutile and brookite. (Du et al., 2020) prepared one-dimensional TiO_2(B) modified by Pd nanowires (Pd/TiO_2(B)) for photo-thermocatalytic benzyl alcohol (BzOH) oxidation. Well-matched heterostructure contact boundary between (1 1 1) planes of Pd nanoparticles and (0 0 3) planes of TiO_2(B) was formed, which might facilitate interfacial photo-generated electrons transfer. The Pd/TiO_2(B) catalyst exhibited enhanced the photo-thermocatalytic activity, attributed to the synergistic effect of Pd, TiO_2(B) and their well-matched heterojunction, and also the synergism between the thermocatalysis and photocatalysis.

(L. Wu et al., 2021) reported the preparation of a photocatalyst (PW_{12}-P-UCNS, where PW_{12} is $H_3PW_{12}O_{40}xH_2O$ and P-UCNS is phosphoric acid-modified unstack graphitic carbon nitride) by incorporating phosphor-tungstic acid on phosphoric acid-functionalized graphitic carbon nitride (g-C_3N_4) nanosheets, Figure 24, for the photooxidation of benzyl alcohol to produce benzaldehyde in aqueous medium at room temperature. Under sunlight, the PW_{12}-P-UCNS photocatalyst showed excellent photocatalytic performance with 58.3% conversion and 99.5% selectivity within two hours and the photocatalyst can be reused up to five times without significant activity loss. A proposed Z-scheme mechanism was suggested according to these reasons; firstly, the incorporated Keggin units facilitate the spontaneous migration of electrons, and thereby accelerate interfacial charge carrier separation; secondly, more abundant $•O_2^-$ radicals (the detected main active species during the current selective photooxidation reaction process) are generated, which is beneficial from the improvement of O_2 adsorption on the surface of P-UCNS.

Figure 24. Illustrated representation of the synthetic process for the PW12-P-UCNS.
Adapted with permission from Ref. (L. Wu et al., 2021) (Copyright 2021, Elsevier)

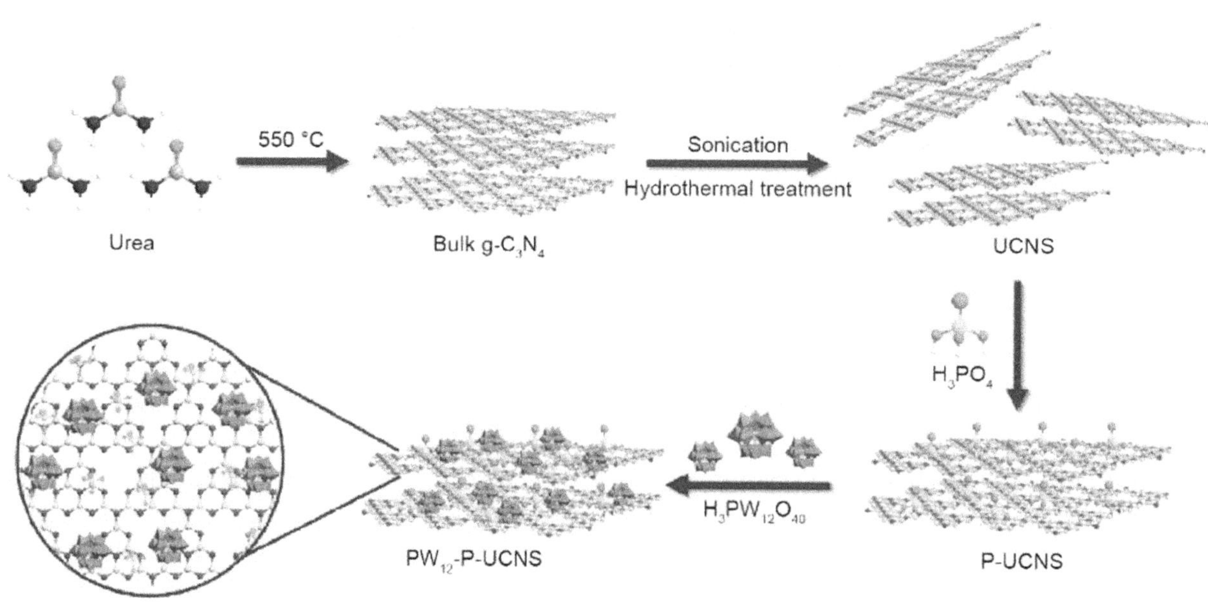

(Mandal et al., 2022) synthesized a new plasmonic Ag NPs coupled mixed ligand Ag(I) based coordination polymer (CP), (Ag@Ag(I)-CP (40%) NH_2) by a one pot synthetic route for photocatalytic oxidation of benzyl alcohol (BA) into benzaldehyde (BD). Under visible light, the Ag@Ag(I)-CP(40% NH_2) achieved enhanced performances toward selective BA oxidation reaction (Conversion rate of BA: k (O_2) = 7.4×10^{-4} min^{-1}; yield: 19.1% BD, and k ($K_2S_2O_8$) = 38.7×10^{-4} min^{-1} with yield: 54.1% BD) which is about 1.4 time, 2.2 time faster compared to that of Ag(I)-CP (40% NH_2) alone, and Ag@Ag(I)-CP, respectively. This enhanced performance is obtained by the synergistic action of two counter parts Ag(I)-CP (40%) NH_2 and Ag NPs. In one side, amine functionalized ligand makes the mixed ligand CP alone a visible light active component and produces a HOCO state (+ 2.4 V *vs.* NHE) which is very selective to initiate hole mediated oxidation of BA to BD ($E^0_{BA/BD}$ = + 1.9 V), whereas in another side Ag NPs (~ 6–7 nm) increases the charge carrier density and lifetime, and promotes the efficiency of adsorption/reduction of oxidants (O_2/$K_2S_2O_8$) in the coupled structure through LSPR and Schottky junction effect. Besides, the coupled structure achieved enhanced stability with respect to reuse than of their mixed ligand CP counterparts. Moreover, it would be an alternate and cheap candidate competing with Ag NPs supported system and precious Au, and Pt NPs metal-based photo-redox systems.

Vanadium oxide supported gold nanocatalysts designated as AuVOx was found to be highly efficient for photocatalytic oxidation of benzyl alcohol and 2-naphthol. The photocatalytic ability of the AuVOx catalyst towards the selective oxidation of 2-naphthol to (Bi naphthol) BINOL was evaluated by (Das et al., 2020) and found to be dependent on the nature of the support. The selectivity of the reaction was found to be significantly altered on hybridization with multi-walled carbon nanotubes (AuVOx-MWCNTs). The well-resolved surface plasmon resonance (SPR) band originated from the gold nanoparticles (Au-NPs) on the vanadium oxides support helped in stimulating the photo-oxidation of benzyl alcohol to benzaldehyde and 2-naphthol to BINOL with almost 100% selectivity. The most notable achievement using the AuVOx and AuVOx-MWCNTs catalyst was in the successful photo-oxidation of 2-naphthol to BINOL in presence of hydrogen peroxide (H_2O_2). The photoactivity of the material and the creation of oxygen defect sites participated in the photochemical processes were estimated from photoluminescence analysis and ESR study. The photochemical reaction was supposed to proceed via superoxide radical generation favored by the oxygen defect sites.

(W. Wang et al., 2022) synthesized Z-scheme heterojunction photocatalysts formed by $FAPbBr_3$ and WO_3 with different morphologies; nanocrystals (NCs) and quantum dots (QDs). Flower-like WO_3 with outstanding activity was exploited to respectively support $FAPbBr_3$ NCs and $FAPbBr_3$ QDs, in order to demonstrate the quantum confinement effect on the photocatalytic benzyl alcohol oxidation. Benefiting from the strong redox potential and morphological engineering, the Z-scheme heterojunction photocatalyst ($FAPbBr_3$ NCs) exhibits an outstanding performance of benzyl alcohol oxidation to benzaldehyde (49.6% conversion rate with 99% selectively). Remarkably, after manipulating the size of $FAPbBr_3$ NCs to quantum dots (QDs), benzoic acid as the main product (65% conversion rate with 90% selectivity) of benzyl alcohol oxidation can be generated on the Z-scheme photocatalyst. As revealed by in-situ electronic paramagnetic resonance (EPR), a new radical will be generated over the $FAPbBr_3$ QDs/WO_3 which alters the product of benzyl alcohol oxidation form benzaldehyde to benzoic acid.

5. Photocatalytic CO_2 Reduction

Carbon dioxide (CO_2) concentrations in the atmosphere are rising as a result of fossil fuel use, raising severe worries about the greenhouse effect and global warming (Figueroa et al., 2008). Solar photocatalytic conversion of CO_2 with H_2O into hydrocarbon fuels by semiconductors has generated a new sustainable development path, as it will assist lower CO_2 levels in the atmosphere while also partially meeting energy demands (Y. Li et al., 2010; Varghese et al., 2009).

The most principal process supporting life on Earth is photosynthesis, in which plants use sunshine to convert CO_2 and water into glucose ($C_6H_{12}O_6$). Researchers have been inspired by this process to investigate various forms of artificial photosynthesis to store solar energy as chemical energy, such as converting CO_2 and water into carbohydrates or hydrocarbons, but the photosynthetic efficiency has been too low for practical applications thus far. Such experiments are still limited to creating short-chain hydrocarbons or carbohydrates, and solar-to-chemical energy conversion efficiencies are one to two orders of magnitude lower than natural photosynthesis's greatest efficiency (approximately 7 percent) (Adachi et al., 1994; Inoue et al., 1979).

Additionally, the rapid charge recombination, poor light absorption and low selectivity still limit the efficiency of photocatalytic CO_2 reduction.

The elementary general mechanism reactions involved during CO_2 photo-reduction to methane are summarized by reactions (Eq. 1) to (Eq. 9) (Habisreutinger et al., 2013):

$CO_2 + e^- \rightarrow {}^\bullet CO_2^-$ Eq. 1

${}^\bullet CO_2^- + H^+ + e^- \rightarrow CO + OH^-$ Eq. 2

$CO + e^- \rightarrow {}^\bullet CO^-$ Eq. 3

${}^\bullet CO^- + H^+ + e^- \rightarrow C + OH^-$ Eq. 4

$C + H^+ + e^- \rightarrow {}^\bullet CH$ Eq. 5

${}^\bullet CH + H^+ + e^- \rightarrow {}^\bullet CH_2$ Eq. 6

${}^\bullet CH_2 + H^+ + e^- \rightarrow {}^\bullet CH_3$ Eq. 7

${}^\bullet CH_3 + H^+ + e^- \rightarrow CH_4$ Eq. 8

$2H_2O + 4h^+ \rightarrow 4H^+ + O_2$ Eq. 9

(Y. Wang et al., 2013) synthesized ordered mesoporous CeO_2-TiO_2 composites with 2D hexagonal structure and varied compositions via a facile nanocasting route using ordered mesoporous SBA-15 as a template. The prepared CeO_2-TiO_2 composites possess high specific surface area and hierarchical porosity with ordered 2D hexagonal meso-structures as elucidated by N2 adsorption-desorption, TEM and XRD. These CeO_2-TiO_2 composites had excellent photocatalytic activity toward methane (CH_4) production via CO_2 reduction with H_2O under simulated solar irradiation as a result of the peculiar composition and structural characteristics. Moreover, the addition of CeO_2 species can effectively extend the spectral response from UV to visible area and enhance the surface chemisorbed oxygen of the ordered mesoporous TiO_2 as confirmed by UV-vis and XPS analysis.

(C. Wang et al., 2020) synthesized cobalt/cobalt oxide (Co/CoO) nanostructured photo-catalyst for artificial photosynthesis via a facile low-cost route. The prepared Co/CoO photocatalyst was used for the conversion of CO_2 into different ranging hydrocarbons from propane (C_3H_8) to pentadecane ($C_{15}H_{32}$) at an average solar-to-chemical energy conversion efficiency>0.1%. The distribution of hydrocarbons follows a Poisson distribution. Experiments using 13C-labeled CO_2 and theoretical calculations reveal strong adsorption of CO_2 on the nanostructured Co/CoO surface. The photon energy of visible light is large enough to excite the adsorbed CO_2 molecules into an unstable state that reacts with dissociated H_2O to form long-chain hydrocarbons. Because CO_2 can be adsorbed from air directly onto the CoO, this catalyst does not require purified CO_2, and thus enables the low-cost production of diesel-range hydrocarbons with solar energy and CO_2 adsorbed from ambient air or unpurified CO_2 sources.

In photo-catalysis, noble metals commonly act as co-catalysts, while the rarity limited their practical applications. In this regard, transition metal sulfide was a good substitute for noble metals owing to their peculiar photo-electrochemical properties. For this reason, (F. Li et al., 2021) shed light on the enhancement of photo-catalytic CO_2 reduction based on bimetallic co-catalyst loaded on g-C_3N_4. They used a facile hydrothermal method to couple bimetallic CoNiSx as a co-catalyst with g-C_3N_4 to form CoNiSx-CN composite with enhanced photocatalytic CO_2 reduction activity with high selectivity to CO production. The production rate of CH_4 and CO in this reaction can reach 0.904 and 11.77 μmol g^{-1} h^{-1}, correspondingly, where the CO yield is more than 10-fold comparable to solely CN. The excellent performance of the composite can be attributed to CoNiSx synergy effect. Due to the potential difference, electrons are more likely to migrate from the conduction band of CN to the conduction band of CoNiSx, inhibiting the recombination of photogenerated electrons and holes, thereby enhancing photocatalytic performance, Figure 25. Furthermore, the thermal performance of CoNiSx can absorb visible light and near-infrared light and convert it into heat. The good thermal conductivity can help increase the temperature of the reaction system and further promote the photocatalytic reaction.

Figure 25. Mechanism of photo-catalytic reduction of CO_2 by CoNiSx-CN.
Adapted with permission from Ref. (F. Li et al., 2021) (Copyright 2021, Elsevier)

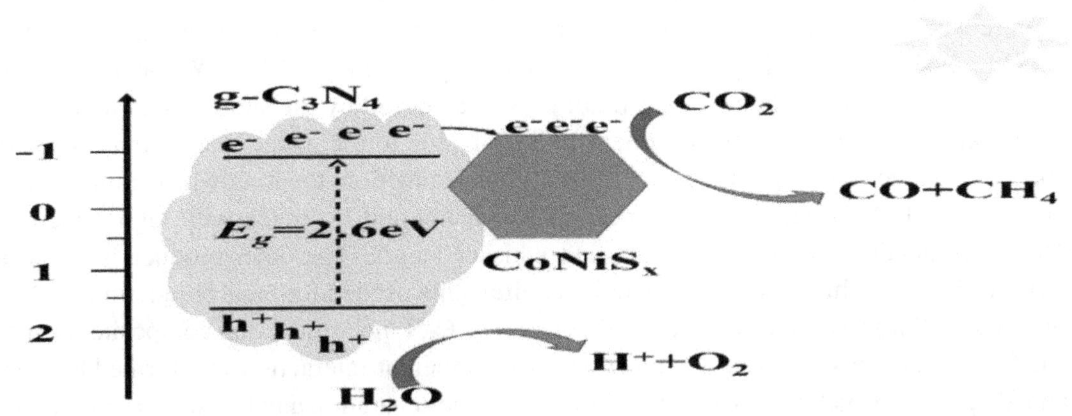

(Moradi et al., 2020) used sol-gel and photo-deposition routes to prepare Pt@Bi-TiO_2 photocatalysts with different Bi (0, 1, 3, 5 wt%) and Pt (0, 1, 1.5, 2 wt%) contents for the photo-reduction of CO_2 in to CH_4. The introduction of Bi into TiO_2 structure and successive loading of Pt on its surface significantly increased the CH_4 yield from CO_2 photo-reduction. The 1.5Pt@3Bi-TiO_2 sample exhibited the highest CH_4 yield (20.6 μmol/g) which was higher than pristine TiO_2 (3.3 μmol/g) by 6.2 times. Physicochemical features of the synthesized photocatalysts indicated that suppression of the recombination of charge carriers, improvement of CO_2 adsorption, oxygen vacancy formation upon Bi doping and co-catalytic features of Pt nanoparticles, enhanced the CO_2 photo-conversion efficiency. The results from this investigation demonstrated that Pt and Bi co-modified TiO_2 was an affordable way of developing CH_4-selective photocatalyst for CO_2 photo-reduction. Moreover, the stability test showed the optimal catalyst (1.5Pt@3Bi-T) could produce 17.9 μmol/g of CH_4 which preserve 87% of the initial methane yield after six successive runs. Besides, a linear correlation between the square root of light intensity and CH_4 yield was found.

(Y. Yang et al., 2022) reported that a mechanical mixture of the as-thermally polymerized melamine (Me)- and 3-amino-1,2,4-triazole (AT)-based carbon nitride (MCN and TCN) can turn into a homojunction material (MTCN) photocatalyst by a facile and effective re-calcination under air. Under visible light, MTCN exhibits the most outstanding performance on tetracycline (TC) degradation and excellent photo-catalytic efficiency for the CO_2 reduction under normal temperature and pressure which can reach 7.75 $\mu mol g^{-1} h^{-1}$ (about 13.8 or 16.1 times higher than the single MCN or TCN). This developed MTCN efficiency is due to the enhanced porosity and photoelectric conversion efficiency. Moreover, the nitrogen-rich structure not only broadens the visible light response range but also provides stronger Lewis basic sites for the adsorption and activation of CO_2 at the surface.

(Nematollahi et al., 2020) used sol-gel method to synthesize a series of Ni (0.5, 1, 1.5 wt%) and Bi (1,3,5 wt%) doped TiO_2 catalysts for CO_2 photoreduction under visible light irradiation. The doping of Ni and/or Bi in the TiO_2 framework increased the specific surface area and enhanced the light absorption toward the visible region with narrower band gap energy compared to pristine TiO_2. Moreover, this doping inhibited the electron-hole recombination rate. Bi doped TiO_2 increased the CO_2 adsorption capacity compared to pristine TiO_2 and Ni doped sample. The maximum CH_4 yield 21.13 $\mu mol/g_{cat}$ (about 6.5 times of pure TiO_2, 3.26 $\mu mol/g_{cat}$) was obtained by TNi_1Bi_3, further doping decreased the CH_4 yield.

(Garay-Rodríguez et al., 2019) used conventional sol-gel method, sono-chemical assisted, and microwave assisted sol-gel method to synthesize $K_2Ti_6O_{13}$ whiskers catalysts with different particle sizes and modified optical, textural, and electrochemical properties for H_2 production and CO_2 photo-reduction. By using pure water as reactant, the highest H_2 production rate of 10.065 μmol g^{-1} is obtained by the long $K_2Ti_6O_{13}$ whiskers prepared by ultrasonic assisted sol-gel method (U-SG). While using a mixture of ethanol-water, the highest rate of 3.2871 μmol g^{-1} was achieved by the shortest and less crystalline $K_2Ti_6O_{13}$ whiskers (C-SG). This enhanced performance is due to the lower charge transfer resistance and the reduced charge recombination rate. For CO_2 photo-reduction, the long whiskers were the most active to transform CO_2 to formaldehyde, methane, methanol and hydrogen with high selectivity for formaldehyde production due to crystallinity and whisker length. The other products were in lower concentrations due to the higher thermodynamic requirements needed for their production.

(Huang et al., 2020) developed a novel 2D/2D Bi_2Se_3/g-C_3N_4 heterojunction composite for the photocatalytic CO_2 reduction selectively. Fortunately, strong inherent interaction was formed between 2D Bi_2Se_3 and 2D g-C_3N_4, which is very beneficial for the charge separation and transfer. In this composite, Bi_2Se_3 acts as a co-catalyst with suitable fermi level and tightly combined with CN so that the charge separation of CN could be apparently improved. Additionally, the absorbed visible and NIR light by Bi_2Se_3 could be largely converted into heat, which further made a positive contribution to the CO_2 reduction. The synergetic effect of highly improved charge separation and strong photo-to-heat conversion efficiency of the 2D/2D Bi_2Se_3/g-C_3N_4 composite led to a superior photocatalytic CO_2 reduction to CO, which is about 6.3 times higher than that of pristine CN under the irradiation of full spectrum (UV, visible and NIR light). Remarkably, the product was nearly 100% CO without other carbon products, and the selectivity for CO was greater than 90% over H_2 evolution.

Methanol photosynthesis via CO_2 reduction is challenging and important because this technology can convert CO_2 by solar- or wind-generated H_2 into liquid fuel. For this purpose, (Z. jun Wang et al., 2019) prepared plasmonic Cu/ZnO catalysts for methanol synthesis as a liquid fuel via CO_2 reduction under visible light (solar energy) and atmospheric pressure. Under visible light irradiation, the methanol production rate was promoted by 54% and increased from 1.38 to 2.13 µmol g^{-1} min^{-1} while the apparent activation energy was reduced by 40% and decreased from 82.4 to 49.4 kJ mol^{-1}. Based on experimental and theoretical studies, the hot electrons were photo-excited by localized surface plasmon resonance (LSPR) on Cu nanoparticles and such photo-excited hot electrons could transfer to ZnO through the metal-support interfaces. The hot electrons on Cu and ZnO synergistically facilitated the activation of reaction intermediates.

CONCLUSION AND FUTURE PROSPECTIVE

Smart nanomaterials have received a lot of interest as one of the promising materials with features tailored to alter in a controlled manner under the effect of external stimuli. Photocatalysts are smart nanomaterials that, depending on their bandgap, act as stimuli when exposed to light, either UV or visible light. These photocatalysts are crucial for a variety of photocatalytic applications, including the treatment of environmental pollutants, hydrogen production, and photosynthesis via alcohol oxidation or CO_2 reduction. Photocatalysis has been shown to be an effective way to reduce pollutants while also producing energy. We can improve the efficiency of photocatalysis using a variety of ways. Coupling semiconductors with other materials like graphene, doping catalysts, producing hetero-structure junctions, and precipitation methods are only a few of them. All of these techniques improve the photocatalysis process' efficiency. Despite its remarkable accomplishments, photocatalysis still faces numerous hurdles in the development of highly effective photocatalytic systems. Charge separation, optical absorption range, and bandgap position are all issues. Furthermore, there are morphological issues that can be managed in a variety of methods. Finally, the photocatalyst's stability is a problem for many solid and practical applications. As a result, new materials with greater visible light absorption must be designed and developed in order to improve the efficiency of photocatalysts and the photocatalytic process. Photocatalysis for the treatment of water and wastewater is gaining popularity around the world. The majority of the studies documented in the literature are still in the laboratory research stage; however, some have been successful in establishing pilot plants. Nanomaterials' diverse features, together with new treatment procedures, give huge opportunities for improving water and wastewater treatment.

REFERENCES

Abou-Hadid, A. F., Usama, El-Behairy, A., Elmalih, M. M., Amdeha, ·E., Ahmed, El Naggar, M. A., Taha, M. H., & Hussein, E. M.(2022). Production of efficient carbon fiber from different solid waste residuals for adsorption of hazardous metals from wastewater samples. *Biomass Conversion and Biorefinery*, *1*, 1–16. doi:10.1007/S13399-022-03097-6

Abualnaja, K. M., ElAassar, M. R., Ghareeb, R. Y., Ibrahim, A. A., & Abdelsalam, N. R. (2021). Development of photo-induced Ag0/TiO_2 nanocomposite coating for photocatalysis, self-cleaning and antimicrobial polyester fabric. *Journal of Materials Research and Technology*, *15*, 1513–1523. doi:10.1016/j.jmrt.2021.08.127

Adachi, K., Ohta, K., & Mizuno, T. (1994). Photocatalytic reduction of carbon dioxide to hydrocarbon using copper-loaded titanium dioxide. *Solar Energy*, *53*(2), 187–190. doi:10.1016/0038-092X(94)90480-4

Ahlawat, J., & Narayan, M. (2020). Introduction to Active, Smart, and Intelligent Nanomaterials for Biomedical Application. *Intelligent Nanomaterials for Drug Delivery Applications*, 1–16. doi:10.1016/b978-0-12-817830-0.00001-1

Amdeha, E. (2021). Recovery of Nanomaterials from Agricultural and Industrial Wastes for Water Treatment Applications. In A. S. H. Makhlouf & G. A. M. Ali (Eds.), *Topics in Mining, Metallurgy and Materials Engineering* (pp. 385–417). Springer. doi:10.1007/978-3-030-68031-2_14

Amdeha, E., Mohamed, R. S., & Dhmees, A. S. (2021). Sonochemical assisted preparation of ZnS–ZnO/MCM-41 based on blast furnace slag and electric arc furnace dust for Cr (VI) photoreduction. *Ceramics International*, *47*(16), 23014–23027. doi:10.1016/j.ceramint.2021.05.015

Arif, M., Muhmood, T., Zhang, M., Amjad Majeed, M., Honglin, Y., Liu, X., & Wang, X. (2022). Highly visible-light active, eco-friendly artificial enzyme and 3D $Bi_4Ti_3O_{12}$ biomimetic nanocomposite for efficient photocatalytic tetracycline hydrochloride degradation and Cr(VI) reduction. *Chemical Engineering Journal*, *434*, 134491. doi:10.1016/j.cej.2021.134491

Bagherzadeh, S. B., Kazemeini, M., & Mahmoodi, N. M. (2021). Preparation of novel and highly active magnetic ternary structures (metal-organic framework/cobalt ferrite/graphene oxide) for effective visible-light-driven photocatalytic and photo-Fenton-like degradation of organic contaminants. *Journal of Colloid and Interface Science*, *602*, 73–94. doi:10.1016/j.jcis.2021.05.181 PMID:34118607

Baig, N., Kammakakam, I., Falath, W., & Kammakakam, I. (2021). Nanomaterials: A review of synthesis methods, properties, recent progress, and challenges. *Materials Advances*, *2*(6), 1821–1871. doi:10.1039/D0MA00807A

Cai, C., Wang, R., Liu, S., Yan, X., Zhang, L., Wang, M., Tong, Q., & Jiao, T. (2020). Synthesis of self-assembled phytic acid-MXene nanocomposites via a facile hydrothermal approach with elevated dye adsorption capacities. *Colloids and Surfaces. A, Physicochemical and Engineering Aspects*, *589*, 124468. doi:10.1016/j.colsurfa.2020.124468

Cai, T., Wang, L., Liu, Y., Zhang, S., Dong, W., Chen, H., Yi, X., Yuan, J., Xia, X., Liu, C., & Luo, S. (2018). Ag_3PO_4/Ti_3C_2 MXene interface materials as a Schottky catalyst with enhanced photocatalytic activities and anti-photocorrosion performance. *Applied Catalysis B: Environmental*, *239*, 545–554. doi:10.1016/j.apcatb.2018.08.053

Cerdan, K., Ouyang, W., Colmenares, J. C., Muñoz-Batista, M. J., Luque, R., & Balu, A. M. (2019). Facile mechanochemical modification of g-C_3N_4 for selective photo-oxidation of benzyl alcohol. *Chemical Engineering Science*, *194*, 78–84. doi:10.1016/j.ces.2018.04.001

Chu, Y., Zheng, X., & Fan, J. (2022). Preparation of sodium and boron co-doped graphitic carbon nitride for the enhanced production of H_2O_2 via two-electron oxygen reduction and the degradation of 2,4-DCP via photocatalytic oxidation coupled with Fenton oxidation. *Chemical Engineering Journal, 431*, 134020. doi:10.1016/j.cej.2021.134020

Das, B., Sharma, M., Baruah, M. J., Mounash, B. P., Karunakar, G. V., & Bania, K. K. (2020). Gold nanoparticle supported on mesoporous vanadium oxide for photo-oxidation of 2-naphthol with hydrogen peroxide and aerobic oxidation of benzyl alcohols. *Journal of Environmental Chemical Engineering, 8*(5), 104268. doi:10.1016/j.jece.2020.104268

Demir, M. E., Chehade, G., Dincer, I., Yuzer, B., & Selcuk, H. (2019). Synergistic effects of advanced oxidization reactions in a combination of TiO_2 photocatalysis for hydrogen production and wastewater treatment applications. *International Journal of Hydrogen Energy, 44*(43), 23856–23867. doi:10.1016/j.ijhydene.2019.07.110

Du, M., Zeng, G., Ye, C., Jin, H., Huang, J., Sun, D., Li, Q., Chen, B., & Li, X. (2020). Solvent-free photo-thermocatalytic oxidation of benzyl alcohol on Pd/TiO_2 (B) nanowires. *Molecular Catalysis, 483*, 110771. doi:10.1016/j.mcat.2020.110771

El-Salamony, R. A., Amdeha, E., El Shafey, A. M., & Al Sabagh, A. M. (2021). Preparation and characterisation of Ce-doped SiO_2 nano-materials as effective photo-catalyst under visible light. *International Journal of Environmental Analytical Chemistry*, 1–16. doi:10.1080/03067319.2020.1865328

Fang, B., Xing, Z., Guo, M., Qiu, Y., Cui, Y., Li, Z., Wang, Y., Chen, P., & Zhou, W. (2021). Phosphorus-doping CdS@NiFe layered double hydroxide as Z-Scheme heterojunction for enhanced photocatalytic and photo-fenton degradation performance. *Separation and Purification Technology, 274*, 119066. doi:10.1016/j.seppur.2021.119066

Feng, X., Yu, Z., Long, R., Li, X., Shao, L., Zeng, H., Zeng, G., & Zuo, Y. (2020). Self-assembling 2D/2D (MXene/LDH) materials achieve ultra-high adsorption of heavy metals Ni^{2+} through terminal group modification. *Separation and Purification Technology, 253*, 117525. doi:10.1016/j.seppur.2020.117525

Figueroa, J. D., Fout, T., Plasynski, S., McIlvried, H., & Srivastava, R. D. (2008). Advances in CO_2 capture technology-The U.S. Department of Energy's Carbon Sequestration Program. *International Journal of Greenhouse Gas Control, 2*(1), 9–20. doi:10.1016/S1750-5836(07)00094-1

Frunza, L., Diamandescu, L., Zgura, I., Frunza, S., Ganea, C. P., Negrila, C. C., Enculescu, M., & Birzu, M. (2018). Photocatalytic activity of wool fabrics deposited at low temperature with ZnO or TiO_2 nanoparticles: Methylene blue degradation as a test reaction. *Catalysis Today, 306*, 251–259. doi:10.1016/j.cattod.2017.02.044

Gao, Y., Yan, S., He, Y., Fan, Y., Zhang, L., Ma, J., Hou, R., Chen, L., & Chen, J. (2021). A photo-Fenton self-cleaning membrane based on NH2-MIL-88B (Fe) and graphene oxide to improve dye removal performance. *Journal of Membrane Science, 626*, 119192. doi:10.1016/j.memsci.2021.119192

Garay-Rodríguez, L. F., Torres-Martínez, L. M., & Moctezuma, E. (2019). Photocatalytic performance of $K_2Ti_6O_{13}$ whiskers to H_2 evolution and CO_2 photo-reduction. *Journal of Energy Chemistry, 37*, 18–28. doi:10.1016/j.jechem.2018.11.014

Gopakumar, D. A., Pai, A. R., Pasquini, D., Shao-Yuan, L., Abdul Khalil, H. P. S., & Thomas, S. (2018). Nanomaterials-State of Art, New Challenges, and Opportunities. *Nanoscale Materials in Water Purification*, 1–24. doi:10.1016/B978-0-12-813926-4.00001-X

Green, M. A., & Bremner, S. P. (2016). Energy conversion approaches and materials for high-efficiency photovoltaics. *Nature Materials*, *16*(1), 23–34. doi:10.1038/nmat4676 PMID:27994249

Gu, S., Zhang, D., Luo, S., & Yang, H. (2019). Construction of direct Z-Scheme photocatalyst by $Mg_{1.2}Ti_{1.8}O_5$ and g-C_3N_4 nanosheets toward photocatalytic H_2 production and disinfection. *International Journal of Photoenergy*, *2019*, 1–9. doi:10.1155/2019/6307858

Habisreutinger, S. N., Schmidt-Mende, L., & Stolarczyk, J. K. (2013). Photocatalytic reduction of CO_2 on TiO_2 and other semiconductors. *Angewandte Chemie International Edition*, *52*(29), 7372–7408. doi:10.1002/anie.201207199 PMID:23765842

Hashem, E. M., Hamza, M. A., El-Shazly, A. N., Abd El-Rahman, S. A., El-Tanany, E. M., Mohamed, R. T., & Allam, N. K. (2021). Novel Z-Scheme/Type-II CdS@ZnO/g-C_3N_4 ternary nanocomposites for the durable photodegradation of organics: Kinetic and mechanistic insights. *Chemosphere*, *277*, 128730. doi:10.1016/j.chemosphere.2020.128730 PMID:33189399

Haw, C., Chiu, W., Abdul Rahman, S., Khiew, P., Radiman, S., Abdul Shukor, R., Hamid, M. A. A., & Ghazali, N. (2016). The design of new magnetic-photocatalyst nanocomposites ($CoFe_2O_4$-TiO_2) as smart nanomaterials for recyclable-photocatalysis applications. *New Journal of Chemistry*, *40*(2), 1124–1136. doi:10.1039/C5NJ02496J

Hosseini, S., & Amoozadeh, A. (2018). Nano-TiO_2-P25-SO_3H as a new and robust photo-catalyst: The acceleration effect of selective oxidation of aromatic alcohols to aldehydes under blue LED irradiation. *Journal of Photochemistry and Photobiology A Chemistry*, *364*, 516–523. doi:10.1016/j.jphotochem.2018.06.035

Huang, Y., Wang, K., Guo, T., Li, J., Wu, X., & Zhang, G. (2020). Construction of 2D/2D Bi_2Se_3/g-C_3N_4 nanocomposite with High interfacial charge separation and photo-heat conversion efficiency for selective photocatalytic CO_2 reduction. *Applied Catalysis B: Environmental*, *277*, 119232. doi:10.1016/j.apcatb.2020.119232

Inoue, T., Fujishima, A., Konishi, S., & Honda, K. (1979). Photoelectrocatalytic reduction of carbon dioxide in aqueous suspensions of semiconductor powders [3]. *Nature*, *277*(5698), 637–638. doi:10.1038/277637a0

Iqbal, K., Halsby, K., Murray, R. D., Carroll, P. V., & Petermann, R. (2019). Glucocorticoid management of adrenal insufficiency in the United Kingdom: Assessment using real-world data. *Endocrine Connections*, *8*(1), 20–31. doi:10.1530/EC-18-0418 PMID:30562160

Jafari, T., Moharreri, E., Amin, A. S., Miao, R., Song, W., & Suib, S. L. (2016). Photocatalytic water splitting - The untamed dream: A review of recent advances. *Molecules (Basel, Switzerland)*, *21*(7), 900. doi:10.3390/molecules21070900 PMID:27409596

Khan, I., Saeed, K., & Khan, I. (2019). Nanoparticles: Properties, applications and toxicities. *Arabian Journal of Chemistry*, *12*(7), 908–931. doi:10.1016/j.arabjc.2017.05.011

Khanna, P., Kaur, A., & Goyal, D. (2019). Algae-based metallic nanoparticles: Synthesis, characterization and applications. *Journal of Microbiological Methods*, *163*, 105656. doi:10.1016/j.mimet.2019.105656 PMID:31220512

Kim, S. J., Koh, H. J., Ren, C. E., Kwon, O., Maleski, K., Cho, S. Y., Anasori, B., Kim, C. K., Choi, Y. K., Kim, J., Gogotsi, Y., & Jung, H. T. (2018). Metallic $Ti_3C_2T_x$ MXene Gas Sensors with Ultrahigh Signal-to-Noise Ratio. *ACS Nano*, *12*(2), 986–993. doi:10.1021/acsnano.7b07460 PMID:29368519

Kuang, P., He, M., Zhu, B., Yu, J., Fan, K., & Jaroniec, M. (2019). 0D/2D NiS_2/V-MXene composite for electrocatalytic H_2 evolution. *Journal of Catalysis*, *375*, 8–20. doi:10.1016/j.jcat.2019.05.019

Kuang, P., Low, J., Cheng, B., Yu, J., & Fan, J. (2020). MXene-based photocatalysts. *Journal of Materials Science and Technology*, *56*, 18–44. doi:10.1016/j.jmst.2020.02.037

Kudo, A., & Miseki, Y. (2009). Heterogeneous photocatalyst materials for water splitting. *Chemical Society Reviews*, *38*(1), 253–278. doi:10.1039/B800489G PMID:19088977

Lewis, N. S. (2016). Research opportunities to advance solar energy utilization. *Science*, *351*(6271), aad1920. doi:10.1126cience.aad1920 PMID:26798020

Li, F., Huang, Y., Gao, C., & Wu, X. (2021). The enhanced photo-catalytic CO_2 reduction performance of g-C_3N_4 with high selectivity by coupling CoNiSx. *Materials Research Bulletin*, *144*, 111488. doi:10.1016/j.materresbull.2021.111488

Li, H., Sun, B., Gao, T., Li, H., Ren, Y., & Zhou, G. (2022). Ti_3C_2 MXene co-catalyst assembled with mesoporous TiO_2 for boosting photocatalytic activity of methyl orange degradation and hydrogen production. *Chinese Journal of Catalysis*, *43*(2), 461–471. doi:10.1016/S1872-2067(21)63915-3

Li, K., Zhang, S., Li, Y., Fan, J., & Lv, K. (2020). MXenes as noble-metal-alternative co-catalysts in photocatalysis. *Chinese Journal of Catalysis*, *42*(1), 3–14. doi:10.1016/S1872-2067(20)63630-0

Li, S., Cai, J., Wu, X., Liu, B., Chen, Q., Li, Y., & Zheng, F. (2018). TiO_2@Pt@CeO_2 nanocomposite as a bifunctional catalyst for enhancing photo-reduction of Cr (VI) and photo-oxidation of benzyl alcohol. *Journal of Hazardous Materials*, *346*, 52–61. doi:10.1016/j.jhazmat.2017.12.001 PMID:29247954

Li, Y., Wang, W. N., Zhan, Z., Woo, M. H., Wu, C. Y., & Biswas, P. (2010). Photocatalytic reduction of CO_2 with H_2O on mesoporous silica supported Cu/TiO_2 catalysts. *Applied Catalysis B: Environmental*, *100*(1–2), 386–392. doi:10.1016/j.apcatb.2010.08.015

Liu, G., Abukhadra, M. R., El-Sherbeeny, A. M., Mostafa, A. M., & Elmeligy, M. A. (2020). Insight into the photocatalytic properties of diatomite@Ni/NiO composite for effective photo-degradation of malachite green dye and photo-reduction of Cr (VI) under visible light. *Journal of Environmental Management*, *254*, 109799. doi:10.1016/j.jenvman.2019.109799 PMID:31710977

Liu, L., & Corma, A. (2018). Metal Catalysts for Heterogeneous Catalysis: From Single Atoms to Nanoclusters and Nanoparticles. *Chemical Reviews*, *118*(10), 4981–5079. doi:10.1021/acs.chemrev.7b00776 PMID:29658707

Lu, N., Jing, X., Zhang, J., Zhang, P., Qiao, Q., & Zhang, Z. (2022). Photo-assisted self-assembly synthesis of all 2D-layered heterojunction photocatalysts with long-range spatial separation of charge-carriers toward photocatalytic redox reactions. *Chemical Engineering Journal, 431*, 134001. doi:10.1016/j.cej.2021.134001

Luo, J., Im, J. H., Mayer, M. T., Schreier, M., Nazeeruddin, M. K., Park, N. G., Tilley, S. D., Fan, H. J., & Grätzel, M. (2014). Water photolysis at 12.3% efficiency via perovskite photovoltaics and Earth-abundant catalysts. *Science, 345*(6204), 1593–1596. doi:10.1126cience.1258307 PMID:25258076

Mandal, S., Nanavati, S. P., Willock, D. J., & Ananthakrishnan, R. (2022). Band gap engineering of amine functionalized Ag(I)-based coordination polymers and their plasmonic Ag^0 coupled novel visible light driven photo-redox system for selective oxidation of benzyl alcohol. *Applied Catalysis B: Environmental, 303*, 120821. doi:10.1016/j.apcatb.2021.120821

Mazloomi, K., & Gomes, C. (2012). Hydrogen as an energy carrier: Prospects and challenges. *Renewable & Sustainable Energy Reviews, 16*(5), 3024–3033. doi:10.1016/j.rser.2012.02.028

Meshram, A. A., Aashish Moses, K., Baral, S. S., & Sontakke, S. M. (2022). Hydrogen production from water splitting of real-time industry effluent using novel photocatalyst. *Advanced Powder Technology, 33*(3), 103488. doi:10.1016/j.apt.2022.103488

Moradi, M., Khorasheh, F., & Larimi, A. (2020). Pt nanoparticles decorated Bi-doped TiO_2 as an efficient photocatalyst for CO_2 photo-reduction into CH_4. *Solar Energy, 211*, 100–110. doi:10.1016/j.solener.2020.09.054

Moscow, S., Kavinkumar, V., Sriramkumar, M., Jothivenkatachalam, K., Saravanan, P., Rajamohan, N., Vasseghian, Y., & Rajasimman, M. (2022). Impact of Erbium (Er) and Yttrium (Y) doping on $BiVO_4$ crystal structure towards the enhancement of photoelectrochemical water splitting and photocatalytic performance. *Chemosphere, 299*, 134343. doi:10.1016/j.chemosphere.2022.134343 PMID:35307389

Mostafa, E. M., & Amdeha, E. (2022). Enhanced photocatalytic degradation of malachite green dye by highly stable visible-light-responsive Fe-based tri-composite photocatalysts. *Environmental Science and Pollution Research International, 1*(46), 1–14. doi:10.100711356-022-20745-6 PMID:35578081

Naguib, M., Kurtoglu, M., Presser, V., Lu, J., Niu, J., Heon, M., Hultman, L., Gogotsi, Y., & Barsoum, M. W. (2011). Two-dimensional nanocrystals produced by exfoliation of Ti_3AlC_2. *Advanced Materials, 23*(37), 4248–4253. doi:10.1002/adma.201102306 PMID:21861270

Naguib, M., Mochalin, V. N., Barsoum, M. W., & Gogotsi, Y. (2014). 25th anniversary article: MXenes: A new family of two-dimensional materials. *Advanced Materials, 26*(7), 992–1005. doi:10.1002/adma.201304138 PMID:24357390

Negi, K., Kumar, M., Singh, G., Chauhan, S., & Chauhan, M. S. (2018). Nanostructured CeO_2 for selective-sensing and smart photocatalytic applications. *Ceramics International, 44*(13), 15281–15289. doi:10.1016/j.ceramint.2018.05.172

Nematollahi, R., Ghotbi, C., Khorasheh, F., & Larimi, A. (2020). Ni-Bi co-doped TiO_2 as highly visible light response nano-photocatalyst for CO_2 photo-reduction in a batch photo-reactor. *Journal of CO_2 Utilization, 41*, 101289. doi:10.1016/j.jcou.2020.101289

Nozari, B., Montazer, M., & Mahmoudi Rad, M. (2021). Stable ZnO/SiO_2 nano coating on polyester for anti-bacterial, self-cleaning and flame retardant applications. *Materials Chemistry and Physics*, *267*, 124674. doi:10.1016/j.matchemphys.2021.124674

Pan, J., Wang, P., Wang, P., Yu, Q., Wang, J., Song, C., Zheng, Y., & Li, C. (2021). The photocatalytic overall water splitting hydrogen production of g-C_3N_4/CdS hollow core–shell heterojunction via the HER/OER matching of Pt/MnOx. *Chemical Engineering Journal*, *405*, 126622. doi:10.1016/j.cej.2020.126622

Qiu, Z., Wang, B., Zhou, X., Yang, S., Gao, Q., Cai, X., Zhang, S., & Fang, Y. (2022). CdS@Mg$(OH)_2$ core/shell composite photocatalyst for efficient visible-light photocatalytic overall water splitting. *International Journal of Hydrogen Energy*, *47*(14), 8729–8738. doi:10.1016/j.ijhydene.2021.12.216

Rakibuddin, M., & Kim, H. (2020). Samarium (III)-doped ZnO/graphitic–C_3N_4 composites for enhanced hydrogen generation from water under visible light photocatalysis. *Journal of Alloys and Compounds*, *832*, 154887. doi:10.1016/j.jallcom.2020.154887

Rane, A. V., Kanny, K., Abitha, V. K., & Thomas, S. (2018). Methods for Synthesis of Nanoparticles and Fabrication of Nanocomposites. *Synthesis of Inorganic Nanomaterials*, 121–139. doi:10.1016/b978-0-08-101975-7.00005-1

Rauf, H., Yasmin, N., Ali, G., Ashiq, M. N., Safdar, M., & Mirza, M. (2022). New insight in photocatalytic degradation of textile dyes over CeO_2/Ce_2S_3 composite. *Physica B, Condensed Matter*, *632*, 413760. doi:10.1016/j.physb.2022.413760

Ravishankar, T. N., Ramakrishnappa, T., Nagaraju, G., & Rajanaika, H. (2015). Synthesis and Characterization of CeO_2 Nanoparticles via Solution Combustion Method for Photocatalytic and Antibacterial Activity Studies. *ChemistryOpen*, *4*(2), 146–154. doi:10.1002/open.201402046 PMID:25969812

Sekar, S., Preethi, V., Srivishnu, V. S., Saravanan, S., & Lee, S. (2022). Highly-efficient photocatalytic activity of TiO_2-AC nanocomposites for hydrogen production from sulphide wastewater. *International Journal of Hydrogen Energy*, *47*(95), 40275–40285. doi:10.1016/j.ijhydene.2022.02.019

Sharmin, F., Chandra Roy, D., & Basith, M. A. (2021). Photocatalytic water splitting ability of Fe/MgO-rGO nanocomposites towards hydrogen evolution. *International Journal of Hydrogen Energy*, *46*(77), 38232–38246. doi:10.1016/j.ijhydene.2021.09.072

She, Z. W., Kibsgaard, J., Dickens, C. F., Chorkendorff, I., Nørskov, J. K., & Jaramillo, T. F. (2017). Combining theory and experiment in electrocatalysis: Insights into materials design. *Science*, *355*(6321). doi:1 doi:0.1126cience.aad4998 PMID:28082532

Tachibana, Y., Vayssieres, L., & Durrant, J. R. (2012). Artificial photosynthesis for solar water-splitting. *Nature Photonics*, *6*(8), 511–518. doi:10.1038/nphoton.2012.175

Tahir, M. B., Iqbal, T., Rafique, M., Rafique, M. S., Nawaz, T., & Sagir, M. (2020). Nanomaterials for photocatalysis. *Nanotechnology and Photocatalysis for Environmental Applications*, 65–76. doi:10.1016/b978-0-12-821192-2.00005-x

Tahir, M. B., Nabi, G., Iqbal, T., Sagir, M., & Rafique, M. (2018). Role of $MoSe_2$ on nanostructures WO_3-CNT performance for photocatalytic hydrogen evolution. *Ceramics International*, *44*(6), 6686–6690. doi:10.1016/j.ceramint.2018.01.081

Tian, J., Leng, Y., Cui, H., & Liu, H. (2015). Hydrogenated TiO_2 nanobelts as highly efficient photocatalytic organic dye degradation and hydrogen evolution photocatalyst. *Journal of Hazardous Materials*, *299*, 165–173. doi:10.1016/j.jhazmat.2015.06.019 PMID:26118828

Tian, Z., Lin, L., Meng, F., & Huang, W. (2009). Combustion synthesis and characterization of nanocrystalline $Ba(Mg_{1/3}Nb_{2/3})O_3$ powders. *Materials Science and Engineering B. Solid-State Materials for Advanced Technology*, *158*(1–3), 88–91. doi:10.1016/j.mseb.2009.01.023

Varghese, O. K., Paulose, M., LaTempa, T. J., & Grimes, C. A. (2009). High-Rate Solar Photocatalytic Conversion of CO_2 and Water Vapor to Hydrocarbon Fuels. *Nano Letters*, *9*(2), 731–737. doi:10.1021/nl803258p PMID:19173633

Wang, C., Ren, H., Zeng, M., Zhu, Q., Zhang, Q., Kan, Z., Wang, Z., Shen, M., Thalavitiya Acharige, M. J., Ruths, M., Ryan, D. K., Li, G., Kolesov, G., Kaxiras, E., & Mazur, E. (2020). Low-cost visible-light photosynthesis of water and adsorbed carbon dioxide into long-chain hydrocarbons. *Chemical Physics Letters*, *739*, 136985. doi:10.1016/j.cplett.2019.136985

Wang, G., Dou, K., Cao, H., Du, R., Liu, J., Tsidaeva, N., & Wang, W. (2022). Designing Z-scheme CdS/WS2 heterojunctions with enhanced photocatalytic degradation of organic dyes and photoreduction of Cr (VI): Experiments, DFT calculations and mechanism. *Separation and Purification Technology*, *291*, 120976. doi:10.1016/j.seppur.2022.120976

Wang, H., Zhang, L., Chen, Z., Hu, J., Li, S., Wang, Z., Liu, J., & Wang, X. (2014). Semiconductor heterojunction photocatalysts: Design, construction, and photocatalytic performances. *Chemical Society Reviews*, *43*(15), 5234–5244. doi:10.1039/C4CS00126E PMID:24841176

Wang, Q., & Domen, K. (2020). Particulate Photocatalysts for Light-Driven Water Splitting: Mechanisms, Challenges, and Design Strategies. *Chemical Reviews*, *120*(2), 919–985. doi:10.1021/acs.chemrev.9b00201 PMID:31393702

Wang, W., Huang, H., Ke, X., Liu, X., Yang, S., Wang, K., Huang, L., Tu, C., Zheng, Z., Luo, D., & Zhang, M. (2022). Morphology- and size-dependent $FAPbBr_3/WO_3$ Z-scheme photocatalysts for the controllable photo-oxidation of benzyl alcohol. *Materials & Design*, *215*, 110502. doi:10.1016/j.matdes.2022.110502

Wang, Y., Li, B., Zhang, C., Cui, L., Kang, S., Li, X., & Zhou, L. (2013). Ordered mesoporous CeO_2-TiO_2 composites: Highly efficient photocatalysts for the reduction of CO_2 with H_2O under simulated solar irradiation. *Applied Catalysis B: Environmental*, *130–131*, 277–284. doi:10.1016/j.apcatb.2012.11.019

Wang, Y., Zhang, Y., Jiang, Z., Jiang, G., Zhao, Z., Wu, Q., Liu, Y., Xu, Q., Duan, A., & Xu, C. (2016). Controlled fabrication and enhanced visible-light photocatalytic hydrogen production of Au at CdS/MIL-101 heterostructure. *Applied Catalysis B: Environmental*, *185*, 307–314. doi:10.1016/j.apcatb.2015.12.020

Wang, Z., Song, H., Pang, H., Ning, Y., Dao, T. D., Wang, Z., Chen, H., Weng, Y., Fu, Q., Nagao, T., Fang, Y., & Ye, J. (2019). Photo-assisted methanol synthesis via CO_2 reduction under ambient pressure over plasmonic Cu/ZnO catalysts. *Applied Catalysis B: Environmental*, *250*, 10–16. doi:10.1016/j.apcatb.2019.03.003

Wellia, D. V., Kusumawati, Y., Diguna, L. J., & Amal, M. I. (2017). *Introduction of Nanomaterials for Photocatalysis*. 1–17. doi:10.1007/978-3-319-62446-4_1

Wieszczycka, K., Staszak, K., Woźniak-Budych, M. J., Litowczenko, J., Maciejewska, B. M., & Jurga, S. (2021). Surface functionalization – The way for advanced applications of smart materials. *Coordination Chemistry Reviews*, *436*, 213846. doi:10.1016/j.ccr.2021.213846

Wu, C., Shen, Q., Zheng, S., Zhang, X., Sheng, J., & Yang, H. (2022). Fabrication of $Bi_2Sn_2O_7$@MIL-100(Fe) composite photocatalyst with enhanced superoxide-radical-dominated photocatalytic activity for ciprofloxacin degradation. *Journal of Molecular Structure*, *1258*, 132657. doi:10.1016/j.molstruc.2022.132657

Wu, L., An, S., & Song, Y. F. (2021). Heteropolyacids-Immobilized Graphitic Carbon Nitride: Highly Efficient Photo-Oxidation of Benzyl Alcohol in the Aqueous Phase. *Engineering*, *7*(1), 94–102. doi:10.1016/j.eng.2020.07.025

Xia, C., Xue, C., Bian, W., Wei, Y., & Zhang, J. (2021). Morphology controlling strategy of ZnS/CeO_2 catalysts to enhance the photocatalytic performance for hydrogen production. *JCIS Open*, *4*, 100030. doi:10.1016/j.jciso.2021.100030

Xu, K., Shen, J., Zhang, S., Xu, D., & Chen, X. (2022). Efficient interfacial charge transfer of BiOCl-In_2O_3 step-scheme heterojunction for boosted photocatalytic degradation of ciprofloxacin. *Journal of Materials Science and Technology*, *121*, 236–244. doi:10.1016/j.jmst.2021.12.070

Yang, J., Lukashuk, L., Li, H., Föttinger, K., Rupprechter, G., & Schubert, U. (2014). High surface area ceria for CO oxidation prepared from cerium t-butoxide by combined sol-gel and solvothermal processing. *Catalysis Letters*, *144*(3), 403–412. doi:10.100710562-013-1162-8 PMID:24532959

Yang, Y., Chen, Y., Li, Z., Tang, S., Li, Y., Fu, Z., Yang, S., Yang, M., & Xie, H. (2022). Homojunction type of carbon nitride as a robust photo-catalyst for reduction conversion of CO_2 in water vapor under visible light. *Chemical Engineering Journal*, *430*, 132668. doi:10.1016/j.cej.2021.132668

Yoshida, M., & Lahann, J. (2008). Smart nanomaterials. *ACS Nano*, *2*(6), 1101–1107. doi:10.1021/nn800332g PMID:19206325

Yue, R., Raisi, B., Rahmatinejad, J., Ye, Z., Barbeau, B., & Rahaman, M. S. (2021). A photo-Fenton nanocomposite ultrafiltration membrane for enhanced dye removal with self-cleaning properties. *Journal of Colloid and Interface Science*, *604*, 458–468. doi:10.1016/j.jcis.2021.06.157 PMID:34273782

Zhang, L., He, Y., Luo, P., Ma, L., Li, S., Nie, Y., Zhong, F., Wang, Y., & Chen, L. (2022). Photocatalytic GO/M88A "interceptor plate" assembled nanofibrous membrane with photo-Fenton self-cleaning performance for oil/water emulsion separation. *Chemical Engineering Journal*, *427*, 130948. doi:10.1016/j.cej.2021.130948

Zhao, R., Wang, M., Zhao, D., Li, H., Wang, C., & Yin, L. (2018). Molecular-Level Heterostructures Assembled from Titanium Carbide MXene and Ni-Co-Al Layered Double-Hydroxide Nanosheets for All-Solid-State Flexible Asymmetric High-Energy Supercapacitors. *ACS Energy Letters*, *3*(1), 132–140. doi:10.1021/acsenergylett.7b01063

Zhu, S., & Wang, D. (2017). Photocatalysis: Basic principles, diverse forms of implementations and emerging scientific opportunities. *Advanced Energy Materials*, *7*(23), 1700841. doi:10.1002/aenm.201700841

KEY TERMS AND DEFINITIONS

2,4-DCP: 2,4-Di ChloroPhenol.
AC: Activated Carbon.
BET: Brunauer–Emmett–Teller.
BJH: Barrett-Joyner-Halenda.
BPA: BisPhenol A.
CIP: Ciprofloxacin.
CN: Graphitic Carbon Nitride; g-C_3N_4.
CR: Congo Red.
CTAB: Cetyl Trimethyl Ammonium Bromide.
TGA: Thermo Gravimetric Analysis.
EDS: Energy-Dispersive-Spectroscopy.
EIS: Electrochemical Impedance Spectroscopy.
FTIR: Fourier-Transform Infra-Red.
GO: Graphene Oxide.
r-GO: Reduced Graphene Oxide.
HOCO: Highest Occupied Crystal Orbital.
HEC: Hydrogen Evolution Cocatalyst.
HRTEM: High-Resolution Transmission Electron Microscopy.
LSPR: Localized Surface Plasmon Resonance.
LUCO: Lowest Unoccupied Crystal Orbital.
MB: Methylene Blue.
MG: Malachite Green.
MO: Methyl Orange.
MOF: Metal Organic Framework.
MWCNTs: Multi Wall Carbon Nano Tubes.
NMs: Nano Materials.
NPs: Nano Particles.
OEC: Oxygen Evolution Cocatalyst.
PL: Photo Luminescence.
RhB: Rhodamine B.
TEM: Transmission Electron Microscope.
UV-Vis: UV-Visible.
DRS: Diffuse Reflectance Spectroscopy.
XRD: X-Ray Diffraction.

Chapter 6
Yttrium Iron Garnet (YIG): A Nano-Material for Tomorrow

Neha Sharma
Keshav Mahavidyalaya, University of Delhi, India

Prashant Kumar
Keshav Mahavidyalaya, University of Delhi, India

ABSTRACT

A cluster of silicate minerals known as garnet is widely used as abrasives and gemstones, has now found an application as smart nonmaterial. The garnet possesses a unique feature of having different chemical composition with similar crystal form and physical properties. Yttrium iron garnet (YIG) also popularly known as iron yttrium oxide, is an example of lab grown garnet. It can have chemical composition of Y3Fe5O1 or Y3Fe2(FeO4). The YIG exhibits a ferrimagnetic nature, with a Curie temperature of 560 K. The chapter focuses on the research around YIG with a mention on fabrication methods and applications.

INTRODUCTION

YIG (Figure 1) has a chemical composition where in an irregular cube structure eight oxygen ions coordinate with the five ions of iron(III), which occupy two octahedral and three tetrahedral sites, with the yttrium(III) ions (Yttrium Iron Garnet, 2015). Magnetic behavior is accounted for by ions of iron, exhibiting different spins in the two coordination sites (Cherepanov et al., 1993). Figure 2 shows the arrangement of atoms, which are present at the line joints.

DOI: 10.4018/978-1-6684-7358-0.ch006

Figure 1. YIG crystal
(source:" https://commons.wikimedia.org/wiki/File:Wiki-YIG.jpg")

Figure 2. Schematic of the different shapes (tetrahydral and octahydral) occupied by cations in YIG

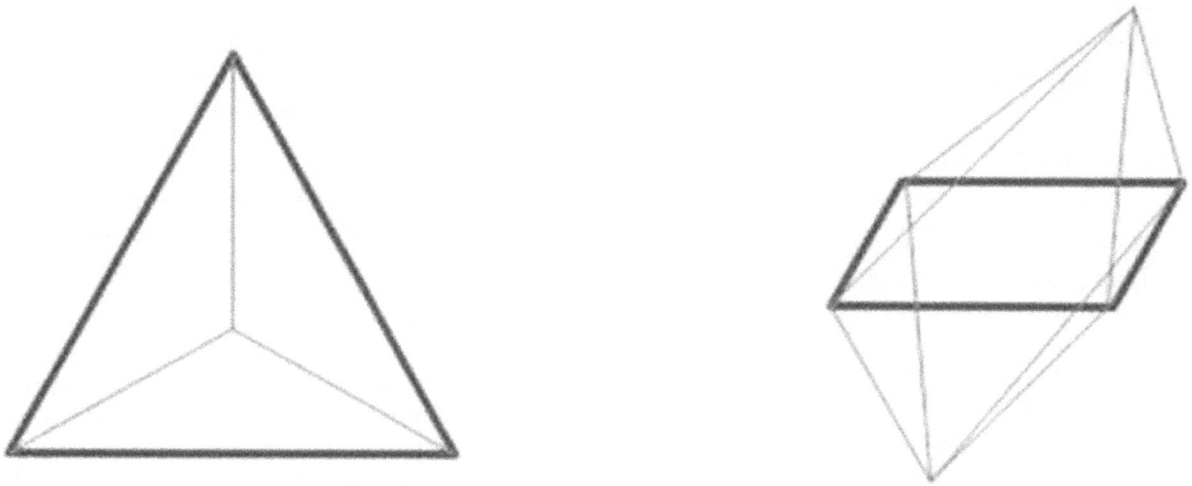

Cations like rare-earth elements metal, alkaline are used as substitutes for specific sites in YIG to obtain magnetic properties (Goulon et al., 2012). The choice of substitute are based on the ionic radii of these atoms. YIG has a high Verdet constant. High values of Verdet constant accounts for number of factors like Faraday effect (Grattan & Meggitt, 1999; Vojna et al., 2020), electron spin resonance with very small value of line-width, microwave frequencies with high quality factor, infrared wavelength with low absorption, as low as 1200nm (Sirohi, 1990). These properties of YIG makes it very useful for applications in superconductors like magneto-optical imaging (MOI) (Magneto-Optical Imaging of Superconductors, 2010). Number of applications have been suggested where YIG can be efficiently used in areas of acoustic transducers and transmitters (Periodic Table of Elements: Yttrium, 2015), filters, bio-antennas, nano-systems, capacitors etc. YIG has many uses in the fields of optical, microwave, and magneto-optical devices. For wavelengths greater than 600nm, YIG is transparent, which makes them more acceptable in the optical devices. More applications like Faraday rotators utilizing lasers based on solid-state, storage of data and optics (nonlinear) (Holm et al., 1984) have also been reported in the literature. The garnet has found use in both biological and engineering research.

SYNTHESIS

YIG powder can be synthesized by

1. co-precipitation (Kimura et al., 1998),

2.microwave synthesis (Ahn et al., 1996),
3. sol–gel (Sanchez et al., 2004) or (and)
4. solid state procedure (Buscaglia et al., 1997).

co-precipitation (Guo et al., 2005; Homg et al., 2008; Xu, 2008), sol–gel (Sanchez et al., 2004; Thiagarajan et al., 2017), techniques include mechanochemical (Sanchez et al., 2004), hydrothermal (Ahn et al., 1996), and microemulsion (Kimura et al., 1998), as well as organic precursor approach (Homg et al., 2008), (Guo et al., 2005), (Xu, 2008). The chemical co-precipitation approach has been shown to be the most effective method due to its low cost and suitability for large production. Additionally, the co-precipitation procedure has many benefits, such as better beginning material mixing and superior chemical homogeneity of the finished product.

3.1 Co-precipitation

The best synthesis technique is co-precipitation, mainly because it is inexpensive and well suited for bulk production. It enables optimal material mixing at the early phases of the process, which results in high chemical homogeneity of the resulting end product. By adding a base under an inert atmosphere at either room temperature or a higher temperature, the process of co-precipitation is a very practical and simple way to make iron oxide nanoparticles (either Fe_3O_4 or -Fe_2O_3) (Gautam & Chattopadhyaya, 2016). Figure 3 shows the detailed synthesis for YIG (Rattan, 2015).

Figure 3. Technique of Co-precipitation for YIG synthesis

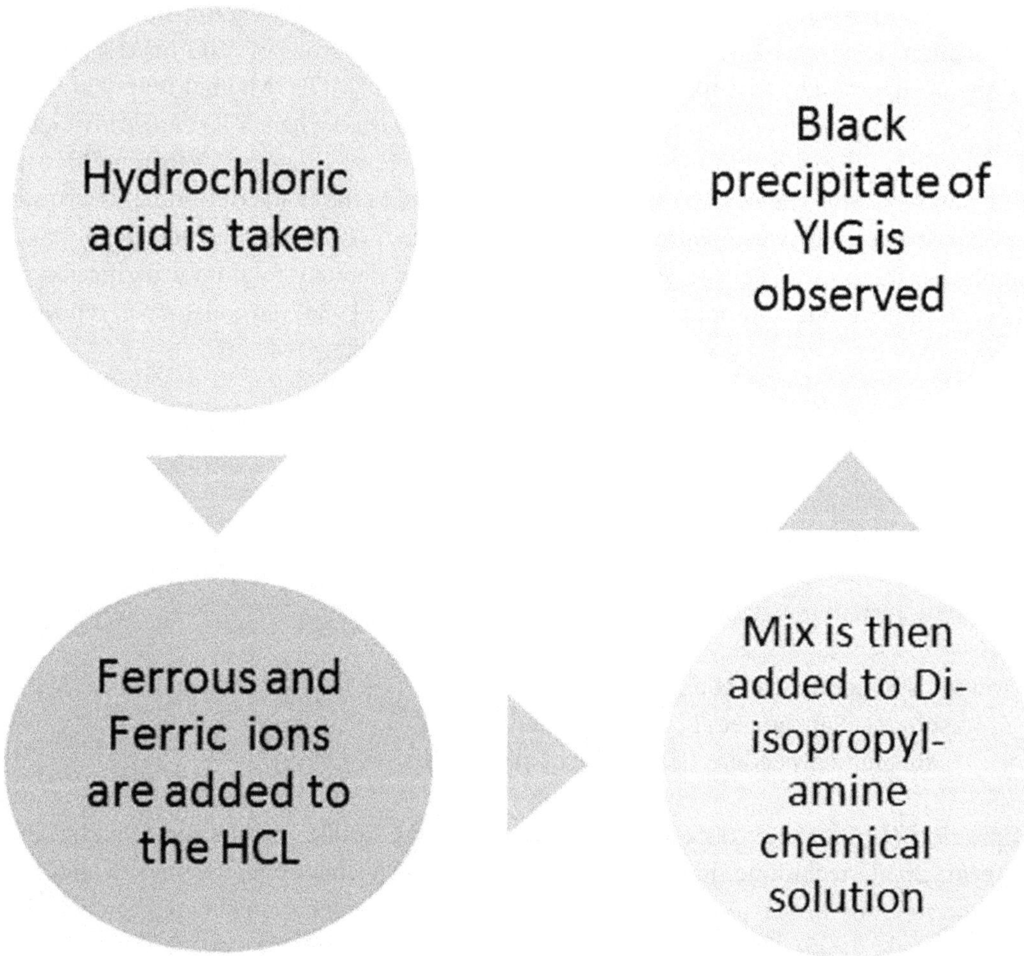

Three procedures of Co-precipitation are recommended namely (Harvey, 2000):

1.inclusion
2.adsorption and
3.occlusion

Inclusion, which is defined as the incorporation in the lattice of the structure of the crystal of the carrier, takes place when a lattice site is occupied by an impurity. This results in crystallographic defect. An important condition for this to happen is that the charge and ionic radius of impurity should be same as that of carrier.

Adsorbate is the term used for an impurity which is bound to the surface of precipitate. It can be either weakly or strongly bounded.

The impurity that has been adsorbed causes occlusion when it becomes physically stuck inside the crystal as it expands.

3.1.1 Models that depict how tracer compounds are distributed between the solution and precipitate ()

Co-precipitation has two models that describe how the tracer component is distributed between the solution and the precipitate.

1. **Doerner Hoskins model-** the logarithmic model This is defined for heterogeneous crystals

It is assumed that no mass exchange occurs between the core of the precipitating crystals and the solution.

The tracer content of the crystal is not uniform when the above assumption is taken into account. The crystals so produced are heterogeneous. Equation 1 gives the proposed model.

$$In\frac{a}{a-x} = \lambda In\frac{b}{b-y} \quad (1)$$

where

a : initial concentration of the tracer
b: initial concentration of the carrier
a – x: concentration of the tracer after separation
b – y: carrier concentrations after separation
x : amount of the tracer on the precipitate
y: amount of carrier on the precipitate and
λ: distribution coefficient

2. **Berthelot-Nernst law** given by equation 2. It is defined for homogeneous crystals

$$\frac{x}{a-x} = D\frac{y}{b-y} \quad (2)$$

where a, b, x and y have same definitions as in equation 1

D: distribution coefficient.

When the concentration of the tracer inside the crystal is uniform, a homogeneous crystal results, and the Berthelot-Nernst law is applicable for such cases. Kinetic effects, such as the existence of mixing and crystallization speed, are crucial for these formations. This paradigm is only applicable in two circumstances.

a. when the original tiny crystals are permitted to re-crystallize; and
b. when internal diffusion, comparable to that in liquids, is conceivable.

If conditions D > 1 and λ > 1 are met for both suggested models, the precipitate is enriched in the tracer. Depending on the co-precipitation system and operational circumstances, either D or a λ may be constant.

3.1.2 Applications of Co-precipitation (Lu et al., 2007)

Co-precipitation finds application in

a. radiochemistry,
b. chemical analysis,
c. Synthesis of magnetic nano-particles
d. water-resources related environmental issues like
    - toxic waste heavy metal transport at defense and industrial locations,
    - waste water treatment technology
    - metal concentrations in aquatic systems,
    - radionuclide migration around waste repositories
    - acid mine drainage etc.

3.1.3 Drawbacks of Co-Precipitation (Harvey, 2000)

Co-precipitation is a crucial step in the study of chemicals. Co-precipitation is a difficult, particularly in gravimetric analysis, which involves precipitating the analyte and measuring its mass to determine its purity or concentration. Excess mass occurs from the presence of unwanted contaminants that co-precipitate with the analyte. Resolving the sample and precipitating it once again or waiting for the precipitate to equilibrate and create larger, purer particles are two ways to solve this issue.

3.2 Microwave Synthesis (Dahiya & Duhan, 2018)

Synthesis using microwaves offers dual advantage of homogeneous heating and speed of the precursor materials. Figure 4 shows the process. Homogeneous heating is accounted by penetration characteristics of microwave radiation. The method is based on the idea of aligning a material's dipoles in an external field while they are being excited by microwave electromagnetic radiations. This process of synthesis is always a part of synthesis strategy, as shown in figure 4, where co-precipitation synthesis is succeeded by microwave synthesis (Liua et al., 2018).

This approach is advised because it allows for precise control over the size and form of the magnetic iron oxide nanoparticles so produced. When molecules are aligned or oriented by an external electrical field, internal heat is generated. Internal heat produced in this way reduces the amount of time and energy needed for processing. This is brought on by the microwaves' uniform heating. Thus, the reaction time in the microwave-assisted synthesis process is shortened. The magnetic characteristics of ferrite nanoparticles can be controlled by fiddling with experimental measurements.

Figure 4. Steps involved in Microwave assisted synthesis of nanoparticles

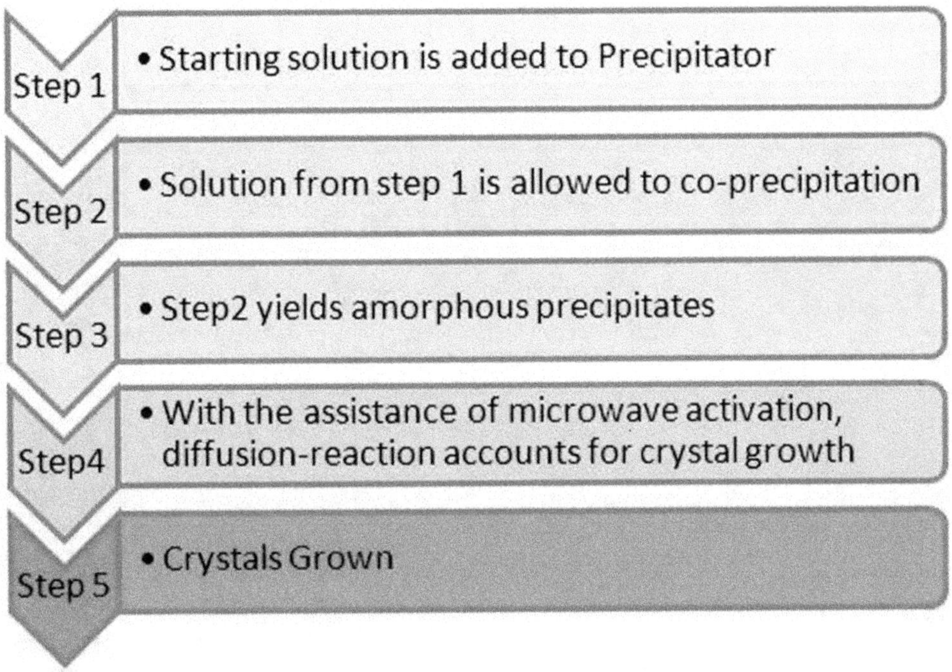

3.3 Sol–gel (Thiagarajan et al., 2017)

High-quality, ultrafine, highly stoichiometric, and homogenous nanostructures can be produced using the cost-effective and highly controlled sol-gel process at low temperatures. The preferred approach for selecting the required metal oxide (MO) nanostructure shape for application-specific structures such as nanofibers, nanoribbons, nanoflakes, nanotubes, nanospheres, and nanorods, among others, is the sol-gel method. Due to their purity, homogeneity, and low temperature synthesis, MO nanostructures made by the sol-gel method are also used in biological applications such as carving natural bone and teeth, medicines, anti-microbial properties, medication transport, etc. Sol-gel is the most popular method for creating MO-based nanostructures with a variety of applications, including sensors and actuators, energy generation and conversion, storage devices, electrical applications, magnetic applications, etc.

Figure 5. Flow diagram for sol-gel processing

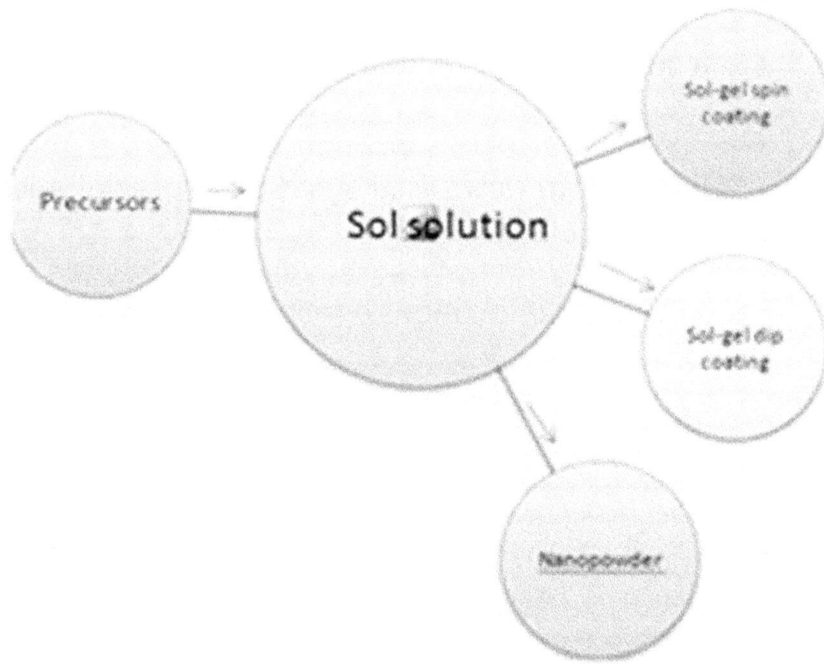

Sol-gel process schematic is shown in Figure 5. It uses a wet chemical process. It is also popularly known by the following names

a. Pechini Method,
b. Self ignition process,
c. Green Combustion,
d. Citrate nitrate combustion and
e. Sol-gel auto combustion.

It involves multiple steps which are both physical and chemical in nature. Sol means colloidal solution and gel means dispersion of liquid molecules in a solid medium.

Steps involved in the sol-gel process are

a. Hydrolysis,
b. Polycondensation,
c. Gelation,
d. Aging,
e. Drying,
f. Densification and
g. Crystallization.

Process of sol-gel can be summarized in following steps

a. It uses either a colloidal or chemical solution to produce an integrated network known as gel.
b. Typical Precursors used are metal chlorides and metal alkoxides.
c. Collides are formed when precursors undergo hydrolysis and polycondensation.
d. The sol forms inorganic continuous network, gel.
e. Metal-hydroxo or metal-oxo polymers are generated in solution due to formation of metal oxides, connecting metal centers with hydroxyl or oxygen respectively.
f. Liquid is removed from the gel by drying
g. Further polycondensation and crystallization is followed by calcination, the thermal treatment.

Principal features of sol-gel are

a. Dopants can be easily introduced in the final product.

b .Easy control of stiochiometry
c. Fine particle size is possible
d. High crystalline features.
e. Products with high purity
f. Better homogeneity
g. Simple equipment
h. Better control of morphology
i. Uniform phase distribution
j. Low processing temperature
k. ultrafine size of particles obtained.

3.4 Solid State Procedure (Ivanov, 2012; Kumar & Swart, 2020)

The traditional procedure for processing using solid-state technique for the synthesis of multi-ferroics nanoparticles involves (figure 6)

a. mixing by milling the carbonates or metal oxides
b. Calcination at high temperature and
c. Interdiffusion of the cations.

Figure 6. Flow diagram of Solid -state technique

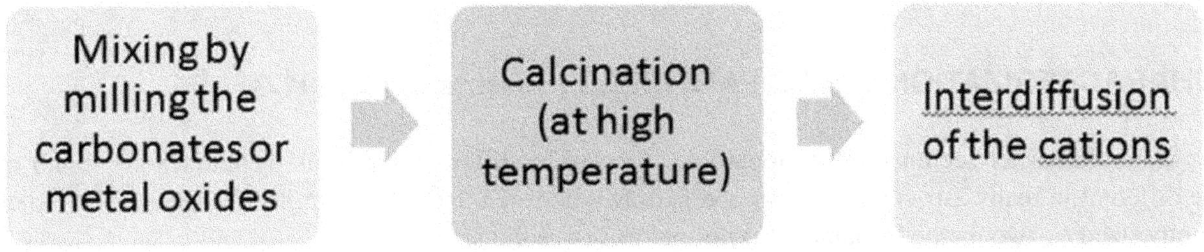

Solid-state diffusion of the ions is by chemical reactions involved, which are characterized by a slow kinetic rate. The degrees of powder particle sized and mixing of importance. At the points of contact between the initial components, reaction commences. This is followed by ionic interdiffusion through the product. The initial reaction is quick due to relatively shorter diffusion paths, but further reaction is very slow. Many unwanted phases appear in the process. This traditional method is widely used because

a. inexpensive,
b. simple and
c. results are acceptable.

Solid-state procedures also come with some drawbacks like

a. poor sintering behavior,
b. non-uniformity of particle size,
c. non-uniformity of particle shape,
d. lack of reproducibility,
e. multiphase character and
f. loss of stoichiometry.

This is frequently employed to create crystalline solids. The solids must be heated to significantly higher degrees because they do not naturally react with one another at room temperature. The solid-state reaction's feasibility and rate depend on the following factors:

a. reaction conditions,
b. reactant structural characteristics,
c. solid surface area,
d. solids' reactivity, and
e. thermodynamic free energy

The initial components used in a typical synthesis procedure are oxide precursors. Using an agate mortar and pestle, the needed quantity of chemicals are pounded to a fine powder. As the ingredients are being mixed, the reaction begins on its own. Heat and water vapour are also released in conjunction with this. Acetone is employed as a mixing medium to produce a homogenous mixture.. After being dry ground, the ground sample is placed in alumina/platinum crucibles and heated at a high temperature for a predetermined amount of time.

CHARACTERIZATION OF YIG PARTICLES (Sharma et al., 2017)

As stated in earlier sections, the nanoparticles formed by different synthesis techniques are found to be different in many aspects. A comparison is drawn between the polycrystalline Yttrium Iron Garnet synthesized by two methods, namely (a) Sol-gel and (b) solid state.

Table 1. Comparison of properties of polycrystalline YIG nanoparticles synthesized by sole-gel and solid-state method

S.no	Property	Sol Gel Method	Solid State Method
1.	Sintering temperature required	750°C	1000°C
2.	The saturation magnetizations	1070 Oe	1125 Oe
3.	Coercivity	Half of that observed for solid-state	Twice of that observed for sol-gel

X-Ray Diffraction, XRD, is used to find the average polycrystalline YIG nanoparticles sizes which are so processed by Sol-gel method and Solid-State method using Scherrer formula (Fiebig, 2005). according to which

$\tau = N\lambda\delta cos\theta$τ=Nλδcosθ

where

N = shape factor
λ = 1.54181 Å, X-ray wavelength (CuKα)(CuKα)
δ = *l*ine broadening at half the maximum intensity (FWHM) (in radians) and
θ = Bragg's angle

The calculations are made using patterns observed in XRD, by studying broadening in peaks results (Sharma et al., 2017) which shows the XRD patterns for Sol gel YIG nanoparticles (a) and Solid state YIG nanoparticles (b), (c) shows room temperature VSM profile of YIG nanoparticles. (d) and (e) exhibit SEM images of YIG NPs produced using the solid state technique and the sol gel, respectively. X-ray powder diffraction in the experiment was performed at 300oK· According to the findings, YIG has a cubic structure with the space group Ia3d and an estimated lattice parameter (estimated) "a" = 12.374. Only the distinctive peaks of Yttrium Iron Garnet were observed in the diffractograms of samples produced at 750 °C and 1000 °C using the sol-gel and solid-state synthesis procedures, respectively.

The dynamic properties of the YIG nanoparticles were studied using the ferromagnetic resonance (FMR) technique. The measurements were carried out in (a) field-sweep and (b) frequency-sweep modes, at various fixed magnetic fields and fixed frequencies for both solid state and sol-gel nanoparticles, the resonance field (Hr) was seen to rise linearly with frequency. It was reported that as compared to nanoparticles manufactured using the solid-state approach, those created via Sol-gel have a lower stop-band bandwidth (frequency linewidth).

Musa et al. (2017) showcased the synthesis of YIG and YAIG ferrite using auto combustion sol-gel technique. In their results they showed that a decrease in saturation magnetization was observed as doping of Al was increased. It was also showed that the magnetic phase change from ferrimagnetic to paramagnetic was observed for Al doping at $x > 1.8$, for $Y_3Fe_{5-x}Al_xO_{12}$,

Magnetic and structural properties of YIG and ($Y_3Al_xFe_{5x}O_{12}$, YAIG) (x = 0.2, 0.6, 1, 1.4, 1.8, and 2.2) nanoparticles was studied. Sample was processed using Sol-gel method. The group used citric acid as combustion fuel and chelating agent. Powder so obtained was heated at 950 °C. X-ray diffraction peaks so observed confirmed the phase formation of garnet. As the level of Al was increased an increase in the crystallite size has been reported from 28.5894 to 28.6170 nm. Also as Al increase from 0.0 to 2.2 for the samples, lattice constant was found to reduce from 12.4674 Å to 12.3233 Å. Fourier Transform Infrared Spectroscopy (FTIR) was used to substantiate the garnet structure, with increasing Al concentration.

Following observations were made

a. It was seen that the primary vibrating modes shifted to a higher wave number.
b. Saturation magnetization, Ms dropped from 20.721 to 0.7586 emu/g while Al increased (0.0 to 2.2).
c. A high Al substitution content on YIG causes the ferrite to behave paramagnetically.
d. A decrease in grain size from 0.64 lm to 0.32 lm
e. When Al is increased from 0.0 to 2.2, the bulk density decreases from 5.058 gcm^3 to 4.233 gcm^3.

Additionally, it was discovered that the diminishing static magnetic field's trends. of YAIG samples were brought on by the addition of Al ions to the YIG crystal lattice.

APPLICATIONS

YIG nanoparticles of microwave absorption make it very popular choice for use with microwave devices. Lot of research is on to explore its applications in magneto-optical devices, optical communication and in passive microwave applications. It finds application in lot of microwave devices like switches, radars, power limiters, filters, attenuators, phase shifters, circulators, isolators and many more YIG film is also used as substrate in microwave integrated circuits.

5.1 Microwave Antenna

In order to transmit or receive a signal, an antenna converts a wave travelling along a transmission line into a plane wave that is propagating in empty space (Pozar, 1998). The material used must have a sufficiently high dielectric permittivity in order to maintain a standing electromagnetic wave within its volume due to total internal reflection at the dielectric interface (Moulson & Herbert, 2003). Other specifications for antenna design can be listed as follows:

Ø light weight
Ø low profile
Ø small size,
Ø low cost,
Ø high radiation efficiency,
Ø large bandwidth,
Ø ease of integration with other passive or active microwave integrated circuit components.

A dielectric resonator which is defined as slice of dielectric, which can be nonconductive, but is polarizable, normally ceramic, which is designed to function as a resonator for radio waves, comes in number of shapes. It can be rectangular, hemispherical, spherical, low profile triangular or circular disk, cylindrical, etc. (Moulson & Herbert, 2003; Pozar, 1998). The use of Sea bed logging (SBL), which relies on an aluminium transmitter and yttrium iron garnet, for detection was demonstrated by Akhtar et al. (2011). A YIG layer with high initial permeability (30.8773), high Q-factor (45.719), and low loss factor was studied by the group (0.0001). Electro-magnetic antenna are designed using high Quality factor materials. According to the group's findings, using an EM antenna in a scale tank enhanced the magnetic field's strength by up to 166%. The research group studied the Aluminum electromagnetic antenna designs with $Y_3Fe_5O_{12}$ and without $Y_3Fe_5O_{12}$ based magnetic feeders. These magnetic feeders, synthesized at a temperature of 1350°C, were used in scale tank. The plots of Magnitude verses offset of Al transmitter and Al transmitter with $Y_3Fe_5O_{12}$ were studied to see the behavior of magnetic feeders using $Y_3Fe_5O_{12}$ in scale tank. According to the research team, the presence of Y3Fe5O12 ferrite in magnetic feeders with aluminium EM antenna was found to boost the magnetic field's strength (Yahya et al., 2011; Yahya et al., 2009).

5.2 Broad Band Microwave Absorber (Vinay Sharma et al., 2017)

Microwave Absorbers, also called noise suppressors, are the devices which attenuate the energy in an electromagnetic wave. These devices are extensively used in military applications like radar cross reductions, antenna pattern shaping etc. Magnetic absorbers are used to reduce electromagnetic interference in wireless electronics at giga-hertz frequencies. They are very thin polymeric materials filled with magnetic nanoparticles (Braginsky,, 1987; Celinski et al., 1997; Wen et al., 2013). These nanoparticles are characterized by both high values of permittivity and permeability. Because of these characteristics, they are very efficient in removing high frequency electromagnetic interference. Bijoy et al demonstrated the application of broad band microwave absorbers based on YIG (Vinay Sharma et al., 2017). A thin nanocomposite layer for complete absorption of incident electromagnetic (EM) radiation covering frequencies from S to K band was proposed by the researchers as part of a process for fabricating thin layer microwave absorbers that operates across a broad frequency range. When the temperature approached the YIG Néel temperature, microwave absorption was seen to disappear. The group reported crystallinity and high purity using microstructural characterization. The nanoparticles, which were deposited using the electrophoretic deposition process, were ferrimagnetic, according to VSM analysis. They reported a 300 micron film thickness and a microwave absorption of 28 dB at 20 GHz.

Figure 7. Applied frequencies versus resonance fields for YIG SS NPs (Vinay Sharma et al., 2017)

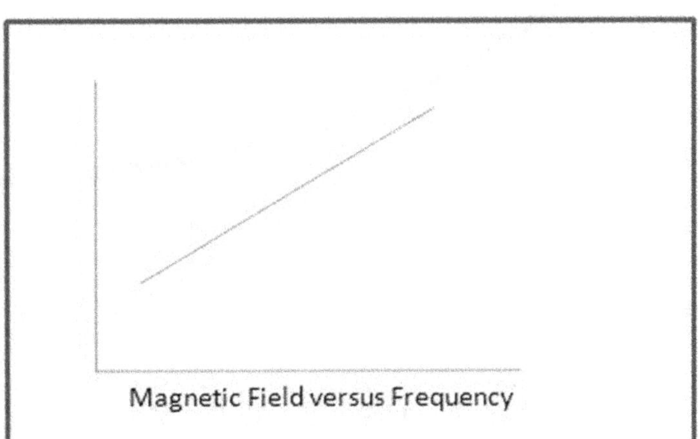

From the study, it was evident that the ferro-magnetic resonance field is directly proportional to operating frequency.. It is observed that at 15 GHz, the nanoparticles synthesized by solid-state method resonate at higher field than those made by sol-gel method.The group reported that the resonance frequency was linearly related to the applied biasing field. Assuming that the nanoparticles are spherical in shape, the expression for ferromagnetic resonance field can be written as

$$H_r = \omega / \gamma - H_{eff}$$

where ω is the Larmor precession angular frequency,

γ : gyro-magnetic ratio.

The effective field, H_{eff} is given by:

$$H_{eff} = H_D + H_K + H_{int}$$

where

H_D: demagnetizing field

H_K : anisotropy field and

H_{int}: interaction field

It has been found that the saturation magnetization of YIG nanoparticles produces a value that is a little bit higher (1.7 kOe) than the value determined using VSM data. For YIG, the interaction field (Hint) is adverse.

Figure 8. Temperature dependence on variable resistance (Vinay Sharma et al., 2017)

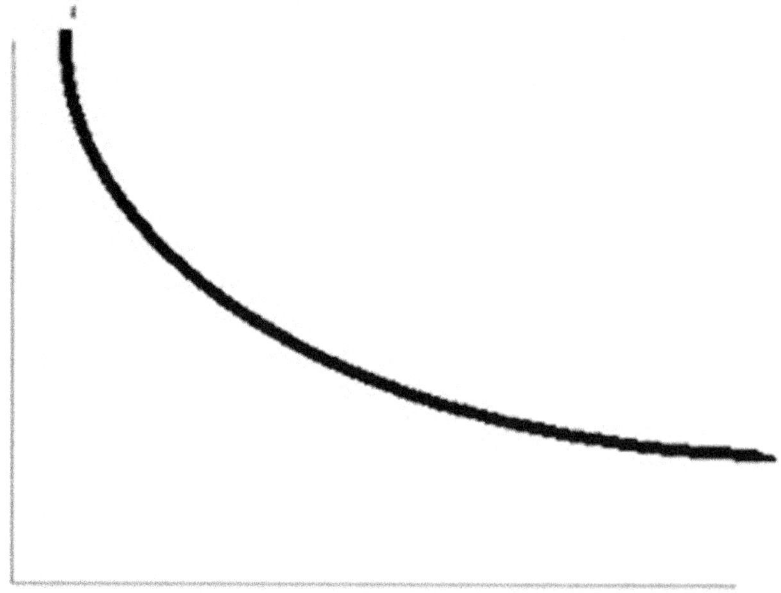

Temperature versus Resistance

The resistance (dc) of nanoparticles was studied as a function of temperature by Bijoy et al. to demonstrate how microwave absorption is temperature-dependent (Figure 8). According to the graph, atomic thermal vibration at the lattice was the reason why the dc resistance decreased from 1G at ambient temperature to 100 at a working temperature of 300 °C. The mechanism of electrons hopping between the ions of Fe2+ and Fe3+ becomes primarily big with an increase in temperature (Fainstein et al., 1999). As a result, the resistance of YIG dramatically decreases at high temperatures. This high temperature drop in resistance is also a result of low concentrations of remnant phases that occur during YIG manufacturing. Similar results were obtained by Wu et al. (2016) who measured the dc conductivity of bulk YIG in the temperature range of 70 K to 530 K. In the operating range of 300 K to 666 K, Sirdeshmukh et al. (1998) found a change on the order of 1012 in the bulk dc conductivity of YIG.

5.3 Modulators (high speed) and Isolators (Firby & Eleezzabi, 2017; Firby & Eleezzabi, 2016)

C.J. Firby and A.Y. Eleezzab presented a design of magneto-plasmonic integrable modulators and isolators. The design recommended was based on magneto-plasmonic waveguide structure. They proposed an efficient architecture of isolator where a Mach-Zehnder inferometer is equipped with a magnetized cerium-substituted YIG waveguides and planar samarium-cobalt biassing magnets. The measurements revealed an isolation of 22.82 dB and an insertion loss of 2.51 dB within a dimension of 6.4 x 10-3 mm^2. A high-speed electrical to optical clock multiplier using bi-substituted YIG in a Mach-Zehnder inferometer was also proposed by the researchers. This clock has an output modulation frequency of 279.9 MHz and a modulation depth of 16.26 dB. The group also gave design for a nonlinear RF magnetoplasmonic modulator for an integrated optical signal. The recommended design had plasmonic Mach-Zehnder Inferometer, constructed using Bi:YIG. A non-reciprocal shift in phase was induced due to transverse component of magnetization due to Bi: YIG. This modulated phase shift in Mach-Zehnder Inferometer leads to modulation of output optical intensity. Presence of harmonics in the driving radio frequency field, splitting around the input frequency was reported. This resulted because of high nonlinear evolution of magnetization in bismuth: yttrium iron garnet. The device so proposed provided a unique method of generation of modulation frequencies within a single structure.

5.4 Power limiters (Stitzer et al., 1997)

S.N. Stitzer et al proposed a frequency-selective power limiter (multistage) in which the process of limiting a part of signal occurs in the three tandem connected stages. In the design presented, the diameter of the three YIG sphere used as power limiters used in three stages were optimized for

Ø lowest threshold,
Ø widest dynamic range and
Ø sharpest selectivity.

Following was observed in the designed phase limiter

- 1.7 dB or less below threshold loss.
- dynamic range of 28 dB

- limited output power less than 5 mw;
- Efficient handling of 3 watts CW,
- X-band with 1 percent bandwidth and
- Third order output inter-modulation signals were more than 20 dB better than a second in-band signal of +33 dBm beating a 0 dBm in-band signal.

The structure proposed found use in high power FW/CW monostatic RADARs, particularly where simultaneous transmission and reception is required,

5. 5 Phase Shifters

A phase shifter is a waveguide which is a two port structure providing variable or fixed phase change of the wave through passing through it. They are commonly used in phased antenna arrays. An ideal phase shifter is matched and lossless. Phase shifters which are tunable find applications in microwave communications like radar systems. They can be synthesized by using ferrites, semiconductors (Koul & Bhat, 1991). Phase shifters based on spin wave propagation in ferrite films, such as epitaxial single-crystal YIG, are used due to their magnetic tunability and planar architecture. High-quality YIG films show relatively little propagation losses for magnetostatic spin waves, less than 30 dB/s., as proven by Adam et al. (2002) and Ishak (1998). Lancaster et al (1998), Kozyrev et al (2003) and Kuylenstierna et al (2006) proposed ferroelectric material based a separate class of microwave phase shifters. The typical aspects of the phase shifters based on ferroelectric are low power consumption and fast electric tunability. Kim et al (1999) and Semenov et al (2006) have reported layered ferrite-ferroelectric based shifters. The results reported by them opened many avenues for microwave devices which were dual-tunable. In the ongoing research, many research groups are investigating several hybrid structure applications in microwave domains. Some of them can be listed as: YIG and lead zirconium titanate layers were used to create a phase shifter that used them as a medium for the transmission of magnetostatic waves. (Fetisov & Srinivasan, 2005a). Based on the ferromagnetic resonance phenomena, the YIG-lead zirconate titanate phase shifter (Tatarenko et al., 2006) and a delay line built on a bilayer of lead magnesium niobate-lead titanate and YIG film (Fetisov & Srinivasan, 2005b) These groups have successfully showcased magnetic and electric tunability by varying permeability and permittivity respectively for ferrites. A. B. Ustinov et al (2007), the research team developed a dual, magnetic and electric field tunable microwave phase changer based on the propagation of hybrid spin-electromagnetic waves in a ferrite-ferroelectric bilayer composed of a single-crystal YIG film and a slab of ceramic barium strontium titanate. They reported that a electrical tunability of the differential phase shift $\Delta\varphi\Delta\varphi$ was achievable if a voltage is applied across the slab.

Yang et al (2014) proposed phase shifters which were voltage tunable on YIG /lead magnesium niobate lead titanate heterostructure. They used bending mode to examine the electric-field tunability of 1-layer and 2-layer PMN-PT structures. Their analysis revealed that the suggested mechanism displayed a phase shift of 119 ° in the presence of an electric field of 11 kV/cm. The system, which utilised partly magnetised ferrite and multiferroics, operated in the absence of an external bias magnetic field. Results showed that compared to conventional multiferroic devices, the design was space-efficient.

CONCLUSION

YIG nanoparticles with unique features have led to a revolution in the microwave and millimeter research areas. YIG naoparticles (YIG-NPs) are coming up as promising material with a wide range of applications in realms of magnanics, optoelectronics, radiofrequency and ferromagnetic devices. Morphology of YIG influence the mechanical reactions where it is used. YIG nanoparticles may be synthesized by a variety of techniques like solid-state process, sol-gel, auto-combustion etc. The application area where they can be very efficiently used can be listed as phase splitters, phase modulators, power limiters, isolators, microwave absorbers etc, A lot of research is on where the specific properties of YIG are explored and applied. In years to come we will see a lot of work done in this regard, thus making YIG a smart nano-material to work on and with.

REFERENCES

Adam, J. D., Davis, L. E., Dionne, G. F., Schloemann, E. F., & Stitzer, S. N. (2002). Ferrite devices and materials. *IEEE Transactions on Microwave Theory and Techniques*, *50*(3), 721–737. doi:10.1109/22.989957

Ahn, R. Y. S., Han, M. H., & Kim, C. O. (1996). Synthesis of yttrium iron garnet precursor particles by homogeneous precipitation. *Journal of Materials Science*, *31*(16), 4233–4240. doi:10.1007/BF00356444

Aichele, T., Lorenz, A., Hergt, R., & Görnert, P. (2003). Garnet layers prepared by liquid phase epitaxy for microwave and magneto-optical applications – a review. *Crystal Research and Technology*, *38*(78), 575–587. doi:10.1002/crat.200310071

Akhtar, M. N., Yahya, N., & Nasir, N. (2011). *Novel EM Antenna based on Y3Fe5O12 magnetic feeders for improved MVO*. IEEE. doi: 10.1109/SIECPC.2011.5876902

AZoNano. (2013). *Yttrium Iron Oxide / Yttrium Ferrite (Y3Fe5O12) Nanoparticles – Properties, Applications*. AZoNano.com.

Bastrukov, S., Khoo, J. Y., Lukiyanchuk, B., & Molodtsova, I. (2012). *Micromagnetic theory of spin relaxation and ferromagnetic resonance in multilayered metallic films, 2012*. arXiv:1210.2609

Belov, L. A., Smolskiy, S. M., & Kochemasov, V. N. (2012). *Handbook of RF, Microwave, and Millimeter-wave Components*. Artech House.

Bonnet, M., Delapalme, A., Fuess, H., & Thomas, M. (1975). Refinement of the structure of yttrium iron garnet (YIG). A case of severe extinction and absorption. *Acta Crystallographica. Section B, Structural Crystallography and Crystal Chemistry*, *31*(9), 2233–2240. doi:10.1107/S0567740875007315

Braginsky, V.B. (1987), Experimental observation of fundamental microwave absorption in high-quality dielectric crystals. *Physics Letters A*, *120*(6), 300-305.

Buscaglia, V., Caracciolo, F., Bottino, C., Leoni, M., & Nanni, P. (1997). Reaction diffusion in the Y_2O_3-Fe_2O_3 system. *Acta Materialia*, *45*(3), 1213–1224. doi:10.1016/S1359-6454(96)00246-7

Celinski, Z., Urquhart, K. B., & Heinrich, B. (1997), Using ferromagnetic resonance to measure the magnetic moments of ultrathin films. *Journal of Magnetism and Magnetic Materials*, *166*(1–2), 6-26.

Cheong, S. W., & Mostovoy, M. (2007). Multiferroics: A magnetic twist for ferroelectricity. *Nature Materials*, *6*(1), 13–20. doi:10.1038/nmat1804 PMID:17199121

Cherepanov, V., Kolokolov, I., & L'Vov, V. (1993). The Saga of YIG: Spectra, Thermodynamics, Interaction and Relaxation of Magnons in a Complex Magnet. *Physics Reports, 229*(3), 84–144.

Dahiya, S., & Duhan, S. (2018). Metal–ferrite nanocomposites for targeted drug delivery. *Applications of Nanocomposite Materials in Drug Delivery.*

Fainstein, A., Winkler, E., Butera, A., & Tallon, J. (1999). Magnetic interactions and magnon gap in the ferromagnetic superconductor RuSr2GdCu2O8. *Physical Review B: Condensed Matter and Materials Physics*, *60*(18), R12597–R12600. doi:10.1103/PhysRevB.60.R12597

Fetisov, Y. K., & Srinivasan, G. (2005a). Ferrite piezoelectric microwave phase shifter: Studies on electric field tunability. *Electronics Letters*, *41*(19), 1066. doi:10.1049/el:20051653

Fetisov, Y. K., & Srinivasan, G. (2005b). Electrically tunable ferrite-ferroelectric microwave delay lines. *Applied Physics Letters*, *87*(10), 103502. doi:10.1063/1.2037860

Fiebig. (2005). Revival of the magnetoelectric effect. *Journal of Physics. D, Applied Physics*, *38*(8), R123–R152. doi:10.1088/0022-3727/38/8/R01

Firby, C. J. & Eleezzabi, A. Y. (2017). Nonlinear optical modulation in a plasmonic Bi:YIG Mach-Zehnder Inferometer. *SPIE Proceedings, Integrated Optics: Devices, Materials and Technologies XXI.*

Firby, C. J., & Eleezzabi, A. Y. (2016). Design of integrated YIG-based isolators and high speed modulators. *SPIE Proceedings, 9746.*

Gautam, R., & Chattopadhyaya, M. (2016). Nanomaterials for Wastewater Remediation. Academic Press. doi:10.1201/9781315368108

Goulon, Rogalev, Wilhelm, Goujon, Yaresko, Brouder, & Ben Youssef. (2012). Site-selective couplings in x-ray-detected magnetic resonance spectra of rare-earth-substituted yttrium iron garnets. *New Journal of Physics, 14*(6), 063001.

Grattan, K. T. V., & Meggitt, B. T. (Eds.). (1999). Optical Fiber Sensor Technology: Volume 3: Applications and Systems. Springer Science & Business Media.

Guo, X. Z., Ravi, B. G., Devi, P. S., Hanson, J. C., Margolies, J., Gambino, R. J., Parise, J. B., & Sampath, S. (2005, August). Synthesis of yttrium iron garnet (YIG) by citrate–nitrate gel combustion and precursor plasma spray processes. *Journal of Magnetism and Magnetic Materials*, *295*(2), 145–154. doi:10.1016/j.jmmm.2005.01.007

Hahn, O. (1936). *Applied Radiochemistry*. Cornell University Press.

Harvey, D. (2000). *Modern Analytical Chemistry*. McGraw-Hill.

Holm, U., Sohlstrom, H., & Brogardh, T. (1984). YIG-Sensor Design for Fiber Optical Magnetic-Field Measurements. *Proceedings - Society of Photo-Optical Instrumentation Engineers*, *514*, 333–336.

Homg, R. Y., Fu, H. P., Di, G. Q., Zheng, Y., & Wei, D. G. (2008, March 15). Facile route to γ-Fe_2O_3/SiO_2 nanocomposite used as a precursor of magnetic fluid. *J. Magn. Magn. Mater. Materials Chemistry and Physics*, *108*(1), 132–141. doi:10.1016/j.matchemphys.2007.09.014

Hongjie, Z., Ji, Z., Yang, B., Zhilun, G., & Longtu, L. (2004). Effect of Bi-substitution on the ielectric properties of polycrystalline yttrium iron garnet. *Journal of Magnetism and Magnetic Materials*, *280*(2-3), 208–213. doi:10.1016/j.jmmm.2004.03.014

Ishak, W. S. (1998). Magnetostatic wave technology: A review. *Proceedings of the IEEE*, *76*(2), 171–187. doi:10.1109/5.4393

Ivanov, S. (2012). Advanced Functional Materials Science and Technology of Atomic, Molecular, Condensed Matter & Biological Systems. Academic Press.

Kawano, K., Kotsugi, M., & Ohkochi, T. (2014). Microstructure effect on magnetization and domain structure in $Ni_{0.49}Zn_{0.49}Co_{0.02}Fe_{1.90}O_x$ ferrite. *Journal of Magnetism and Magnetic Materials*, *361*, 57–60. doi:10.1016/j.jmmm.2014.02.048

Kim, W. J., Chang, W., Quadri, S. B., Wu, H. D., Pond, J. M., Kirchoefer, S. W., Newman, H. S., Chrisey, D. B., & Horwits, J. S. (1999). Electrically and magnetically tunable microwave device using (Ba, Sr) TiO3/Y3Fe5O12 multilayer. *Applied Physics. A, Materials Science & Processing*, *71*, 1564.

Kimura, T., Takizawa, H., Uheda, K., Endo, T., & Shimada, M. (1998). Microwave Synthesis of Yttrium Iron Garnet Powder. *Journal of the American Ceramic Society*, *8*(11), 2961–2964. doi:10.1111/j.1151-2916.1998.tb02720.x

Kohara, Y., Yamasaki, Y., Onose, Y., & Tokura, Y. (2010). Excess-electron induced polarization and magnetoelectric effect in yttrium iron garnet. *Physical Review. B*, *82*(10), 104419. doi:10.1103/PhysRevB.82.104419

Koul, S., & Bhat, B. (1991). *Microwave and Millimeter Wave Phase Shifters. Artech House*.

Kozyrev, A., Ivanov, A., Soldatenkov, O., Tumarkin, A., Ivanova, S., Kayadanov, T., Perkins, J. D., Alleman, J., Ginley, D. S., Sengupta, L., Chiu, L., & Zhang, X. (2003). Millimeter-Wave Loaded Line Ferroelectric Phase Shifters. *Integrated Ferroelectrics*, *55*(1), 847–852. doi:10.1080/10584580390259290

Kumar, V., & Swart, H. C. (2020). Luminescence properties of rare-earth doped oxide materials. *Spectroscopy of Lanthanide Doped Oxide Materials*.

Kuylenstierna, D., Vorobiev, A., Linner, P., & Gevorgian, S. (2006). Composite right/left handed transmission line phase shifter using ferroelectric varactors. *IEEE Microwave and Wireless Components Letters*, *16*(4), 167–169. doi:10.1109/LMWC.2006.872145

Lancaster, M. J., Powell, J., & Porch, A. (1998). Thin-film ferroelectric microwave devices. *Superconductor Science and Technology*, *11*(11), 1323–1334. doi:10.1088/0953-2048/11/11/021

Liua, J., Yua, P., Jina, Q., Chong, Z., Ming, Z., & Harrisb, V. G. (2018). Microwave-accelerated rapid synthesis of high-quality yttrium iron garnet nano powders with improved magnetic properties. *Mater. Res. Lett.*, *6*(1), 36–40

Lu, A.-H., Salabas, E. L., & Schüth, F. (2007). Magnetic Nanoparticles: Synthesis, Protection, Functionalization, and Application. *Angewandte Chemie International Edition*, *46*(8), 1222–1244. doi:10.1002/anie.200602866 PMID:17278160

Luk, K.-M., & Leung, K.-W. (2003). Dielectric Resonator Antennas. Research Studies Press Ltd.

Magneto-Optical Imaging of Superconductors. (2010). Department of Physics, University of Oslo.

Moulson, A. J., & Herbert, J. M. (2003). *Electroceramics: Materials - Properties – Aplications* (2nd ed.). John Wiley & Sons Inc. doi:10.1002/0470867965

Musa, Azis, R. S., Osman, N. H., Hassan, J., & Zangina, T. (2017). Structural and magnetic properties of yttrium iron garnet (YIG) and yttrium aluminum iron garnet (YAlG) nanoferrite via sol-gel synthesis. *Results in Physics*, *7*, 1135–1142. doi:10.1016/j.rinp.2017.02.038

Nazarov, A. V., Ménard, D., Green, J. J., Patton, C. E., Argentina, G. M., & van Hook, H. J. (2003). Near theoretical microwave loss in hot isostatic pressed (hipped) polycrystalline yttrium iron garnet. *Journal of Applied Physics*, *94*(11), 7227–7234. doi:10.1063/1.1622996

Periodic Table of Elements: Yttrium. (2015), Los Alamos National Laboratory.

Pozar, D. M. (1998). *Microwave Engineering* (2nd ed.). John Wiley & Sons Inc.

Rattan, R. K. (2015). Nanotechnology in Soil Science. Soil Science: An Introduction, 775-807.

Ristic, M., Nowik, I., Popovic, S., Felner, I., & Music, S. (2003). Influence of synthesis procedure on the YIG formation. *Materials Letters*, *57*(16-17), 2584–2590. doi:10.1016/S0167-577X(02)01315-0

Sanchez, D., Ramos, C. A., Rivas, J., Vaqueiro, P., & Lopez-Quintela, M. A. (2004). Ferromagnetic resonance and magnetic properties of single-domain particles of $Y_3Fe_5O_{12}$ prepared by sol-gel method. *Physica B, Condensed Matter*, *354*(1-4), 104–107. doi:10.1016/j.physb.2004.09.028

Semenov, A. A., Karmanenko, S. F., Demidov, V. E., Kalinikos, B. A., Srinivasan, G., Slavin, A. N., & Mantese, J. V. (2006). Ferrite-ferroelectric layered structures for electrically and magnetically tunable microwave resonators. *Applied Physics Letters*, *88*(3), 033503. doi:10.1063/1.2166489

Sharma, V., Saha, J., Patnaik, S., & Kuanr, B. K. (2017). Synthesis and characterization of yttrium iron garnet (YIG) nanoparticles - Microwave material. *AIP Advances*, *7*(5), 056405. doi:10.1063/1.4973199

Sirdeshmukh, L., Krishna Kumar, K., Bal Laxman, S., Rama Krishna, A., & Sathaiah, G. (1998). Dielectric properties and electrical conduction in yttrium iron garnet (YIG). *Bulletin of Materials Science*, *21*(3), 219–0. doi:10.1007/BF02744973

Sirohi, R. S. (1990). *Optical Components, Systems and Measurement Techniques*. CRC Press.

Stitzer, S. N., Carter, P. S., & Goldie, H. (1997). A High Power X-Band Frequency Selective Passive YIG Limiter. In *IEEE MTT-S International Microwave Symposium Digest*. IEEE.

Sun, K., Zhang, Z., Fan, R., Chen, M., Cheng, C., Hou, Q., Zhang, X., & Liu, Y. (2015). Random copper/yttrium iron garnet composites with tunable negative electromagnetic parameters prepared by *in situ* synthesis. *RSC Advances*, *5*(75), 61155–61160. doi:10.1039/C5RA09882C

Tatarenko, A. S., Srinivasan, G., & Bichurin, M. I. (2006). Magnetoelectric microwave phase shifter. *Applied Physics Letters, 88*(18), 183507. doi:10.1063/1.2198111

Thiagarajan, S., Sanmugam, A., & Vikraman, D. (2017). Facile Methodology of Sol-Gel Synthesis for Metal Oxide Nanostructures. *Recent Applications in Sol-Gel Synthesis.*

Townshield, A., & Jackwerth, E. (1989). Precipitation of major constitutuents for trace preconcentration: Potential and problems. *Pure and Applied Chemistry, 61*(9), 1643–1656. doi:10.1351/pac198961091643

Ustinov, A. B., Srinivasan, G., & Kalinikos, B. A. (2007). Ferrite-ferroelectric hybrid wave phase shifters. *Applied Physics Letters, 90*(3), 031913. doi:10.1063/1.2432953

Vinay Sharma, J., Saha, S., Patnaik, B., & Kuanr, K. (2017), YIG based broad band microwave absorber: A perspective on synthesis methods. *Journal of Magnetism and Magnetic Materials, 439*, 277-286.

Vojna, D., Slezák, O., Yasuhara, R., Furuse, H., Lucianetti, A., & Mocek, T. (2020). Faraday Rotation of Dy2O3, CeF3 and Y3Fe5O12 at the Mid-Infrared Wavelengths. *Materials, 13*(23), 324. doi:10.3390/ma13235324

Wen, B., Cao, M.-S., Hou, Z.-L., Song, W.-L., Zhang, L., Lu, M.-M., Jin, H.-B., Fang, X.-Y., & Wang, W.-Z. (2013). Temperature dependent microwave attenuation behavior for carbon-nanotube/silica composites. Carbon, 65, 124-139.

Wu, H., Huang, F., Lu, X., Xu, T., Lu, X., Ti, R., Jin, Y., & Zhu, J. (2016). Grain size and Fe2+ concentration-dependent magnetic, dielectric, and magnetodielectric properties of Y3Fe5O12 ceramics. *Physica Status Solidi, 213*(1), 146–153. doi:10.1002/pssa.201532582

Xu, H.T. (2008). Synthesis of Ni nanoparticles and their catalytic effect on the decomposition of ammonium perchlorate. *J. Mater. Process. Technol., 208*(1–3), 494-498.

Yahya, N. Akhtar, M.N., Nasir, N. Shafie, A., Jabeli, M. S., & Koziol, K. (2011). CNT Fibres /Aluminium -NiZnFe2O4 Based EM Transmitter for Improved Magnitude vs. Offset (MVO) in a Scaled Marine Environment. *J. Nanosci. Nanotechnol.*

Yahya, N., Zhu, T. W., Rusop, M., & Soga, T. (2009). Development of Y3Fe5O12 nano-magnetic feeder for em source of an intelligent horizontal twin dipoles. *AIP Conference Proceedings, 1136*, 269–276. doi:10.1063/1.3160145

Yang, X. (2014), Voltage Tunable Multiferroic Phase Shifter With YIG/PMN-PT Heterostructure. *IEEE Microwave and Wireless Components Letters, 24*(3), 191-193. doi:10.1109/LMWC.2013.2292924

Yiyan, S. M. W. (2013). Yttrium Iron Garnet Nano Films: Epitaxial Growth, Spin-Pumping Efficiency, and Pt-Capping-Caused Damping. *Solid State Physics, 64*, 157–191. doi:10.1016/B978-0-12-408130-7.00006-X

Yttrium Iron Garnet. (2015). American Elements.

Chapter 7
Synthesis Nanoparticals of SiO2 From Rice Husk and Its Industrial Application

Seitkhan Azat
Al-Farabi Kazakh National University, Kazakhstan

Ulan Zhantikeyev
https://orcid.org/0000-0002-1200-2340
Al-Farabi Kazakh National University, Kazakhstan

Kydyr Askaruly
https://orcid.org/0000-0002-8998-0409
Al-Farabi Kazakh National University, Kazakhstan

Kalampyr Bexseitova
https://orcid.org/0000-0002-1200-2340
Al-Farabi Kazakh National University, Kazakhstan

Yerbolat Sailaukhanuly
Al-Farabi Kazakh National University, Kazakhstan

Kainaubek Toshtay
Al-Farabi Kazakh National University, Kazakhstan

Zhandos Tauanov
Al-Farabi Kazakh National University, Kazakhstan

Darkhan Ybyraiymkul
Al-Farabi Kazakh National University, Kazakhstan

ABSTRACT

Rice, which provides a major source of food for billions of people, covers 1% of the surface of the earth. Rice husk (RH), the outer covering of paddy rice, is an important byproduct during the milling process, and residue ash is generated after the burning of RH that is called rice husk ash (RHA), which is a primary waste material in the agricultural industry. The main components in RH are lignin, cellulose, and hemicellulose, which are generally named lignocellulose. RH also contains ca. 15 to 28 wt% of silica. The high content of silica in RH presents opportunities for the preparation of value-added silicon-based

DOI: 10.4018/978-1-6684-7358-0.ch007

materials. Since the 1970s, various silicon-based materials, including silica, silicon carbide, silicon nitride, silicon tetrachloride, zeolite, and silicates, have been successfully synthesized using RH as the silicon source. This field of research has been significantly advanced and expanded in the past decade spurred by the global attention on sustainable and renewable resources.

1. INTRODUCTION

Rice, which provides a major source of food for billions of people, covers 1% of the surface of the earth. Rice husk (RH), the outer covering of paddy rice, is an important by-product during the milling process and residue ash is generated after the burning of RH and the ash is called rice husk ash (RHA), which is a primary waste material in the agricultural industry (Azat et al., 2019).

The main components in RH are lignin, cellulose, and hemicellulose, which are generally named lignocellulose. RH also contains ca. 15 to 28 wt% of silica (Kerimkulova et al., 2019; Askaruly et al., 2019; Azat et al., 2019). The high content of silica in RH presents opportunities for the preparation of value-added silicon-based materials. Since the 1970s, various silicon based materials, including silica, silicon carbide, silicon nitride, silicon tetrachloride, zeolite, and silicates, have been successfully synthesized using RH as the silicon source (Sunb & Gong, 2001). This field of research has been significantly advanced and expanded in the past decade spurred by the global attention on sustainable and renewable resources.

Extraction of silica from rice husk is an emerging trend in the current research field. Large amount of rice husk (RH) are treated as waste and disposed off at the landfill site. These waste materials can also cause fire, which may lead to severe environmental pollutions. The airborne particles produced from dust may induce respiratory disease to human beings (Bock & Su, 1970). The burning of rice husk results in the formation of Rice Husk Ash (RHA) with major SiO_2 content with 10 to 20% of carbon and traces of other organic components depending on the burning conditions, the furnace type, the rice variety, the climate and the geographical area (Brindley & Zussman, 1959).

Moreover, the commonly used silica precursor like tetraethyl orthosilicate (TEOS) is more expensive, and hence rice husk ash (RHA) and other waste sources having silica are used as an alternative (JB d'Espinose de la Caillerie et al., 1995; Mansurov et al., 2013). Acid leaching of the rice husk ash was carried out to remove soluble elemental impurities and hence it increases the purity of the silica content. The organic compounds in Rice Husk and other waste materials can be decomposed under Sintering conditions. The metallic impurities can be transferred to soluble ions by simple acid treatment (Azat et al., 2014; Tauanov et al., 2020). With the aforementioned method, sintering the sources at high temperature requires high energy and the utilization of acid especially the volatile hydrochloric acid not only corrupts the reaction equipment but also brings an extra financial cost and pollutes environment. Hence, sulfuric acid treatment was selected for acid treatment and the purification treatment was done in two steps. The first step is leaching with acid solution to remove the soluble metallic impurities and the second one is sintering the leached sample in order to reduce the total carbon content. The expected results of these two-unit processes are an overall increase in the purity of the silica content.

The development of technology for the production of silica from rice husks is gaining popularity due to the growing consumption of amorphous silicon dioxide in many countries, including Kazakhstan, which has the widest application in various industries. At present, silicon dioxide materials are prepared using several methods, including reactions in the vapor phase, sol-gel and thermal decomposition methods. Nevertheless, their high cost of preparation limits their wide application (Kapizov, et al., 2020).

Considering the above, the relevant task is search of new, alternative sources of silica and development of methods of obtaining from them silicon-containing materials.

2. STRUCTURE AND COMPONENTS OF RICE HUSK (RH)

Critical economic and environmental situations of the current days encourage companies and researchers to develop and improve technologies intended to reduce or minimize industrial wastes. As a consequence, much effort has been expended in different areas, including the agricultural production.

Rice is the second largest produced cereal in the world. Its production is geographically concentrated in Asia with more than 90 percent of world output. The United States and Brazil are the most important non-Asian producers and Italy ranks first in Europe. According to information from the Food and Agriculture Organization of the United Nations, world rice production in 2016 was 748.0 million tons in the global rice market (FAO, 2018). While in Kazakhstan, according to the information of Ministry of Agriculture of the Republic of Kazakhstan, the main regions in the territory of the republic involved in rice production are: Kyzylorda Region (78,4 thousand ha), Almaty Region (11,1 thousand ha) and South Kazakhstan Region (3 thousand ha) (IPAD, 2022; Ministry, 2022). Rice growing in Kazakhstan is developed mainly in the land areas of the lower reaches of the Syr Darya River, in the territory of the Kyzylorda Region. The location of this region below sea level, state of the soil, ameliorative and agricultural conditions have become the reason for the creation of a special irrigation engineering system for watering of rice fields.

In most varieties rice is composed by approximately 20% of rice husk, which contains a fibrous materials and silica (from 1 ton of paddy it is possible to produce 220 kg of rice husk); however, the amount of each component depends on the climate and geographic location of rice crop. Therefore, due to its high percentage in the grain composition, the husk is considered a by-product in the mills and creates disposal and pollution problems (Baimenov, et al., 2020; Sultanov, et al., 2020).

RH is a major agro-industrial by-product of rice milling. It is the natural protective outer shell that develops during the growth of rice, and its main components are palea and lemma (Daulbayev, et al., 2021). Representative dissected structure of mature rice grain, respective constituent structures is shown in Figure 1. These two components interlock with each other after the pollen grains are shed and provide an enclosure to rice grain. The lemma and palea remain firmly attached to rice grain providing shelter during its growth and are separated during rice milling. Structure and internal organization of lemma are different from palea; however, their surface morphology is similar (Ndazi, et al., 2007). In addition, morphology of the outer surface of lemma/palea is different from the inner ones, which houses the rice grain. The inner surface of this protective shell (layer) is comparatively smooth owing to the presence of wax and natural fats, while the outer surface is highly irregular (Daulbayev et al. 2021; Ndazi et al. 2007; Luduena et al., 2011). The presence of waxy materials in the inner surface of RH, which houses the rice grain, provides great asylum to rice grain (paddy) against termites and other harmful microorganisms (Chowdhury et al., 2011). On the other hand, the presence of these waxy materials is considered to be responsible for the poor affinity/interactions between RH particles and the polymeric matrices owing to their ability to hide the reactive sites available for interaction (Mohanty et al., 2001).

Figure 1. Representative dissected structures of mature rice grain

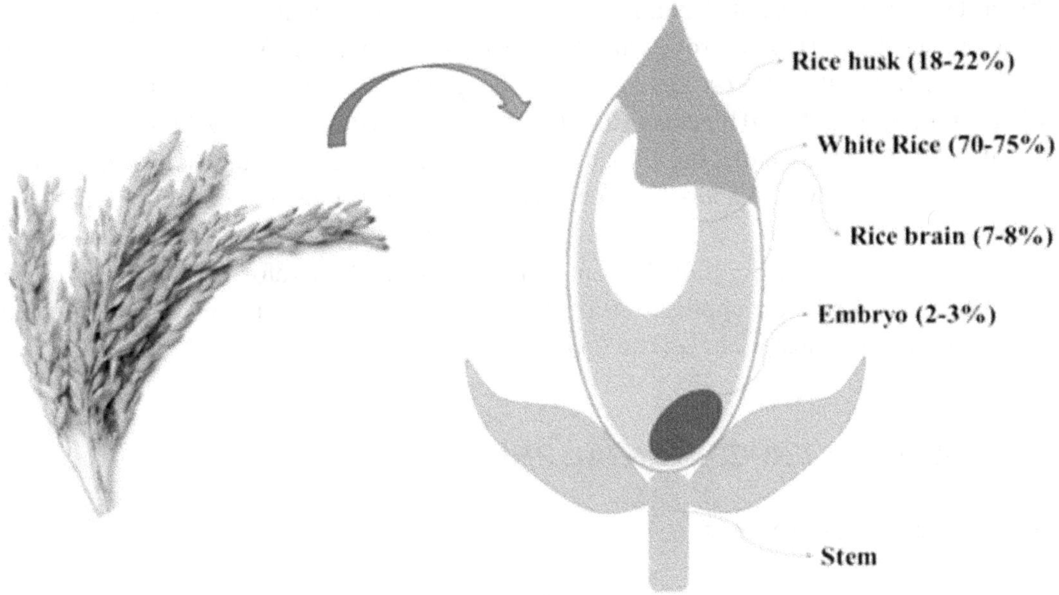

Major constituents of RH are cellulose, hemicellulose, lignin, and silica, and a typical composition of these constituents is tabulated in Table 1.

Table 1. Typical composition of rice husk (Buggenhout, et al., 2013; Kieling, 2009; Johar, et al., 2012; Arjmandi, et al., 2015; Majeed, et al., 2015)

Composition	Percentage
Cellulose	25-35
Hemicellulose	18-21
Lignin	26-31
Silica	15-17
Soluble	2-5
Moisture content	5-10

Reports have been published on the composition, properties, and intended uses of RHs since at least as early as 1871 (Houston, 1972) . Silicon enters the rice plant through its root in a soluble form, probably as a silicate or monosilicic acid, and then moves to the outer surface of the plant, where it becomes concentrated by evaporation and polymerization to form a cellulose silica membrane. There is quite general agreement that the silica is predominantly in inorganic linkages, but some of the silica is also bonded covalently to the organic compounds. This portion of the silica cannot be dissolved in alkali and can withstand very high temperatures (Patel & Karera, 1991). Characterizations by SEM, energy-dispersive X-ray analysis, AES, etc., suggest that the silica is mainly localized in the tough interlayer (epidermis) of the RH and that it also fills in the spaces between the epidermal cells (Houston, 1972; Ding, 1992; Krishnarao & Godkhindi, 1972; Sharma, et al., 1984).

Harvesting silica or silicon from agricultural wastes including RH can not only take full advantage of their highest potential value, but also minimize the related environmental issues from the valorization of RH (Ayatzhan et al., 2021; Jandosov et al., 2022; Satayeva et al., 2022). Thermochemical treatment such as pyrolysis, gasification is considered as the most promising approach that can converted RH to biofuels (e.g., bio-oil, syngas) and biochars simultaneously (Flavio et al., 2010; Akhmetzhan et al., 2021; Toshtay et al., 2021). Subsequently, bio-oil can be converted to renewable biofuels (e.g., biodiesel) by using the RH-derived catalysts. Vapors from organic matters decomposition can be upgraded to value-added syngas for energy application or chemical synthesis by using the RH silica materials, which are used for gas cleaning (Sapei et al., 2008; Soltani et al., 2015), catalytic reforming (Vassilis et al., 2021; Azat et al., 2020). In general, the applications of RH-derived biochars mainly include soil remediation, pollutants removal, silicon materials, and so on (Figure 2).

3 METHODOLOGY & EXPERIMENTAL PROCEDURE

3.1 Samples for Experiments

As research objects served the shells of rice (Rice Husks), selected from different regions of the Republic of Kazakhstan (Figure 2). Samples of rice husk with a length of about 5-10 mm were pre-washed with water and dried in air. Prepared samples were treated and analyzed by using a number of physicochemical methods.

Figure 2. Rice husks from different regions of Kazakhstan (objects of experiments)

1. Almaty region (Balkhash) Rice variety: Regul	**2.Kyzylorda region (Syrdarya)** Rice variety: Lider	**3. South Kazakhstan region (Zhetysay)** Rice variety: 42 or Uzbekski

3.2 Extraction of Silica Nano Particles

The versatile material RHA has generated great amount of interest in researchers world-wide. Researchers have worked on various techniques for the synthesis of silica nanoparticles using different templates to control the shape and size of the particles. However in this present study, a simple, template free different methods such as pre-acid wash treatment (hydrochloric acid, citric acid) and thermally treatment was used to prepare silica nano particles (Figure 3). The aim of acid pre-treatment is to improve the purity of silica product. It proves to be an effective way in substantially removing most of the metallic impurities and producing ash silica completely white in colour. The acid treatment also gives a high surface area for the silica when it is precipitated. In addition the quality of producing nano-silica powder depends on the calcinations or combustions temperature and duration of calcinations or combustions.

WRHA which predominantly contains silica dissolves in sodium hydroxide solution after pre-treatment. Sodium silicate formed is used as precursor for silica synthesis. The principal reaction can be represented as

$$NaOH + SiO_2 \rightarrow Na_2SiO_3 + H_2O$$

$$Na_2SiO_3 + 2HCl \rightarrow H_2SiO_3^- + 2NaCl$$

$$H_2SiO_3 \rightarrow SiO_2 + H_2O$$

Reaction of hydrochloric acid with sodium silicate promotes a silanol (R_3Si–OH) groups formation and condensation of which leads to the formation of extended three-dimensional Si–O–Si network.

Figure 3. Appearance of raw rice husk, white rice husk ash and extracted silica from treated rice husk in 2 M HCl acid followed by combustion at 600 °C

3.3 Pre-Treatment with Hydrochloric Acid

Rice husks (Almaty region, Kyzylorda region, South Kazakhstan region) which is an agricultural residue was used as the main raw material for synthesizing silicon dioxide.

The samples of rice husks which selected in the Almaty region in the village of Bakanas were previously washed with water for the purification of the composition from foreign substances. Then the initial raw materials were dried in the laboratory drying oven at the temperature of 90 °C for 2 hours (for complete evaporation of the water in the composition).

The sample (50 g) of the raw material was treated with the 2M hydrochloric acid solution (500 ml) and dried at 90 °C for 2 hours (Figure 4 a).

Figure 4. Treatment of the raw material with 2M hydrochloric acid solution, a), Calcination of rice husks in the muffle furnace at the temperature of 600 °C, b)

The all prepared samples were continuously washed with distilled water until pH 7 and separated from solution via filtration. Thereupon, RH was dried at 105 °C for overnight and then calcinated at 600 °C for 4 hours in a mufñe furnace (AAF series, Carbolite (UK)) to produce white rice husk ash (WRHA) (Figure 4 b).

The main purpose of heat treatment (calcination) is production pure silicon dioxide from rice husks by burning of organic compounds and carbon in the composition of raw materials at high temperature (Figure 5a).

Figure 5. Combustion of carbon in the composition of RH at high temperature a), formation of the white rice husk ash, b)

After the end of the process, all organic compounds in the rice husk are burned completely (at the temperature of 600 °C for 4 hours) and eventually the ash of white rice husks is formed (Figure 5b).

As can be seen from Figure 6a, the mass of the white rice husk ash obtained after the combustion process was 8.707 grams.

Figure 6. Mass of obtained white rice husk ash, a) treatment of the WRHA with 2M NaOH solution, b)

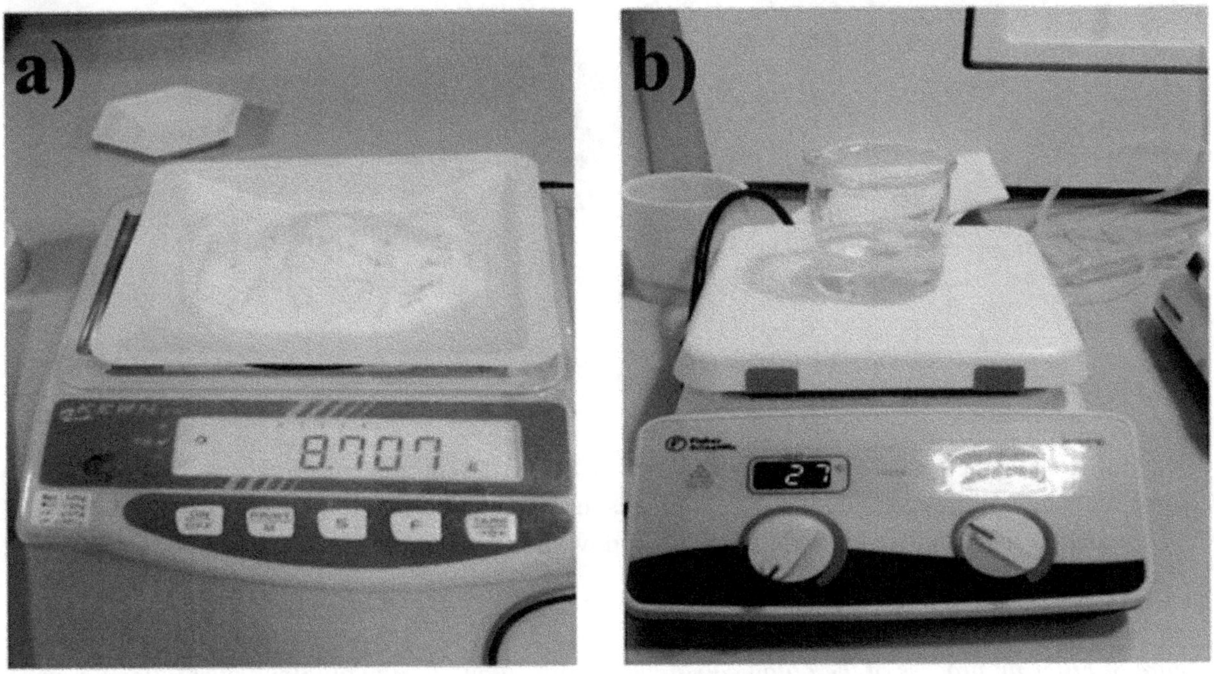

Subsequently the WRHA was mixed with 100 ml of 2M NaOH at 90 °C at continuous vigorous stirring for 2 hours in order to extract the solid silica into water soluble sodium silicate (Figure 6b).

The water soluble sodium silicate solution was ðltered via the vacuum pump to remove insoluble residues (Figure 7a).

Figure 7. Installation of vacuum filtration, a) Formation of the final product (Silicon dioxide), b)

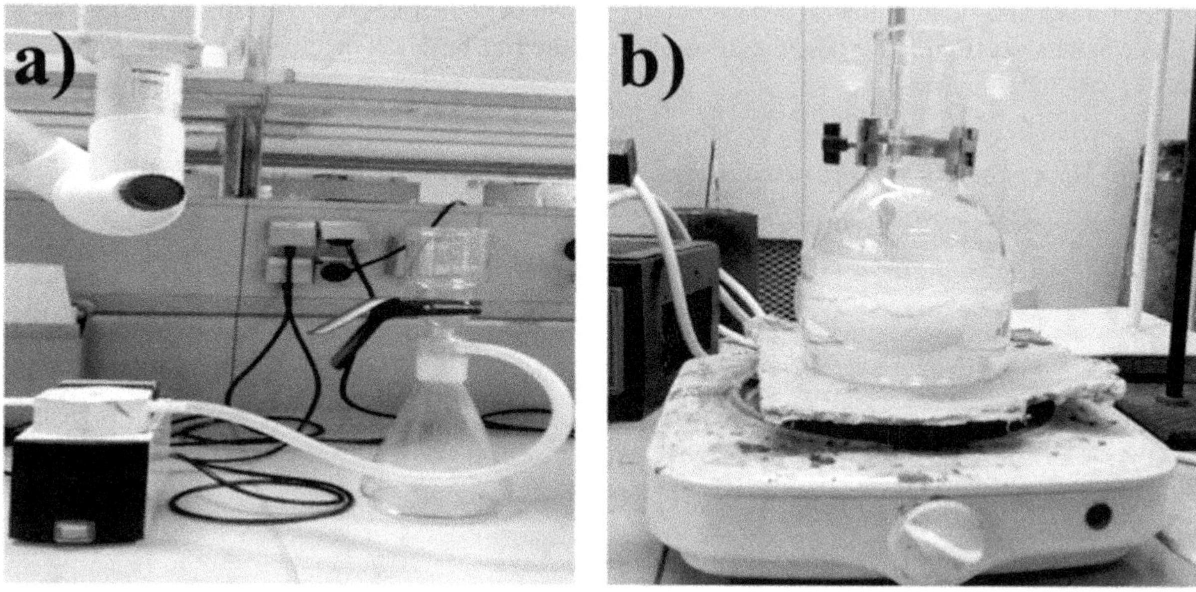

After filtration, the water soluble sodium silicate solution (the filtrate) converted into insoluble silicic acid by reaction with concentrated HCl (30 minutes, under continuous stirring) (Figure 7b).

The final product is passed through the filter and washed with hot water to remove by-products (the NaCl residue in its composition) (Figure 8a,b).

Figure 8. Filtering process of the obtained silicon dioxide, a and b), the obtained pure silicon dioxide after washing with hot water c)

Fltration via the vacuum pump | Fltration via filter paper | Pure silica

The following figure 8c shows the final finished product after washing with hot water. Finally, the obtained pure silicon dioxide was placed in the laboratory drying oven at 105 °C for the duration of a night. After completion of the final product drying, were carried out various physicochemical investigation of the obtained silicon dioxide, the details of the characterizations are given in chapter 3 followed by results and discussion.

Scheme of rice husk conversion into the silica by pre-treatment with hydrochloric acid is presented on a Figures 9.

Figure 9. Scheme of the silicon oxide production from the raw rice husk by pre-treatment with hydrochloric acid

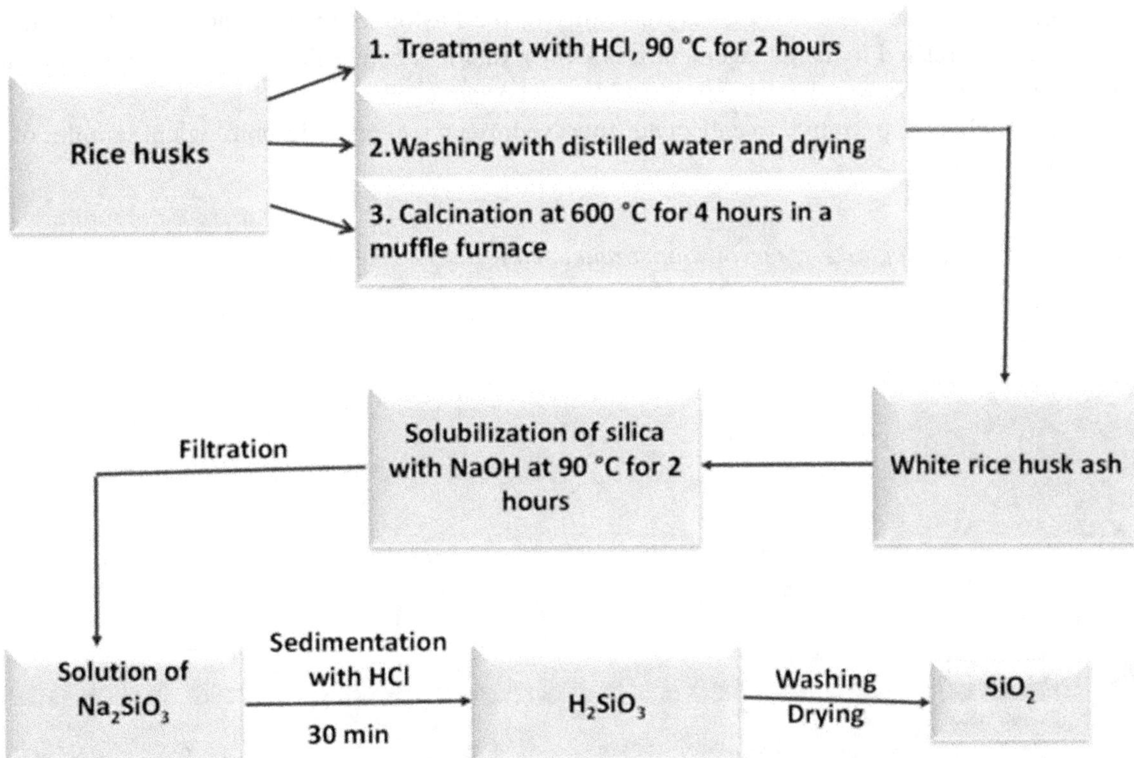

As a whole, as shown by the experiment, from 50 g of rice husks it is possible to obtain 8.71 g of white rice ash, 6g of pure silicon oxide, which means that from 1 ton of the rice husks it is possible to obtain 120 kg of silicon oxide, and from the remains of rice husks in Kazakhstan, approximately in a year 8000 tons of silicon dioxide.

4 HYDRIDE SILICA COMPOSITES FOR RAPID AND FACILE REMOVAL OF AQUEOUS MERCURY

Collection Of "Real" Water Samples and Lake Sediments From the Lake-Reservoir Bylkildak, Pavlodar

The lake-reservoir Bulkuldak was built in 1973 (simultaneously with the site of the former Pavlodar Chemical Plant, succeed to the JSC "Kaustik") with the sole purpose of collecting industrial, household, drainage and storm sewage from the entire territory of the Northern industrial zone of the city of Pavlodar (Figure 10). The reservoir is under the responsibility of the Pavlodar city Akimat, and JSC "Kaustik" has no obligations related to its management.It is known that it is contaminated not only with mercury, but also with other heavy metals, sulfates, chlorides and organic substances.

It was an excellent opportunity to collect aqueous, sediment, soil and microbiological samples of the heavily polluted area.

Figure 10. The Bulkuldak lake-reservoir appearance, view on the Pavlodar prom zone

Experimental Part

An oxidation-reduction (redox) reaction is a type of chemical reaction that involves a transfer of electrons between two species (molecule, atom, or ion changes by gaining or losing an electron). Redox reactions are comprised of two processes, a reduction and an oxidization, that always occur simultaneously. The reduced species gains electrons and the oxidation number decreases, while the oxidized loses electrons and the oxidation number increases.

Methods for the study of reduction-oxidation reactions of nanoparticles in composites were investigated. The silica oxide (obtained from the rice husk ash as described previously) were prepared by modification of with Triethoxysilane (TES, Sigma Aldrich 390143. 95%) into siloxanes by reaction with acetic acid, and conducted procedure for silver nanoparticles formation on silica surface and tested Mercury adsorption onto hydride silica composites.

Synthesis of Surface Modified Silica Substrate

1. Batch (3 g) of the silica oxide was added into a round bottom flask equipped with a reflux condenser. The flask was placed in a water bath with constant temperature (90 ^{0}C) and solution of modifier (0.4 ml triethoxysilane (TES, Sigma Aldrich, 390143, 95%) in 60 ml of the glacial acetic acid) was added under continuous stirring (Figure 11). After 2 h of reaction, the mixture was cooled to room temperature and filtered. Obtained solid was dried at 90 ^{0}C. Resulting modified silica samples were used in reaction of the silver nanoparticles formation.

Figure 11. Modification of SiO_2 by triethoxysilane

Typical procedure for silver nanoparticles formation on silica surface was as follows: 5 samples of modified silica (0.5 g. each) were immersed into silver nitrate (Sigma Aldrich, S0139, ≥99.9%). 1 mmol•L^{-1} and 10 mmol•L^{-1} to make two series of samples) solutions at different volume (2.5, 5, 10, 15, 20 ml, Figure 12 (a,b)), all experiments were carried out in the light shielded conditions to prevent the light degradation of the silver nitrate. Silver nanoparticles are formed on the surface of silica through the chemical reduction of silver ions into zero-valence state as result of reaction with silicon-hydride groups on the silica surface (Figure 13). The obtained samples were filtered and dried for 12 h at 105 °C in the bench oven.

In our case, aqueous mercury solution interacts with silver metal (Ag^0) at a stoichiometric ratio of 1:2 (Eq. (1)), resulting in zero valent mercury. An oxidation-reduction (redox) reaction: reduction – gain of electrons and oxidation – loss of electrons, so we can show next type of redox reaction:

Figure 12. Modification of silica with silver nitrate solution of a different volume: silver nitrate solutions concentration of 1 mmol·L⁻¹ (a) and silver nitrate solutions concentration of 10 mmol·L⁻¹ (b)

$$Hg^{2+} + 2Ag^0_{NPs} \rightarrow Hg^0 + 2Ag^+ \quad (1)$$

$$2Ag^0 - 2e^- \rightarrow 2Ag^+ \quad | \quad Reducer$$

$$Hg^{2+} + 2e^- \rightarrow Hg^0 \quad | \quad Oxidizer$$

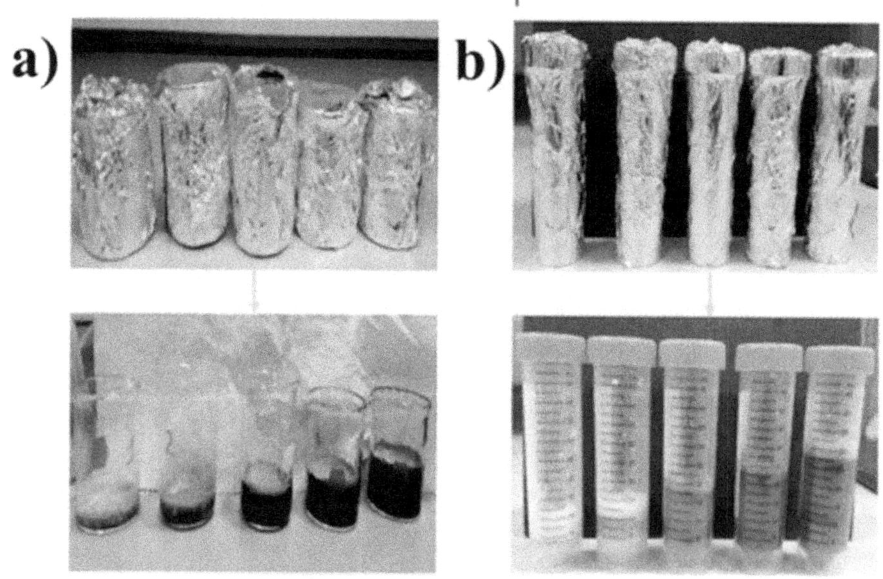

Figure 13. Silver nanoparticles generation on the surface of modified silica substrate

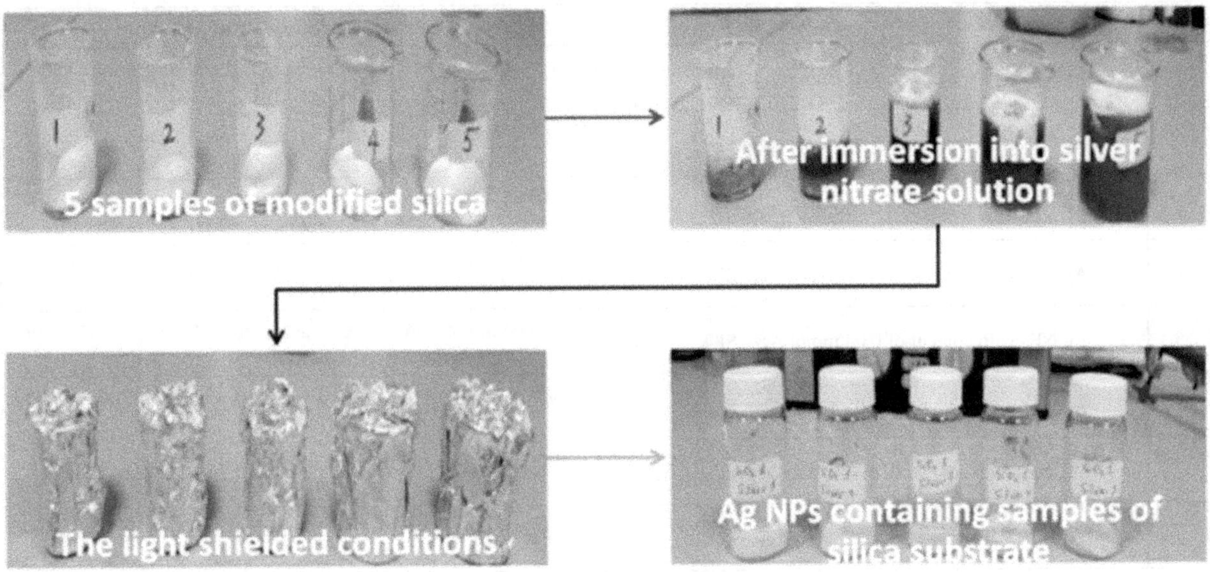

Incineration Of Silver Nanoparticles on Silica Substrate with Aqueous Solution of Mercury (Modular Solution)

Silver nanoparticles deposited on the silica surface were tested in reactions with mercury chloride ($HgCl_2$), for each experiment 0.1 g of silver containing silica was placed in a conical flask and 10 ml of $HgCl_2$ (Sigma Aldrich, M6529, ≥99.5%) solution (100 mg/l) was added. The mixture was continuously stirred at ambient temperature for 1.5 hours. After reaction, the mixture centrifuged and solution analysed for mercury content.

Different concentrations of silver nanoparticles on silica substrate were synthesized and calculated stoichiometric ratio of Hg^{II} to Ag^0: 0.005 mmol Ag/g SiO_2, 0.01 mmol Ag/g SiO_2, 0.02 mmol Ag/g SiO_2, 0.03 mmol Ag/g SiO_2, 0.04 mmol Ag/g SiO_2, 0.05 mmol Ag/g SiO_2, 0.1 mmol Ag/g SiO_2, 0.2 mmol Ag/g SiO_2, 0.3 mmol Ag/g SiO_2, 0.4 mmol Ag/g SiO_2.

All the obtained samples were assigned with the terms given in Table 2.

Table 2. Different silver containing silica samples for interactions with mercury nitrate solution

N	Sample name of silver containing silica	Interactions with mercury nitrate solution
1	Ag NPs with content of 0.005 mmol Ag/g SiO_2	0.1 g of 0.005 mmol of Ag/g SiO_2 was mixed with 10 mL of 100mg/l of $HgCl_2$
2	Ag NPs with content of 0.01 mmol Ag/g SiO_2	0.1 g of 0.01 mmol of Ag/g SiO_2 was mixed with 10 mL of 100mg/l of $HgCl_2$
3	Ag NPs with content of 0.02 mmol Ag/g SiO_2	0.1 g of 0.02 mmol of Ag/g SiO_2 was mixed with 10 mL of 100mg/l of $HgCl_2$
4	Ag NPs with content of 0.03 mmol Ag/g SiO_2	0.1 g of 0.03 mmol of Ag/g SiO_2 was mixed with 10 mL of 100mg/l of $HgCl_2$
5	Ag NPs with content of 0.04 mmol Ag/g SiO_2	0.1 g of 0.04 mmol of Ag/g SiO_2 was mixed with 10 mL of 100mg/l of $HgCl_2$
6	Ag NPs with content of 0.05 mmol Ag/g SiO_2	0.1 g of 0.05 mmol of Ag/g SiO_2 was mixed with 10 mL of 100mg/l of $HgCl_2$
7	Ag NPs with content of 0.1 mmol Ag/g SiO_2	0.1 g of 0.1 mmol of Ag/g SiO_2 was mixed with 10 mL of 100mg/l of $HgCl_2$
8	Ag NPs with content of 0.2 mmol Ag/g SiO_2	0.1 g of 0.2 mmol of Ag/g SiO_2 was mixed with 10 mL of 100mg/l of $HgCl_2$
9	Ag NPs with content of 0.3 mmol Ag/g SiO_2	0.1 g of 0.3 mmol of Ag/g SiO_2 was mixed with 10 mL of 100mg/l of $HgCl_2$
10	Ag NPs with content of 0.4 mmol Ag/g SiO_2	0.1 g of 0.4 mmol of Ag/g SiO_2 was mixed with 10 mL of 100mg/l of $HgCl_2$

Incineration Of Silver Nanoparticles on Silica Substrate with Aqueous Solution Of Mercury (Real Solution)

The interaction of the noble nanoparticles containing nature sourced materials (silica's obtained from rice husk) with "real" mercury-contain aqueous samples of the lake-reservoir Balkyldak was analyzed using Lumex RA-915+ mercury analyzer (Table 3).

Removal of mercury from aqueous solution analyzed by RA - 915M Mercury Analyzer with pyrolysis attachment (PYRO-915^{+}). The initial concentration of the mercury in aqueous sample was 14836.6 ng/l.

Table 3. Residual mercury concentration and adsorption rates (%) given in respect to the silver nanoparticles contain in nature sourced materials (silica's obtained from rice husk) and time of interaction with "real" mercury containing aqueous samples of the lake –reservoir Bylkyldak

N	Samples and time contacts	Concentration of mercury in water after sorption ng/l	Adsorption rates (%)
1	0.2 mmol Ag/g SiO_2, 15 min	3059.3	79.4
2	0.2 mmol Ag/g SiO_2, 30 min	8354.2	43.7
3	0.2 mmol Ag/g SiO_2, 60 min	5553.6	62.6
4	0.2 mmol Ag/g SiO_2, 90 min	1037.6	93
5	0.2 mmol Ag/g SiO_2, 120 min	766.0	94.8
6	0.3 mmol Ag/g SiO_2, 15 min	2525.5	83
7	0.3 mmol Ag/g SiO_2, 30 min	3292.1	77.8
8	0.3 mmol Ag/g SiO_2, 60 min	2167.2	85.4
9	0.3 mmol Ag/g SiO_2, 90 min	1921.3	87
10	0.3 mmol Ag/g SiO_2, 120 min	1760.7	88.1
11	0.4 mmol Ag/g SiO_2, 15 min	5775.1	61
12	0.4 mmol Ag/g SiO_2, 30 min	3074.5	79.3
13	0.4 mmol Ag/g SiO_2, 60 min	5704.6	61.5
14	0.4 mmol Ag/g SiO_2, 90 min	2441.1	83.5
15	0.4 mmol Ag/g SiO_2, 120 min	4935.9	66.7

RESULTS AND DISCUSSION

Silver Nanoparticles Generation

The size and purity of obtained nanoparticles were confirmed via the X-RAY phase analysis (XRD) using the "Dron-6" (Burevesnik).This method allows to determine structural and dimensional features, and to correlate data of this features according to synthesis conditions. Analysis of nanocomposite speciation AgNPs/SiO2 with silver nanoparticles (AgNPs) content equal to 0.2, 0.3 and 0.4 mmol/g. Figures 14a,b,c showed, that spectra of all samples contain five major peaks at 38°, 44.2°, 64.3°, 77.3° and 81.4° 2θ, respectively. Presence of given peaks confirms successful formation of silver nanoparticles, that have face-centered cubic crystalline structure. Average diameters for silver nanoparticles were calculated using XRD-graphs (Figure 14) and their sizes in respect to the NPs loading are: 0.2 mmol/g – 50 nm, 0.3 mmol/g – 80 nm, and 0.4 mmol/g - 140 nm, respectively.

Figure 14. X-ray diffraction patterns for silver nanoparticles on silica subtrate synthesized from 0.2 mmol Ag/ g SiO_2 concentration of silver, a), X-ray diffraction patterns for silver nanoparticles on silica subtrate synthesized from 0.3 mmol Ag/ g SiO_2 concentration of silver, b), X-ray diffraction patterns for silver nanoparticles on silica subtrate synthesized from 0.4 mmol Ag/ g SiO_2 concentration of silver c).

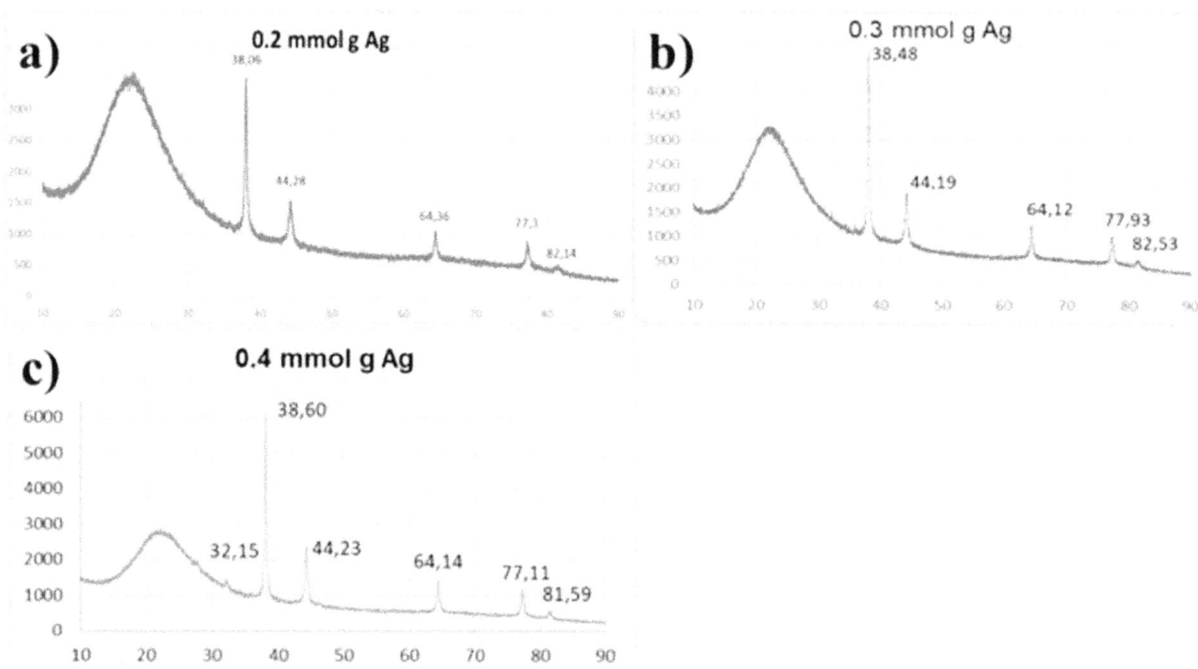

The Amount of Deposited AG Nanoparticles on the Surface of Silica

Silver nanoparticles formation on silica surface was as follows: 5 samples of modified silica (1.1 g. each) were immersed into silver nitrate (10 mmol.L^{-1}) aqueous solution at ambient temperature with different volume (5.5, 11, 22, 33, 44 ml) of silver nitrate; all experiments were carried out in the light shielded conditions to prevent the light degradation of the silver nitrate. Silver nanoparticles are formed on the surface of silica through the chemical reduction of silver ions into zero-valence state as result of reaction with silicon-hydride groups on the silica surface (Table 4). The obtained samples were filtered and dried for 12 h at 105 °C in the bench oven.

Table 4. The amount of deposited Ag nanoparticles on the surface of silica

Sample N°	Description	Ag NPs mass. On silica surface, g
N°1	Ag NPs with content of 0.05 mmol Ag/g SiO2	0.008
N°2	Ag NPs with content of 0.1 mmol Ag/g SiO2	0.017
N°3	Ag NPs with content of 0.2 mmol Ag/g SiO2	0.034
N°4	Ag NPs with content of 0.3 mmol Ag/g SiO2	0.051
N°5	Ag NPs with content of 0.4 mmol Ag/g SiO2	0.067

Surface Modification of Silica from Rice Husk (Almaty Region)

Fourier Transform Infrared Spectroscopy (FTIR) was performed using Agilent technologies, Cary 600 series FTIR spectrometer in transmission (T) mode at mid IR, wavenumbers range 500-4000 cm^{-1} to conduct IR measurements (Figure 15). Band intensities can be also expressed in absorbance mode, in which same results are expected. The powder was then dispersed in a matrix of potassium bromide (KBr) in the ratio of ~1:10. The absorption coefficient of KBr is far less than 400 cm^{-1}, therefore no absorption peak of KBr appeared in the range of measurement.

Silicon hydride groups anchored to the surface of silica particles possess weak reducing properties, which are sufficient for generating "chemically pure" zero-valent silver by the reduction of silver cations according to Equation (2).

Figure 15. IR spectra of Silica and modified silica

$$\equiv SiH + Ag^{+} + 2H_2O \rightarrow \equiv SiOH + Ag^{0} + H_3O^{+} + 1/2\ H_2 \quad (2)$$

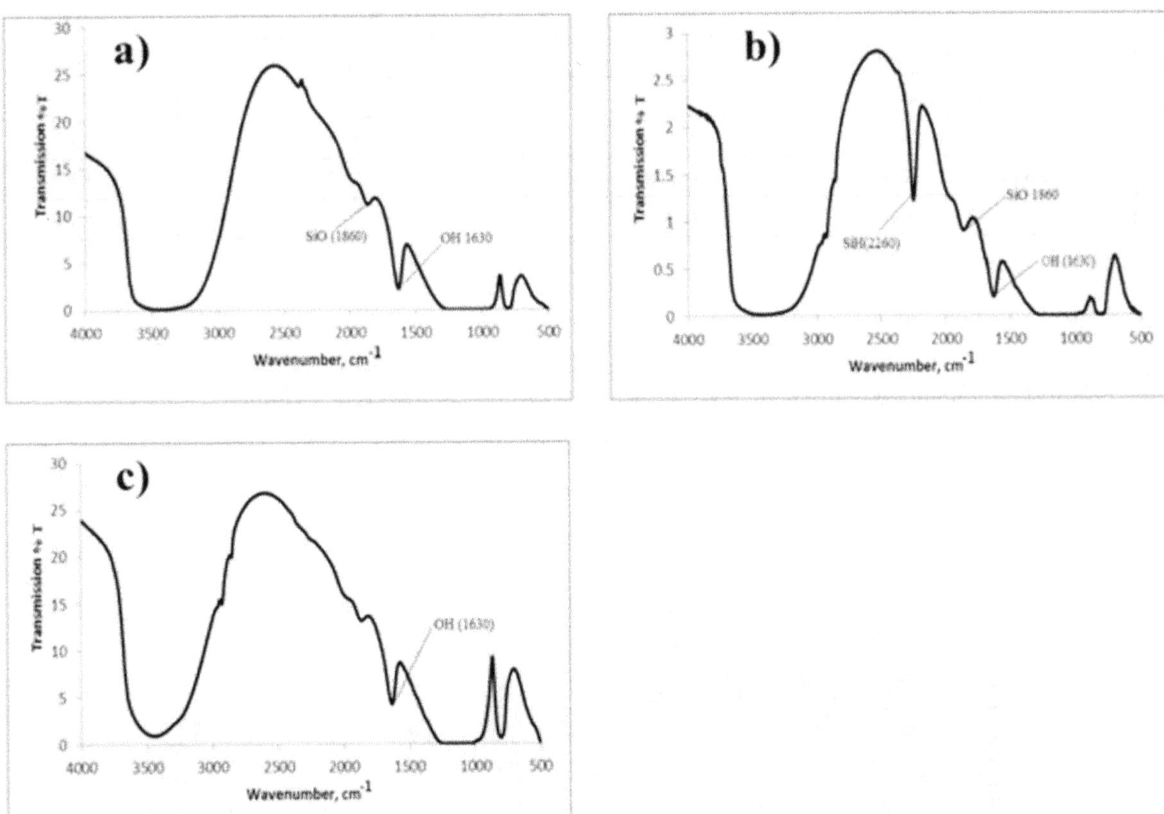

In the IR spectra of Silica, stretching vibrational bands of Si-O can be observed at 1860 cm^{-1} (Figure 15a) contained an intense band with an absorption maximum at 2260 cm^{-1}. This band corresponds to Si–H bond stretching vibrations in surface chemical compounds (Figure 15b). A broad absorption band at 3300–3800 cm^{-1} is evidence of the presence of adsorbed water and perturbed ° SiOH groups in the surface layer. After modiðed silica was brought in contact with a solution of silver nitrate, and silver nanoparticles were reduced in the surface layer of silica matrices (Figure 15c).

The produced silver nanoparticles immobilized on silica support were tested on removal the mercury (II) ions from aqueous solution and calculated stoichiometric ratio of Hg^{II} to Ag^0 (Table 5).

Removal of Hg^{2+} from aqueous solution analyzed by RA-915M Mercury Analyzer with pyrolyzer PYRO-915$^+$. The concentration of Hg^{2+} in water before sorption 98243.1 ng/g (Modular solution).

Table 5. Removal of Hg^{2+} from aqueous solution

N	Sample description	Stoichiometric ratio of Hg^{II} to Ag^0	Concentration Of Hg^{2+} in water after sorption ng/g	Adsorption capacity (%)
1	0.005 mmol Ag/g SiO_2	7.4:1	21014.8	79
2	0.01 mmol Ag/g SiO_2	3.7:1	9741.3	90
3	0.02 mmol Ag/g SiO_2	1.85:1	2927.9	97
4	0.03 mmol Ag/g SiO_2	1.2:1	376.6	99.6
5	0.04 mmol Ag/g SiO_2	0.925:1	797.6	99.1
6	0.05 mmol Ag/g SiO_2	0.74:1	21431.5	78
7	0.1 mmol Ag/g SiO_2	0.37:1	20431.	78
8	0.2 mmol Ag/g SiO_2	0.185:1	17156.3	83
9	0.3 mmol Ag/g SiO_2	0.12:1	15557.3	84
10	0.4 mmol Ag/g SiO_2	0.09:1	14650.9	85

Table 6. Samples of modified silica with silver nitrate solution at different volume

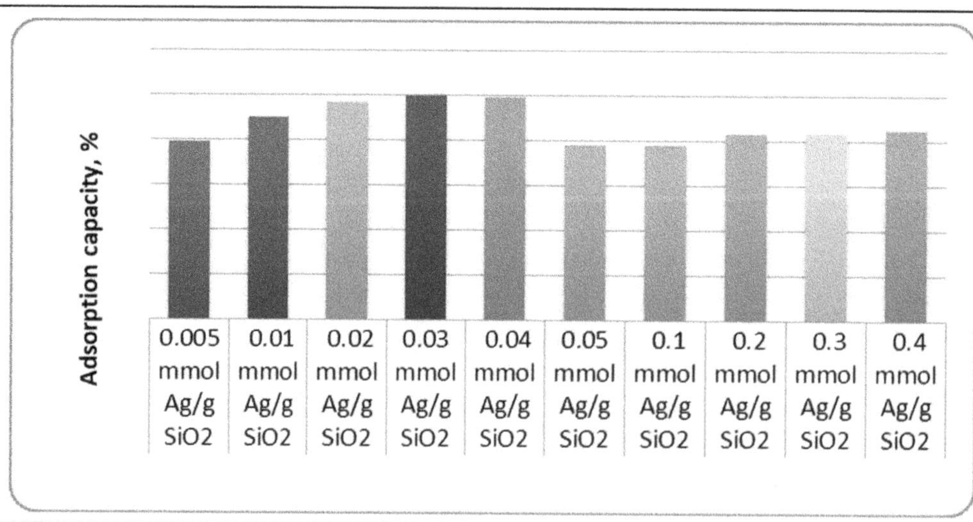

Mercury adsorption onto hydride silica composites is a fast and efficient process, allowing the loading of up to 0.02, 0.03, 0.04 mmol of Ag/g SiO_2 adsorbent with a stoichiometric molar ratio 1.85:1, 1.2:1, 0.925:1 between Hg^{II} and Ag^0 on the silica surface.

It has been found that the occurrence of silver nanoparticles of various sizes on the surface of silicon composite material depends upon the concentration and volume of silver nitrate solution (*Table 6*). Dependence-silicon composite sorption of mercury in water is shown in the following figure (Figure 16).

Figure 16. Samples of modified silica 1mmol·L⁻ (a) and 10mmol·L^{-1} (b) silver nitrate solution at different volume for removal of Hg^{2+} from aqueous solution

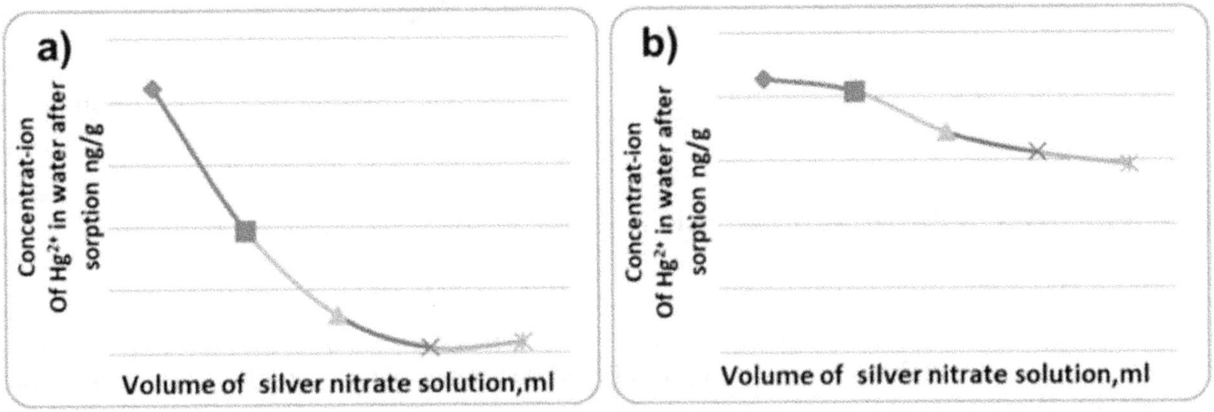

It was determined that the mercury sorption increases depending on the volume of silver nitrate 1 mmol•L^{-1} and 10 mmol•L^{-1} solutions used for silica modification.

Effect of Silver NPs Concentration

Transmission Electron Microscope (TEM) was employed to provide further insights into the influence of the concentration on the size and morphology of silver nanoparticles. Figure 17 show that all the generated nanoparticles have near-spherical shape with varying sizes. It is also seen from the images that the particles are not aggregated with each other agglomerated but spread over the surface of the silica. The reason is that Ag NPs appears only at those sites where the silicon hydride groups are present due to highly reactive properties of the latter component. Moreover, the surface density of the SiH groups are small which also prevents the agglomeration and the stability of the generated nanoparticles results in the higher surface area of the obtained material compared with bulk scale silver.

Figure 17. TEM images of silver nanoparticles synthesized based on different reaction times a) Ag NPs with content of 0.05 mmol Ag/g SiO_2 b) Ag NPs with content of 0.2 mmol Ag/g SiO_2

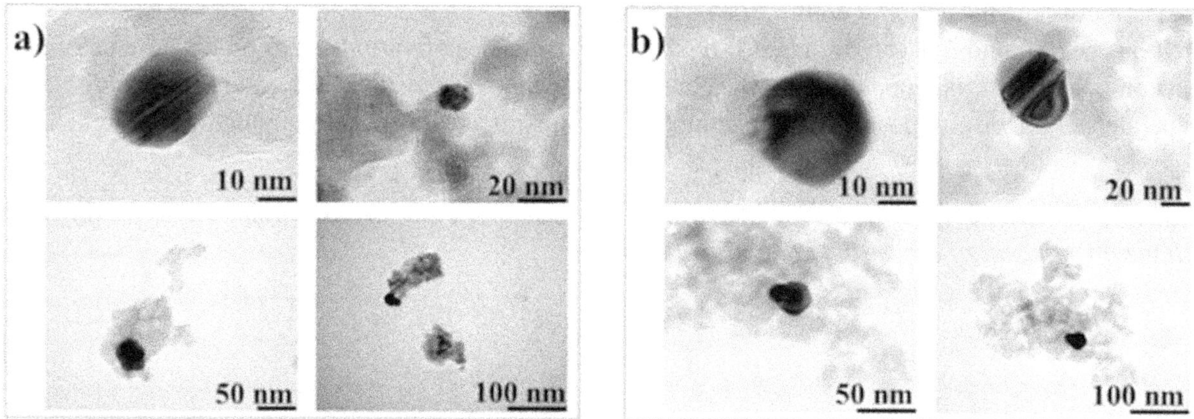

Effect of sizes of Silver Nanoparticles

Interesting picture was observed when the silver containing silica samples were dried in the oven for 12 h. All the samples had the yellow color (Figure 18), meaning that the surrounding aqueous environment affected each samples differently. It is strongly believed that this is due to the differences in the sizes of nanoparticles.

Figure 18. Dried silver containing silica samples synthesized based on the reaction time

CONCLUSION

1. The method of conversion of the rice husk into nano-SiO_2 was developed and confirmed. Pre-treatment methods of rice husk are studied and the composition and content of rice husks components are deter-

mined. The main substance in the ash is amorphous silicon dioxide, the content of which differs in the studied varieties from 90.1-99.5%. The obtained data indicate the existence of the dependence of the studied parameters on the plant variety and the reagent used for the treatment.

2. Different samples of amorphous silica were obtained from the rice husks by chemical reagent treatment. Their elemental composition, structure and physicochemical properties are established and the dependence of the parameters on the method of production and on the plant variety is shown.

3. The surface characteristics of silicon dioxide samples obtained from the rice husk of different region of Kazakhstan are studied. The specific surface area is determined by the BET method (~295 m^2/g). Data about the pore sizes (there are mainly micropores) are investigated. Structure of the obtained nano-SiO_2 was confirmed using an XRD and SEM analyses. The broad diffused peak with maximum of intensity at 24-theta/deg was observed in XRD spectra, indicating amorphous structure and nanoscale of the obtained silica.

4. The dependence of morphology and surface characteristics of silicon dioxide samples obtained from rice husks on the method of production and on the conditions of processing raw materials is established.

5. The use of RHA as partial replacement of geopolymer concrete has been investigated in this thesis.

6. The adsorption of ionic mercury (II) from aqueous solution on functionalized hydride silicon materials was investigated.

In summary, the amorphous silica-rich RH could become a potential resource of low cost precursors for the production of high value-added silica/silicon based materials for practical applications.

«This research has been/was/is funded by the Ministry of Science and Higher Education of the Republic of Kazakhstan (Grant No. AP14869499 Plasma-Assisted water treatment for degradation of persistent polymeric and organic pollutants: from micro- to Nanoparticles to molecules 2022 - 2024) ».

REFERENCES

Akhmetzhan, A., Abeu, N., Longinos, S. N., Tashenov, A., Myrzakhmetova, N., Amangeldi, N., & Toktarbay, Z. (2021). Synthesis and Heavy-Metal Sorption Studies of N, N-Dimethylacrylamide-Based Hydrogels. *Polymers*, *13*(18), 3084.

Arjmandi, R., Hassan, A., Majeed, K., & Zakaria, Z. (2015). Rice husk filled polymer composites. *International Journal of Polymer Science*, 2015.

Askaruly, K., Azat, S., Sartova, Z., Yeleuov, M., Kerimkulova, A., & Bekseitova, K. (2020). Obtaining And Characterization Of Amorphous Silica From Rice Husk. *Journal of Chemical Technology & Metallurgy*, *55*(1).

Ayatzhan, A., Tashenov, A., Nurgeldi, A., Zhanar, O., Zhexenbek, T., Kaldibek, A., & Nuraje, N. (2021). P (DADMAAC-co-DMAA): Synthesis, thermal stability, and kinetics. *Polymers for Advanced Technologies*, *32*(7), 2669–2675.

Azat, S., Arkhangelsky, E., Papathanasiou, T., Zorpas, A. A., Abirov, A., & Inglezakis, V. J. (2020). Synthesis of biosourced silica–Ag nanocomposites and amalgamation reaction with mercury in aqueous solutions. *Comptes Rendus. Chimie*, *23*(1), 77–92.

Azat, S., Busquets, R., Pavlenko, V. V., Kerimkulova, A. R., Whitby, R. L., & Mansurov, Z. A. (2014). Applications of activated carbon sorbents based on greek walnut. []. Trans Tech Publications Ltd.]. *Applied Mechanics and Materials*, *467*, 49–51.

Azat, S., Korobeinyk, A. V., Moustakas, K., & Inglezakis, V. J. (2019). Sustainable production of pure silica from rice husk waste in Kazakhstan. *Journal of Cleaner Production*, *217*, 352–359. doi:10.1016/j.jclepro.2019.01.142

Azat, S., Sartova, Z., Bekseitova, K., & Askaruly, K. (2019). Extraction of high-purity silica from rice husk via hydrochloric acid leaching treatment. *Turkish Journal of Chemistry*, *43*(5), 1258–1269. doi:10.3906/kim-1903-53

Baimenov, A., Berillo, D., Azat, S., Nurgozhin, T., & Inglezakis, V. (2020). Removal of Cd2+ from water by use of super-macroporous cryogels and comparison to commercial adsorbents. *Polymers*, *12*(10), 2405.

Bock, J. A. N., & Su, G. J. (1970). Interpretation of the infrared spectra of fused silica. *Journal of the American Ceramic Society*, *53*(2), 69–73. doi:10.1111/j.1151-2916.1970.tb12012.x

Brindley, G. W., & Zussman, J. (1959). Infra-red absorption data for serpentine minerals.

Buggenhout, J., Brijs, K., Celus, I., & Delcour, J. A. (2013). The breakage susceptibility of raw and parboiled rice: A review. *Journal of Food Engineering*, *117*(3), 304–315.

Chowdhury, S., Mishra, R., Saha, P., & Kushwaha, P. (2011). Adsorption thermodynamics, kinetics and isosteric heat of adsorption of malachite green onto chemically modified rice husk. *Desalination*, *265*(1-3), 159–168.

d'Espinose de la Caillerie, J. B., Kermarec, M., & Clause, O. (1995). 29Si NMR observation of an amorphous magnesium silicate formed during impregnation of silica with Mg (II) in aqueous solution. *Journal of Physical Chemistry*, *99*(47), 17273–17281. doi:10.1021/j100047a036

Daulbayev, C., Sultanov, F., Korobeinyk, A. V., Yeleuov, M., Azat, S., Bakbolat, B., & Mansurov, Z. (2021). Bio-waste-derived few-layered graphene/SrTiO3/PAN as efficient photocatalytic system for water splitting. *Applied Surface Science*, *549*, 149176.

Ding, M. (1992). Rice husk silicon and its applications. *Inorg. Chem. Ind*, *24*(6), 36–42.

FAO. (2018). FAO Rice Market Monitor. FAO. https://www.fao.org/economic/est/publications/rice-publications/rice-market-monitor-rmm/en/

Houston, D. F. (1972). Rice: Chemistry and Technology (No. HOU 633.18 (BR 125.9)). American Association of Cereal Chemists.

Inglezakis, V. J., Azat, S., Tauanov, Z., & Mikhalovsky, S. V. (2021). Functionalization of biosourced silica and surface reactions with mercury in aqueous solutions. *Chemical Engineering Journal*, *423*, 129745.

Jandosov, J., Alavijeh, M., Sultakhan, S., Baimenov, A., Bernardo, M., Sakipova, Z., & Berillo, D. (2022). Activated Carbon/Pectin Composite Enterosorbent for Human Protection from Intoxication with Xenobiotics Pb (II) and Sodium Diclofenac. *Molecules (Basel, Switzerland)*, *27*(7), 2296.

Johar, N., Ahmad, I., & Dufresne, A. (2012). Extraction, preparation and characterization of cellulose fibres and nanocrystals from rice husk. *Industrial Crops and Products*, *37*(1), 93–99.

Kapizov, O., Azat, S., Askaruly, K., Zhantikeyev, U., Tauanov, Z., Bergeneva, N. S., & Satayeva, A. R. (2020). Perspectives of the Silicon Dioxide Production from Rice Husk in Kazakhstan: An Overview. *Eurasian Chemico-Technological Journal*, *22*(4), 285–293.

Kerimkulova, A. R., Azat, S., Velasco, L., Mansurov, Z. A., Lodewyckx, P., Tulepov, M. I., & Imangazy, A. M. (2019). Granular rice husk based sorbents for sorption of vapors of organic and inorganic matters. *J. Chem. Technol. Metall*, *54*, 578–584.

Kieling, A. G. (2009). *Influência da segregação no desempenho de cinzas de casca de arroz como pozolanas e material adsorvente* [*Influence of segregation on the performance of rice husk ash as pozzolans and absorbant material*].

Krishnarao, R. V., & Godkhindi, M. M. (1992). Distribution of silica in rice husks and its effect on the formation of silicon carbide. *Ceramics International*, *18*(4), 243–249.

Ludueña, L., Fasce, D., Alvarez, V. A., & Stefani, P. M. (2011). Nanocellulose from rice husk following alkaline treatment to remove silica. *BioResources*, *6*(2), 1440–1453.

Majeed, K., Hassan, A., & Bakar, A. A. (2014). Influence of maleic anhydride-grafted polyethylene compatibiliser on the tensile, oxygen barrier and thermal properties of rice husk and nanoclay-filled low-density polyethylene composite films. *Journal of Plastic Film & Sheeting*, *30*(2), 120–140.

Mansurov, Z. A., Jandosov, J. M., Kerimkulova, A. R., Azat, S., Zhubanova, A. A., Digel, I. E., Savistkaya, I. S., Akimbekov, N. S., & Kistaubaeva, A. S. (2013). Nanostructured carbon materials for biomedical use. *Eurasian Chemico-Technological Journal*, *15*(3), 209–217. doi:10.18321/ectj224

Mohanty, A. K., Misra, M., & Drzal, L. T. (2001). Surface modifications of natural fibers and performance of the resulting biocomposites: An overview. *Composite Interfaces*, *8*(5), 313–343.

Ndazi, B. S., Karlsson, S., Tesha, J. V., & Nyahumwa, C. W. (2007). Chemical and physical modifications of rice husks for use as composite panels. *Composites. Part A, Applied Science and Manufacturing*, *38*(3), 925–935.

Patel, M., & Karera, A. (1991). SiC whisker from rice husk: Microscopic study. Powder Metallurgy International;(Germany, FR), 23(1).

Sapei, L., Nöske, R., Strauch, P., & Paris, O. (2008). Isolation of mesoporous biogenic silica from the perennial plant Equisetum hyemale. *Chemistry of Materials*, *20*(5), 2020–2025.

Satayeva, A., Baimenov, A., Azat, S., Zhantikeyev, U., Seisenova, A., & Tauanov, Z. (2022). Review on coal fly ash generation and utilization for resolving mercury contamination issues in Central Asia: Kazakhstan. *Environmental Reviews*, (ja).

Sharma, N. K., Williams, W. S., & Zangvil, A. (1984). Formation and structure of silicon carbide whiskers from rice hulls. *Journal of the American Ceramic Society*, *67*(11), 715–720.

Silva, F. S., Cristale, J., André, P. A., Saldiva, P. H., & Marchi, M. R. (2010). PM2. 5 and PM10: The influence of sugarcane burning on potential cancer risk. *Atmospheric Environment*, *44*(39), 5133–5138.

Soltani, N., Bahrami, A., Pech-Canul, M. I., & González, L. A. (2015). Review on the physicochemical treatments of rice husk for production of advanced materials. *Chemical Engineering Journal*, *264*, 899–935.

Sultanov, F., Daulbayev, C., Azat, S., Kuterbekov, K., Bekmyrza, K., Bakbolat, B., & Mansurov, Z. (2020). Influence of Metal Oxide Particles on Bandgap of 1D Photocatalysts Based on SrTiO3/PAN Fibers. *Nanomaterials (Basel, Switzerland)*, *10*(9), 1734.

Sun, L., & Gong, K. (2001). Silicon-based materials from rice husks and their applications. *Industrial & Engineering Chemistry Research*, *40*(25), 5861–5877. doi:10.1021/ie010284b

Tauanov, Z., Azat, S., & Baibatyrova, A. (2022). A mini-review on coal fly ash properties, utilization and synthesis of zeolites. *International Journal of Coal Preparation and Utilization*, *42*(7), 1968–1990.

Toshtay, K., Auyezov, A., Korkembay, Z., Toktassynov, S., Seytkhan, A., & Nurakyshev, A. (2021). Partial hydrogenation of sunflower oil on platinum catalysts: Influence of process conditions on the mass content of geometric isomers. *Molecular Catalysis*, *513*, 111819.

Chapter 8
Development and Characterization of As-Deposited CdTe Thin Films in a Non-Aqueous Medium

Sudeshna Surabhi
National Institute of Advanced Manufacturing Technology, Ranchi, India

Kumar Anurag
National Institute of Advanced Manufacturing Technology, Ranchi, India

S. R. Kumar
National Institute of Advanced Manufacturing Technology, Ranchi, India

ABSTRACT

Cadmium telluride (CdTe) is a binary II-VI direct band gap semiconducting material. Cadmium telluride is a promising electrochemical and photovoltaic material for thin film solar cells. It shows both p and n type conductivity. A nanocrystalline thin film of CdTe was deposited on nickel plate by chemical bath deposition method containing 0.01M Cadmium acetate and 0.02M tellurium dioxide. The structural, compositional, and optical analysis were studied by X-ray diffraction (XRD), energy-dispersive X-ray spectroscopy (EDS), scanning electron microscope (SEM), fourier transform infrared spectroscopy (FTIR) and photoluminescence (PL).The diffraction peak observed at 2=23.59° with (111) plane indicate the crystalline phase of CdTe film. The average crystalline size is measured to be 7nm.Compositional analysis reveals the presence of both Cd and Te elements. The EDS spectroscopy shows the ratio of Cd and Te is1:2 in case of as deposited film. The photoluminescence peak is observed at 471nm.

DOI: 10.4018/978-1-6684-7358-0.ch008

1. GENERAL INTRODUCTION

The properties and quality of the deposited thin film depends upon its growth parameters and deposition methods. These properties of the thin film need to be understood in order to make use of the grown samples for specific applications. The optimization of growth parameters leads to optimal conditions and hence enhance the performance of the desired device. The characterization techniques help in prediction and analysis for various properties of the deposited thin film and help to control or modify deposition process. Hence the deposition methods and characterization techniques play a crucial part in the growth of any material for device fabrication.

1.1 Thin Films

A thin film is a layer of material that is placed on a substrate, which can be made of plastic, metal, ceramic, or a semiconductor. Thin film is described as having a thickness of less than 1 micrometer and a fraction of a nanometer. So basically, a system solid or liquid possessing at most 2-D order of periodicity is known as a thin film. This 2-D order system accounts for the significant differences in physical and chemical properties of the materials in comparison to their bulk form. In recent times researchers are focusing much on thin film technology to use them for various applications owing to their interesting electrical, optical and magnetic properties. The specific properties of thin films can be tuned by controlling different growth parameters. Thin films are widely used in fundamental research and solid state technology. Due to their affordable production costs, they are widely utilized in photovoltaic conversion of solar energy and other optoelectronic devices. The silicon based solar cells, although have good efficiency and used in several applications, they are not commercially viable yet. The issue of silicon's high cost was identified early on, and it was also discovered that cheaper solar cells and other optoelectronic devices could only be made if materials were combined with less expensive technologies, specifically thin films. For the deposition of various layers, a range of substrates (flexible or rigid, metal or insulator) can be used (contact, buffer, absorber, reflector etc.). The schematic of thin film is shown in figure 1.

Figure 1. Schematic of thin films

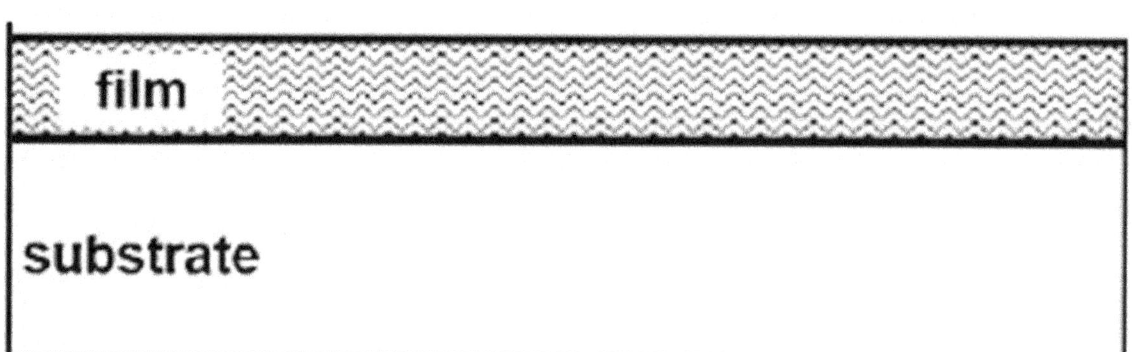

1.2 Selection of Deposition Method

There is no single technique which is ideally suitable for the deposition of thin film having all the desirable properties with no containments. The selection of deposition technique is important in order to obtain reliable, good quality films. The following criteria may be adopted in choosing the deposition technique.

§ The process should be able to deposit desired material on suitable substrate.
§ Deposition over a wide range of preparative parameters.
§ Possible to deposit over a large area.
§ Controlling of growth rate, structural parameters and defects.
§ The stoichiometric composition of the film should be maintained.
§ The films obtained should be reproducible. Among the different deposition methods discussed previously, chemical bath deposition satisfied most of the above said criteria for thin film deposition.

1.3 Cadmium Telluride (CdTe)

In recent years, the compound of chalcogenide family is more attracted towards thin film as it has high efficiency and was used to produce solar cells. The compounds of chalcopyrite families had two sub classes – (i) Group II–VI and (ii) Group III–V compound. Among these materials, cadmium telluride (CdTe) is the member of group II–VI compound, and it is best material used in biological labelling (Lee et al.,2004) light emitting devices(Gaponik et al.,2000) solar cells(Gupta et.al.,2006) etc. Due to their critical technical applicability in the domains of electronics, electrochemistry, and photovoltaic material for thin-film solar cells, cadmium telluride (CdTe) thin films have gained much research(Candless et.al., 2003; Britt & Ferekides, 1993; Russell et.al., 2015; Echendu et.al., 2016). Thin film solar cells CdTe have emerged as a promising candidate for window layer. They provide a huge selection of options for device design and fabrication. Long-term research has been conducted on the II-VI family semiconductor called cadmium telluride (CdTe). The earliest studies on CdTe production and applications were conducted between 1890 and 1920. CdTe has mostly been investigated as a quantum dot and a polycrystalline thin film over the past ten years. Cd-based compounds are crucial II-IV family semiconductor materials. The appeal of Cd-based binary and ternary compounds comes from their potential use in solar cells, X-ray detectors, and other devices. These Cd-based device types are widely used in a variety of industries. The following are some of the characteristic features of II-VI compound semiconductors which make them suitable for wide applications in photovoltaic technology as well as other optoelectronic devices (Caglar et.al., 2006; Liu et.al., 2016).

- They are wide direct band gap semiconductors. The band gap ranges from ultraviolet to Infrared region,
- The possibility to tune band gap.
- Some materials such as CdS has high transmission/emission coefficient and some materials such as CdTe has high absorption coefficient.
- Materials having a thickness of roughly 1 micrometer are sufficient to absorb 99% of solar light with photon energies greater than the band gap.
- Hence, they are suitable for thin film optical devices and the use of II-VI semiconductors has an economically viable approach to utilize solar energy for terrestrial applications.

CdTe is binary semiconducting material. Energy bandgap is an essential parameter for semiconductors which gives the idea about electronic and optical properties. At room temperature, the visible light spectrum range can be turned into electricity by solar cells thanks to the direct band gap of CdTe, which has an energy band gap of 1.45 eV. Thin film photovoltaics' (PV) low cost promise has been the argument for the technology for decades. It was believed that low-cost, large-area deposition techniques might be created to produce highly efficient devices, and monolithic integration techniques could bring down the price of module production. With a market share of between 5 and 6%, Cadmium telluride (CdTe) is now the most commercially viable thin-film PV technology after having accomplished many of these goals.

Due to the thin film nature of solar cells, numerous researchers have attempted to employ various methodologies to find the optimal electrical quality for photovoltaic applications. CdTe thin film have been fabricated using number of different methods, such as chemical vapour deposition (Chou et.al.,1986; Deivanayaki et.al., 2010) UHV sublimation(Lin Ming et.al.1998), MOVPE(metal-organic vapour phase epitaxy) (Nishino et.al., 1996), hot wall epitaxy (Korenstein et.al., 1992; Tatsuoka et.al., 1993), chemical bath deposition, close spaced vapour transport, R.F. sputtering, molecular beam epitaxy, electrodeposition, thermal vapour evaporation, etc.(Lerma- Stoelo et.al., 2001; Padam & Malhotra, 1989; Klochko et.al., 2009 ; Khrypunov et.al., 2011; Deivanayak et.al., 2010; Garadkar et.al., 2010; Wang et.al., 2015; Shenouda et.al., 2015; Chauhan et.al., 2014; Zhu et.al., 2019; Ishikawa et.al., 2016; Oluyamo et.al., 2020). Physical and chemical methods can both be used to deposit CdTe. Both of these methods have various benefits and drawbacks; in brief, the physical methods are expensive, and the production process is halted due to the vacuum chamber, whilst the chemical methods are less expensive but less effective. To accomplish this, a high efficiency to cost ratio must be established. Of all the methods, coating a big surface area with materials using CBD is the most straightforward, affordable, and practical (Chi et.al.,2017). The easiest and most practical method for depositing thin films of binary, ternary, etc. is chemical bath deposition (CBD). It is a slow process that involves dual mechanisms of ion by ion condensation and cluster by cluster condensation that lead to the regulated precipitation of the desired molecule on the substrate in the presence of an appropriate complexing agent. The process for depositing thin films is called chemical bath deposition (CBD), also referred to as solution growth. In 1884, PbS and PbSe were the only materials used to create the first CBD thin films. CBD films began to focus on solar energy applications around 1980. A controlled precipitation of a substance from the solution onto a suitable substrate is a component of the chemical bath deposition process. The method's capacity is to blanket a big area in an affordable and repeatable manner. The window layer for thin film solar cells is frequently grown using this process, which is appropriate for generating thin films of many different kinds of materials. The fundamental idea behind CBD is that a molecule must have a higher ionic product than its solubility product in order to precipitate it out of a solution. When a cationic solution (metal) and an anionic solution (chalcogenide) are combined, rapid precipitation happens when the ions join on the substrate and in the solution if the ionic product is greater than the solubility product of the reactive species. However, if the same reaction takes place in the presence of a complexing agent, such as NH4OH, metal ions are complexed and their concentration decreases enough to cause the regulated precipitation of the desired chemical compound, which leads to the deposition of thin film on the substrate. The preparation of films by CBD is governed by factors such as

Ø Bath temperature
Ø Nature and concentration of the precursor
Ø Adsorption and deposition time

Ø Nature of the substrate
Ø Rinsing time

In CBD, the compound's solubility product is crucial for the development of thin films. Homogeneity, stoichiometry, and adhesiveness are just a few of the qualities that heavily rely on solubility. A saturated solution with A+ and B+ ions in contact with the undissolved solid AB is created when a sparingly double salt, AB, is dissolved in water. This equilibrium between the solid phase and the ion in the solution is represented by the equation.

$AB \rightarrow A^{+} + B^{-}$

From the law of action

$$K = \frac{\left[C_A^{+} \times C_B^{-}\right]}{\left[C_{AB}\right]} \tag{1}$$

Where $[C_A^{+}]$ and $[C_B^{-}]$ are the concentration of A^{+} and B^{-}, $[C_{AB}]$ is the concentration of AB in the solution. Let $C_{AB} = K^1$, then the above equation becomes

$KK^1 = [C_A^{+}] \times [C_B^{-}]$

Where K and K^1 are constants

Let $KK^1 = K_s$

Therefore, $K_s = [C_A^{+}]\ X\ [C_B^{-}]$

K_s is called the solubility product. $I_p = C_A^{+}\ X\ C_B^{-}$ is called the ionic product.

In the case of saturated solution, the two products are equal ($K_s = I_p$)

When ionic product is less than the solubility product precipitation does not occur hence no film deposition takes place. However, when ionic product exceeds the solubility product ($I_P > K_S$) the solution is super saturated, super saturated the ions combines in the solution as well as at the substrate to form a nuclei.

The CBD approach can be used to create nanocrystalline CdTe thin films in non-aqueous media. It has been noted that the aqueous system has restrictions on temperature, the region of the film's solubility, hydrogen gas evolution, etc. Work on the non-aqueous medium for the deposition of binary and ternary compounds has been chosen to circumvent the aforementioned issue. The non-aqueous bath provides more options for deposition sources, wider temperature ranges, and freedom from the omnipresent hydrogen evolution process, which is sometimes a pain in the neck when it comes to making stress- and pinhole-free deposits (Kumar et.al., 1991).

The findings of structural, morphological, elemental, and optical investigations of CdTe nanocrystalline thin films are presented in this chapter. The following techniques are used to confirm the structure and formation of the prepared samples: XRD, SEM, EDS, FTIR, UV-Visible Spectroscopy, and PL Spectroscopy.

2. EXPERIMENTAL PROCEDURE

This chapter explains the chemical bath deposition method used to create the thin film semiconducting material CdTe on nickel substrates and blue star glass plates. Figure 2 displays a graphic representation of the experimental setup. X-ray diffraction analysis, scanning electron microscopy (SEM) analysis, energy-dispersive X-ray spectroscopy, FTIR analysis, and photoluminescence were used to characterize the produced thin films.

Figure 2. Graphical representation of experimental setup

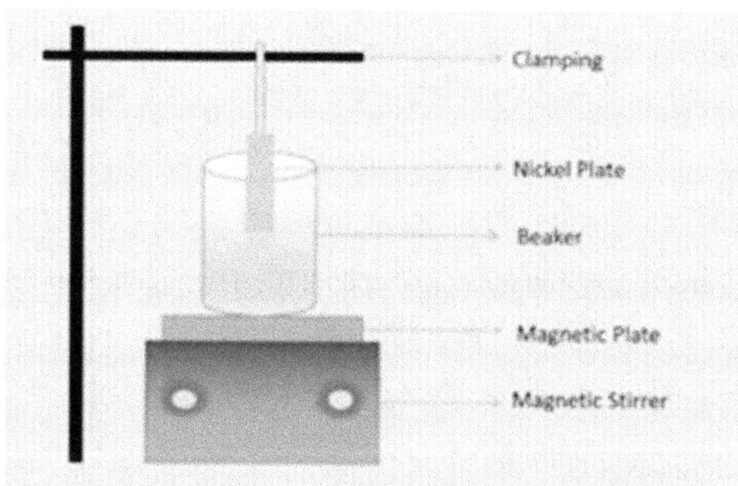

2.1 Chemical Bath Deposition of CdTe

Figure 3 depicts the experimental equipment utilized to deposit CdTe film. It is made up of a substrate mounting mechanism and a reaction cell made of Borosil glass. To agitate the electrolyte, a magnetic stirrer with a hot plate was employed. Glass plate and nickel plate were utilized as the substrate for this work. The substrate was mounted on a holder, and a stand was used to attach it vertically. The substrate was effectively cleaned before deposition using a soap solution, followed by water and distilled water. In order to dye the substrates, an oven was used.

Figure 3. Experimental setup of CdTe film by CBD method

2.2 Preparation of Bath

The electrolyte was prepared using 40ml of ethylene glycol as well as 0.1M cadmium acetate (AR grade). The electrolyte's temperature was kept constant at 160°C. For one and a half hours, a magnetic stirrer was used to agitate the mixture moderately. Two blue star glass plates were dipped in the solution vertically in a parallel position with no spacing between them. Tellurium dioxide with an AR grade of 0.02M was added to the solution after dipping both of the glass substrates in it. A 15-minute deposition was conducted. Nickel plates were used for the same procedure. Figure 3 depicts the same experiment's set-up. The films were then dried at room temperature after being washed with distil water for 10 seconds to get rid of counterions and organic contaminants. The CdTe coatings on the nickel substrate were physically stable and had excellent adhesion. The film's initial colour was a brownish gray. To evaluate the solid-state properties, the films structural, morphological, compositional, optical, and electrical characteristics were scrutinized.

3. RESULTS AND DISCUSSIONS

3.1 Structural Analysis

By using an X-ray diffraction, the structural characterization of the CdTe thin films was thoroughly examined (XRD). The as-deposited CdTe film's XRD spectra are displayed in Figure 4. The spectra prove the existence of CdTe by demonstrating that peaks are formed. Along with prominent peaks from the tetragonal TeO_2 compound's (222), (123), and (321) planes, sharp peaks of cubic CdTe are also seen. The planes (111), (221), and (331) all provide views of the CdTe peak locations. The observed CdTe peak "d" values are in good agreement with the reference "d" values, as indicated in table 1. For various planes, smaller hexagonal Cd peaks are also seen. Using Scherrer's relation, the crystallite size of the as cast CdTe thin films were determined.

$$D = \frac{0.94\lambda}{\beta \cos \Delta} \tag{2}$$

Where, λ is wavelength of incident X-ray(=0.154nm), β is full width at half maxima of (111) peak in radian and θ is diffracted angle(Surabhi et.al., 2022).

Figure 4. XRD Spectra of as deposited CdTe film(Surabhi et.al., 2021b)

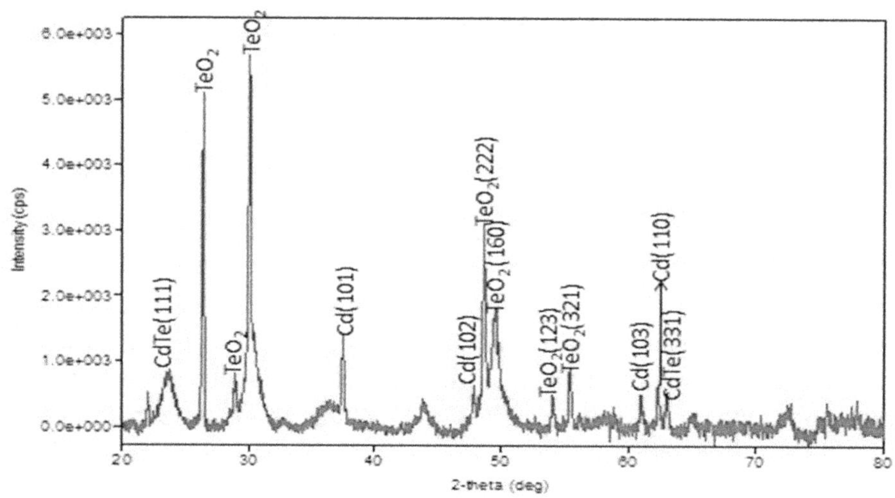

The observed value 2θ and β (full width at half maximum) for the CdTe (111) peak are 23.59° and 0.0223, respectively. Accordingly, the crystallite size of the deposited CdTe was found to be 7nm for a wavelength (λ) of 0.154nm, which corresponds to CuKα radiations. In light of this, the current work uses chemical bath deposition in a non-aqueous bath to create a nanocrystalline CdTe coating. In this regard, it is significant to highlight that the CdTe produced by the aqueous electrodeposition technique has poor crystalline character when utilizing titanium or tin-oxide substrate, and annealing is necessary to generate CdTe through electrodeposition from an aqueous bath using titanium substrate. Since both nickel and CdTe have cubic crystal structures, it is clear that this similarity in crystal structure encourages the growth of the CdTe deposit in the crystalline form.

Table 1. Comparision of 2θ, 'd', FWHM and crystal planes of observed CdTe film and the standard Value(Surabhi et.al., 2021b)

Observed Value			Standard Value		
2θ (degree)	"d" Value (A°)	FWHM (degree)	2θ (degree)	"d" Value (A°)	Miller Planes
23.59	3.769	1.28	24.072	3.701	CdTe (111) Cubic
26.35	3.378	0.168	26.668	3.340	TeO_2
28.86	3.091	0.34	28.785	3.099	TeO_2
30.02	2.974	0.29	30.012	2.975	TeO_2
37.58	2.391	0.53	38.610	2.330	Cd(101) Hexagonal
47.90	1.897	0.25	48.104	1.890	Cd(102)
48.72	1.867	0.177	48.459	1.877	TeO_2(222)
49.55	1.838	0.62	49.380	1.844	TeO_2
54.01	1.696	0.150	54.143	1.690	TeO_2(123)
55.40	1.657	0.163	54.907	1.670	TeO_2(321)
60.99	1.518	0.190	61.345	1.510	Cd(103)
62.27	1.489	0.178	62.728	1.480	Cd(110)

3.2 Surface Morphology

Scanning electron microscopy has been used to examine the film's surface morphology. The deposited samples were examined using a scanning electron microscope under a high vacuum and an accelerating voltage of 20kV. The material's surface morphology plays a significant role in the efficiency of the device's solar energy conversion. Figure 5 depicts the photography of the CdTe film as it had been deposited. We can infer from the figure that the grains are tightly clustered, spherical, and well-connected without pinholes, pits, or voids, as seen by the contrast of the film. The SEM image, which was captured at magnifications of 5KX and 10KX, reveals that the film has a consistent, crack-free surface with grains of spherical shape dispersed throughout (Surabhi et.al., 2021a). The micrograph also shows how uniform and adhesive the films are (Samanta et.al., 2015). Clear clustering of tiny crystals can be seen in the as-deposited CdTe layers (Ojo & Dharmadasa, 2017). It should be noted that in CdTe-based solar cells, the richness of Cd in the CdTe layer has been shown to be more favorable in the fabrication of high frequency devices than Te-rich absorber layer (Dharmadasa et al., 1986; Burst et al., 2016; Reese et al., 2015; Diso et al., 2016). The film's average grain size has been determined to be 1.08 μm.

Figure 5. Scanning Electron Microscopy (SEM) of as-deposited CdTe film at (a) 5KX and (b) 10KX (Surabhi et.al., 2021b)

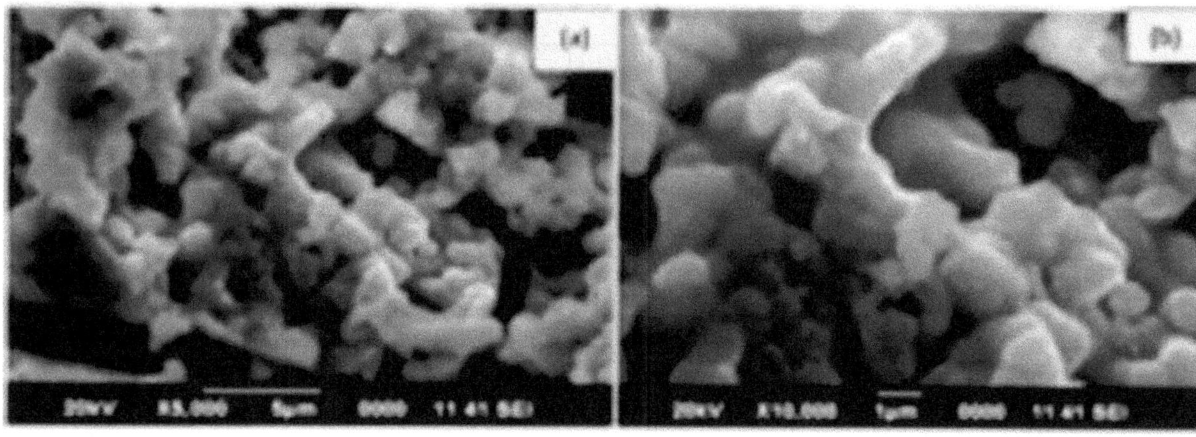

3.3 Energy Dispersive Spectroscopy

With the aid of energy dispersive spectroscopy (EDS), the elemental analysis of the as-cast CdTe nano-crystalline thin film deposited on nickel substrate is carried out and is displayed in figure 6. The spectra's distinct peaks indicate the presence of cadmium and tellurium. It was discovered through EDS analysis that all of the films are almost stoichiometric. The atomic and weight percentage of Cd and Te are shown in Table 2. By boosting the growing parameters, the divergence in stoichiometry can be reduced.

Figure 6. EDS Spectra of as-deposited CdTe film (Surabhi et.al., 2021b)

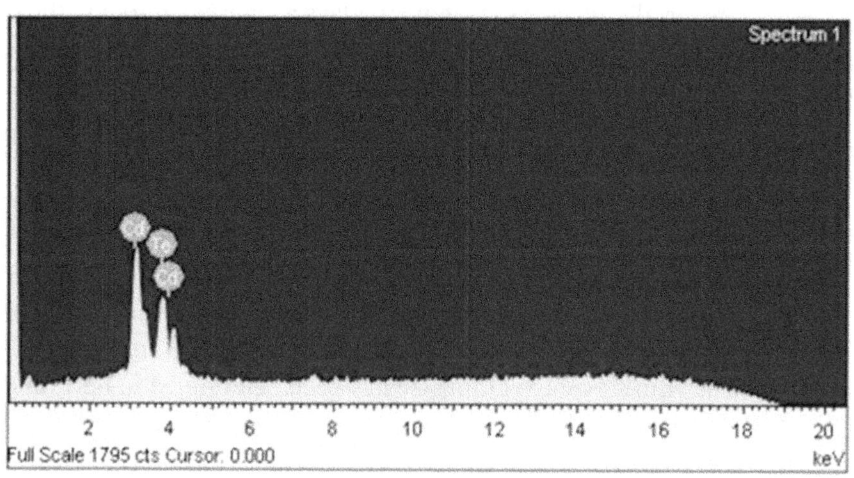

Table 2. The weight and atomic percentage of as deposited CdTe film(Surabhi et.al., 2021b)

Element	Weight%	Atomic%
Cd	48.35	51.52
Te	51.65	48.48
Total	100.00	

3.4 FTIR Analysis

The study of any changing signal into its constituent frequency components is known as fourier spectroscopy. Fourier transforms are very effective in spectroscopy. Infrared spectroscopy, also known as Fourier transform infrared (FTIR), nuclear magnetic resonance (NMR), and electron spin resonance (ESR) spectroscopy are among the spectroscopies that can be affected by the Fourier transform. The analytical method of Fourier Transform Infrared Spectroscopy (FTIR) is used to distinguish between organic and inorganic materials. The Fourier transform of an optical interferogram can be used to get the absorption, reflection, emission, or photoacoustic spectrum, which is included in FTIR spectroscopy. An important system in organic chemistry that can identify functional groups in a molecule is infrared or IR spectroscopy. The idea behind infrared spectroscopy is that molecules vibrate at specific frequencies (Meir, 2005). When IR radiation is incident on a sample, it absorbs radiation at frequencies that are similar to its molecular vibration frequencies and emits other frequencies. These frequencies (4000 to 200 cm^{-1}) fall in the IR section of the electromagnetic spectrum. By using an infrared spectrometer to identify the frequencies of absorbed radiation, an infrared spectrum—a graph of absorbed energy against frequency—can be created. A specific molecule can be recognized because dissimilar materials have different vibrations and produce different infrared spectra. In a nonlinear molecule with N atoms, the atoms move in 3N–6 vibrational patterns that are also known as fundamental vibrations or normal modes. Only IR active chemical vibrations occur in the IR spectrum (or absorb IR light). A change in the vibrating molecule's dipole moment is required for a mode to be IR active. As a result, the symmetric vibration is not visible in the IR spectrum. The vibrations in a molecule with a center of symmetry that are symmetric to the centre are IR inactive. But since all of these asymmetric molecule vibrations are IR active, they all show up in the spectrum. As a result, all chemical groups found in the substance are detected simultaneously. The FTIR spectra of CdTe film are displayed in Figure 7. The compound's bonding is ascertained using this analysis. Due to C-O stretching, the IR bond of CdTe occurs at 1041 cm^{-1} and 1084 cm^{-1} (Choti et.al., 2017). The peaks seen at 3352cm^{-1} indicate the presence of the stretching mode of -OH, which was brought on by the environment (Kumar et.al., 2015). The presence of C-H stretching vibration is shown by the peaks that formed between 2800 cm^{-1} and 2900 cm^{-1}. (Tang et.al., 2005). The C-H stretching reveals the existence of species on the nanocrystal surface (Surabhi et.al., 2021b). The C-O stretching vibration caused by the ethylene glycol is thought to be responsible for the peak at 1200cm^{-1} to 1400cm^{-1} (Surabhi et.al., 2021b). The curves' bands at 1041 and 1084 cm^{-1} match to CdTe. The coordinating interactions are responsible for these FTIR frequency changes. It establishes the CdTe film's calibre and consistency. The mixture's component amounts are also determined by it. Table 3 displays the stretching band of several functional groups present in the as-deposited CdTe thin film

Figure 7. FTIR Spectra of as-deposited CdTe Thin Film(Surabhi et.al., 2021b)

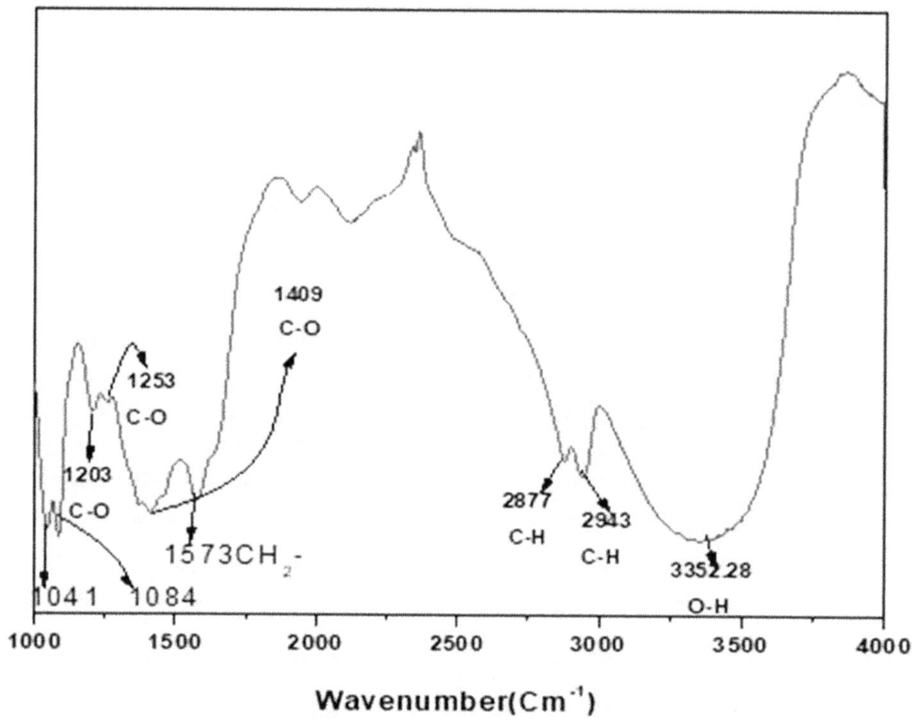

Table 3. Different functional groups present in the as deposited CdTe thin films (Surabhi et.al., 2021b)

S.No.	Positions(Cm^{-1})	Assignments
1.	1041	CdTe
2.	1084	CdTe
3.	1203	C-O
4.	1253	C-O
5.	1409	C-O
6.	1573	CH_2^-
7.	2877	C-H
8.	2943	C-H
9.	3352.28	O-H

3.5 Photoluminescence Study of as-deposited CdTe film

Studies on photoluminescence are conducted using a spectrophotometer with a wavelength range of 250 nm to 750 nm. The photoluminescence (PL) spectra of the chemical bath deposited CdTe thin film on nickel substrate are shown in Figure 8. In order to describe the optical characteristics of CdTe, the photoluminescence spectrum was obtained. The recombination of the surface states is where the photoluminescence comes from. CdTe has a peak with the highest intensity at 471nm. The band edge exciton peak correlates to emission that peaks at 471 nm. This characterization makes use of a 250nm xenon lamp source as the excitation wavelength. In this study, a sample of CdTe nanoparticles was made using less reagent. This is as a result of the emission created having less recombination with electrons and material traps at lower concentrations. As a result, the amount of produced CdTe nanoparticles is inversely related to the intensity of PL spectra (Felix-Olvera et.al., 2020). The particle size distribution of CdTe nanocrystalline thin films was responsible for the broadening of emission peaks (471nm) (Mahdi et.al., 2012; Tiwart et.al., 2008, 2009).

Figure 8. Photoluminescence spectra of as deposited CdTe film(Surabhi et.al., 2021b)

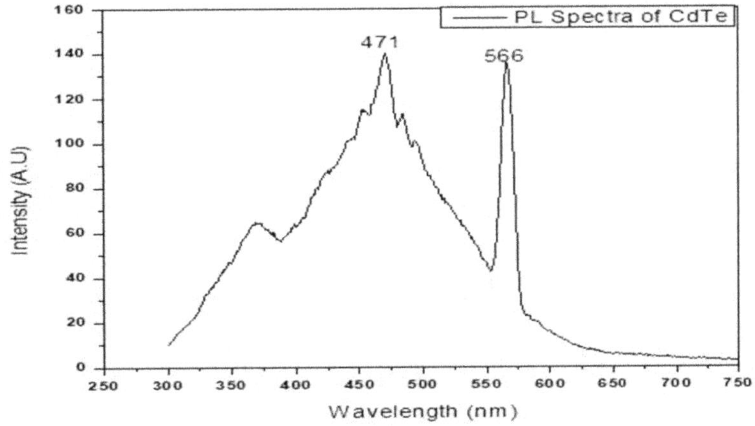

3.6 UV-Visible Analysis

To determine the absorbance spectra of a substance in solution or as a solid, ultraviolet-visible (UV-Visible) spectroscopy is employed. The absorption of electromagnetic radiation or light energy that excites electrons in a compound or substance from its ground state to its initial singlet excited state is what is seen under a spectroscope. The UV-Visible region of the electromagnetic spectrum has a wavelength range of 800–200 nm and has an energy range of 1.5–6.2 eV. The principle underpinning absorbance spectroscopy is the Beer-Lambert equation. A is absorbance, ε is the molar absorptivity of the substance or molecule in solution, b is the cuvette's path length (often 1 cm), and c is the solution's concentration at a single wavelength.

$A = \varepsilon bc$

This section discusses the optical properties of CdTe thin films in the situation of as cast. The optical absorption spectra of the CdTe thin film as it was deposited are shown in Figure 9. Analysis of UV-Visible absorption spectra is a crucial technique since it provides knowledge on the sample's energy gap. The band gap energy affects the film's emission characteristics. The plot$(\alpha h\upsilon)^2$ against $h\upsilon$ to energy axis. Using the Taucs method, analysis of the UV-Visible absorption spectra provides data on the particle's band gap. For h> Eg, the Taucs formula is given as $\alpha h\vartheta = A(h\vartheta\text{-}Eg)^{1/2,}$ where h is the Planck's constant, ϑ is the vibrational frequency, Eg is the band gap, and A is the proportionality constant (Ubale et.al., 2006). While an electron in the valence band will absorb a photon from the incident light and transmit from the valence band to the conduction band when the wavelength of the incident light fulfils $\lambda < hc/Eg$, this will cause a rapid increase in optical absorption (Corregidor et.al., 2002). In as cast case, the predicted energy gap is determined to be 1.43 eV, which is in good agreement with the reported values of CdTe thin layer coatings by other authors (Isshiki & Wang, 2017; Ameen et.al., 2017).

Figure 9. Band gap of as deposited CdTe thin film

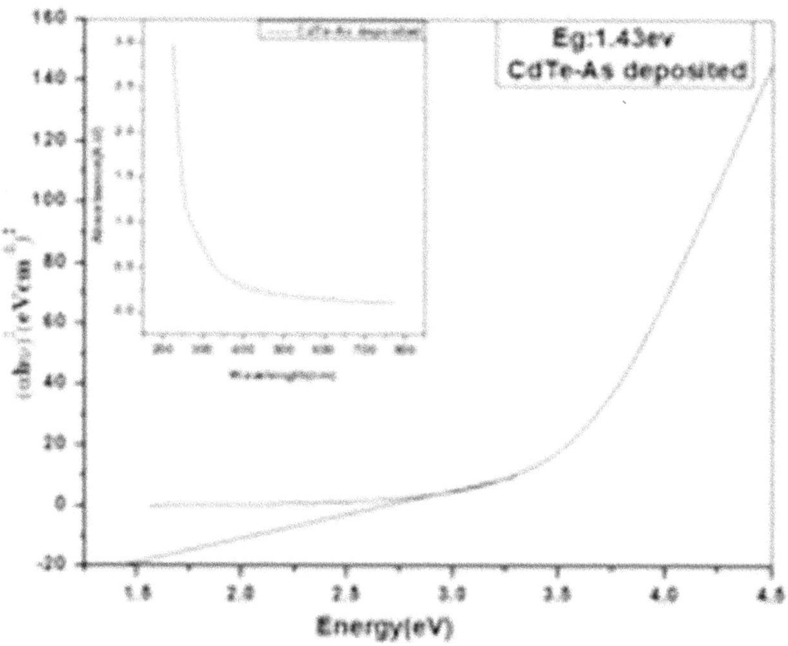

3.7 Resistivity Measurement

Electrical resistivity, which measures a material's resistance to current flow and is the inverse of conductivity, is a fundamental attribute of materials. The doping of the material, manufacturing, and ambient elements like temperature and humidity all affect a material's resistivity. The properties of a device built of that material, such as the series resistance, threshold voltage, capacitance, and other parameters, can be influenced by the material's resistivity. It is common practice in both research and fabrication environments to determine a material's resistivity. There are numerous ways to determine a material's resistivity, but the method used will depend on the material's kind, degree of resistance, form, and thickness. A four-point collinear probe is one of the most used tools for determining the resistivity of various thin, flat materials, such as semiconductors or conductive coatings. The four-point probe approach includes making contact with an unidentified material's resistance using four evenly spaced probes. A voltmeter monitors the voltage differential between the inner and outer probes while an ADC current is forced between the outer two probes. Among the equipment utilized for this test are a collinear probe and a DC current source. Using the four probe method, the CdTe film's resistivity was measured. By using the formula, ρ =V/I×2πS, where V is the potential difference between the inner probes in volts, I is the current flowing through the outer pair in amperes, and S is the probe distance in centimeters, the resistivity was computed (0.200 cm). In the 298 to 333K temperature range, the resistivity of CdTe was assessed. At normal temperature, CdTe was found to have a resistance of 7.7 106 cm. As temperature rises, resistivity decreases in semiconductors. The electrical resistivity of as cast CdTe is shown in Figure 10. The resistance drops from 7.7 106 cm to 0.319 106 cm as the temperature rises because the excited electrons in the valence band leap into the conduction band, increasing conductance, which in turn causes the resistance to decrease. This variation demonstrates how intrinsic nature differs from extrinsic nature. Since resistance and resistivity are inversely proportional, resistivity also drops (Xu et.al., 2008).

Figure 10. Resistivity of as-grown CdTe thin films

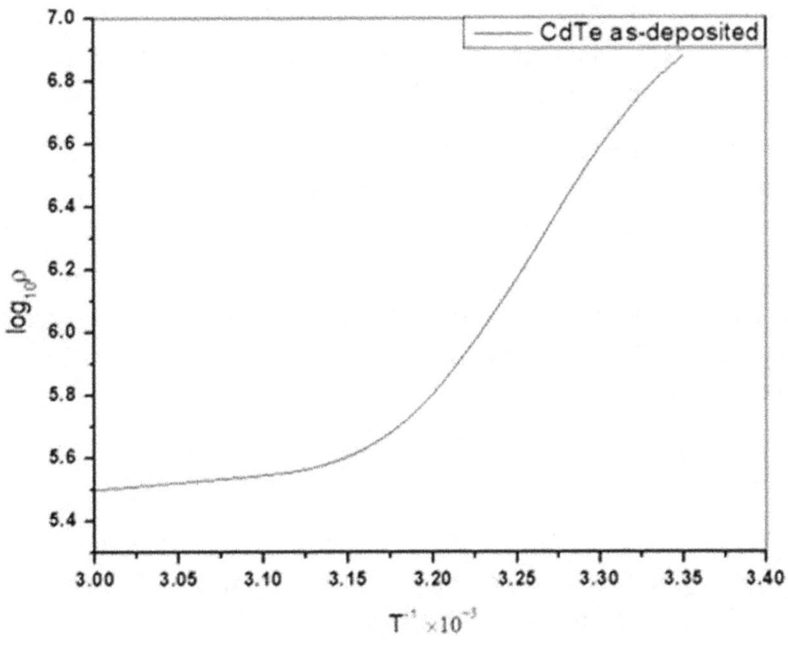

4. CONCLUSION

By employing non-aqueous media and chemical bath deposition (CBD), CdTe nanocrystalline films are created. When compared to aqueous precursor, CBD enables the creation of films of cadmium telluride on a nickel plate more consistently. The construction of a nickel substrate coating method that maintains stability during the deposition process is essential to the approach's success. XRD, SEM, EDS, FTIR, UV-visible, and PL were used to describe the sample as it was deposited. The XRD results indicate that the cubic structure of the CdTe film has a plane at (111). SEM pictures depict surfaces that are homogenous, smooth, compact, and void-free. The presence of CdTe is verified by the FTIR spectra at 1041 cm^{-1} and 1084 cm^{-1}. PL detects the excitation peak of CdTe. The peak of Cd and Te may be seen in the

EDS spectrum with an observed Cd:Te atomic ratio of over 1.27. The CdTe film that has been shown is well suited for photovoltaic, solar cell, and sensor applications.

ACKNOWLEDGMENT

We thank CIF,BIT Mesra for their help with all of the characterizations.

REFERENCES

Ali, A., Rahman, K. S., Ali, L., Akhtaruzzaman, M., Sopian, K., Radiman, S., & Amina, N. (2017). A computational study on the energy bandgap engineering in performance enhancement of CdTe thin film solar cells. *Results in Physics*, *7*, 1066–1072. doi:10.1016/j.rinp.2017.02.032

Britt, J., & Ferekides, C. (1993). Thin Film CdS/CdTe solar cell with 15.8% efficiency. *Applied Physics Letters*, *62*(22), 2851–2852. doi:10.1063/1.109629

Burst, J. M., Duenow, J. N., Albin, D. S., Colegrove, E., Reese, M. O., Aguiar, J. A., Jiang, C. S., Patel, M. K., Al-Jassim, M. M., Kuciauskas, D., Swain, S., Ablekim, T., Lynn, K. G., & Metzger, W. K. (2016). CdTe solar cells with open-circuit voltage breaking the 1 V barrier. *Nature Energy*, *1*(3), 16015. doi:10.1038/nenergy.2016.15

Caglar, M., Zor, M., Ilican, S., & Caglar, Y. (2006). Effect of indium incorporation on the optical properties of spray pyrolyzed Cd0.22Zn0.78S thin films. *Czechoslovak Journal of Physics*, *56*(3), 277–287. doi:10.100710582-006-0088-4

Candless, B., Hegedus, A. S. S, & Sites, J.R.. (2003). Handbook of Photovoltaic Science and Engineering, 617-662. John Wiley and Sons.

Chauhan, K. R., Burgess, I. J., Chang, G. S., & Mukhopadhyay, I. (2014). Preparation of CdTe thin film by electrodeposition in butyl methyl imidazolium bath at 80 C. *Journal of Electroanalytical Chemistry (Lausanne, Switzerland)*, *713*, 70–76. doi:10.1016/j.jelechem.2013.11.032

Chi, K., Li, Q., Meng, X., Liu, L., & Fu, W. (2017). Morphological control and characterization of CdTe microstructure arrays synthesized by one- step electrodeposition. *Materials Letters*, *194*, 78–80. doi:10.1016/j.matlet.2017.02.023

[] Choti, C., Sharma, V., Tarachand, Deshpandey, U., Okram, G. S. (2017). Stable Dispersibility study of Cadmium Telluride Nanoparticles in Deionized water. *SF Nano Res Let., 1.*

Chou, R. L., Lin, M. S., & Chou, K. S. (1986). Characteristic of CdTe Grown on Silicon by Low Pressure metal Organic Chemical Vapour Deposition. *Appl. Phys. Leets.*, *48*(8), 523–525. doi:10.1063/1.96494

Corregidor, V., Saucedo, E., Fornaro, L., Castaño, J. L., & Dieguez, E. (2002). Defects in CdTe polycrystalline films grown by physical vapour deposition. *Materials Science and Engineering B*, *91–92*, 525–528. doi:10.1016/S0921-5107(01)01063-7

Deivanayaki, S., Jayamurugan, P., Mariappan, R., & Ponnuswamy, V. (2010). Optical and structural characterization of CdTe thin films by chemical bath deposition technique. *Chalcogenide Letters*, *7*, 159–163.

Deivanayaki, S., Jayamurugan, P., Mariappan, R., & Ponnuswamy, V. (2010). Optical and structural characterization of CdTe thin films by chemical bath deposition technique. *Chalcogenide Letters*, *7*, 159–163.

Dharmadasa, I. M., Herrenden-Harker, W. G., & Williams, R. H. (1986). Metals on cadmium telluride: Schottky barriers and interface reactions. *Applied Physics Letters*, *48*(26), 1802–1804. doi:10.1063/1.96792

Diso, D. G., Fauzi, F., Echendu, O. K., Olusola, O. I., & Dharmadasa, I. M. (2016). Optimisation of CdTe electrodeposition voltage for development of CdS/CdTe solar cells. *Journal of Materials Science*, *27*, 12464–12472.

Echendu, O. K., Okeoma, K. B., Oriku, C. I., & Dharmadasa, I. M. (2016). *Electrochemical Deposition of CdTe Semiconductor Thin Films for Solar Cell Application Using Two-Elecytrode and Three Electrode Configuration: A Comparative Study* (Vol. 2016). Advance in Materials Science and Engineering.

Gaponik, N. P., Talapin, D. V., Rogach, A. L., & Eychumuller, A. (2000). Electrochemical synthesis of CdTe Cdnanocrystal/polypyrrole composites for optoelectronic applications. *Journal of Materials Chemistry*, *10*(9), 2163–2166. doi:10.1039/b001958p

Garadkar, K. M., Pawar, S. J., Hankare, P. P., & Patil, A. A. (2010). Effect of annealing on chemically deposited polycrystalline CdTe thin films. *Journal of Alloys and Compounds*, *491*(1-2), 77–80. doi:10.1016/j.jallcom.2009.10.146

Guptaa, P. (2006). High efficiency ultra-thin sputtered CdTe solar cells.Solar Energy Materials &amp. *Solar Cells*, *90*(15), 2263–2271. doi:10.1016/j.solmat.2006.02.029

Ishikawa, R., Yamaguchi, T., Ohtaki, Y., Akiyama, R., & Kuroda, S. (2016). Thin film growth of a topological crystal insulator SnTe on the CdTe (111) surface by molecular beam epitaxy. *Journal of Crystal Growth*, *4531*, 124–129. doi:10.1016/j.jcrysgro.2016.08.027

Isshiki, M., & Wang, J. (2017). II-IV Semiconductors for Optoelectronics: CdS, CdSe, CdTe. In S. Kasap & P. Capper (Eds.), *Springer Handbook of Electronic and Photonic Materials. Springer Handbooks.* Springer. doi:10.1007/978-3-319-48933-9_33

Jeisthardt, R. M., & Topic, M. Sites, J. R. (2015). Characteristics of CdTe grown on Si by low pressure metalorganic chemical vapour deposition, 5. IEE.

[] Khrypunov, G., Klochko, N., Volkova, N., Lyubov, V. and Li, T. (2011). Single-phase cadmium telluride thin films deposited by electroless electrodeposition. *Japanese Journal of Applied Physics, 50*, 05FH01- 05FH02.

Klochko, N. P., Volkova, N. D., Kharchenko, M. M., & Kopach, V. R. (2009). Contact electrodeposition of CdTe thin films. *Functional Materials*, *16*, 190–191.

Korenstein, R., Madison, P., & Hallock, P. (1992). Growth of (111) CdTe on GaAs/Si and Si substrates for HgCdTe epitaxy. *Journal of Vacuum Science & Technology. B, Microelectronics and Nanometer Structures: Processing, Measurement, and Phenomena: An Official Journal of the American Vacuum Society*, *10*(4), 1370. doi:10.1116/1.585870

Kumar, S. R., Gore, R. B., Kulkarni, S. K., & Pandey, R. K. (1991). Rapid Thermal Annealing of Cu-In/Se stacked layers. *Semiconductor Science and Technology*, *6*(9), 940–941. doi:10.1088/0268-1242/6/9/019

Kumar, S. R., Kumar, S., Sharma, S. K., & Roy, D. (2015). Structure Composition and Optical properties of Non aqueous Deposited ZnCdS Nanocrystalline Film. *Materials Today: Proceedings*, *2*(9), 4563–4568. doi:10.1016/j.matpr.2015.10.071

Lee, J., Govorov, A. O., Dulka, J., & Kotov, N. A. (2004). Bioconjugates of CdTe Nanowires and Au Nanoparticles: Plasmon–Exciton Interactions, Luminescence Enhancement, and Collective Effects. *Nano Letters*, *4*(12), 2323–2330. doi:10.1021/nl048669h

Lerma, M. S., Zingaro, R. A., & Castillo, S. J. (2001). Optical and structural characterization of CdTe thin films by chemical bath deposition technique, Journal of Oregano metallic. *Chemistry (Weinheim an der Bergstrasse, Germany)*, *623*, 81–86.

Lin, C. M., Chuu, D., Xu, J. A., Huang, E., Chou, W. C., Hu, J. Z., & Pei, J. H. (1998). High-pressure phase transitions in $Zn_{1-x}M_xSe$ (M=Cd, Fe, and Mn). *J.Phys.Rev.B*, *58*(1), 16–19. doi:10.1103/PhysRevB.58.16

Liu, I. P., Chen, L. Y., & Lee, Y. (2016). L. *Journal of Power Sources*, *325*, 706–713. doi:10.1016/j.jpowsour.2016.06.095 PMID:12200862

Mahdi, M. A., Hassan, Z., Nag, S. S., Hassan, J. J., & Bakhori, S. K. (2012). Structural and optical properties of nanocrystalline CdS thin films prepared using microwave-assisted chemical bath deposition. *Thin Solid Films*, *520*(9), 3477–3484. doi:10.1016/j.tsf.2011.12.059

Meir, R. (2005). Vibrational spectroscopy: A 'vanishing' discipline. *Chemical Society Reviews*, *34*, 734–752. PMID:16100614

Nishino, H., & Nishijima, Y. (1996). CdTe (111) B/Si (100) structure grown by metalorganic vapor phase epitaxy with Te adsorption and annealing. *Journal of Crystal Growth*, *167*(3-4), 488–494. doi:10.1016/0022-0248(96)00288-6

Ojo, A. A., & Dharmadasa, I. M. (2017). Analysis of electrodeposited CdTe thin films grown using Cadmium chloride precursor for applications in solar cells. *Journal of Materials Science Materials in Electronics*, *28*(19), 14110–14120. doi:10.100710854-017-7264-0

Oluyamo, S. S., Faremi, A. A., Olusola, O. I. O., & Odusote, Y. A. (2020). Tunability of conductivity type and energy band gap of CdTe thin film in the electrodeposition technique. *Materials Today: Proceedings*, *38*, 558–563. doi:10.1016/j.matpr.2020.02.962

Olvera-Felix, C., Ramirez-Bon, R., Ochoa-Landi'N, R., Ruvalcba-Manzo, S. G., & Castillo, S. J. (2020). Optical and Structural Characterization of CdTe Nanoparticles synthesized Using Chemical Bath Deposition Technique. *Journal of Electronic Materials*, *49*(2), 1257–1265. doi:10.100711664-019-07719-0

Padam, G. K., & Malhotra, G. L. (1989). preparation and study of CdTe thin films grown by the solution method. *Materials Research Bulletin*, *24*(5), 595–601. doi:10.1016/0025-5408(89)90107-4

Reese, M. O., Perkins, C. L., Burst, J. M., Farrell, S., Barnes, T. M., Johnston, S. W., Kuciauskas, D., Gessert, T. A., & Metzger, W. K. (2015). Intrinsic surface passivation of CdTe. *Journal of Applied Physics, 118*(15), 155305. doi:10.1063/1.4933186

Samanta, S., Shinde, M. S., Salunke, D., & Patil, R. S. (2015). Studies on growth and characterization of CdTe thin film deposited by Chemical Bath Deposition. *Applied Scientific Research, 11*, 10–15.

Shenouda, A. Y., Sayed, E., & Sayed, M. E. (2015). Electrodeposition, characterization and photo electrochemical properties of CdSe and CdTe. *Ain Shams Engineering Journal, 6*(1), 341–346. doi:10.1016/j.asej.2014.07.010

Surabhi, S., Anurag, K., & Kumar, S. R. (2021). Effect of annealing on the structural, compositional and optical properties of CdTe films. *Materials Today: Proceedings, 45*, 4477–4482. doi:10.1016/j.matpr.2020.12.988

Surabhi, S., Anurag, K., & Kumar, S. R. (2022). Structural, morphological, electrical and optical properties of Zn doped CdTe thin films. *Chalcogenide Letters, 19*(2), 143–152. doi:10.15251/CL.2022.192.143

Surabhi, S., Anurag, K., Rajpal, S., & Kumar, S. R. (2021). A new route for preparing CdTe thin films by chemical bath deposition. *Materials Today: Proceedings, 44*, 1463–1467. doi:10.1016/j.matpr.2020.11.635

Tang, H., Yan, M., Zhang, H., Xia, M., & Yang, D. (2005). Preparation and characterization of water-soluble CdS nanocrystals by surface modification of ethylene diamine. *Materials Letters, 59*(8-9), 1024–1027. doi:10.1016/j.matlet.2004.11.049

Tatsuoka, H., Kuwabara, H., Nakanishi, Y., & Fujiyasu, H. (1993). CdTe (111) growth on misoriented Si (100) substrates by hot-wall epitaxy. *Journal of Crystal Growth, 129*(3-4), 686–690. doi:10.1016/0022-0248(93)90504-P

[] Tiwary, C. S., Sarkar, R., Kumbhakar, P., Mitra, A. K. (2008). Synthesis and optical characterization of monodispersed Mn2+ doped CdS nanoparticles. *Phys. Lett. A, 327*, 5825-5830.

Tiwary, C. S., Sarkar, R., Kumbhakar, P., Mitra, A. K., & Chattopadhyay, K. (2009). Synthesis of wurtzite-phase ZnS nanocrystal and its optical properties. *Journal of Luminescence, 129*(11), 1366–1370. doi:10.1016/j.jlumin.2009.07.004

Ubale, A. U., Dhokne, R. J., Chikhlikar, P. S., Sangawar, V. S., & Kulkarni, D. K. (2006). Characterization of nanocrystalline cadmium telluride thin films grown by successive ionic layer adsorption and reaction (SILAR) method. *Bulletin of Materials Science, 29*(2), 165–168. doi:10.1007/BF02704610

Wang, J., Li, Q., Mu, Y., Li, S., Yang, L., Lv, P., Su, S., Liu, T., Fu, W., & Yang, H. (2015). Fabrication of CdTe thin films grown by the two-step electrodeposition technique on Ni foils. *Journal of Alloys and Compounds, 636*, 97–101. doi:10.1016/j.jallcom.2015.02.094

Xu, Y., Jie, W., Sellin, P., Wang, T., Liu, W., Zha, G., Veeramani, P., & Mills, C. (2008). Study on temperature dependent resistivity of indium-doped cadmium Zinc telluride. *Journal of Physics. D, Applied Physics, 42*(3), 35105. doi:10.1088/0022-3727/42/3/035105

Zhu, X., Lian, Q., Tang, K., Bai, W., & Chu, J. (2019). Epitaxial processing optimization and photoluminescence spectra of CdTe thin films grown on highly dissimilar SrTiO3 (001) by molecular beam epitaxy. *Thin Solid Films, 6691*, 551–557. doi:10.1016/j.tsf.2018.11.047

Chapter 9
Effect of Carbon-Based Composites in Fast-Charging Battery Technologies

Kefayat Ullah
University of Swat, Pakistan

Bakht Mand Khan
University of Swat, Pakistan

Fawad Khan
University of Swat, Pakistan

Won Chun Oh
Hanseo University, South Korea

ABSTRACT

The need for electric vehicles and other portable devices globally, advancements in charging technologies, are inflection points in the evolution of upcoming batteries. The proper engineering of electrodes are essential for stability, lifetime and cyclability. Several materials have high theoretical capacity i.e Na, Al, Li are desirable for next generation fast charging batteries. However there are some drawbacks such as low columbic efficiency, dendrite growth and volume expansion restricting their practical applications. We organized many scheme and strategies for the engineering of these metal electrodes and electrolytes. The effect of many composite structure were found to improve the stability and cyclability. The carbon derivatives specially graphene, graphene oxide, carbon nanotube's, carbon fiber, graphite and their composites were studies and found to be the most suitable candidates for making the anode composite for better electrochemical performance of fast charging batteries.

DOI: 10.4018/978-1-6684-7358-0.ch009

1 INTRODUCTION

With the rapidly increasing demands for electric vehicles, portable devices and energy-storage systems, high-energy-density rechargeable batteries are needed for their wide range of applications. Fast charging, as one of the basic technologies of rechargeable batteries, has attracted the attention of many consumers and has a promising market potential(G. L. Zhu et al., 2019).Furthermore, the charging duration for normal rechargeable batteries often take several hours, significantly limit their usage and efficiency in the vehicle industry. To lower the time of charging in electric vehicles, It is necessary to have a battery that can charge quickly. A fast-charging battery may be charged approximately 60% of their maximum capacity in 6 hours or less. Unfortunately, high energy density is difficult to attain in conventional electrodes because of weak electrical conductivity and complex ion-diffusion paths. As a result, increasing power density and achieving faster charge/discharge of batteries is necessary(Q. Huang et al., 2021). In order to fulfill the requirements of ever-growing industries and minimize the charging time, the need of fast charging batteries are more urgent (S. Wang et al., 2021). The carbon derivatives specially graphene, graphene oxide, carbon nanotube's, carbon fiber, graphite and their composites were commonly applied as conductive additives to increase the electron transport of electrodes because of its outstanding electrical conductivity (R. Wang et al., 2020).

Carbon-based materials have enormous potential to use as an anode materials for different batteries due to its benefits such as natural availability, chemical inertness and low cost. Carbon-based materials have a variety of structures including graphene, carbon nanotubes, graphite and carbon fiber and morphologies (0D, 1D, 2D, and 3D, where D stands for dimension), allowing them to satisfy the requirement for extremely efficient Li^+, Na^+, and Al^+ storage (L. Zhang, Wang, Lu, & Xiang, 2021). This chapter focuses on various carbon derivatives including graphene, carbon nanotube's, carbon fiber, graphite and their composite. These materials are tested as an anode materials for fast charging batteries. These materials has ability to improve the electrochemical performance, fast charging rate, life time and capacity of batteries.

2 LITHIUM-ION BATTERIES

Lithium-ion batteries (LIBs) are fast growing rechargeable batteries which store electric charge using lithium ions (Li^+) as carriers. The cathode, anode, separator and electrolytes are the main components of LIBs (Figure 1). Lithium-ions migrate via electrolyte from cathode to anode during charging, and lithium-ions go in the reverse manner during discharging mechanism. In recent times, LIBs are now employed in portable electronics, electric cars, and power grids. Whittingham, Yoshino, and Goodenough were awarded the Nobel Prize in Chemistry for their significant contributions in the invention of LIBs(Or, Gourley, Kaliyappan, Yu, & Chen, 2020; G. Wang, Yu, & Feng, 2021). Despite advances, mileage anxiety as well as time-consuming charging remain technological barriers to the widespread use of electric-vehicles. Although raising the pack of battery size allows for longer-distance driving, the resultant rise in battery charging time defies customer satisfaction. Increasing consumer comfort by reducing charging time without losing mileage. As a result, quick charging has emerged as a long-term major aim for battery development in the industry of electric vehicles(L. Li et al., 2021). Fast-charging batteries are more important than ever before. Current LIBs with consist of graphite anodes, on the other hand, are unable to achieve the necessary fast-charging aim of 80% capacity in 15 minutes(Jiang et al., 2021).

Figure 1. Components and working mechanism of LIBs. Reproduced from ref (Or et al., 2020) under the terms of the Creative Commons CC BY license

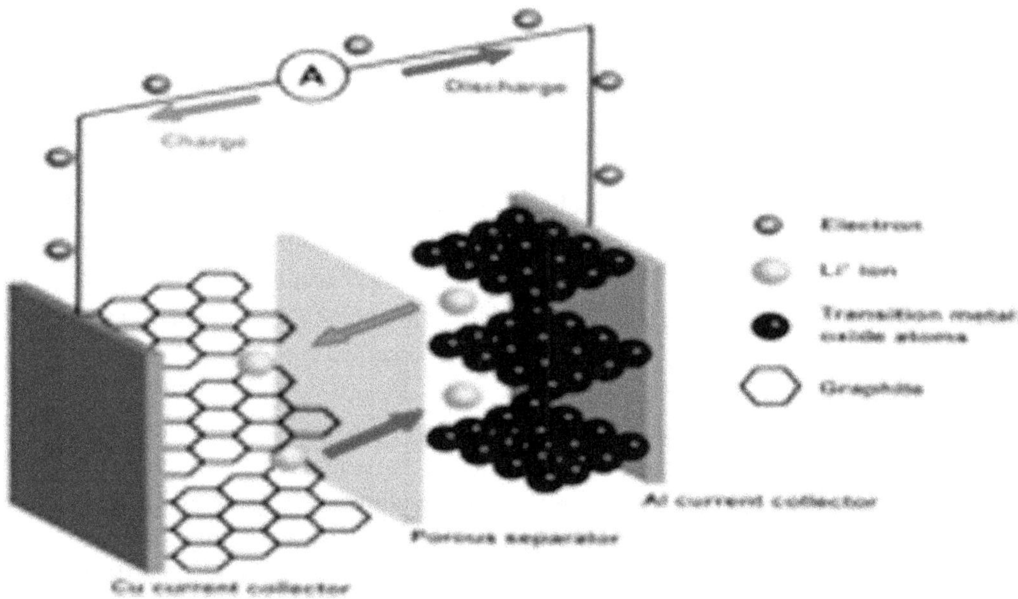

2.1 Carbon-Based Materials for Fast Charging Lithium-Ion Batteries

Carbon-based composites have great potential for rapid charging LIBs because they can benefit from both carbon's high electrical conductivity as well as the remarkable energy storage capabilities of noncarbon components in a synergistic manner. The production of a carbon-based composite as a anode with better lithium ion storing characteristics has made significant progress. Efforts to build carbon-based composites with an unique two-dimensional nanosheet structure have gained a lot of attention(Pang et al., 2020).

2.1.1 Graphene Based Lithium-Ion Fast Charging Batteries

Several Carbon-based composite materials are increasingly being investigated as anode materials due its excellent physical and chemical stability, including graphene, carbon nanotubes, natural graphite, graphitized carbon, and amorphous carbon. Among these options, graphene is considered to be best negative electrode for LIBs due to its unique properties including large theoretical capacity, large specific-area, superior conductivity, and excellent electrochemical stability(Mand Khan, Chun Oh, Nuengmatch, & Ullah, 2023; Pang et al., 2020). Graphene is a single layer of carbon atoms with a honey comb like structure made up of sp^2 hybridization, shows remarkable features including large specific surface-area, thermal conductivity, good mechanical strength, outstanding electrical conductivity, charge density, carrier mobility and good chemical stability, enabling it to be used in energy storage technology(Hussain, Ihrar, Hussain, Oh, & Ullah, 2020; Sui et al., 2020; Zheng et al., 2021).

Because of its unusual features, graphene has drawn the attentions of researchers for possible commercial uses since Novoselov and Geim first isolated graphene from graphite by scotch tape in 2004. LIB anodes based on this unique Mn_2O_3/N-RGO material shows large specific surface-area, high electron transport capability and shorter electron transport channel. These characteristics may help to improve the LIBs cyclability, since they have such a large capacity about 1350 mAh/g and a very rapid charging speed t a10 A/g, as well as at 0.5 A/g it exhibits discharge capacity about 516 mAh/g after 200 cycles(L. Li et al., 2021).

In Li-ion batteries, doping heteroatoms including phosphorous, boron and nitrogen with pure graphene lattice can effectively improve the properties of graphene such as electrical properties, semiconductor properties, band structure, magnetic properties and surface reactivity. Because of these properties graphene have large specific-capacity, good capacity retention, high rate of capability and increased charging speed (Xiang et al., 2021). The composites of graphene with polymers, metal and metal oxides are considered one of the best materials in batteries technology(Saba & Jawaid, 2018).

Xin Zhao et al. used a facile electrochemically exfoliated graphene (EG) technique to make a good quality EG anode. The EG anode performs good charge and discharge capability LIBs. It shows an excellent lithium storage capacity and great stability . A very large capacity about 356 mAh/g was obtained at 1 A/g. A capacity of 100% was retained after 6500 cycles(X. Zhao et al., 2020).

Three-dimensional surface-porous graphene negative electrode (anode), which was manufactured at 550 ºC by a reactions between Carbon dioxide and Sodium liquid, was successfully developed by Lei Wang and his colleagues. At 0.2 C the 3D-G anode shows a large first discharge capacity about 972 mAh/g and a reversible capacity about 277 mAh/g was attained at 150 C, showing its rapid charge/discharge capability in a very less time approximately 16 seconds. Moreover, even after a 1000 cycles it shows a discharge capacity about 280 mAh/g as well as 91% capacity retention was achieved at 50 C(L. Wang et al., 2020).

Thanapat Autthawong and his group produced and studied TiO_2(B)/N-doped graphene (TNG) composite anode materials using a simple chemical approach and heat treatment procedure. The TNG composites anode are reported to exhibit good anode material properties, including significant specific capacity even at a large current density, adaptive quick charging capabilities, a long lifetime, and a high degree of safety. Surprisingly, the TNG-24h combined with TiO_2(B)-nanorods gave a specific capacity about 350 mAh/g at 10C over 1000 cycles, owing to the TiO_2(B) nanorods' large crystal voids as well as pseudo-capacitive tunnels, which encourage Li-ion diffusion and transport. Figure 2 shows the electrochemical performance of TNG composite anode(Autthawong et al., 2020).

Figure 2. TNG electrode electrochemical measurements: (a) rate cycle capability at current densities range from 1C to 50C. (b) Cycle stability of the obtained TNG composites anode at 10C for 1000 cycles, and (c and d) long-term cycling of TNG-24h anode at 20C and 50C, respectively. Reproduced from ref(Autthawong et al., 2020) under the terms of the Creative Commons CC BY license.

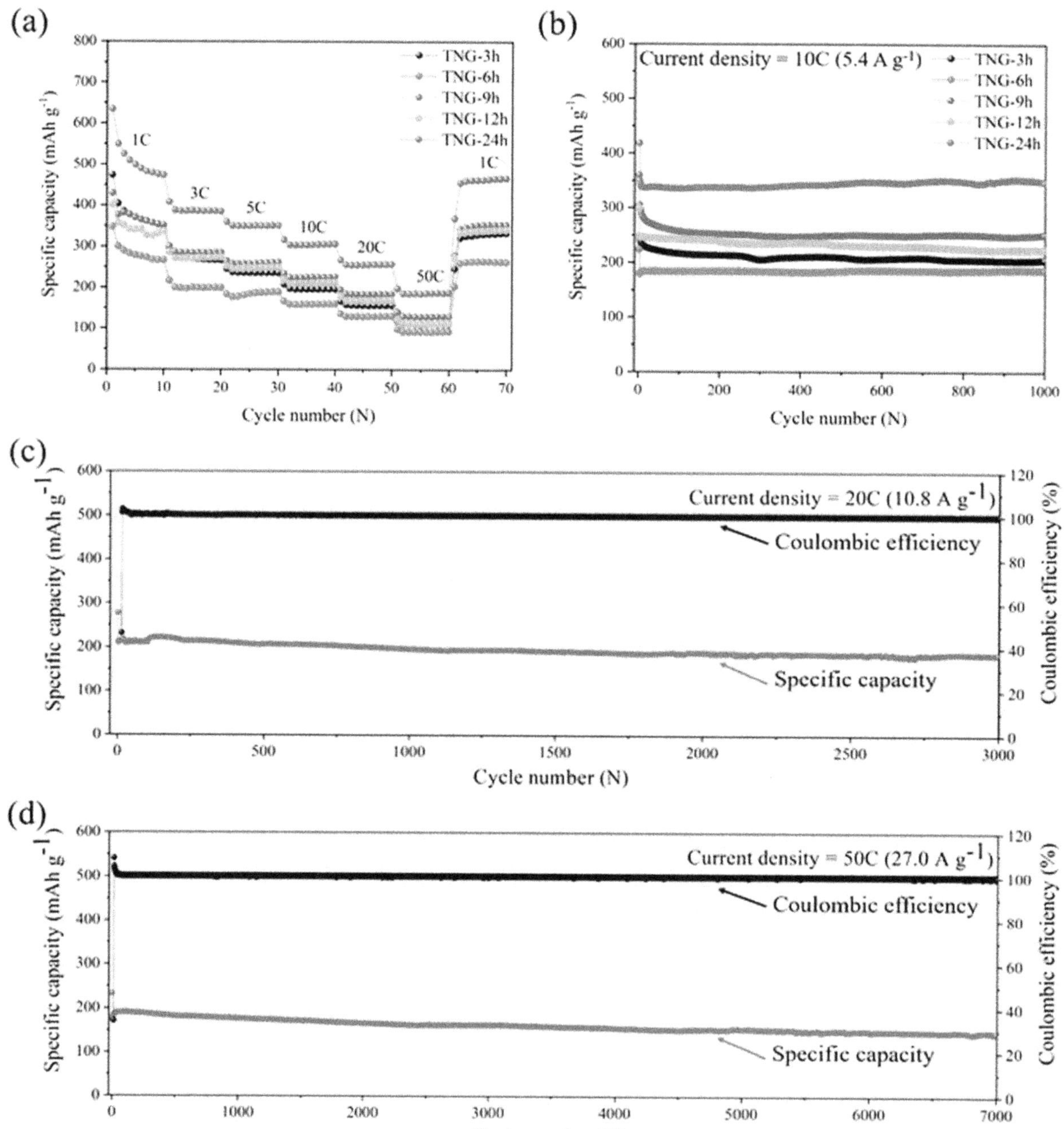

Junke Zhu and his co-workers use two-step calcination technique for producing hollow like Co_3O_4 nanoparticles encapsulated in an N, S-combined rGO structure. In the initial stage, pyrolysis of Co-based MOFs/GO precursors in inert environment at temperature of 800 °C yields core-shell like of $Co@Co_3O_4$

encapsulated in N and S-co-combined with rGO. The hollow like Co_3O_4 nanoparticles having average size of particles is about 25 nm as well as a wall thickness approximately 4-5nm are generated by a subsequent calcination procedure in air at temperature of 250 °C using nanoscale Kirkendall-effect. The $HoCo_3O_4$/NS-RGO framework, both micropores as well as mesopores are formed. After 600 cycles at rate of 1 A/g, the prepared composite show a large specific capacity about 1590 mAh/g. and also exhibits excellent rate of performance ranging from 0.2-10 A /g due to the porous structure of Co_3O_4 and the combining of N and S atoms with RGO(J. Zhu et al., 2020).

Baicen Wan et al. effectively created a composite composed of few-layer of TMOs with graphene nanosheets containing interfacial Co-O-C bonds using a simple precipitation technique. The CoO/RGO shows a large first reversible capacity approximately 2189.4 mAh/g even at 100 mA/g, and it exhibits outstanding reversible retention (70%) at 2 A/g after 300 cycles. At 5 A/g, it shows good reversible cyclic stability about 1167 mAh/g having capacity retention about 76%. It is proposed that building interfacial Co-O-C interconnections can maintain the network structure even against agglomeration while simultaneously providing rapid paths for ion transport as well as charge transfer(Madhusanka et al., 2020).

Xiaoyu Dong et al. created an aerogel-structure SiO_2/NGA composite using one-pot hydrothermal technique and examined its frameworks and lithium storage ability. The SiO_2/NGA nanocomposite may alleviate the circumstances that lead to fragment as well as reduce the volume change of SiO_2 during charge /discharge operations due to its 3-dimensional carbon framework. At 100 mA/g after 800cycles, the prepared composite exhibits good reversible capacity about 1052 mAh/g when the its electrochemical performance was tested. Furthermore, following the rate of performance test, its specific capacity has obtained up to 1000 mAh/g while rate was reduced to 100 mA/g, confirming the electrochemical reaction's outstanding reversibility(Dong et al., 2020).

Siling Gu and his co-workers developed a unique graphene nanosheet attach with Fe_3O_4 nanoparticles to make its composite using a simple freeze-drying of graphene and Fe^+ solution supported by solvent thermal preparation technique. Fe_3O_4 nanoparticles size in a very unique architecture may be regulated ranges of 20-100 nm by changing the mass concentration of graphene to Fe_3O_4 NPs, while Fe_3O_4 nanoparticles exhibit a homogeneous dispersion state. Graphene nanosheets significantly reduce volume expansion, inhibit Fe_3O_4 NP aggregation, and subsequently retain the structural stability of the prepared GNSs/Fe_3O_4 NPs electrode during the Lithium insertion/extraction mechanism. Furthermore, graphene nanosheets included in GNSs-Fe_3O_4 composite offer the essential conductive pathways for the electronic transmission even without inclusion of any extra conductive agent. After 120 cycles at 100 mA/g, the obtained composite electrode shows has a large reversible capacity about 1145 mAh/g and at 0.5 A/g an outstanding rate capability about 650 mAh/g was achieved(Gu & Zhu, 2020).

S.Y. Kim et al. synthesize rGO/Si/GA composite as a anode using spray dying and uniformly inserted it into a 3-dimensional, porous like GA structure by aerogel production method. Si-based composite enclosed in numerous graphene layers was clearly employed as an anode in LIBs even without the addition of in-active components like a binder, metal substrate and conducting material. Because of its multilayer shape, the rGO/Si/GA composite with many graphene shells has an extraordinarily first discharge capacity about 1217 mAh/g, after 200 cycles it shows outstanding cyclic stability about 462 mAh/g at 1.0 C, and at large rate 10 C it displays better rate capability about 819 mAh/g. It was concluded that, this easy method for constructing Si anodes covered in many graphene shells will help in the growth of advanced LIBs used for electric vehicles(S. Y. Kim, Kim, & Yang, 2021).

2.1.2 Carbon Nanotubes Based Lithium-Ion Fast Charging Batteries Supported By Silicon

Silicon based anode materials were intensively explored by researchers as the greatest viable anode material for LIBs due to their abundant availability, large theoretical capacity and low working voltage platform (Ruan, Wang, & Ying, 2021). The use of Si-carbon anode composites for next-generation LIBs has been proposed, with the goal of effectively combining Si high specific capacity with carbon's good mechanical flexibility and electrical conductivity(Bai, Zhang, Lin, & Zhang, 2021). Owing to their excessive energy density and extensive lifecycle, LIBs were widely used in modern mobile EVs and portable devices to date. Traditional graphite anodes capacity (372 mAh/g) is deficient in the fulfillment of LIBs demand for higher energy-density. Due its natural abundance, large theoretical capacity about 3579 mAh/g, Si anodes are considered to be the most promising site for LIBs . Recent research has attempted to minimize Si significant volume change during cycling by designing different nanostructures like nanosheets, nanotubes, nanowires and nanorods(Zeng et al., 2021). However, substantial instabilities resulting from the considerable volume shift about (300%) during charge-discharge process have persistently hampered the commercialization of Si anodes(Seok et al., 2021).

Many methodologies have been proposed to date in order to overcome the aforementioned issues and increase the commercialization of silicon materials. Qian et al. created nano-silicon which exhibits large reversible capacity about 1150 mAh/g at 1 A/g after 100 cycles. In comparison to a pure silicon anode, Zuo and his workers synthesized Si/C composite having a 3-D hierarchical mesoporous composition that can retain a capacity about 710 mAh/g after 300 cycles at 0.2 A/g(Luo et al., 2020). When compared to carbon black conductive additive, Lim et al. found that CNTs can enhance anode material cycling performance(Batool et al., 2020).

A viable but difficult road to high energy-density LIBs is to use nano - structured Si materials to construct an freestanding membrane having excellent mechanical properties and continuous conducting network (LIBs). Physical vapour deposition (PVD) can be employed to develop freestanding film anode having good volumetric/areal capacity as well as good energy density by using unique 2D porous amorphous Silicon nanoflakes anchored by MWCNTs. When used directly as the anode electrode, the Si-MWCNT film shows a remarkable specific capacity about 1556 mAh/g(Z. Wang et al., 2020).

The cyclic performance of silicon flakes partly encapsulated by SiO_2 and coated by conductive nanocarbon films and CNTs as an anode materials for LIBs has so far been impressive. On a silicon flake, the SiO_2 surface layer helps to improve stability of a silicon-based anode. Lithium ions and electrons can move swiftly on the bare silicon surface. Electrical connections between current collector and silicon flakes are provided by nanocarbon sheets CNTs and nanocarbon sheets. Even after volume fluctuations during repeated charging and draining cause silicon to split into many smaller pieces, Such carbon nanostructures enable electrical conductivity between the Si particles and flakes that surround them. After more than 100 charging and discharging cycles, It was discovered that a charge capacity more than 2000 mAh/g could be found(Tzeng, Chen, & He, 2020).

Carbon nanotube networks in the all-carbon binder also contribute to the mechanical and conductivity qualities of the material. Because of its integrated structural pattern, the electrochemical performances this electrodes with CNTs-reinforced all-carbon binder are substantially enhanced in terms cycling stability and rate. When silicon is utilized as an active component, the carbon binder technique is also useful. It was also experienced the Li storage characteristics of electrode containing silicon (Sie-C-CNT), and discovered that at 0.5 C after cycling for 300 times, the prepared Si-C-CNT exhibits large capacity about 902.4 mAh/g, having good capacity retention of 74.6 percent, representing the advantage of carbon binder in the electrode-based on alloy anode material (125.7 mAh/g after 225 cycles)(Fan et al., 2021).

2.1.3 Graphite Based Lithium-Ion Fast Charging Batteries

Since the introduction of lithium-ion batteries to the market, graphite has dominated as a stable electrode material, and its use in future graphite-based energy storage devices could result in a surge in demand(Divya, Natarajan, Lee, & Aravindan, 2020). Because of its high specific capacity, lithium-ion batteries turn out to be the industry standard for energy storage. To increase the wellbeing of Li-ion devices, Yazami et al. accept graphite as a conductor of Li ions at the negative electrode(P.-H. Guo et al., 2021). The growing popularity of electric, hybrid, and plug-in hybrid vehicles has sparked interest in silicon as a component in LIBs anodes. Silicon having a theoretical specific capacity of 3590mAh/g (equivalent to $Li_{15}Si_4$), which is about ten times that of the extensively used and commercially leading graphite anodes (372 mAh/g)(Dhillon, Hernández, Wagner, Svensson, & Brandell, 2021).

Guo et al. fabricated a high capacity and nominal cost $SiO_x/C@$ graphite (SCG) hybrid anode. At 0.5 A/g the prepared composite shows a large cycling stability about 842 mAh/g after 300 cycles and after 300 cycles at 1 A/git shows a rate capability about 562 mAh/g(X. Wang, J. Wang, Z. Chen, K. Yang, Z. Zhang, Z. Shi, T. Mei, J. Qian, J. Li, & X. Wang, 2020). Despite great efforts, lithium-ion batteries rely on silicon anodes and composites, the issues having low conductivity, side reactions and drop in energy-density have yet to be addressed. A novel electrochemical cell design has been introduced with an anode built of silicene on a graphite substrate (Galashev, Ivanichkina, Katin, & Maslov, 2020) . Figure 3 shows schematic diagram of LIB cell with a silicene anode.

Figure 3. LIB cell with a silicene anode schematic diagram. Reproduced from ref(Galashev et al., 2020), under the terms of the Creative Commons CC BY license)

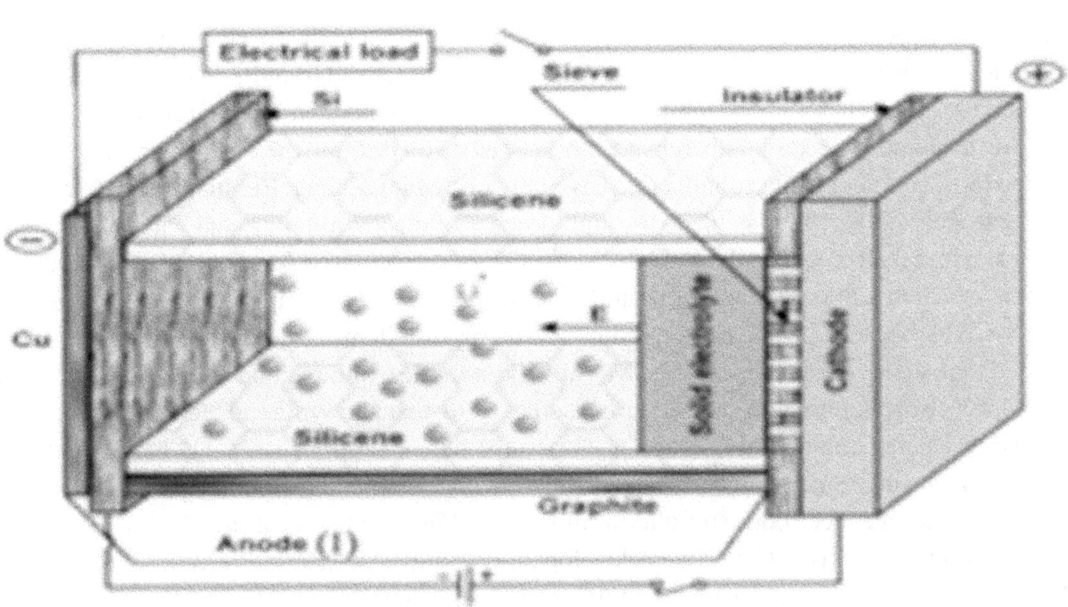

Lithium adsorption calculations on single-layer of silicene using density functional theory (DFT) suggest a capacity (1196 mAh/g) for silicene-anode with no breaking of Si-Si links. During lithium intercalation, the charge capacity of double layer silicene can reach (1384 mAh/g)(Galashev et al., 2020). Using functional materials to modify graphite-based anodes appears to be a viable option for achieving LIBs that charge quickly without losing energy density. Kim et al. have been noted an improved rate capability hybrid negative electrode (anode) made of a nanolayer of Silicon and having edge-site graphite, claiming that, nanolayer facilitates fast Lithium diffusion on graphite core while reducing the composite material's energy-density loss(S.-M. Lee et al., 2021).

Graphite was improved by moderate oxidation and ball-milling, the amount of edges and attaching sites between amorphous MoSx and graphite increased. Because of its oxygen functional groups, MoS_x is a key constituent of various large capacity materials, remained stable in a large C-rate situation. The MoS_x/B-rGO anode showed good cycling stability and specific capacity about 1239 at 0.12 A/g and at 2.60 A/g it shows a capacity about 403 mAh/g. Also after 100cycles it deliver a capacity about 1016 mAh/g (Y.-K. Lee et al., 2020).

The possibility of quick charging with typical graphite anodes has been investigated through modelling & experiments. In this work, we use a laser patterning technique to accurately create a channels of pores in post-calendered-graphite anodes. It creates a highly-ordered laser-patterned electrode (HOLE) design having vertical arrays of channels running the length of the electrode. At 4C and 6C after 100 cycles, a capacity retention was noted > 97% and >93% at fast-charging rate of 4C and 6 C over 100 cycles using the HOLE design, compared to 69% and 59% for patterned electrodes. At 4C the capacity retention of the obtained HOLE cells was very large approximately 91% and at 6 C it show capacity retention about 86% after 600 cycles. Moreover, the HOLE design enables cells to reach more than 90% of their complete capacity while fast charging(Chen et al., 2020).

2.1.4 CNFs Based Lithium-Ion Fast Charging Batteries

The fast development of high-performance LIBs have been prompted by the rising demand for effective energy storage for different electronic devices and electric vehicles. Graphite, which is used as a commercial anode for LIBs, has a good stability. As a result, advancements in the search for new LIB anode materials are desperately needed. The transportation of electron through the electrochemical mechanism is facilitated by 3D interconnected conductive CNF network(Manj, Zhang, Rehman, Luo, & Yang, 2020). In LIBs electrodes, CNFs which is made through process of electrospinning and carbon cloth could be utilize as a flexible matrix, particularly CNFs having large surface-area and good electronic conductivity. In a few research works, carbon cloth and electrospun CNFs were applied as an anode for LIBs, particularly CNFs because of its characteristics including excellent and large specific-surface-area. It was reported that, at 0.1 A/g the binder-free $MoSe_2$/CNFs anode exhibits good capacity 610.8 mAh/g(Ni et al., 2021). Electrospinning is an cost-effective and easy-to-manage synthetic technique for continuous one-dimensional nanostructures. For good-performance battery applications, 1D nanostructures have proven to be promising. Surprisingly, nanofiber mats with flexible and lightweight qualities might be used as new binder-free electrodes straight out of the box(Ding, Liang, & Zhao, 2021).

Cobalt embedded CNFs (Co-CNFs) were developed and synthesized using an electrospinning approach, while Co1-xS-CNFs were created using a solvothermal vulcanization process. This design had numerous noteworthy benefits: (I) The skeleton of the CNFs served as a support material, enhancing the electrode's conductivity and promoting electron transit. Furthermore, (II) the Co1-xS particles incorporated in the

CNFs may help to minimize volume expansion in charging/discharging mechanism. The synergistic effect of the CNFs superior electrical conductivity and at 100 mA/g due to the large capacity of Co1-xS resulted in a good discharge capacity about 620 mAh/g and it deliver better cycling stability. After 500 cycles at 2000 mA/g the capacity was retained up to 252 mAh/g. The Co1-x S-CNFs electrodes shows maximum reversible capacity as well as better cycle stability as cathode material(G. Zhu et al., 2021).

After 200 cycles of fical electrospinning, rice-panicle-like Fe_2O_3-CNFs had a large specific-capacity about 1252 mAh/g at 10 A/g. The electrospun Co_3O_4-carbon porous fibers demonstrated reversible capacity retention about 1100 mAh/g after 500 cycles. Cobalt vacancies riches Co_2AlO_4/carbon nanofibers, which are synthesized by electrospinning, are another excellent anode material for LIBs. Co_2AlO_4 nanoparticles were strongly bound to carbon nanofibers, whereas Co_2AlO_4 nanosheets developed on nanofiber surface. When used as an anode materials, the obtained composites demonstrated excellent cyclic stability about 500 mA/g and it exhibits a discharge-capacity about 627.4 mAh/g after 500 cycles(Shi et al., 2020).

W. Weng, et al. produced a CNF/Sb-SnO_2/C composite by electrospinning it first, then progressively coating it with Sb-SnO_2 and carbon layers using a hydrothermal technique. This material is believed to be an ideal anode material for LIBs. Between the outer layer and CNF core, the layer of Sb-SnO_2 was sandwiched. At 0.1 A/g, the obtained composite exhibits a very large specific-capacity about 1161 mAh/g for initial cycle, at 2 A/g it exhibits a capacity about 411 mAh/g, and at 2 A/g it shows an good cyclic stability about 347 mAh/g after 100 cycles (Weng, Kurihara, Wang, & Shiratori, 2019).

3 SODIUM ION BATTERIES

Sodium ion batteries (SIBs) have drawn a lot of attention in electrochemical energy storage technologies due to its natural availability and relatively inexpensive cost(Wu et al., 2022). SIB is made up positive and negative sodium (Na) electrodes that are ionically linked by immersing them in an electrolyte, that is usually a sodium (Na) salt dissolved in solvent As sodium does not make an alloy with aluminium, then conventional current collector (copper) is replaced by an Aluminium. As a result, a low cost aluminium is more inexpensive anode current collector than copper. SIBs working principle is similar to LIBs, with sodium ions (Na^+) from cathode to anode during charging mechanism, and during discharging it return to cathode(Wasalathilake, Li, Xu, & Yan, 2020). However, because the radius of Na^+ is (1.02) which is almost 35% greater than that of Li^+ (0.76), several LIBs anode materials are not suited for SIBs. For example, conventional graphite used in LIBs as an anode material having a capacity about 372 mAh/g, but it only delivers a capacity of just only 35 mAh/g and short cycle life in SIBs because of Na^+ not being able to insert into graphite. To address the needs of large and growing electrochemical energy-storage technologies, it is critical to create high-capacity alternate anode materials(Feng, Maça, & Etacheri, 2020; Q. Guo et al., 2021; Xu, Sun, & Wei, 2020). Thus carbon based materials is considered to be best anode materials for SIBs because of its beneficial physical and chemical properties.

3.1 Carbon-Based Materials For Fast Charging Sodium-Ion Batteries

Carbon-based materials have gotten a lot of interest as potential anode materials for fast charging SIBs. It is critical to develop carbon-based anodes for the commercial use of SIBs in energy storage systems(Yasin et al., 2020). Carbon based materials have now been intensively investigated in recent years for composite production due to their wide availability, outstanding thermal and physical qualities, and inexpensive cost(Thakur et al., 2021).

3.1.1 Graphene Based Sodium Ion Fast Charging Batteries

Currently, graphene, an developing 2D material, has been intensively investigated as a possible negative electrode for SIBs due to its remarkable physical and chemical properties such as outstanding electronic conductivity, chemical stability and strong mechanical property(Ambade et al., 2021). A 3-dimensional graphene-based framework including vertically oriented graphene nanosheets, in particular, has been recognized as an excellent approach for improving conductivity, improving active reaction sites, and efficiently reducing aggregation(Cui et al., 2020; Kefayat, Bakht Mand, Amin Ur, & Won Chun, 2022). From a commercial perspective, graphene-based materials fulfill all anode material characteristics, including outstanding cycle stability, nontoxicity, low cost, durability and safety. Therefore, different graphene-based composite anode materials have been widely researched by researchers in order to improve battery performance(Thakur et al., 2021).

C. Zheng et al. developed a Fe_2O_3 - holey graphene (FHG) hybrid as a anode material for SIBs, under which holey graphene limits Fe_2O_3 growth, resulting in evenly distributed NPs, and improves ion diffusion throughout electrochemical processes. Furthermore, the incorporation of additional hole edges as well as functional groups facilitates in the anchoring of the NPs by establishing Fe-O-C bond, that successfully inhibits aggregation of NPs during production and electrochemical operations, providing maximum NP usage as well as great cyclic stability. As a consequence, the obtained FHG hybrid exhibits improved sodiation/desodiation mechanisms, resulting in increased electrochemical cycle stability and enhance rate performance of SIBs. At 0.1 A/g, It displays a large specific capacity about 923 mAh/g and at 2 A/g it shows a capacity about 318 mAh/g. However, at 2 A/g, a capacity is maintained up to 214 mAh/g after 100 cycles(Xiao et al., 2021).

Xu Yang and his co-workers used freeze-drying and ball milling to produce a sandwich like SnO/rGO composite anode. Upon rGO nanosheets, SnO particles are homogeneously distributed. Reduced graphene oxide helps to enhance electrical conductivity and also reduce volume change of SnO during intercalation/de-intercalation mechanism. The SnO/rGO composite demonstrates outstanding electrochemical performance when it use as a anode material in SIBs. At 0.05 A/g, It exhibits a reversible capacity about 399.4 mAh/ and it also shows a capacity about 132.3 mAh/g at 5 A/g . Furthermore, after 1200 cycles, it shows a charge capacity about 109.5 mAh/g and a good capacity retention was obtained up to 70.62% at 4 A/g (Yang et al., 2021).

W. Zhao et al. used a sonochemical technique to synthesize a novel Sb_2S_3-graphene composite anode with an amorphous structure. After 50 cycles, the composites may give first discharge-capacity about 1867.1 mAh/g . The amorphous samples outperformed the crystalline equivalent in terms of reversible capacity, thermodynamic reversibility and cycle stability when used as an anode materials in SIBs. Regardless of the quick response kinetics, diffusion network, as well as homogeneous dispersion on conductive matrix, the Sb_2S_3-graphene composites have good durability and accommodation to particularly reduce volume changes(W. Zhao et al., 2021).

3.1.2 Carbon Nanotubes Based Sodium Ion Fast Charging Batteries

Although Li-ion batteries has been effectively used in portable devices for decades, the depletion and high cost of lithium has prompted a quest for alternate energy storage solutions. Li has been replaced with other elements including Al, Na, Ca and Mg by researchers. Because sodium and lithium are both alkaline metals, sodium ion batteries (SIBs) were actually developed as a result of the success of LIBs(Vo et al., 2020). Na-ion batteries work in the same way as LIBs, and the redox potential of Na^+ is just 0.3 V making it a viable option to LIBs, especially for extensive energy storage. Although SIBs have numerous obvious disadvantages, Na^+ have a higher ionic radius (1.02) than Li^+ (0.76), resulting in a smaller energy-density and the insertion–extraction is really tough. As a result, an appropriate host electrode materials having bigger tunnel structure as well as ion diffusion channel must be investigated(Feng et al., 2020; Xu et al., 2020).

For superior sodium storage capabilities, an efficient design of carbon-based materials with adequate nanoarchitecture and improved physicochemical properties is required. By using polypyrrole as a high N-carbon precursor, researchers were able to create N-rich up to 15.7% N-CNT with a long aspect ratio using a template approach. The N-CNT displays good Na^+ storage capability with benefit to reduce volume expansion, quick ion and electron transfer at high rate due to its hollow structure contain wide cavity and lengthy aspect ratio, showing good performance and outstanding cycling stability. Furthermore, the high nitrogen concentration generates a lot of active sites along with extrinsic defects, which helps to enhance specific capacity and rate performance. At 4000 mA/g after 5000 cycles the obtained N-CNT-600 electrode deliver a suitable reversible capacity about 132 mAh/g for SIBs(M. Li et al., 2020).

By tightly attaching SnS nanoparticles on porous CNT film (SnS/CNT), the capacity of SIBs anode can be increased. The well-dispersed SnS nanoparticles attached on CNT network significantly improve contact surface-area between Na^+ and active material by developing chemical-interactions between CNT network and SnS nanoparticles. By attaching SnS nanoparticles at porous CNT film (SnS/CNT), the capacity of SIBs anode can be increased. The contact surface area among sodium ions and active material can be increase greatly with CNT network having attached dispersed SnS nano-particles on it. After 100 cycles SnS/CNT anode has outstanding cycle performance having capacity about 666 mAh/g and at 1 A/g a capacity about 615 mAh/g was obtained after 500 cycles(Luu et al., 2020).

Tin sulphides (SnS_2 as well as SnS) consider to be a replacement for Na^+ owing to maximum spacing of 4.33 (SnS) for SnS and 5.90 (SnS_2) Vs 1.02 for sodium ion. SnS material in smaller size that is in the range of quantum dot could help overcome these problems by suppressing volume changes while also facilitating Na^+ transport. At 100 mA/g they obtained SnS QDs@NC electrode shows a capacity about 281 mAh/g and outstanding stability during the cycles, after 500 cycles having capacity retention (75%) at optimum current density (1000 mA/g) (Veerasubramani, Park, Choi, & Kim, 2020).

Sulfur-mediated interface engineering was established by Wang et al. to effectively enhance the kinetic characteristics of SnS nanosheet anode, obtained by carbon deposition/doping synchronous reduction process. The Sulphur in raw SnS_2 generates a sulfur-combined amorphous layers of carbon on the reduced SnS nanosheet. The sulfur-mediated layer interface improve reversibility as well as kinetics properties, boost ion/electron transport, and retain the configurational completeness of the SnS nanosheet anodes. As a result, at 1.0 A/g after 1 300 cycles a sodium-storage capacity about 349 mAh/g was achieved. The simple procedure and outstanding results point to interface tuning as a viable option for producing SnS-based anode material for batteries applications(Cheng et al., 2020).

Lu et al. used nitrogen doped CNT coated porous carbon as an anode for NIB, which have the benefits of both porous carbon and CNT. As a result, the synergistic impact of a porous structure of carbon with enough nitrogen content and a one-dimensional shape can always enhance Na^+ storage. MN-BCNT performs well as an anode in NIB, at 50 mA/g it showing a reversible specific capacity about 278 mAh/g(Shende, Chandran, & Ramaprabhu, 2020).

3.1.3 Graphite Based Sodium Ion Fast Charging Batteries

The idea of sodium-ion batteries was developed almost simultaneously with that LIBs in 1970s for LIBs and 1980s for SIBs. SIBs, on the other hand, vanished from view after the commercialization of LIBs in1990s, because of its lower specific-capacity, poorer energy-density and poor rate and cycle performance. The basic and most simple reason that SIBs behave worse compare to LIBs is that sodium is heavier, larger and has a larger standard potential than lithium, resulting in lower capacity and a narrower redox potential range(Xie, Xu, Guo, & Titirici, 2020). The goal of recent Na-ion battery research has been to develop acceptable electrode material choices. For the reason that (I) sodium-ion instability in interstitial locations in anode materials (II) repeated sodium-ion intercalation causes structural stability and (iii) larger ionic-radius of sodium ion (1.02) as compare to Li^+ (0.76), because of this conventional anode materials used in LIBs are unsuitable for Na-ion batteries(Y. J. Kim, Pyo, Kim, & Ryu, 2021). SIBs can charge to 4C and have a far easier cycling tendency than LIBs for the reason that SIBs can drain the battery to 0 V, leaving it free of electric current leakage; consequently, these batteries are very safe to store or export. However, the size of the battery which is significantly bigger than LIBs containing the same volt and loses charge quicker than LIBs are frequent issues that we have yet to fix, making them rather average as charge carriers (Karuppasamy et al., 2020). Furthermore, the considerable volume change that occurs during the intercalation and deintercalation of the bigger Na^+ causes severe structural deformation, which leads to cycling inconsistency and rate performance. Consequently, much work has gone into developing novel electrode materials for SIBs(Sheng, Zhang, Ji, Tong, & Tang, 2017). Anode material in Na-ion batteries must have low Na^+ intercalation and deintercalation voltages, rather below 0.5 V vs Na/Na^+ electrode having capacities greater than 250 mAh/g . The Na^+ deintercalation voltage of the negative electrode is essential because Na^+ are deintercalated from negative electrode and intercalated into positive electrode during discharge process of Na^+ cell (Abraham, 2020).

Jache et al proposed employing solvent co-intercalation chemistry to use up graphite as an negative electrode for Na-ion battery demonstrated Na-ion insertion using a graphite anode in an electrochemical cell having sodium triflate (NaOTf) in diglyme electrolyte, with and a coulombic efficiency (99.87%) and a reversible specific capacity (100 mAh/g) (Figure 4)(Park, Xu, & Kang, 2020).

Because of the lower stabilizing energy and larger ionic-radius sodium in graphite, traditional graphite anodes for LIBs do not accept Na^+ accommodation into the interlayer area. By widening the axial slab gap of graphite, we propose a viable technique for dramatically enhancing Na capacity. Using self-assembly negative graphene oxide flakes and Aluminium cation pillars, then a chemical interaction between Al-GO materials, we were able to successfully generate new graphite materials. Rebuilt graphite anodes with Al pillars have an extraordinary high capacity about 780 mAh/g was very large as compared to rGO materials having capacity about 210 mAh/g(Pyo et al., 2020).

The molybdenum pentachloride (MoI_5)–graphite intercalation compound (GI) has the good electronic conductivity, highest bulk density and excellent chemical stability among the transition-metal chloride-GIs. Intercalated MoI_5-single layers can influence a significant charge-transfer chemical combining effect having electronic conductivity about 2.8105 S cm^{-1} as well as enhance the graphite interlayer-distance with tap density about 1.55 g cm^{-3}, creating GIs a promising electrode material for batteries. Because of this the MoI_5-GI anode exhibits exceptional gravimetric capacities for sodium-ion batteries of 275 mAh/g and 426 mAh cm^{-3}, respectively(J. Zhu et al., 2020) .

Figure 4. Electrolyte linking structure, intercalated graphite, and co-intercalated graphite schematic diagrams, respectively. The insertion of dissolved ions is known as co-intercalation. Reproduced from ref (Park et al., 2020) under the terms of the Creative Commons CC BY license.

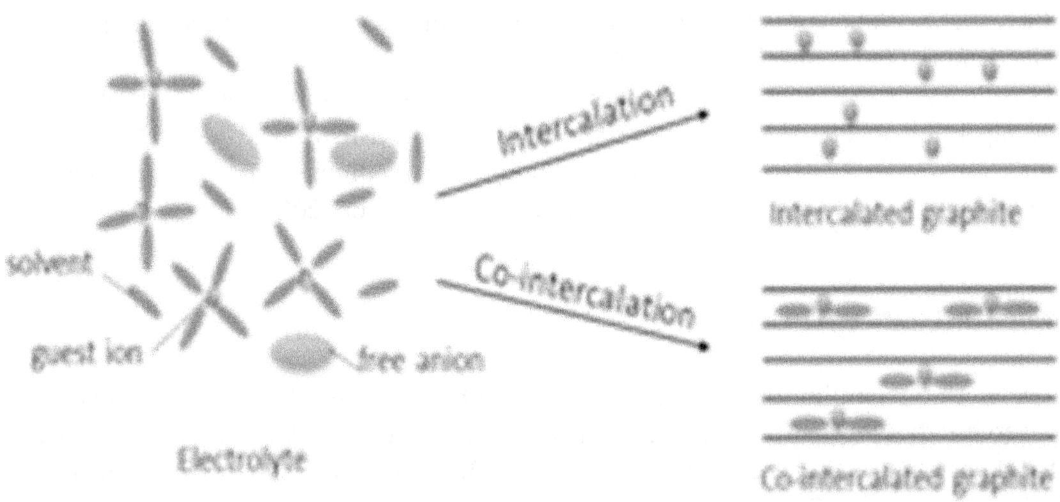

3.1.4 CNFs Based Sodium Ion Fast Charging Batteries

In electric vehicles and portable electronic-devices, LIBs are the most extensively utilized energy-storage technologies. However, due to lithium's limited deposits and unequal geographical distribution, alternative alkali metal-ion batteries including Na^+ and K^+, which have the benefits such as low cost, richness, and global distribution, have high expectations. Because of a paucity of good-performance negative electrode, SIBs have now been hindered in their development as a new technology(Tian et al., 2021). The performance of carbon nanofiber as an negative electrode material can be dramatically improved by covering it with a thin layer of oxygen-vacancy-rich TiO_2. At 10 A/g after 20,000cycles, the obtained (TiO_2/PBC) composite anode material shows reversible capacity about 180.2 mAh/g with 110.7% capacity retention. A 3-dimensional cross-linked conductive matrix, exposed ion transport pathways, and, most significantly, a TiO_2 coating containing large oxygen vacancies, that expands the lattice as well as so facilitates rapid ion transport, are all factors that contribute to TiO_2/PBC's excellent sodium storage performance (L. Huang et al., 2021).

The carbon substrate's hierarchical porous structure helped to improve the active materials' Li^+/Na^+ storage, as did the carbon wrappings made as of polyacrylonitrile (PAN) nanofibers. In lithium ion batteries (LIBs), After 180 cycles at 200 mA/g, the obtained Sn-CNF anode exhibits discharge capacity about 610.8 mAh/g, and at 100 mA/g it shows a capacity about 360.5 mAh/g after 100 successive cycles in SIBs.(S. Zhu, Huang, Wang, & Xu, 2021).

The search for more effective CNFs based anodes for SIBs manufactured using easy and cost-effective processes is still ongoing. A plasma-controlled approach is established for modifying nitrogen riches electrospinning CNFs as an free-standing anode material for SIBs. After the rate capability test, at 0.1 A/g the optimized sample exhibits reversible capacity (199 mAh/g) and at 500 mA/g an outstanding lasting stability having reversible capacity about 150 mAh/g was achieved after 2000 cycles (Yan et al., 2021).

Carbon anode innovation is important for the commercialization of Na-ion batteries. The use of a simple chemical vapour deposition process to make hybrid CNFs has significant implications as an anode for Na^+ storage. The developed hybrid CNFs contain linear and spiral shapes and have diameters ranges from 50-150 nm. The CNFs anode has impressive electrochemical characteristics, with an first Coulombic efficiency about 87.6% and good cycling stability, with a capacity retention about 75.4% was obtained after 100th cycle(Xue et al., 2021).

In half-cells, Fe_7Se_8-NCFs exhibits exceptional stability about 232 mAh/g at 0.1 A/g after 400 cycles and a rate capability about 153 mAh/g was achieved at 2 A/g. Furthermore, Fe_7Se_8-NCFs paired with $Na_3V_2(PO_4)_3$ (abbreviated as NVP) offer the possibility of full sodium-ion batteries, which have good reversible-capacity and rate-capability(L. Hu, Shang, Wang, & Zhou, 2021).

4 ALUMINIUM ION BATTERIES

Rechargeable Aluminium-ion batteries is considered to be excellent electrochemical energy storage device due to its advantages including large availability, long cycle life, low cost, high specific-energy, well-defined charge and discharge plateaus and easy to handle in ambient environments (Fard et al., 2020; J. Kim, Raj, & Lee, 2021). Aluminium (Al) is the most common metallic element exist in the Earth, consists of three electron per atom make up good charge-density, resulting in large theoretical capacities about 2978 mAh/g and 8034 mAh cm^{-3}, this high numbers are favorable for energy-storage applications. Furthermore, because one of the most important factors influencing practical as well as large-scale usage of technology is cost. One important benefit of Al-ion batteries is the less expensive of Al metal, which is due to both the large abundance of Aluminium as well as maturity of aluminium smelting technology compared to other metals (Akgenc, Sarikurt, Yagmurcukardes, & Ersan, 2021; Faegh, Ng, Hayman, & Mustain, 2021; X. Wang, J. Wang, Z. Chen, K. Yang, Z. Zhang, Z. Shi, T. Mei, J. Qian, J. Li, X. J. J. o. A. Wang, et al., 2020; Yoo, Heeney, Glöcklhofer, & Choi, 2021).

4.1 Carbon-Based Materials for Aluminium Ion Batteries

Carbonaceous materials are used as a cathode materials, aluminium foil as a anode, and ionic liquid is used electrolyte for making AIBs. During charging and discharge process, $AlCl_4$ intercalates and deintercalates carbon cathode materials, resulting in high rate performance and good cycle. . Presently, research focuses mostly on graphene-based materials having large specific-surface area and excellent electron transport abilities) and on graphite materials (which have layered and porous structure), both of which facilitate $AlCl_4$ insertion/deintercalation (Akgenc et al., 2021).

4.1.1 Graphene Based Fast Charging Aluminum Ion Batteries

Owing to its special physical characteristics, including such large surface-area and good conductivity, graphene has proven itself as a popular option as an cathode material for Al-ion batteries(Huang et al., 2020; Kong et al., 2021).

Graphene based cathode materials shows great cycling stability (thousands of cycles), quick charging (over 60 C), nearly 100% Coulombic efficiency, and very good discharge voltage-plateau values about 1.8 to 2.0 V. N. Li et al. produced graphene fiber-based aluminium-ion battery, which shows a large specific capacity about 115 mAh/g at 50 A/g, excellent rate performance, and as well as outstanding cyclic stability after 1000 cycles. The newly constructed AIB had a large energy density about 58 Wh kg^{-1} and a subsequent power-density about 26 kW kg^{-1}, both of which were equivalent to traditional energy-storage devices. The obtained wire-shaped graphene based Al-ion batteries can retain its electrochemical behavior across a variety of bending states as well as long bending cycles, indicating high mechanical flexibility and good stability (N. Li et al., 2021).

Lu et al. used Ar^+ -plasma etching and CVD technique to produce graphene nanoribbons upon highly porous 3-dimensional graphene (GNHPG) foam, establishing nanovoids and additional spaces for $AlCl_4$ – intercalation and deintercalation throughout the 3-dimensional graphene. At 5 A/g, the obtained GNHPG shows large capacity of 123 mAh/g having an excellent columbic efficiency greater than 98% and no capacity decrease after 100 cycles. Moreover, the Al/GNHPG displayed good rate capabilities at various current densities ranging from 2- 8 A/g, as well as superior high-low temperature of (0, 40, 60, and 80 °C) characteristics in terms of cycle life and capacity(Z. Hu et al., 2020).

4.1.2 Carbon Nanotubes Based Aluminum-Ion Fast Charging Batteries

Aluminum (Al) is abundant in the earth's crust, with 82000 ppm. During the electrochemical process, Al can exchange three electrons, resulting in a large gravimetric capacity about 2980 mAh/g as well as large volumetric-capacity about 8046 mA h cm^{-3}. Because of its minimal chemical reactivity, Al can also be handled in the open air. As a result, aluminium-based batteries are seen as a viable alternative or complement to LIBs(S. Li et al., 2021). Since 2014, AIBs have made significant progress as a novel electrochemical energy storage device. However, there are still certain issues with cathode materials, such as carbon-based materials' restricted capacity, metal chalcogenides and oxides lower rate-capacity, and cycling stability(Zhou et al., 2020). The majority of earlier research centered on increasing the electrochemical parameters (like cycle stability, discharge voltage and rate capability) of AIBs with traditional planer and button-type(N. Li et al., 2021).

AIBs work on the same principles as Li-ion batteries, with the main concept being that a reactive species can be intercalated inside the carbon. The key claim for these cells is that AIBs can replace the traditional one-electron lithium process with a three-electron process at the anode(Faegh et al., 2021). The development of cathode materials able to storing multivalent aluminium containing ions carrier is one of the most difficult challenges to overcome. Due to multi-valency of aluminium or aluminium-complex ions causes significant Coulombic contact with anionic structures, finding such materials with appropriately large ionic conductivity is a non-trivial task in practice(Yoo et al., 2021).

Because of the obvious large size of $AlCl_{4-}$ anions, improving the reversible capacity of $AlCl_{4-}$ intercalation and deintercalation in typical cathodes of AIBs is generally challenging. The composite films out of polyaniline and single-walled CNTs (PANI/SWCNTs) and also protonated the PANI nanorods in them. At 1 A/g the Al/PANI(H^+) battery has large discharge capacity about 200 mAh/g Furthermore, PANI(H^+)/SWCNT cathode have a long cycling life, having just only 0.003% capacity loss per cycle after 8000 cycles and having coulombic efficiency about 100%. Moreover, PANI(H^+) has good mechanical characteristics(S. Wang, Huang, Yao, Zhang, & Niu, 2020).

According to Bhauriyal et al. an Al-SWCNT cell with a tube structure can deliver 1.96 V discharge voltage and 275 mAh/g specific capacity. The development of the SWCNT cathode has been hampered by a lack of experimental assistance. At 200 mA/g the prepared Al-GMWCNT cell shows good specific capacity about 4 mAh/g. After thousand charge/discharge cycles, discharge capacity is almost unaffected and remains about 58 mAh/g (Coulombic efficiency: 99.5%), indicating that the Al-G-MWCNT cell has outstanding cycling stability (Han et al., 2020).

At 50 mA/g, GCNT/S as a positive electrode in Al-ion battery exhibits specific capacity about 507 mAh/g and a steady cyclic stability after 600 cycles. When compared to GCNT, this represents a 7.2 times increase within specific capacity. This specific capacity value is larger than that of carbon-based materials, paving the path for the creation of large-energy density Al-ion batteries(Saroja, Kamaraj, & Ramaprabhu, 2020).

4.1.3 Graphite Based Aluminium Ion Fast Charging Batteries

The current state-of-the-art lithium-ion batteries have spectacular success in applications like as mobile electronic gadgets and electric/hybrid automobiles(Tu, Wang, Zhu, & Jiao, 2020). Due to the limitations of LIBs in terms of cost, energy-density, and safety in recent years, other innovative secondary batteries have gotten a lot of attention. Aluminum metal's volumetric energy density about 8.064 Ah cm^{-3} is roughly four times that of Li metal in many secondary batteries. Furthermore, aluminium metal has a lower activity than lithium metal, making it easier to store with greater security. Above all, the storage capacity of aluminium metal in the crust is three orders of magnitude greater than lithium, lowering the cost of aluminium batteries significantly(Li, Li, Li, & Zhang, 2020).

Aluminum is the greatest rich metal element exist in the earth when compared to other metals. Aluminum's mass capacity about 2.98 Ah/g is comparable to Li which is 3.86 Ah/g in terms of energy density and charge storage capacity, having its volumetric capacity about 8.04 Ah cm^{-3} which is three times that of lithium(Y. Zhang et al., 2020). Because metal aluminium is the primary negative electrode material in aluminium batteries, the performance of cathode materials limits the advancement of aluminium batteries(Huo et al., 2020). As a result, some research on carbon-based materials as positive electrode for Al-ion batteries has been published in recent years, such as 3-dimensional graphene, few-layer graphene, graphite. The experiment was reported by Dai et al at Stanford University, and it revealed that Al-ion batteries having graphitic-foam cathodes shows excellent electrochemical behavior. The battery's discharge voltage is higher, plateauing is ~2.0V, and it can sustain 7500 cycles with no losing capacity. The specific discharge capacity of graphitic-foam cathode is lower and is maintained at 66 mAh/g(Tan et al., 2020).

Dai et al. investigated a AIB with a 3D graphitic-foam positive electrode and an $AlCl_3$/[EMIm]Cl ionic liquid electrolyte, which showed outstanding cycling performance along with discharge voltage plateaus approximately 2 volts, having a capacity about 70mAh/g. Jiao group investigated an Al-graphite battery based on $AlCl_3$/[EMIm]Cl polymeric-electrolyte having discharge capacity about 120 mAh/g (near to graphite's theoretical value)(Y. Zhang et al., 2021). However, at room temperature, there is still the issue of a very small electrochemical window about 1–2.45 V vs. Al/Al_3^+. As a result, overcharging and safety risks may reduce the use of such kind of batteries. The authors of this research suggested that organic compounds be added to the graphitic cathode to improve protection whenever the battery is overcharged. After 200 cycles with CuPc added to graphitic cathode, there was no substantial capacity depreciation (lower than 10%), and the specific capacity loss was less than 25% for initial 1,000 cycles. When the CuPc/graphite composite cathode was employed, after 1000 charging/discharging cycles, the swelling phenomenon not happen (Bian et al., 2020). When deposited/dissolved at moderate current densities onto a suitable substrate, one of the primary advantages of aluminium is that it does not form dendrites. Dendrites develop only above current densities of 100 mA cm^{-2} when aluminium is deposited from $AlCl_3$-[EMIm]Cl. This is because plating occurs quicker than diffusion can feed species from the electrolyte, resulting in preferential plating onto projecting regions of the electrode, resulting in dendritic development. Fortunately, current densities in batteries are significantly lower (80 C may be attained at 10 mA cm^{-2})(Craig, Schoetz, Cruden, & de Leon, 2020).

When graphite is employed as an potential active material for ALBs, the $AlCl_4$- ions have being regarded the intercalated species and reversibly intercalated/deintercalated between the layers of graphite. The major carrier species is $AlCl_4$- ions. After the shallow charge intercalation and deintercalation, some $AlCl_4$- and $Al_2 Cl_7$- ions persist in the interlayer, causing exfoliation and eventual disintegration of structure of graphite. To better graphite cyclability as well as capacity as a component of battery, it would be beneficial to avoid big anions from intercalating into the structure(Yamagata, Takahara, Wang, Mizoguchi, & Yagi, 2020).

CONCLUSION

We have looked to the use of carbon-based nanomaterials in different batteries. Carbon-based nanomaterials were identified as promising materials for different batteries applications. A variety of carbon-based materials including graphene, CNTs, CNFs and graphite have been used as effective electrode materials for fast charging batteries. Although carbon-based nanomaterials are effective for batteries applications, greater surface area materials with a higher reaction rate are required to meet the need for high-power electrochemical cells. Furthermore, the nanoparticles used in batteries have a low density. As a result, only materials that conduct the necessary reaction in a very selective manner are suitable for usage at the nanoscale. New low-cost anode material development is predicted to improve the cycle performance, safety and rate capability of both small and large batteries. In conclusion carbon nanomaterials based fast charging batteries have been investigated for so long and provide the dominant energy storage approach for EVs and portable electronic-devices because of its low cost, long cycle life and remarkable energy-density. Among all batteries, Lithium ion batteries is one of the best electrochemical device because of its low cost, large capacity, safety, high energy-density and excellent cycle life. Carbon based materials has ability to further enhance the fast charging capability, electrochemical performance, storage capacity of next generation LIBs.

REFERENCES

Abraham, K. (2020). How comparable are sodium-ion batteries to lithium-ion counterparts? *ACS Energy Letters*, *5*(11), 3544–3547.

Akgenc, B., Sarikurt, S., Yagmurcukardes, M., & Ersan, F. (2021). Aluminum and lithium sulfur batteries: A review of recent progress and future directions. *Journal of Physics Condensed Matter*, *33*(25), 253002.

Ambade, R. B., Veerasubramani, G. K., Ambade, S. B., Christy, M., Eom, W., Shin, H., ... Han, T. H. (2021). Photonic split-second induced mesoporous TiO2-Graphene architectures for efficient sodium-ion batteries. *Carbon*, *178*, 332–344.

Autthawong, T., Chimupala, Y., Haruta, M., Kurata, H., Kiyomura, T., Yu, A.-s., ... Sarakonsri, T. (2020). Ultrafast-charging and long cycle-life anode materials of TiO 2-bronze/nitrogen-doped graphene nano-composites for high-performance lithium-ion batteries. *RSC Advances*, *10*(71), 43811–43824.

Bai, X., Zhang, H., Lin, J., & Zhang, G. (2021). UV-ozone contributions towards facile self-assembly and high performance of silicon-carbon fiber materials as lithium-ion battery anodes. *Journal of Colloid and Interface Science*, *598*, 339–347. doi:10.1016/j.jcis.2021.04.044

Batool, S., Idrees, M., Kong, J., Zhang, J., Kong, S., & Dong, M. (2020). *Assessment of the electro-chemical behaviour of silicon@ carbon nanocomposite anode for lithium-ion batteries*. Academic Press.

Bian, Y., Su, L., Yu, Z., Lv, Z., Chen, H., Zhou, Y., . . . Cao, H. (2020). *Graphite/copper phthalocyanine composite cathode for overcharge protection and gas evolution suppression in aluminum-ion batteries at room temperature*. Academic Press.

Chen, K.-H., Namkoong, M. J., Goel, V., Yang, C., Kazemiabnavi, S., Mortuza, S., ... Sakamoto, J. (2020). Efficient fast-charging of lithium-ion batteries enabled by laser-patterned three-dimensional graphite anode architectures. *Journal of Power Sources*, *471*, 228475.

Cheng, Y., Wang, Z., Chang, L., Wang, S., Sun, Q., Yi, Z., & Wang, L. (2020). Sulfur-mediated interface engineering enables fast SnS nanosheet anodes for advanced lithium/sodium-ion batteries. *ACS Applied Materials & Interfaces*, *12*(23), 25786–25797.

Craig, B., Schoetz, T., Cruden, A., & de Leon, C. P. (2020). Review of current progress in non-aqueous aluminium batteries. *Renewable & Sustainable Energy Reviews*, *133*, 110100.

Cui, Z., He, S. A., Liu, Q., Guan, G., Zhang, W., Xu, C., ... Zou, R. (2020). Graphene-Like Carbon Film Wrapped Tin (II) Sulfide Nanosheet Arrays on Porous Carbon Fibers with Enhanced Electrochemical Kinetics as High-Performance Li and Na Ion Battery Anodes. *Advancement of Science*, *7*(18), 1903045.

Dhillon, S., Hernández, G., Wagner, N. P., Svensson, A. M., & Brandell, D. (2021). Modelling capacity fade in silicon-graphite composite electrodes for lithium-ion batteries. *Electrochimica Acta*, *377*, 138067.

Ding, X., Liang, D., & Zhao, H. (2021). *Enhanced electrochemical Performance Promoted by Tin in silica Anode materials for Stable and High-capacity Lithium-ion Batteries*. Academic Press.

Divya, M., Natarajan, S., Lee, Y.-S., & Aravindan, V. (2020). Achieving high-energy dual carbon Li-ion capacitors with unique low-and high-temperature performance from spent Li-ion batteries. *Journal of Materials Chemistry. A, Materials for Energy and Sustainability*, *8*(9), 4950–4959.

Dong, X., Zheng, X., Deng, Y., Wang, L., Hong, H., & Ju, Z. (2020). SiO 2/N-doped graphene aerogel composite anode for lithium-ion batteries. *Journal of Materials Science*, *55*, 13023–13035.

Faegh, E., Ng, B., Hayman, D., & Mustain, W. E. (2021). Practical assessment of the performance of aluminium battery technologies. *Nature Energy*, *6*(1), 21–29.

Fan, X., Wang, Z., Cai, T., Yang, Y., Wu, H., Cao, S., ... Zhang, W. (2021). An integrated highly stable anode enabled by carbon nanotube-reinforced all-carbon binder for enhanced performance in lithium-ion battery. *Carbon*, *182*, 749–757. doi:10.1016/j.carbon.2021.06.065

Fard, L. S., Peighambardoust, N. S., Jang, H. W., Dehghan, A., Saligheh, N. N. K., Iranpour, M., & Rajabi, M. I. (2020). The rechargeable aluminum-ion battery with different composite cathodes: A review. *Journal of Composites and Compounds*, *2*(4), 138–146.

Feng, W., Maça, R. R., & Etacheri, V. (2020). High-energy-density sodium-ion hybrid capacitors enabled by interface-engineered hierarchical TiO2 nanosheet anodes. *ACS Applied Materials & Interfaces*, *12*(4), 4443–4453.

Galashev, A. Y., Ivanichkina, K. A., Katin, K. P., & Maslov, M. M. (2020). Computer test of a modified silicene/graphite anode for lithium-ion batteries. *ACS Omega*, *5*(22), 13207–13218.

Gu, S., & Zhu, A. (2020). Graphene nanosheets loaded Fe3O4 nanoparticles as a promising anode material for lithium ion batteries. *Journal of Alloys and Compounds*, *813*, 152160. doi:10.1016/j.jallcom.2019.152160

Guo, P.-H., Cheng, H.-M., Chen, J.-S., Wu, Y.-S., Yang, C.-C., Zeng, F.-Y., ... Chen, P.-T. (2021). Novel Triaxial Raman Scanning Platform for Evaluating Integrity of Graphite Electrodes in Li-Ion Batteries. *IEEE Access: Practical Innovations, Open Solutions*, *9*, 81895–81901.

Guo, Q., Shao, H., Zhang, K., Chen, G., Kong, W., Feng, X., ... Dong, C. (2021). CoP Nanoparticles Intertwined with Graphene Nanosheets as a Superior Anode for Half/Full Sodium-Ion Batteries. *ChemElectroChem*, *8*(11), 2022–2027.

Han, M., Lv, Z., Hou, L., Zhou, S., Cao, H., Chen, H., ... Cai, M. (2020). Graphitic multi-walled carbon nanotube cathodes for rechargeable Al-ion batteries with well-defined discharge plateaus. *Journal of Power Sources*, *451*, 227769.

Hu, L., Shang, C., Wang, X., & Zhou, G. (2021). Fe 7 Se 8 encapsulated in N-doped carbon nanofibers as a stable anode material for sodium ion batteries. *Nanoscale Advances*, *3*(1), 231–239.

Hu, Z., Zhang, H., Wang, H., Zhang, F., Li, Q., & Li, H. (2020). Nonaqueous aluminum ion batteries: Recent progress and prospects. *ACS Materials Letters*, *2*(8), 887–904.

Huang, H., Zhou, F., Lu, P., Li, X., Das, P., Feng, X., ... Wu, Z.-S. (2020). Design and construction of few-layer graphene cathode for ultrafast and high-capacity aluminum-ion batteries. *Energy Storage Materials*, *27*, 396–404.

Huang, L., Zeng, L., Zhu, J., Sun, L., Yao, L., Deng, L., & Zhang, P. (2021). Oxygen-vacancy-rich TiO2-coated carbon nanofibers for fast sodium storage in high-performance sodium-ion hybrid capacitors. *Journal of Power Sources*, *493*, 229678.

Huang, Q., Ni, S., Jiao, M., Zhong, X., Zhou, G., & Cheng, H. M. (2021). Aligned Carbon-Based Electrodes for Fast-Charging Batteries: A Review. *Small*, *17*(48), 2007676.

Huo, X., Liu, J., Li, J., Zhang, B., Zhang, Y., Yu, Y., & Kang, F. (2020). Hexagonal composite CuSe@C as a positive electrode for high-performance aluminum batteries. *ACS Applied Energy Materials*, *3*(11), 11445–11455.

Hussain, S. Z., Ihrar, M., Hussain, S. B., Oh, W. C., & Ullah, K. (2020). A review on graphene based transition metal oxide composites and its application towards supercapacitor electrodes. *SN Applied Sciences*, *2*, 1–23.

Jiang, L. L., Yan, C., Yao, Y. X., Cai, W., Huang, J. Q., & Zhang, Q. (2021). Inhibiting solvent co-intercalation in a graphite anode by a localized high-concentration electrolyte in fast-charging batteries. *Angewandte Chemie International Edition*, *60*(7), 3402–3406.

Karuppasamy, K., Jothi, V. R., Nichelson, A., Vikraman, D., Tanveer, W. H., Kim, H.-S., & Yi, S.-C. (2020). Nanostructured transition metal sulfide/selenide anodes for high-performance sodium-ion batteries. In Nanostructured, Functional, and Flexible Materials for Energy Conversion and Storage Systems (pp. 437-464). Elsevier.

Kefayat, U., Bakht Mand, K., Amin Ur, R., & Won Chun, O. (2022). Perspective Chapter: Graphene Based Nanocomposites for Supercapacitor Electrodes. In S. Zoran (Ed.), Updates on Supercapacitors. Rijeka: IntechOpen.

Kim, J., Raj, M. R., & Lee, G. (2021). High-defect-density graphite for superior-performance aluminum-ion batteries with ultra-fast charging and stable long life. *Nano-Micro Letters*, *13*(1), 171.

Kim, S. Y., Kim, C. H., & Yang, C.-M. (2021). Binder-free silicon anodes wrapped in multiple graphene shells for high-performance lithium-ion batteries. *Journal of Power Sources*, *486*, 229350. doi:10.1016/j.jpowsour.2020.229350

Kim, Y. J., Pyo, S., Kim, S., & Ryu, W.-H. (2021). Group VI metallic pillars for assembly of expanded graphite anodes for high-capacity Na-ion batteries. *Carbon*, *175*, 585–593.

Kong, Y., Tang, C., Huang, X., Nanjundan, A. K., Zou, J., Du, A., & Yu, C. (2021). Thermal Reductive Perforation of Graphene Cathode for High-Performance Aluminum-Ion Batteries. *Advanced Functional Materials*, *31*(17), 2010569.

Lee, S.-M., Kim, J., Moon, J., Jung, K.-N., Kim, J. H., Park, G.-J., ... Lee, J.-W. (2021). A cooperative biphasic MoO x–MoP x promoter enables a fast-charging lithium-ion battery. *Nature Communications*, *12*(1), 39.

Lee, Y.-K., Lee, M., Lee, G. W., Eom, K., Lee, M.-H., Lee, S., & Joh, H.-I. (2020). Amorphous MoSx embedded within edges of modified graphite as fast-charging anode material for rechargeable batteries. *Applied Surface Science*, *509*, 145352. doi:10.1016/j.apsusc.2020.145352

Li, L., Zhang, D., Deng, J., Gou, Y., Fang, J., Cui, H., ... Cao, M. (2021). Carbon-based materials for fast charging lithium-ion batteries. *Carbon*, *183*, 721–734.

Li, M., Liu, T., Bi, X., Chen, Z., Amine, K., Zhong, C., & Lu, J. (2020). *Cationic and anionic redox in lithium-ion based batteries*. Academic Press.

Li, N., Yao, Y., Lv, T., Chen, Z., Yang, Y., Cao, S., & Chen, T. (2021). High-performance wire-shaped aluminum ion batteries based on continuous graphene fiber cathodes. *Journal of Power Sources, 488*, 229460.

Li, S., Liu, Y.-M., Zhang, Y.-C., Song, Y., Wang, G.-K., Liu, Y.-X., ... Guo, X.-D. (2021). A review of rational design and investigation of binders applied in silicon-based anodes for lithium-ion batteries. *Journal of Power Sources, 485*, 229331.

Li, Z., Li, J., Li, X., & Zhang, W. (2020). Pseudocapacitance effect in Al-C batteries with expanded graphite positive electrode at different temperatures. *Journal of Power Sources, 467*, 228323.

Luo, H., Wang, Q., Wang, Y., Xu, C., Wang, B., Wang, M., ... Zhang, Y. (2020). Nano-silicon embedded in MOFs-derived nitrogen-doped carbon/cobalt/carbon nanotubes hybrid composite for enhanced lithium ion storage. *Applied Surface Science, 529*, 147134. doi:10.1016/j.apsusc.2020.147134

Luu, T. H. T., Duong, D. L., Lee, T. H., Pham, D. T., Sahoo, R., Han, G., ... Lee, Y. H. (2020). Monodispersed SnS nanoparticles anchored on carbon nanotubes for high-retention sodium-ion batteries. *Journal of Materials Chemistry. A, Materials for Energy and Sustainability, 8*(16), 7861–7869. doi:10.1039/C9TA13136A

Madhusanka, S., Sandaruwan, R., Athar, M., Zaib, M., Munasinghe, H. M., Arachchige, B., . . . Gunawardhana, N. (2020). *TiO2 Microparticles/Reduced Graphene Oxide Composite as Anode Material for Lithium Ion Battery*. Academic Press.

Mand Khan, B., Chun Oh, W., Nuengmatch, P., & Ullah, K. (2023). Role of graphene-based nanocomposites as anode material for Lithium-ion batteries. *Materials Science and Engineering B, 287*, 116141. doi:10.1016/j.mseb.2022.116141

Manj, R. Z. A., Zhang, F., Rehman, W. U., Luo, W., & Yang, J. (2020). *Toward understanding the interaction within Silicon-based anodes for stable lithium storage*. Academic Press.

Ni, X., Cui, Z., Luo, H., Chen, H., Liu, C., Wu, Q., & Ju, A. (2021). Hollow multi-nanochannel carbon nanofibers@ MoSe2 nanosheets composite as flexible anodes for high performance lithium-ion batteries. *Chemical Engineering Journal, 404*, 126249.

Or, T., Gourley, S. W., Kaliyappan, K., Yu, A., & Chen, Z. (2020). Recycling of mixed cathode lithium-ion batteries for electric vehicles: Current status and future outlook. *Carbon Energy, 2*(1), 6–43.

Pang, H., Yu, P., Xu, F., Zhang, W., Peng, J., Hu, H., ... Liang, Y. (2020). A general strategy for metal oxide nanoparticles embedded into heterogeneous carbon nanosheets as high-rate lithium-ion battery anodes. *Journal of Materials Chemistry. A, Materials for Energy and Sustainability, 8*(47), 25382–25389. doi:10.1039/D0TA08198A

Park, J., Xu, Z.-L., & Kang, K. (2020). Solvated ion intercalation in graphite: Sodium and beyond. *Frontiers in Chemistry, 8*, 432.

Pyo, S., Eom, W., Kim, Y. J., Lee, S. H., Han, T. H., & Ryu, W.-H. (2020). Super-expansion of assembled reduced graphene oxide interlayers by segregation of Al nanoparticle pillars for high-capacity Na-ion battery anodes. *ACS Applied Materials & Interfaces, 12*(21), 23781–23788.

Ruan, L. F., Wang, J. W., & Ying, S. M. (2021). *Research progress and application of modified silicon-based anode materials for lithium-ion batteries.* Paper presented at the Materials Science Forum.

Saba, N., & Jawaid, M. (2018). Energy and environmental applications of graphene and its derivatives. In *Polymer-based Nanocomposites for Energy and Environmental Applications* (pp. 105–129). Elsevier.

Saroja, A. P. V. K., Kamaraj, M., & Ramaprabhu, S. (2020). Strongly coupled sulfur nanoparticles on graphene-carbon nanotube hybrid electrode for multifunctional sodium and aluminium ion storage. *Journal of Alloys and Compounds, 818*, 152864.

Seok, J. Y., Kim, S., Yang, I., Park, J. H., Lee, J., Kwon, S., & Woo, K. (2021). Strategically Controlled Flash Irradiation on Silicon Anode for Enhancing Cycling Stability and Rate Capability toward High-Performance Lithium-Ion Batteries. *ACS Applied Materials & Interfaces, 13*(13), 15205–15215. doi:10.1021/acsami.0c22983

Shende, R. C., Chandran, P., & Ramaprabhu, S. (2020). In situ nitrogen-doped, defect-induced carbon nanotubes as an efficient anode for sodium-ion batteries. *Nanotechnology, 31*(23), 235403.

Sheng, M., Zhang, F., Ji, B., Tong, X., & Tang, Y. (2017). A novel tin-graphite dual-ion battery based on sodium-ion electrolyte with high energy density. *Advanced Energy Materials, 7*(7), 1601963.

Shi, Q., Zhang, Y., Chen, K., Yuan, S., Chang, T., Tian, F., ... Yang, S. (2020). Cobalt vacancies assisted ion diffusion in Co 2 AlO 4 carbon nanofibers for enhancing lithium battery performance. *Dalton Transactions (Cambridge, England), 49*(29), 10127–10137.

Sui, D., Xu, L., Zhang, H., Sun, Z., Kan, B., Ma, Y., & Chen, Y. (2020). A 3D cross-linked graphene-based honeycomb carbon composite with excellent confinement effect of organic cathode material for lithium-ion batteries. *Carbon, 157*, 656–662.

Tan, B., Han, S., Luo, W., Chao, Z., Fan, J., & Wang, M. (2020). Synthesis of RGO-supported layered MoS2 with enhanced electrochemical performance for aluminum ion batteries. *Journal of Alloys and Compounds, 841*, 155732.

Thakur, A. K., Ahmed, M. S., Oh, G., Kang, H., Jeong, Y., Prabakaran, R., ... Hwang, J.-Y. (2021). Advancement in graphene-based nanocomposites as high capacity anode materials for sodium-ion batteries. *Journal of Materials Chemistry. A, Materials for Energy and Sustainability, 9*(5), 2628–2661.

Tian, S., Cai, T., Wang, D., Kong, D., Ren, H., & Xing, W. (2021). *A superior cathode of sodium-ion battery: A V2O5 nanosheets anchored on carbon nanofibers.* Paper presented at the IOP Conference Series: Earth and Environmental Science.

Tu, J., Wang, J., Zhu, H., & Jiao, S. (2020). The molten chlorides for aluminum-graphite rechargeable batteries. *Journal of Alloys and Compounds, 821*, 153285.

Tzeng, Y., Chen, R., & He, J.-L. (2020). Silicon-based anode of lithium ion battery made of nano silicon flakes partially encapsulated by silicon dioxide. *Nanomaterials (Basel, Switzerland), 10*(12), 2467.

Veerasubramani, G. K., Park, M.-S., Choi, J.-Y., & Kim, D.-W. (2020). Ultrasmall SnS quantum dots anchored onto nitrogen-enriched carbon nanospheres as an advanced anode material for sodium-ion batteries. *ACS Applied Materials & Interfaces, 12*(6), 7114–7124.

Vo, T. N., Kim, D. S., Mun, Y. S., Lee, H. J., Ahn, S.-k., & Kim, I. T. (2020). Fast charging sodium-ion batteries based on Te-PC composites and insights to low-frequency limits of four common equivalent impedance circuits. *Chemical Engineering Journal, 398*, 125703.

Wang, G., Yu, M., & Feng, X. (2021). Carbon materials for ion-intercalation involved rechargeable battery technologies. *Chemical Society Reviews, 50*(4), 2388–2443. doi:10.1039/D0CS00187B

Wang, L., Deng, J., Deng, J., Fei, Y., Fang, Y., & Hu, Y. H. (2020). Ultra-fast and ultra-long-life Li ion batteries with 3D surface-porous graphene anodes synthesized from CO 2. *Journal of Materials Chemistry. A, Materials for Energy and Sustainability, 8*(26), 13385–13392.

Wang, R., Yu, J., Islam, F., Tahmasebi, A., Lee, S., & Chen, Y. (2020). State-of-the-Art Research and Applications of Carbon Foam Composite Materials as Electrodes for High-Capacity Lithium Batteries. *Energy & Fuels, 34*(7), 7935–7954. doi:10.1021/acs.energyfuels.0c01802

Wang, S., Cheng, Y., Xue, H., Liu, W., Yi, Z., Chang, L., & Wang, L. (2021). Multifunctional sulfur-mediated strategy enabling fast-charging Sb 2 S 3 micro-package anode for lithium-ion storage. *Journal of Materials Chemistry. A, Materials for Energy and Sustainability, 9*(12), 7838–7847.

Wang, S., Huang, S., Yao, M., Zhang, Y., & Niu, Z. (2020). Engineering active sites of polyaniline for AlCl2+ storage in an aluminum-ion battery. *Angewandte Chemie International Edition, 59*(29), 11800–11807.

Wang, X., Wang, J., Chen, Z., Yang, K., Zhang, Z., & Shi, Z. (2020). *Yolk-double shell Fe3O4@ C@ C composite as high-performance anode materials for lithium-ion batteries*. Academic Press.

Wang, X., Wang, J., Chen, Z., Yang, K., Zhang, Z., Shi, Z., ... Wang, X. (2020). Yolk-double shell Fe3O4@ C@ C composite as high-performance anode materials for lithium-ion batteries. *Journal of Alloys and Compounds, 822*, 153656.

Wang, Z., Li, Y., Huang, S., Liu, L., Wang, Y., Jin, J., ... Schmidt, O. G. (2020). PVD customized 2D porous amorphous silicon nanoflakes percolated with carbon nanotubes for high areal capacity lithium ion batteries. *Journal of Materials Chemistry. A, Materials for Energy and Sustainability, 8*(9), 4836–4843. doi:10.1039/C9TA12923E

Wasalathilake, K. C., Li, H., Xu, L., & Yan, C. (2020). Recent advances in graphene based materials as anode materials in sodium-ion batteries. *Journal of Energy Chemistry, 42*, 91–107. doi:10.1016/j.jechem.2019.06.016

Weng, W., Kurihara, R., Wang, J., & Shiratori, S. (2019). Electrospun carbon nanofiber-based composites for lithium-ion batteries: Structure optimization towards high performance. *Composites Communications, 15*, 135–148. doi:10.1016/j.coco.2019.07.005

Wu, Z., Zhang, X., Deng, L., Zhang, Y., Wang, Z., Shen, Y., & Shao, G. (2022). Atomic Layer Coated Al2O3 on Nitrogen Doped Vertical Graphene Nanosheets for High Performance Sodium Ion Batteries. *Energy & Environmental Materials, 5*(1), 285–294. doi:10.1002/eem2.12169

Xiang, Y., Xin, L., Hu, J., Li, C., Qi, J., Hou, Y., & Wei, X. (2021). Advances in the applications of graphene-based nanocomposites in clean energy materials. *Crystals, 11*(1), 47.

Xiao, Z., Wang, R., Jiang, D., Qian, Z., Li, Y., Yang, K., ... Wu, F. (2021). *Recent Developments of Two-Dimensional Anode Materials and Their Composites in Lithium-Ion Batteries*. Academic Press.

Xie, F., Xu, Z., Guo, Z., & Titirici, M.-M. (2020). Hard carbons for sodium-ion batteries and beyond. *Progress in Energy*, *2*(4), 042002.

Xu, Y., Sun, X., & Wei, C. (2020). Improving the electrochemical performance of lithium Si batteries by multilayer porous carbon nanosheets/multi-walled carbon nanotubes composite inert nano-Ag. *Ionics*, *26*, 1149–1158.

Xue, Z., Xiong, Q., Zou, C., Chi, H., Hu, X., & Ji, Z. (2021). Growth of carbon nanofibers through chemical vapor deposition for enhanced sodium ion storage. *Materials Research Bulletin*, *133*, 111049.

Yamagata, S., Takahara, I., Wang, M., Mizoguchi, T., & Yagi, S. (2020). EQCM analysis of intercalation species into graphite positive electrodes for Al batteries. *Journal of Alloys and Compounds*, *846*, 156469.

Yan, X., Liang, S., Shi, H., Hu, Y., Liu, L., & Xu, Z. (2021). Nitrogen-enriched carbon nanofibers with tunable semi-ionic CF bonds as a stable long cycle anode for sodium-ion batteries. *Journal of Colloid and Interface Science*, *583*, 535–543.

Yang, X., Liang, H.-J., Zhao, X.-X., Yu, H.-Y., Wang, M.-Y., Nie, X.-J., & Wu, X.-L. (2021). A sandwich nanocomposite composed of commercially available SnO and reduced graphene oxide as advanced anode materials for sodium-ion full batteries. *Inorganic Chemistry Frontiers*, *8*(2), 396–404. doi:10.1039/D0QI01033B

Yasin, G., Arif, M., Mehtab, T., Shakeel, M., Mushtaq, M. A., Kumar, A., ... Song, H. (2020). A novel strategy for the synthesis of hard carbon spheres encapsulated with graphene networks as a low-cost and large-scalable anode material for fast sodium storage with an ultralong cycle life. *Inorganic Chemistry Frontiers*, *7*(2), 402–410.

Yoo, D.-J., Heeney, M., Glöcklhofer, F., & Choi, J. W. (2021). Tetradiketone macrocycle for divalent aluminium ion batteries. *Nature Communications*, *12*(1), 2386.

Zeng, Y., Huang, Y., Liu, N., Wang, X., Zhang, Y., Guo, Y., ... Zhang, Q. (2021). N-doped porous carbon nanofibers sheathed pumpkin-like Si/C composites as free-standing anodes for lithium-ion batteries. *Journal of Energy Chemistry*, *54*, 727–735. doi:10.1016/j.jechem.2020.06.022

Zhang, L., Wang, W., Lu, S., & Xiang, Y. (2021). Carbon anode materials: A detailed comparison between Na-ion and K-ion batteries. *Advanced Energy Materials*, *11*(11), 2003640.

Zhang, Y., Zhang, B., Li, J., Liu, J., Huo, X., & Kang, F. (2021). SnSe nano-particles as advanced positive electrode materials for rechargeable aluminum-ion batteries. *Chemical Engineering Journal*, *403*, 126377.

Zhang, Y., Zhang, Y., Zhang, K., Jia, K., Liu, G., & He, X. (2020). *Nitrogen-doped graphene nanosheet coated nanospherical Fe3O4 from zeolitic imidazolate frameworks template as anode of lithium ion batteries*. Academic Press.

Zhao, W., Li, M., Qi, Y., Tao, Y., Shi, Z., Liu, Y., & Cheng, J. (2021). Ultrasound sonochemical synthesis of amorphous Sb2S3-graphene composites for sodium-ion batteries. *Journal of Colloid and Interface Science*, *586*, 404–411. doi:10.1016/j.jcis.2020.10.104

Zhao, X., Li, H., Han, F., Dai, M., Sun, Y., Song, Z., ... Niu, L. (2020). Electrochemical exfoliation of graphene as an anode material for ultra-long cycle lithium ion batteries. *Journal of Physics and Chemistry of Solids*, *139*, 109301. doi:10.1016/j.jpcs.2019.109301

Zheng, X., Hu, Q., Zhou, X., Nie, W., Li, C., & Yuan, N. (2021). Graphene-based fibers for the energy devices application: A comprehensive review. *Materials & Design*, *201*, 109476. doi:10.1016/j.matdes.2021.109476

Zhou, Q., Zheng, Y., Wang, D., Lian, Y., Ban, C., Zhao, J., & Zhang, H. (2020). Cathode materials in non-aqueous aluminum-ion batteries: Progress and challenges. *Ceramics International*, *46*(17), 26454–26465.

Zhu, G., Tang, C., Jiang, M., Du, A., Zhang, H., & Yang, J. (2021). Regulating the interfacial behavior of carbon nanotubes for fast lithium storage. *Electrochimica Acta*, *388*, 138591. doi:10.1016/j.electacta.2021.138591

Zhu, G. L., Zhao, C. Z., Huang, J. Q., He, C., Zhang, J., Chen, S., ... Zhang, Q. (2019). Fast charging lithium batteries: Recent progress and future prospects. *Small*, *15*(15), 1805389.

Zhu, J., Tu, W., Pan, H., Zhang, H., Liu, B., Cheng, Y., ... Zhang, H. (2020). Self-templating synthesis of hollow Co3O4 nanoparticles embedded in N, S-dual-doped reduced graphene oxide for lithium ion batteries. *ACS Nano*, *14*(5), 5780–5787.

Zhu, S., Huang, A., Wang, Q., & Xu, Y. (2021). MOF-derived porous carbon nanofibers wrapping Sn nanoparticles as flexible anodes for lithium/sodium ion batteries. *Nanotechnology*, *32*(16), 165401.

Chapter 10
Nanoparticles as Fillers in Composites for X-Ray and Gamma-Ray Shielding:
A Review

Srilakshmi Prabhu
CHRIST University (Deemed), India

Bubbly S. G.
https://orcid.org/0000-0003-1234-0662
CHRIST University (Deemed), India

Shivappa B. Gudennavar
https://orcid.org/0000-0003-1234-0662
CHRIST University (Deemed), India

ABSTRACT

In last few decades, nanomaterials have gained enormous attention in the scientific industry due to their tunable physico-chemical and biological properties with enhanced performance over their bulk counterparts. In particular, nanoparticles have been extensively investigated for their usefulness in X-ray and gamma-ray shielding applications. Various elements and compounds, with high atomic numbers and effective atomic numbers respectively, have the potential to form nanoparticles that offer remarkable enhancement in the shielding performance. Composites, obtained by doping different nanoparticles into structural matrices (concrete, glass, or polymers), not only possess striking thermo-mechanical properties but also are effective shielding materials to replace conventional lead shields. This review is an attempt to throw light on various aspects of nanoparticles and their influence on shielding effectiveness. The authors also summarize the experimental findings so as to highlight the potential underlying the radiation-matter interaction mechanism in nanostructured systems.

DOI: 10.4018/978-1-6684-7358-0.ch010

INTRODUCTION

Significant advancements in nuclear science and technology have hastened the worldwide usage X-rays and -rays in a plethora of industries, including medical diagnostic imaging and radiation therapy, food sterilizing and cultivation, nuclear research laboratories, aeronautical, and space exploration, among others. Prolonged exposure to these radiations poses serious threat to human health, causing genetic damage, leukemia and even death. Although limiting the exposure time and keeping a safe distance from radiation source is instrumental in achieving low dose accumulation, the use of personal protective equipment (PPE) would be the most feasible way to reduce the radiation exposure. Along with personnel protection, it is essential to address the challenges presented by nuclear research labs, aerospace, or observational astronomy, wherein high energy radiations often damage the electronic equipment and instrumentation facilities. Hence, radiation protection and shielding are inevitable for medical personnel and electronic instruments closely associated with radiation field.

From the physics of interactions of radiations with matter, it is known that the attenuation of X-rays and γ-rays increases significantly with increase in atomic number (Z) of the interacting medium. Accordingly, high-Z and high-density materials are preferred for radiation shielding and the conventionally employed materials are lead (Pb), bismuth (Bi), tungsten (W), tin (Sn) and antimony (Sb), manufactured mostly in the in the shape of sheets, fibres, or thin films. Among them, lead with high-Z (82), high density (11.34 g cm^3) and higher interaction cross-section has been most widely used for shielding. Similar shielding properties are provided by bismuth and tungsten, however at greater mass thicknesses than Pb. As a result, efforts have been made to develop non-toxic and non-carcinogenic materials to replace lead and minimize the weight and thickness of PPE. Currently, the Today's radiation protective aprons typically have a shielding efficiency of 0.25 to 0.5 mmPb lead equivalent corresponding to 3 mm thickness of overall shielding material. However, increasing the density of the shielding material by reducing porosity and enhancing the particle packing density, is the most practical and simplest technique to reduce the thickness of the shielding material. Reducing the size of the material's constituent particles, or using nanomaterials, is another way to achieve this.

Nanomaterials, especially nanoparticles, are in high demand for a range of practical applications. The design and development of high-performance materials is often facilitated by novel aspects of nanoparticles such small size effect, surface and boundary effect, increased surface-to-volume ratio, and quantum size impact. Nanoparticles offer tunable physicochemical properties, such as soft magnetic properties, tensile strength, and ductility, as well as high melting point, good diffusivity, better wettability, high catalytic activity, greater luminous efficiency, higher electrical and thermal conductivity, compared to their bulk counterparts. Thus, nanoparticles constitute a dynamic area of research and a socio-economic sector that contributes to almost every field of science including physics, chemistry, materials science, biomedical and aerospace engineering, and computer science. One of the most significant advances, in the recent years, has been towards human health with promising results, notably in the field of radiation shielding and protection. Particle size reduction results in shorter interparticle distances, which enhances the material's packing density. In contrast to micron-sized particles, which create holes in the system as seen in Figure 1, nanoparticles occupy every interstitial space. Consequently, attenuation of incident photon energy in nanomaterials is greater than their micro counterparts, mostly due to increased probability of interaction of photons with nanoparticles of the attenuation medium. Therefore, nanoparticles can be either used independently or incorporated as additives/fillers into structural materials like concrete, glass and polymers and molded into desired shape and size according to the end-user application so as to achieve enhanced shielding performance.

Figure 1. Representative internal structure of materials constituted of micro and nanoparticles

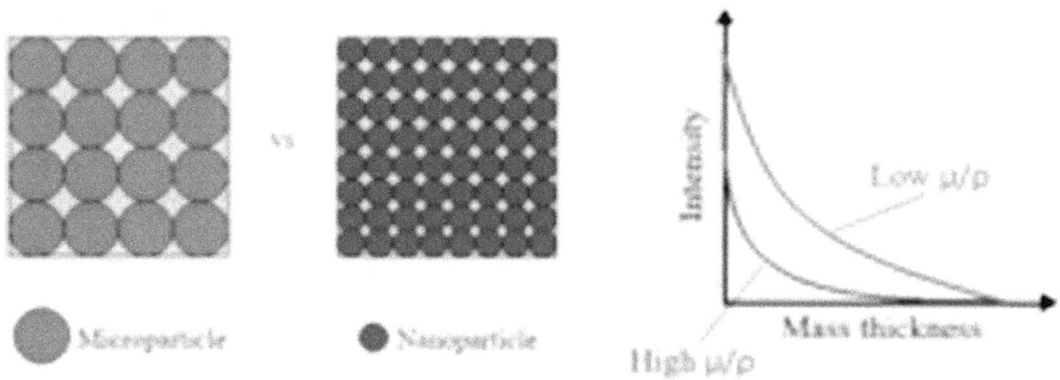

Nanoparticles, when used as fillers, are more effective in blocking pinholes/voids in composites through which radiations could be transmitted, therefore offering better attenuation than micro composites. Recent studies on developing lightweight shielding materials for use in medical industry mainly focus on polymer nanocomposites. The choice of particle size and the base matrix that disperses it are very important in the manufacturing of shielding sheets because nanoparticles typically exhibit high dispersion forces. Filler size-dependent radiation studies on polymer composites have repeatedly shown that nanocomposites perform better in terms of physico-mechanical, thermal, and radiation attenuation properties (Kim et al., 2014; Li et al., 2017; Mahmoud et al., 2018a, 2018b; El-Khatib et al., 2019; Muthamma et al., 2021; Prabhu et al., 2021a; Abbas et al., 2022; Prabhu et al., 2022). The desire to achieve shielding performance on par with lead-based materials necessitates the search for new kind of non-toxic and multifunctional nanoparticle-based additives. In view of this, we present a brief overview of the most cutting-edge research on employment of nanoparticles in developing X- and γ-ray shielding materials. The spotlight on recent trends and advancement in nanoparticles used for shielding purposes in this article may provide future directions for the radiation researchers.

MASS ATTENUATION COEFFICIENT FOR X-RAYS AND γ-RAYS

The shielding efficiency of the materials depends on energy (E) of the incident photons, atomic number (Z) and density (ρ) of the interacting medium. Shielding properties of materials for X-rays and γ-rays are usually characterized by the radiation interaction parameter called mass attenuation coefficient, μ/ρ (cm^2/g), the calculation of which requires the knowledge of mechanism of radiation-matter interaction. X-ray and γ-rays, on interaction with materials, are either completely absorbed or scattered from their path. The most dominant absorption and scattering processes are photoelectric absorption and pair production, and Compton scattering, respectively. It is vital to know that each of these processes becomes dominant in different energy regions as they have dissimilar dependence of interaction cross-sections on atomic number (Z) and incident photon energy (E). The incident photon interacts with inner shell electron ejecting that electron as photoelectron by the process of photoelectric absorption. Photoelectric interaction predominates in the lower energy region below 500 keV with its cross-section for K-shell electron varying as ~Z^5 and $E^{-3.5}$. During the Compton scattering process, the incident photon ejects an valence electron while the incident photon gets scattered with decreased energy. With no dependence on Z, Compton scattering dominates for γ-ray energies above 500 keV and the scattered photon depending on its energy, will further trigger either photoelectric or Compton scattering. Cross-section for pair production process is directly proportional to Z^2 and the process occurs for incident photon energies above 1.022 MeV. With a poly-energetic beam, which is frequently employed in healthcare, the low-energy part of the spectrum will interact with matter by photoelectric absorption, but the medium- and high-energy parts are more likely to do so through Compton scattering. According to the exponential decay law, a medium with a thickness of x (cm) determines the intensity of a transmitted beam of photons (I) as follows,

$$I = I_0 e^{-\mu x} \tag{1}$$

where, I_0 is the incident photon intensity and μ is the photon linear attenuation coefficient (cm^{-1}). If thickness is considered in terms of mass thickness (ρx, g cm^{-2}), the attenuation coefficient is called mass attenuation coefficient (μ/ρ, cm^2 g^{-1}). The fractional attenuation of incident photons per mass thickness of material traversed is measured as μ/ρ and it is the sum of the various partial attenuation coefficients such as photoelectric absorption coefficient (μ/ρ_{pe}), coherent scattering coefficient (μ/ρ_{coh}), Compton scattering coefficient (μ/ρ_{incoh}), absorption coefficient for the pair production in the field of the nucleus (μ/ρ_{pair}) and triplet production in the field of the atomic electrons (μ/ρ_{trip}) (see for eg, Prabhu et al., 2021b). Accordingly,

$$\mu / \rho = \mu / \rho_{pe} + \mu / \rho_{coh} + \mu / \rho_{incoh} + \mu / \rho_{pair} + \mu / \rho_{trip} \tag{2}$$

μ/ρ of a multi-elemental substance is given by the mixture rule,

$$\mu / \rho = \sum_i w_i (\mu / \rho)_i \tag{3}$$

where, $(\mu/\rho)_i$ is the mass attenuation coefficient of the ith element in the mixture/compound, and $w_i = \frac{n_i A_i}{\sum_i n_i A_i}$ is the weight fraction of the ith element, n_i and A_i are the number of atoms and atomic mass of ith element, respectively. From Equation (3), it is clear that the shielding ability of materials depends on relative abundance of each element and their attenuation properties.

μ/ρ is directly proportional to the photon total interaction cross-section per atom, σ (b/atom) as given by the relation,

$$\frac{\mu}{\rho} = \frac{\sigma N_A}{A} \tag{4}$$

where, N_A is the Avogadro constant and A is the atomic weight of the element. Atomic weight has to be replaced by molecular weight in mixture/composite materials. σ is also expressed as the sum of partial interaction cross sections given by,

$$\sigma = \sigma_{pe} + \sigma_{coh} + \sigma_{incoh} + \sigma_{pair} + \sigma_{trip} \tag{5}$$

where, σ_{pe}, σ_{coh}, σ_{incoh}, σ_{pair} and σ_{trip} are the photoelectric absorption cross section per atom, coherent and incoherent scattering cross sections per atom, pair and triplet production absorption cross sections per atom in the fields of the nucleus and atomic electrons, respectively. In general, each of these processes has different interaction cross-sections and different dependence on Z of the medium and E of incident photon. Above all, the attenuation tends to increase significantly with increase in Z of the absorbing medium, thus justifying the use of high-Z fillers. In the next section, we present a brief overview of nanoparticle-based materials used for radiation shielding applications.

Nanostructured Materials for Radiation Shielding

International Organization for Standardization has defined 'nanomaterial' as a 'material with any external nanoscale dimension or having internal nanoscale surface structure'. A nanometer (nm), that represents 10^{-9} meter in length, is an International System of Units unit. In principle, nanomaterials are described as materials with length in the range 1 – 100 nm in at least one dimension. They may be in the form of nanoparticles, nanofibers, nanotubes, nanocomposites and nanostructured materials. Based on dimensionality, Pokropivny and Skorokhod (2007) classified nanomaterials as 0D, 1D, 2D and 3D. This scheme of classification is dependent only on the movement of electron along the different dimensions in the nanomaterials. In the following section, we discuss the correlation between dimension of nanoparticles and shielding performance.

0D Nanomaterials

0D nanomaterials are negligibly small (nano-scale) in size when measured in all three dimensions and are popularly known as quantum dots. This class of nanomaterials are mostly spherical, quasi-spherical, cubical, or polygonal in shape with a diameter of less than 100 nm. Carbon and graphene quantum dots, polymer dots, fullerenes, magnetic nanoparticles, inorganic quantum dots and up-conversion nanoparticles are some advanced 0D nanomaterials (Wang et al., 2020) as represented in Figure 2. Several studies have reported the X-/γ-ray shielding properties of 0D nanomaterials.

Figure 2. 0D nanomaterials

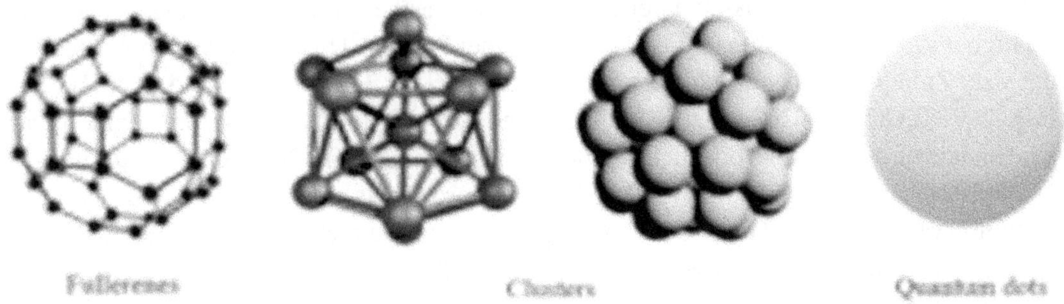

Moharram et al. (2020) have synthesized nanoparticles (NPs) of lead oxide (PbO) and iron oxide (Fe_2O_3) with particle size of about 50 nm using sol-gel method and investigated their γ-ray shielding properties. NPs were reported to exhibit greater shielding ability compared to their micro-counterparts, mostly attributed to higher-surface area of NPs and consequent increase in radiation interaction probability. A similar study on magnesium oxide (MgO) and zinc oxide (ZnO) NPs synthesized by solid-solid reaction method by Rashad et al. (2020) has showed higher μ/ρ values for NPs than their micro counterparts. ZnO (50 nm) NPs showed greater attenuation ability than MgO (70 nm) at all the energies in the range 0.356 – 1.332 MeV, which was attributed to the relatively higher atomic number of Zn over Mg. Recently, barium sulphate ($BaSO_4$) NPs have garnered great deal of research interests due to their excellent X-ray absorption properties. Fang et al. (2021) employed precipitation method in a rotating packed bed to prepare transparent and stable $BaSO_4$ nano-dispersion, with particle size around 10 - 17 nm. The authors claimed that the $BaSO_4$ nano-dispersion could be used for making transparent composite materials and X-ray coatings.

Zhou et al. (2017) have investigated the γ-ray and neutron co-shielding ability of bismuth borate ($Bi_6B_{10}O_{24}$) NPs. The μ/ρ value of $Bi_6B_{10}O_{24}$ NPs (100 – 300 nm) synthesized by facile sol-gel was found to be 0.199 cm^2g^{-1} at 662 keV. $Bi_6B_{10}O_{24}$ NPs exhibited neutron shielding ability of ~ 61% for ^{252}Cf neutron source. This was attributed to the presence of boron (B) which is a good moderator material, thus, proving that these nanofillers can be used for dual purpose shielding applications. Another interesting study involving the synthesis of semi-solid based Bi_2O_3 NPs (< 100 nm) for use in X-ray shielding (Shirkhanloo et al., 2017) demonstrated that these semi-solid Bi_2O_3 NPs showing 56.79% attenuation have better X-ray shielding ability than 0.5 mm Pb (40 and 42%) for diagnostic X-ray energies (50 – 150 kV). These semi-solid based Bi_2O_3 NPs find applications in preparation of ointments and other pharmaceuticals required in dental X-rays.

Very recently, spinel soft magnetic NPs of nickel ferrite ($NiFe_2O_4$) were synthesized by combustion method for investigation of their X-/γ-ray shielding, photoluminescence and anti-microbial properties (Reddy et al., 2021). Highly crystalline NPs of $NiFe_2O_4$ (36.69 - 52.49 nm) were found to display interesting properties such as wide band gap (~ 5.2 eV) and relatively good antimicrobial activity compared to streptomycin. With photoluminescence emission spectra peaks at 448, 540 and 628 nm, these NPs were found to be useful in white light LED displays. Currently, nano-spinel ferrites are technologically trending materials due to their extensive applications in biomedical instrumentation, wireless mobile communication, phase shifters, transformers, inductors, capacitors, gyrators, and phase shifters. Yet another study aimed to exploit the γ-ray shielding (122–1330 keV) properties of various spinel ferrites *viz.*

nickel ferrite ($NiFe_2O_4$), manganese ferrite ($MnFe_2O_4$), zinc ferrite ($ZnFe_2O_4$), copper ferrite ($CuFe_2O_4$) and cobalt ferrite ($CoFe_2O_4$) (Lokhande et al., 2021). From their study, they observed that $ZnFe_2O_4$ displayed higher μ/ρ value as compared to the other spinel ferrites. This result was correlated with their crystallite size: $ZnFe_2O_4$ possesses a lowest crystallite size (~32.31 nm) compared to $NiFe_2O_4$ (~36.25 nm), $MnFe_2O_4$ (~33.87 nm) $CuFe_2O_4$ (~39.75 nm) and $CoFe_2O_4$ (~52.45 nm), thus demonstrating the positive influence of particle size on shielding properties. Interestingly, $NiFe_2O_4$ crystals as reported by Reddy et al. (2021) and Lokhande et al. (2021) showed approximately same shielding properties, mostly due to similar crystallite sizes. These results are suggestive of the fact that the shielding efficiency can be improved significantly with decrease in dimension of fillers from micro to nano-scale.

1D Nanomaterials

1D nanomaterials like nanorods, nanowires, nanotubes, and nanofilaments have one dimension outside the nanoscale. These nanostructures possess superior conductivity, low percolation, high aspect ratio, transparency, stretchability, enhanced efficiency, optimum energy storage, ultra-low thermal conductivity and high charge mobility (Han & Ho, 2014; Gong & Cheng, 2017; Akpan et al., 2019). They have also been investigated for their shielding properties.

Figure 3. 1D nanomaterials

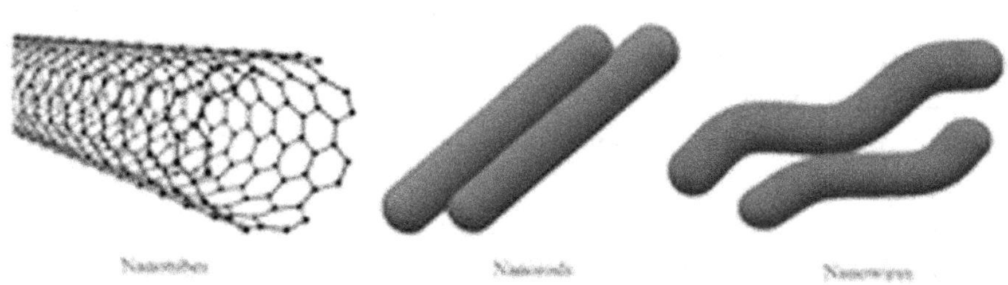

Badran et al. (2017) have synthesized lithium (Li) doped hydroxyapatite (HAp) 1D nanorods of length (87.8 – 1766.5 nm) by sol-gel technique followed by microwave-hydrothermal treatment for use as bone substitute materials. Their microstructure not only resembled that of the human bone, but low dielectric constant values of these nanostructures could also expedite the rate of both healing fractures in bones and osteobonding. Additionally, 40 wt% Li doped HAp nanocomposites were found to have μ value of 2.190 cm^{-1} at 662 keV suggesting their potentiality as fillers for γ-ray attenuation. In the same year, Shkir et al. (2017) have reported the microwave-assisted synthesis of W doped HAp nanorods at low temperatures. The study showed that 40 wt% W doped HAp composites could attain μ value of 2.610 cm^{-1} at 662 keV, that is around 19% higher compared to that of 40 wt% Li doped HAp nanocomposites (Badran et al., 2017). This could be attributed to two reasons, firstly, the length of W-HAp nanorods (10 nm) were lesser than the corresponding Li-HAp nanorods, thus demonstrating positive influence of nanometer effects. Secondly, the effective atomic number (Z_{eff}) and effective electron density (N_{eff}) of W-HAp nanorods is much greater than Li-HAp, attributed to higher atomic number of W ($Z = 74$) over Li ($Z = 3$). Additionally, W doped HAp was found to be a good catalyst in the presence of Enterococcus faecalis bacteria indicating its relevance in waste water treatment and environmental clean-up. Another study reported by Yahia et al. (2017) employed microwave-assisted synthesis of tellurium (Te) doped HAp nanorods. With just 0.24 wt% Te loading, HAp composites formed 5 nm thick nanosheets showing μ value of 2.512 cm^{-1} at 662 keV. These composites were also found to have excellent antimicrobial properties, thus making it a promising candidate for biomedical applications.

Xu et al. (2018) synthesized 1D lead tungstate ($PbWO_4$) nanofibers using (3-Mercaptopropyl) trimethoxysilane (KH590) as surfactant employing a facile precipitation method. This work was intended to develop high performance filling materials for better γ-ray shielding by exploiting the synergistic effect of two high atomic number elements, namely Pb and W. Addition of KH590 largely influences the crystal phase and preferential growth of $PbWO_4$ NPs. The authors claim that self-assembly, spontaneous alignment and fusion of NPs are responsible for the formation of these lead tungstate nanofibers. The attenuation studies revealed that the radiation shielding ability of $PbWO_4$ crystals decreased with increase in dosage of KH590. This was attributed to increase in the size of $PbWO_4$ crystals with KH590 content suggesting that nanoscale size effects could possibly enhance the probability of interaction of γ-rays with matter.

2D Nanomaterials

2D nanomaterials are those materials where only one dimension exists in nanoscale. This class of materials exhibits plate-like shapes and includes graphene, nano prisms, nanowalls, nanosheets, nanofilms, nanolayers, and nano-coatings. These materials have better fire-resistant properties and high degree of reinforcement compared to 1D and 3D nanofillers (Akpan et al., 2019). These have also been studied for their shielding effectiveness.

Figure 4. 2D nanomaterials

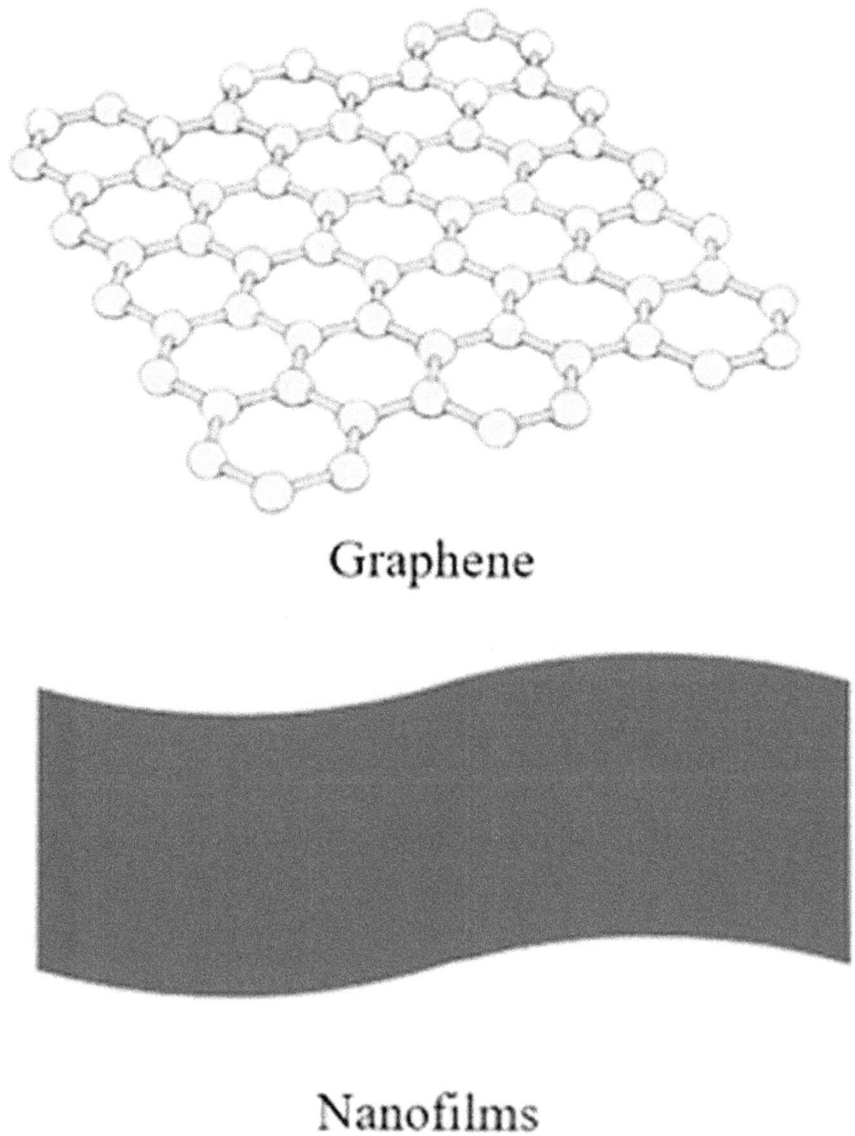

Zhang et al. have fabricated 2D lead borate single crystal ($Pb_3B_{10}O_{16}[OH]_4$) nanosheets by a facile solvothermal method for attenuation of both neutron and γ-rays (Zhang et al., 2017). It was found that these extremely thin nanosheets of size 20 – 100 nm exhibited significantly better shielding ability than their micron counterpart for γ-ray energies in the range 59 – 105 keV. The authors claim that the chemically bonded atoms of B and Pb in the nanosheets enable better attenuation of neutron and γ-rays when compared to PbO and B_2O_3, owing to improved dispersibility and matrix-filler interfacial adhesion. Also, 2D fillers, in comparison to other shapes, are instrumental to achieve low percolation threshold for creating firm network structures and exhibiting enhanced performance. Despite the fact that Pb-based fillers have high toxicity, they remain the primary choice so as to achieve best shielding properties. In this context, Al-Balushi et al. (2017) have synthesized series of ZnO nanorod thin films (ZONRs-TFs)

doped with various concentrations (1, 2, and 3 wt.%) of lead acetate (LA), lead nitrate (LN), and bismuth nitrate (BN) employing chemical bath deposition. Studies on the X-ray attenuation in the energy range of 50 – 100 kVp revealed that the trend in the shielding properties of ZONRs-TFs was: 1 wt.% LA-doped TF > 1 wt.% LN-doped TF > 3 wt.% BN-doped TF > undoped TFs.

In the same year, Jia et al. (2016) have synthesized non-lead based nanosheets of cerium (Ce) doped barium tantalates (Ce-$Ba_2Ta_4O_{15}$) to exploit the radiation shielding properties of tantalum. With increase in Ce doping from 0 – 16 mol%, the shielding efficiency increased from 29.7% to 38.6% at 59.5 keV. This is particularly interesting as significant enhancement in shielding efficiency is observed due to doping of relatively medium-Z element (Ce) and this is attributed to the dominance of photoelectric effect by the K-edge absorption at low energies (< 100 keV). Ce with its K-edge at 40.4 keV in combination with Ta having K- and L-I edges, at 67.4 keV and 11.7 keV, respectively, can show a shielding effect in the Ta feeble absorbing area, thus making up for the shortcoming of the latter. Therefore, doping crystal lattice of $Ba_5Ta_4O_{15}$ with Ce enhances the shielding performance at lower energies.

Recently, an interesting work has been carried out to bring out the synergistic effect of combining nanoclay with inorganic metal oxide for γ-ray shielding application (Allam et al., 2021). Spherical nanoparticles of nano-mercuric oxide (n-HgO) was intermixed with platelet-like nano-bentonite (n-bent) clay, forming intercalated structure of n-HgO above the n-bent sheets. The nanocomposite containing 40 wt% HgO and 60 wt% n-bent displayed the best shielding properties in the γ-ray energy region of 356 – 1332 keV.

Perovskites have been widely used for space applications. As materials for space applications need to withstand harsh environmental conditions, there have been several studies on the radiation shielding abilities of these materials. The following section gives details on the perovskite materials, especially perovskite nanostructures explored for X-/γ-ray shielding applications.

Perovskite Nanostructures

Perovskites have emerged as the most promising and efficient low-cost energy materials for optoelectronic and photovoltaic device applications owing to their attractive properties including favourable band gap, appropriate charge diffusion lengths, inexpensive preparation techniques, low-temperature processability, high absorption coefficient, stable structure, and compositional flexibility. In addition, perovskites can also detect radiations such as X-rays, γ-rays and charged particles (Wei et al., 2017; Chen et al., 2018; Graham et al., 2019; Xu et al., 2019). Nazarenko et al. (2017) have employed formamidinium (FA)-lead halide perovskites ($FAPbX_3$ where X = I or Br/I) for the sensitive detection of γ-rays. With very high mobility-lifetime products of up to 1.2×10^{-1} $cm^2 V^{-1}$, such high-Z semiconductor single crystals exhibited increased sensitivity to γ-rays from ^{241}Am and ^{137}Cs sources, thus enabling single-photon counting and inexpensive dosimetry.

The high photoconversion efficiency (~ 25.5%) and excellent radiation resistance properties of perovskites make them most suitable materials for use in space applications, not only to power the space vehicles but also to provide shielding for instruments overboard (Tu et al., 2021). Theoretical investigations on the γ-ray shielding effectiveness of a series of inorganic halide cubic perovskites, $CsBX_3$ (where B is either of Sn, Ge or Pb and X is either of I, Br, Cl) by Oto et al. (2019) showed highest radiation shielding properties for $CsPbBr_3$, probably attributed to the presence of high Z constituent, Pb. Further, attenuation studies on Ce doped bismuth tungstate (Bi_2WO_6) tuned perovskite crystals by Xu et al. (2019a) showed 10 mol% Ce-doped Bi_2WO_6 to have ~ 49% higher shielding efficiency than Bi_2WO_6 at 59.5 keV.

Considering perovskite nanostructures in particular, Xu et al. (2019b) examined the radiation shielding ability of highly transparent organic-inorganic methylammonium lead halide perovskite single crystals $CH_3NH_3PbX_3$ (X = Br, I, Cl). Among them, $CH_3NH_3PbCl_3$, $CH_3NH_3PbBr_{0.02}Cl_{2.98}$, and $CH_3NH_3PbBr_{1.65}Cl_{1.35}$ exhibited better shielding efficiency up to 90% for X-rays from 30 kV X-ray tube. Interestingly, the shielding ability of these perovskite nanocrystals were about ten times more than bone equivalent plastic, polyethylene terephthalate plastic and colourless glass. With very high stability even under high dose irradiation of γ-rays, increased visible light transparency with a tunable cut-off edge and refractive index, these single nanocrystals can be used as both photovoltaic device and radiation shield, particularly useful in satellites. Recently, Hemeda et al. (2021) investigated the γ-ray shielding properties of nanometer-sized $PbZr_xTi_{1-x}O_3$ powder at 662 keV. $PbZr_{0.52}Ti_{0.48}O_3$ showed the highest μ/ρ value of 0.0741 cm^2g^{-1} at 662 keV, which is ~62% and ~16% higher than bismuth borate and n-bent/n-HgO nanoparticles (Zhou et al., 2017; Allam et al., 2021). The potentiality of nanostructured perovskites for radiation shielding remains less explored. With combined opto-electronic and photon shielding properties, these perovskites can widen their perspectives in space applications.

Extensive research has been conducted to examine the effect of nanoparticle reinforcement on shielding abilities of different kinds of structural matrices (e.g., glass, concrete, polymers, etc) forming composite materials. Among them, polymer matrix composites (PMCs) have gained enormous interest owing to their excellent properties of light-weight, low cost, ease of processability, chemical and corrosion resistance. The incorporation of suitable nanofillers not only introduces shielding properties in PMCs but also enhances their thermal stability and mechanical strength. In the next section, we have briefly compared the radiation shielding properties of various polymer composites as function of filler/particle size.

Polymer Nanocomposites

With the increasing demand for developing light-weight high performance composites with multifunctionalities, different kinds of polymers including engineering thermoplastic and thermosetting polymers, smart polymers, elastomers, silicon- and vinyl-based polymers, biopolymers and so on, have been explored for shielding applications (Prabhu et al., 2023). Varieties of functional fillers such as lead (III) oxide, bismuth (III) oxide, bismuth oxychloride, tungsten (III) oxide, tungsten carbide, tantalum (V) oxide, gadolinium (III) oxide, samarium (III) oxide and so forth have been mostly used. In particular, several researchers have compared the physico-mechanical, thermal and shielding properties of nanocomposites with their micro counterparts so as to understand the effect of filler size on various properties. Table 1 presents various polymer composites and their μ/ρ values at 662 keV.

Kim et al. (2014) have compared the shielding efficiency of nano and micro-W particle reinforced ethylene propylene monomer (EPM)/ high-density polyethylene (HDPE) blend composites in the energy range 0.3 – 1 .2 MeV. High energy ball milling was adopted to carry out pulverization of metallic W-particles and simultaneous encapsulation of the formed nano-W particles with low density polyethylene (LDPE). 25 μm W particles and freshly formed 300 nm shell structured nano-W/PE particles were then filled into polymer mixture consisting of 70 wt% EPM and 30 wt% HDPE. At all the energies, nanocomposites showed higher attenuation than the micro-composites at a given filler loading. These nano-W/PE loaded composites with low filler content (7.5 and 15 wt%) were found to be much better than other highly loaded composites studied in the consequent years (Hou et al., 2018; Ahmed et al., 2020; Wu et al., 2020). This could be possibly attributed to enhanced degree of dispersion of fillers in the matrix as nano-W is more compatible compared to micro-W, with HDPE matrix, suggesting that filler-matrix interfacial adhesion is crucial in determining the shielding efficiency.

Hexagonal boron nitride (h-BN) nanosheet, a 2D nanomaterial, having superior physical properties such as high temperature stability, large optical absorption and neutron capture interface, corrosion resistance, large negative electron affinity and good thermal conductivity, has been used as filler in polymers to develop shields (Wang et al., 2017; Irim et al., 2018). High performance ternary composites of HDPE loaded with hybridised nano-h-BN and nano-Gd_2O_3 have been developed by Irim et al. (2018) for shielding against both γ-rays and neutrons. The mechanical studies revealed that the increasing nano-Gd_2O_3 loading at a given h-BN content did not affect the tensile strength and modulus of the composites but the mechanical properties were enhanced with increasing h-BN content at a given loading of nano-Gd_2O_3. Shielding ability of HDPE composites for both the γ-rays and neutrons increased with h-BN content and in particular, composites containing highest h-BN (11 wt%) and 3 wt% nano-Gd_2O_3 showed highest μ/ρ value of 0.1123 cm^2g^{-1} at 662 keV.

Mahmoud et al. (2018a) have synthesized composites of virgin HDPE loaded with micro and nano-sized PbO by compression mold technique for shielding. Thermal studies revealed an enhancement in the thermal stability of nano composites over their micro counterparts. At a given filler loading, the higher attenuation showed by nanocomposites compared to micro composites, was attributed to several reasons: PbO nanoparticles with spherical shape provided larger interfacial area between the individual particles in composites compared to large irregular flakes of bulk PbO; Nanoparticles with surface area approximately three times greater than that of bulk PbO particles not only ensured their symmetrical distribution within matrix but also enhanced interfacial adhesion; Bulk PbO fillers created voids at filler-matrix interface resulting in debonding of fillers from matrix. The same group conducted a comparative study on reinforcement of micro and nano PbO particles of same size but incorporated recycled-HDPE (r-HDPE) (Mahmoud et al., 2018b), producing very similar but better results in all aspects. The μ/ρ values of r-HDPE were much higher than HDPE at all energies and hence consequently r-HDPE/micro/nano composites were better attenuators compared to HDPE composites.

Shielding studies using epoxy composites with micro/nano- Bi_2O_3 particles as fillers have revealed that 14 wt% nano-Bi_2O_3 (20 nm) loaded epoxy composites displayed shielding ability on par with 30 wt% micro-Bi_2O_3/epoxy composites (Muthamma et al., 2021). All of the composites demonstrated thermal stability up to 400 °C, with nanocomposites demonstrating thermal stability substantially greater than their micro counterparts. Tensile strength of nano-Bi_2O_3/epoxy composites with 14 wt% filler content was 326 MPa, which is about 38% higher than that of 30 wt% micro Bi_2O_3/epoxy composites. In a subsequent study, it was found that the decoration of micro-Bi_2O_3 particles on surface of graphene oxide (GO) would change the microstructure of overall composite forming BGO (abbreviation for Bi_2O_3 decorated GO) nanocomposites (Prabhu et al., 2022). 15 wt% BGO loaded epoxy nanocomposites displayed comparable shielding with 14 wt% n-Bi_2O_3 loaded epoxy composites (Muthamma et al., 2021). Even though the size of nano-Bi_2O_3 particles (20 nm) was significantly smaller than that of BGO nanocomposites (42 nm), the formation of a continuous, interconnected BGO/epoxy network ensured uniform distribution of fillers throughout the matrix, thereby increasing radiation attenuation and compensating for differences in filler sizes (Muthamma et al., 2021; Prabhu et al., 2022). Yet again, another study explored the effect of Ta_2O_5 filler size and loading on thermal stability, tensile, viscoelastic and shielding properties of thermo-mechanical and shielding properties of epoxy matrix (Prabhu et al., 2021a). 30 wt% nano-Ta_2O_5 (75 nm) loaded epoxy composite demonstrated dramatical enhancement over the corresponding micro counterpart in terms of thermo-tensile and viscoelastic characteristics. The μ/ρ values of this nano-composite were, 0.154, 0.114, 0.096, 0.069, 0.065 and 0.063 cm^2 g^{-1} at 356, 511, 662, 1173, 1274 and 1332 keV energies, respectively, 6 – 13% higher than the micro-composites. Interestingly, 20 wt% nano-Ta_2O_5/epoxy composites showed significantly higher shielding properties than 30 wt% micro-Ta_2O_5/epoxy composites. These results unequivocally show that radiation attenuation capabilities equivalent to those of fully loaded micro-composites may be obtained with minimal loadings of nanoparticles in matrix.

Enhancement in shielding ability of nanocomposites over micro-composites is also accompanied with improved thermal and mechanical properties of the composite system. Dong et al. (2012) demonstrated an enhancement of ~ 16% and 57%, respectively, in bending strength and tensile strength of epoxy/WO_3 nanocomposites over their micro counterparts. Another study on epoxy/Gd_2O_3 nanocomposites showed equivalent flexural strength but a 15% higher flexural modulus compared to micro-composites (Li et al., 2017). Both the studies showed nanocomposites to be better than micro-composites in terms of radiation attenuation efficiency. In general, Table 1 provides clear evidence for the fact that nanocomposites provide γ-ray shielding efficiency as high as micro-composites at a given energy of 662 keV from ^{137}Cs source.

On similar grounds, significant efforts have been made to develop lead-free radioprotective materials for protection against X-rays employed for medical purposes, mainly in the energy region of 40 - 120 keV. According to Cornish and Li (2017), medical radiation attenuation (RA) gloves that are commercially available are typically made of hevea natural rubber (HNR) or guayule natural rubber (GNR), both of which do not contain lead. RA gloves like RadiaXon, Ansell, and Kiran, for instance, have fillings made of Bi-based materials. To be utilized as effective RA gloves, all materials must meet the ASTM D7866 criteria for minimal radiation attenuation (Table 2). At any given X-ray energy, polymer nanocomposites displayed upper-hand in shielding effectively compared to the micro-composites. For instance, Nikeghbal et al. (2020) developed polyvinyl chloride (PVC)-based shields against the ionizing effect of cone beam computed tomography (CBCT) X-ray exposure. 60 wt% of each of micro (4 μm for WO_3 and 0.8 μm for Bi_2O_3) and nano metal oxide powders (60 nm for WO_3 and 80 nm for Bi_2O_3) were added to PVC matrix to develop X-ray shields. Yet again, their study illustrated that nanocomposites were more effective in the X-ray attenuation. For nano-Bi_2O_3/PVC and nano-WO_3/PVC composites, half value layer (HVL), i.e., the thickness of the shielding material required to reduce the intensity of incident radiations by one half, was found to be around 0.035 and 0.039 cm respectively, at 110 kVp. Although lead has a lower HVL (0.300 cm) for 110 kVp X-ray than the nano-PVC composites, the shields made of latter can provide equivalent X-ray attenuation at much lighter weight than lead aprons. Recently, light-weight and low-cost LDPE-based X-ray shielding materials were developed by Alshahri et al. (2021). In their study, it was found that the LDPE composites containing 15% wt of nano-Bi_2O_3, with particle size less than 100 nm, could block about 80% of X-rays at 47.9 keV. Table 3 summarizes the X-ray shielding characteristics of several Bi-based polymer nanocomposites at different X-ray tube energy (tube voltage in kVp).

This is because the reduction in filler size, increased surface-to-volume ratio and thus the increasing number of particles at the interface results in increased probability of radiation interaction and attenuation. Taking into consideration all the above-mentioned factors, efficient designing of shielding materials can provide full-fledged radiation protection. This goes hand-in-hand with the scientific advancement of nanotechnology and opportunity to exploit the properties of nanostructures to demonstrate their potential benefits. Working with nanotechnology in future for radiation protection would be highly beneficial for developing high performance multi-functional radiation shielding materials.

CONCLUSION

This article provides a brief insight about the state-of-the-art nanostructured materials used as X- and γ-ray shielding materials since the advent of nanotechnology. The highlights on the recent trends in nanostructured materials, can assist the radiation researchers to further identify the gap in this research area and navigate through the most appropriate solution. Some of the crucial factors to be considered when using nanoparticles as shielding materials are summarized below:

1. Smaller the size of nanoparticles, greater is the interfacial area and better is the probability of radiation-matter interaction and attenuation. This effect becomes particularly predominant in composites at lower nanoparticle loadings.
2. Combination of medium - or high-Z inorganic metal oxide nanoparticles with carbon-based nanomaterials, such as graphene oxide, carbon nanotubes, etc., offers synergistic radioprotection over wider range of photon energy radiations compared to their utilization individually.
3. New kind of nanoparticles based on perovskites, carbon nanotubes, nanoclay, etc. display multifunctional properties. Along with shielding, these nanoparticles impart exceptional material properties to the end-product.

Thus, factors such as size and shape of nanofillers, composition and energy range of deployment are decisive in the development of advanced nanomaterials with the goal of providing full-fledged radiation protection and material-property enhancement.

Table 1. Comparison of shielding efficiency of polymer nanocomposites at 662 keV

Polymer	Micro fillers	μ/ρ ($cm^2\ g^{-1}$)	Nanofillers	μ/ρ ($cm^2\ g^{-1}$)
Epoxy (Dong et al., 2012)	50 wt % m-WO_3 (1 μm)	0.075	50 wt % n-WO_3 (50 nm)	0.078
EPM/HDPE (Kim et al., 2014)	7.5 wt% m-W (25 μm)	0.192	7.5 wt% n-W (300 nm)	0.234
HDPE (Mahmoud et al., 2018a)	50 wt% m-PbO (0.7 μm)	0.104	50 wt% n-PbO (10 nm)	0.114
r-HDPE (Mahmoud et al., 2018b)	50 wt% m-PbO (0.7 μm)	0.133	50 wt% n-PbO (10 nm)	0.143
SR (Verdipoor et al., 2018)	50 wt% m-PbO (1 μm)	0.074	50 wt% n-PbO (100 nm)	0.085
	50 wt% m-Bi_2O_3 (1 μm)	0.075	50 wt% n-Bi_2O_3 (100 nm)	0.085
	50 wt% m-WO_3 (1 μm)	0.076	50 wt% n-WO_3 (100 nm)	0.084
PVA (Kazemi et al., 2019)	50 wt% m-WO_3 (10 μm)	0.053	50 wt% n-WO_3 (30 nm)	0.060
HDPE (El-Khatib et al., 2019)	40 wt% m-CdO (0.58 – 0.95 μm)	0.082	40 wt% n-CdO (50 nm)	0.093
Epoxy (Prabhu et al., 2021a)	30 wt% m-Ta_2O_5 (4 μm)	0.087	30 wt% n-Ta_2O_5 (80 nm)	0.096
SR (Abbas et al., 2022)	30 wt% m-Bi_2O_3 (15 μm)	0.089	30 wt% n-Bi_2O_3 (30 nm)	0.104

Table 2. ASTM D7866 specification for radiation attenuation gloves

Energy (kVp)	60	80	100	120
Minimum attenuation (%)	22	29	18	15

Table 3. X-ray attenuation efficiencies of different Bi-based polymer nanocomposites at varied energies (tube voltage in kVp)

Tube Voltage (kVp)	Polymer composite	X-ray attenuation (%)
30	38 wt% n-Bi_2O_3/PLA (Noor Azman et al., 2013)	100% at 1 mm
100	8 wt% n-Bi_2O_3/P(VDF-TrFe) (Fontainha et al., 2016)	20% at 0.5 mm
25-35	18 wt% n-Bi_2O_3/PVA (Noor Azman et al., 2016)	100% at 0.25 mm
100	20 wt% n-Bi_2O_3/epoxy–PVA (Abunahel et al., 2018)	10% at 2 mm
110	60 wt% n-Bi_2O_3/PVC (Nikeghbal et al., 2020)	98.06% at 2 mm
110	60 wt% n-WO_3/PVC (Nikeghbal et al., 2020)	97.14% at 2 mm
60	15 wt% n-Bi_2O_3/LDPE (Alshahri et al., 2021)	79.16% at 16 mm
120	15 wt% n-Bi_2O_3/LDPE (Alshahri et al., 2021)	40.24% at 16 mm
60	25 wt% n-Bi_2O_3/n- WO_3/GO/HDPE (Abdolahzadeh et al., 2022)	18.99% at 6 mm
80	25 wt% n-Bi_2O_3/n- WO_3/GO/HDPE (Abdolahzadeh et al., 2022)	16.32% at 6 mm
120	25 wt% n-Bi_2O_3/n-WO_3/GO/HDPE (Abdolahzadeh et al., 2022)	12.63% at 6 mm

ACKNOWLEDGMENT

One of the authors (SP) acknowledges Department of Science and Technology, Government of India, New Delhi for the DST-Inspire Fellowship.

REFERENCES

Abbas, M. I., El-Khatib, A. M., Dib, M. F., Mustafa, H. E., Sayyed, M. I., & Elsafi, M. (2022). The influence of Bi_2O_3 nanoparticle content on the γ-ray interaction parameters of silicon rubber. *Polymers, 14*(5), 1048. doi:10.3390/polym14051048 PMID:35267871

Abdolahzadeh, T., Morshedian, J., & Ahmadi, S. (2022). Preparation and characterization of nano WO_3/Bi_2O_3/GO and $BaSO_4$/GO dispersed HDPE composites for X-ray shielding application. *Polyolefins Journal, 9.*

Abunahel, B. M., Ramli, R. M., Quffa, K. M., & Azman, N. Z. N. (2018). Effect of nanofibrous porosity on the X-ray attenuation properties of electrospun n-Bi_2O_3/epoxy–polyvinyl alcohol (PVA) nanofiber mats. *Applied Physics. A, Materials Science & Processing, 124*(8), 1–9. doi:10.100700339-018-1975-9

Ahmed, B., Shah, G. B., Malik, A. H., & Rizwan, M. (2020). Gamma-ray shielding characteristics of flexible silicone tungsten composites. *Applied Radiation and Isotopes, 155*, 108901. doi:10.1016/j.apradiso.2019.108901 PMID:31655354

Akpan, E. I., Shen, X., Wetzel, B., & Friedrich, K. (2019). *Design and synthesis of polymer nanocomposites. Polymer Composites with Functionalized Nanoparticles*. Elsevier.

Al-Balushi, M. A., Ahmed, N. M., Zyoud, S. H., Mohammed Ali, M. K., Akhdar, H., Aldaghri, O. A., & Ibnaouf, K. H. (2022). Ionization radiation shielding effectiveness of lead acetate, lead nitrate, and bismuth nitrate-doped zinc oxide nanorods thin films: A comparative evaluation. *Materials (Basel), 15*(1), 3. doi:10.3390/ma15010003 PMID:35009152

Allam, E. A., El-Sharkawy, R. M., El-Taher, A., Shaaban, E. R., Massoud, E. E. S., & Mahmoud, M. E. (2022). Enhancement and optimization of gamma radiation shielding by doped nano HgO into nanoscale bentonite. *Nuclear Engineering and Technology, 54*(6), 2253–2261. doi:10.1016/j.net.2021.12.026

Alshahri, S., Alsuhybani, M., Alosime, E., Almurayshid, M., Alrwais, A., & Alotaibi, S. (2021). LDPE/bismuth oxide nanocomposite: Preparation, characterization and application in X-ray shielding. *Polymers, 13*(18), 3081. doi:10.3390/polym13183081 PMID:34577982

Badran, H., Yahia, I. S., Hamdy, M. S., & Awwad, N. S. (2017). Lithium-doped hydroxyapatite nanocomposites: Synthesis, characterization, gamma attenuation coefficient and dielectric properties. *Radiation Physics and Chemistry, 130*, 85–91. doi:10.1016/j.radphyschem.2016.08.001

Chen, Q., Wu, J., Ou, X., Huang, B., Almutlaq, J., Zhumekenov, A. A., Guan, X., Han, S., Liang, L., Yi, Z., Li, J., Xie, X., Wang, Y., Li, Y., Fan, D., Teh, D. B. L., All, A. H., Mohammed, O. F., Bakr, O. M., & Liu, X. (2018). All-inorganic perovskite nanocrystal scintillators. *Nature, 561*(7721), 88–93. doi:10.103841586-018-0451-1 PMID:30150772

Cornish, K., & Li, Z. (2017). Guayule medical radiation attenuation glove. *Rubber & Plastics News, Technical Note, October, 2*, 20-25.

Dong, Y., Chang, S. Q., Zhang, H. X., Ren, C., Kang, B., Dai, M. Z., & Dai, Y. D. (2012). Effects of WO_3 particle size in WO3/epoxy resin radiation shielding material. *Chinese Physics Letters, 29*(10), 108102. doi:10.1088/0256-307X/29/10/108102

El-Khatib, A. M., Abbas, M. I., Elzaher, M. A., Badawi, M. S., Alabsy, M. T., Alharshan, G. A., & Aloraini, D. A. (2019). Gamma attenuation coefficients of nano cadmium oxide/high density polyethylene composites. *Scientific Reports*, *9*(1), 1–11. doi:10.103841598-019-52220-7 PMID:31690761

Fang, L., Sun, Q., Duan, Y.-H., Zhai, J., Wang, D., & Wang, J. X. (2020). Preparation of transparent $BaSO_4$ nanodispersions by high-gravity reactive precipitation combined with surface modification for transparent X-ray shielding nanocomposite films. *Frontiers of Chemical Science and Engineering*, *15*(4), 902–912. doi:10.100711705-020-1985-y

Fontainha, C. C., Baptista-Neto, A. T., & Faria, L. O. (2016). Polymer-based nanocomposites of P (VDF-TrFE)/Bi. *Journal of Materials Science*, 4.

Gong, S., & Cheng, W. (2017). One-dimensional nanomaterials for soft electronics. *Advanced Electronic Materials*, *3*(3), 1600314. doi:10.1002/aelm.201600314

Graham, E., Gooding, D., Gruszko, J., Grant, C., Naranjo, B., & Winslow, L. (2019). Light yield of perovskite nanocrystal-doped liquid scintillator. *Journal of Instrumentation: An IOP and SISSA Journal*, *14*(11), P11024. doi:10.1088/1748-0221/14/11/P11024

Han, N., & Ho, J. C. (2014). One-dimensional nanomaterials for energy applications. *Nanocrystalline Materials,* 75-120.

Hemeda, O. M., Eid, M. E. A., Sharshar, T., Ellabany, H. M., & Henaish, A. M. A. (2021). Synthesis of nanometer-sized $PbZr_xTi_{1-x}O_3$ for gamma-ray attenuation. *Journal of Physics and Chemistry of Solids*, *148*, 109688. doi:10.1016/j.jpcs.2020.109688

Hou, Y., Li, M., Gu, Y., Yang, Z., Li, R., & Zhang, Z. (2018). Gamma ray shielding property of tungsten powder modified continuous basalt fiber reinforced epoxy matrix composites. *Polymer Composites*, *39*(S4), E2106–E2115. doi:10.1002/pc.24469

İrim, Ş. G., Wis, A. A., Keskin, M. A., Baykara, O., Ozkoc, G., Avcı, A., Doğru, M., & Karakoç, M. (2018). Physical, mechanical and neutron shielding properties of h-BN/Gd_2O_3/HDPE ternary nanocomposites. *Radiation Physics and Chemistry*, *144*, 434–443. doi:10.1016/j.radphyschem.2017.10.007

Jia, X., Zhou, Y., Zheng, J., Li, Y., Zou, H., & Xie, R. (2016). Cerium doped barium tantalates: Fabrication, characterization, and investigation of gamma radiation attenuation. *Journal of Alloys and Compounds*, *688*, 679–684. doi:10.1016/j.jallcom.2016.07.188

Kim, J., Seo, D., Lee, B. C., Seo, Y. S., & Miller, W. H. (2014). Nano-W dispersed gamma radiation shielding materials. *Advanced Engineering Materials*, *16*(9), 1083–1089. doi:10.1002/adem.201400127

Li, R., Gu, Y., Wang, Y., Yang, Z., Li, M., & Zhang, Z. (2017). Effect of particle size on gamma radiation shielding property of gadolinium oxide dispersed epoxy resin matrix composite. *Materials Research Express*, *4*(3), 035035. doi:10.1088/2053-1591/aa6651

Lokhande, R. M., Vinayak, V., Mukhamale, S. V., & Khirade, P. P. (2021). Gamma radiation shielding characteristics of various spinel ferrite nanocrystals: A combined experimental and theoretical investigation. *RSC Advances*, *11*(14), 7925–7937. doi:10.1039/D0RA08372K PMID:35423294

Mahmoud, M. E., El-Khatib, A. M., Badawi, M. S., Rashad, A. R., El-Sharkawy, R. M., & Thabet, A. A. (2018a). Fabrication, characterization and gamma rays shielding properties of nano and micro lead oxide-dispersed-high density polyethylene composites. *Radiation Physics and Chemistry, 145*, 160–173. doi:10.1016/j.radphyschem.2017.10.017

Mahmoud, M. E., El-Khatib, A. M., Badawi, M. S., Rashad, A. R., El-Sharkawy, R. M., & Thabet, A. A. (2018b). Recycled high-density polyethylene plastics added with lead oxide nanoparticles as sustainable radiation shielding materials. *Journal of Cleaner Production, 176*, 276–287. doi:10.1016/j.jclepro.2017.12.100

Moharram, B. M., Nagy, M. E., Shaat, M. K., El Sayed, A. R., Fayiz, M., Dwidar, S. A., & Dorrah, M. E. (2020). Performance of lead and iron oxides nanoparticle materials on shielding properties for γ-rays. *Radiation Physics and Chemistry, 173*, 108880. doi:10.1016/j.radphyschem.2020.108880

Muthamma, M. V., Prabhu, S., Bubbly, S. G., & Gudennavar, S. B. (2021). Micro and nano Bi2O3 filled epoxy composites: Thermal, mechanical and γ-ray attenuation properties. *Applied Radiation and Isotopes, 174*, 109780. doi:10.1016/j.apradiso.2021.109780 PMID:34052516

Nazarenko, O., Yakunin, S., Morad, V., Cherniukh, I., & Kovalenko, M. V. (2017). Single crystals of caesium formamidinium lead halide perovskites: Solution growth and gamma dosimetry. *NPG Asia Materials, 9*(4), e373–e373. doi:10.1038/am.2017.45

Nikeghbal, K., Zamanian, Z., Shahidi, S., Spagnuolo, G., & Soltani, P. (2020). Designing and fabricating nano-structured and micro-structured radiation shields for protection against CBCT exposure. *Materials (Basel), 13*(19), 4371. doi:10.3390/ma13194371 PMID:33008078

Noor Azman, N. Z., Musa, N. F., Nik Ab Razak, N. N., Ramli, R. M., Mustafa, I. S., Abdul Rahman, A., & Yahaya, N. Z. (2016). Effect of Bi_2O_3 particle sizes and addition of starch into Bi_2O_3–PVA composites for X-ray shielding. *Applied Physics. A, Materials Science & Processing, 122*(9), 1–9. doi:10.100700339-016-0329-8

Noor Azman, N. Z., Siddiqui, S. A., Haroosh, H. J., Albetran, H. M., Johannessen, B., Dong, Y., & Low, I. M. (2013). Characteristics of X-ray attenuation in electrospun bismuth oxide/polylactic acid nanofibre mats. *Journal of Synchrotron Radiation, 20*(5), 741–748. doi:10.1107/S0909049513017871 PMID:23955038

Oto, B., Gulebaglan, S. E., Madak, Z., & Kavaz, E. (2019). Effective atomic numbers, electron densities and gamma rays buildup factors of inorganic metal halide cubic perovskites $CsBX_3$ (B = Sn, Ge; X = I, Br, Cl). *Radiation Physics and Chemistry, 159*, 195–206. doi:10.1016/j.radphyschem.2019.03.010

Pokropivny, V. V., & Skorokhod, V. V. (2007). Classification of nanostructures by dimensionality and concept of surface forms engineering in nanomaterial science. *Materials Science and Engineering C, 27*(5-8), 990–993. doi:10.1016/j.msec.2006.09.023

Prabhu, S., Bubbly, S. G., & Gudennavar, S. B. (2021a). Thermal, mechanical and γ-ray shielding properties of micro-and nano-Ta_2O_5 loaded DGEBA epoxy resin composites. *Journal of Applied Polymer Science, 138*(44), 51289. doi:10.1002/app.51289

Prabhu, S., Bubbly, S. G., & Gudennavar, S. B. (2022). Bismuth (III) oxide decorated graphene oxide filled epoxy nanocomposites: Thermo-mechanical and photon attenuation properties. *Advanced Composite Materials*, 1–27. doi:10.1080/09243046.2022.2128264

Prabhu, S., Bubbly, S. G., & Gudennavar, S. B. (2023). X-ray and γ-ray shielding efficiency of polymer composites: Choice of fillers, effect of loading and filler size, photon energy and multifunctionality. *Polymer Reviews (Philadelphia, Pa.)*, *63*(1), 246–288. doi:10.1080/15583724.2022.2067867

Prabhu, S., Jayaram, S., Bubbly, S. G., & Gudennavar, S. B. (2021b). A simple software for swift computation of photon and charged particle interaction parameters: PAGEX. *Applied Radiation and Isotopes*, *176*, 109903. doi:10.1016/j.apradiso.2021.109903 PMID:34419873

Rashad, M., Tekin, H. O., Zakaly, H. M., Pyshkina, M., Issa, S. A., & Susoy, G. (2020). Physical and nuclear shielding properties of newly synthesized magnesium oxide and zinc oxide nanoparticles. *Nuclear Engineering and Technology*, *52*(9), 2078–2084. doi:10.1016/j.net.2020.02.013

Reddy, B. C., Manjunatha, H. C., Vidya, Y. S., Sridhar, K. N., Pasha, U. M., Seenappa, L., Mahendrakumar, C., Sadashivamurthy, B., Dhananjaya, N., Sankarshan, B. M., Krishnaveni, S., Sathish, K. V., & Gupta, P. S. D. (2021). Synthesis and characterization of multi-functional nickel ferrite nano-particles for X-ray/gamma radiation shielding, display and antimicrobial applications. *Journal of Physics and Chemistry of Solids*, *159*, 110260. doi:10.1016/j.jpcs.2021.110260

Shirkhanloo, H., Safari, M., Amini, S. M., & Rashidi, M. (2017). Novel semisolid design based on bismuth oxide (Bi_2O_3) nanoparticles for radiation protection. *Nanomedicine Research Journal*, *2*, 230–238.

Shkir, M., Kilany, M., & Yahia, I. S. (2017). Facile microwave-assisted synthesis of tungsten-doped hydroxyapatite nanorods: A systematic structural, morphological, dielectric, radiation and microbial activity studies. *Ceramics International*, *43*(17), 14923–14931. doi:10.1016/j.ceramint.2017.08.009

Tekin, H. O., Singh, V. P., Kara, U., Manici, T., & Altunsoy, E. E. (2016). Investigation of nanoparticle effect on radiation shielding property using Monte Carlo method. *CBU Journal of Science*, *12*(2), 196–200. doi:10.18466/cbujos.15586

Tu, Y., Wu, J., Xu, G., Yang, X., Cai, R., Gong, Q., & Huang, W. (2021). Perovskite solar cells for space applications: Progress and challenges. *Advanced Materials*, *33*(21), 2006545. doi:10.1002/adma.202006545 PMID:33861877

Verdipoor, K., Alemi, A., & Mesbahi, A. (2018). Photon mass attenuation coefficients of a silicon resin loaded with WO_3, PbO, and Bi_2O_3 micro and nano-particles for radiation shielding. *Radiation Physics and Chemistry*, *147*, 85–90. doi:10.1016/j.radphyschem.2018.02.017

Wang, J., Ma, F., Liang, W., & Sun, M. (2017). Electrical properties and applications of graphene, hexagonal boron nitride (h-BN), and graphene/h-BN heterostructures. *Materials Today Physics*, *2*, 6–34. doi:10.1016/j.mtphys.2017.07.001

Wang, Z., Hu, T., Liang, R., & Wei, M. (2020). Application of zero-dimensional nanomaterials in biosensing. *Frontiers in Chemistry*, *8*, 320. doi:10.3389/fchem.2020.00320 PMID:32373593

Wei, H., DeSantis, D., Wei, W., Deng, Y., Guo, D., Savenije, T. J., Cao, L., & Huang, J. (2017). Dopant compensation in alloyed $CH_3NH_3PbBr_{3-x}Cl_x$ perovskite single crystals for gamma-ray spectroscopy. *Nature Materials*, *16*(8), 826–833. doi:10.1038/nmat4927 PMID:28671663

Wu, Y., Cao, Y., Wu, Y., & Li, D. (2020). Mechanical properties and gamma-ray shielding performance of 3D-printed poly-ether-ether-ketone/tungsten composites. *Materials (Basel)*, *13*(20), 4475. doi:10.3390/ma13204475 PMID:33050304

Xu, Q., Shao, W., Li, Y., Zhang, H., Ouyang, X., Zhang, X., Nie, J., Ouyang, X., & Liu, B. (2019). Visual clarity methylammonium lead trichloride perovskite single crystals for X and gamma rays protection. *Journal of Alloys and Compounds*, *810*, 151896. doi:10.1016/j.jallcom.2019.151896

Xu, Q., Shao, W., Li, Y., Zhang, X., Ouyang, X., Liu, J., Liu, B., Wu, Z., Ouyang, X., Tang, X., & Jia, W. (2019b). High-performance surface barrier X-ray detector based on methylammonium lead tribromide single crystals. *ACS Applied Materials & Interfaces*, *11*(10), 9679–9684. doi:10.1021/acsami.8b21605 PMID:30793585

Xu, Y. C., Song, C., Ding, X. Y., Zhao, Y., Xu, D. G., Zhang, Q. P., & Zhou, Y. L. (2019a). Tailoring lattices of Bi_2WO_6 crystals via Ce doping to improve the shielding properties against low-energy gamma rays. *Journal of Physics and Chemistry of Solids*, *127*, 76–80. doi:10.1016/j.jpcs.2018.12.007

Xu, Y. C., Zhang, Q. P., Liu, J. H., Wu, Y., Liu, L. P., Xu, D. G., & Zhou, Y. L. (2018). $PbWO_4$ nanofibers for shielding gamma radiation: Crystal growth, morphology and performance evaluation. *CrystEngComm*, *20*(40), 6197–6206. doi:10.1039/C8CE01224E

Yahia, I. S., Shkir, M., AlFaify, S., Ganesh, V., Zahran, H. Y., & Kilany, M. (2017). Facile microwave-assisted synthesis of Te-doped hydroxyapatite nanorods and nanosheets and their characterizations for bone cement applications. *Materials Science and Engineering C*, *72*, 472–480. doi:10.1016/j.msec.2016.11.074 PMID:28024611

Zhang, Q. P., Sun, N., Zheng, J., Wang, S. W., Wu, Y., Luo, D. L., & Zhou, Y. L. (2017). Fabrication of lead borate single crystal nanosheets for attenuating both neutron and gamma radiations. *Advanced Engineering Materials*, *19*(3), 1600650. doi:10.1002/adem.201600650

Zhou, D., Zhang, Q. P., Zheng, J., Wu, Y., Zhao, Y., & Zhou, Y. L. (2017). Co-shielding of neutron and γ-ray with bismuth borate nanoparticles fabricated via a facile sol-gel method. *Inorganic Chemistry Communications*, *77*, 55–58. doi:10.1016/j.inoche.2017.01.034

KEY TERMS AND DEFINITIONS

Effective Atomic Number: An energy dependent parameter used to differentiate different materials based on total radiation interaction probability.

Effective Electron Density: The number of electrons per unit mass of the interacting medium representing the radiation interaction probability.

Half Value Layer: The thickness of the shielding material required to reduce the intensity of radiations to half of the incident radiation intensity.

Linear Attenuation Coefficient: The fraction of beam of X-rays or γ-rays that is absorbed or scattered per unit thickness of the absorber.

Mass Attenuation Coefficient: The fraction of beam of X-rays or γ-rays that is absorbed or scattered per unit mass thickness of the absorber.

Nanocomposite: A material composed of two or more component, of which at least one has a nanoscale dimension, such as nanoparticles dispersed throughout another medium.

Nanomaterial: Material with one or more external dimensions, or an internal structure, at nanoscale and which could exhibit novel characteristics compared to the same material at a larger scale.

Nanoparticle: Particle with one or more dimensions of the order of 100 nm or less.

Nanostructure: Structure with one or more dimensions at the nanoscale.

Packing Density: The ratio of the atomic volume within a unit cell to the volume of the unit cell.

Surface-to-Volume Ratio: The ratio of the amount of surface area per unit volume of an object.

Chapter 11
Piezoelectric Smart Materials and Commercialization

Rahul Saha
https://orcid.org/0000-0002-6573-2347
University of Calcutta, India

Ridipt Mishra
https://orcid.org/0000-0002-5353-4765
University of Calcutta, India

Avijit Dhara
https://orcid.org/0000-0001-9786-7791
University of Calcutta, India

Abir Chakravorty
University of Calcutta, India

Subhasis Roy
https://orcid.org/0000-0003-4197-535X
University of Calcutta, India

ABSTRACT

This chapter provides an exhaustive review of piezoelectric smart materials and their applications for commercialization with the addition of recent research and breakthroughs. Piezoelectric smart materials convert electrical signals into mechanical stresses and vice versa. They are remarkably used in energy harvesting from different processes and diverse human activities. Piezoelectric materials are convenient, readily available, adaptable, and have a speedy recurrence reaction time. Smart piezoelectric actuators embedded into adaptive structures are significant. Recent progress in their applications is evident in the versatile smart wings present in an aircraft, since they correlate with the overall effectiveness. In addition, a wide range of industrial applications, for example, diesel fuel injectors, quick responding solenoids, optical lenses, mechanical technology, vibrational damping, etc., utilize piezoelectric actuators. Further, this chapter aims to address the latest improvements in piezoelectric materials in connection to the realm of nanotechnology.

DOI: 10.4018/978-1-6684-7358-0.ch011

1. INTRODUCTION

The global exigencies in non-renewable energy sources have drawn the attention of researchers worldwide in the search for an alternative energy harvesting technology. Conventional power sources such as batteries experience many issues such as limited lifespan, power efficiency, and periodic recharging, which leads to its replacement in portable, remote devices (Safaei, Sodano, & Anton 2019).

Numerous initiatives by now have been made in energy conversion from various forms of energy to electrical energy. As of late, the point of examination in this field has exchanged its way towards anthropogenic energy scrounging, or the harvesting of energy discreetly through numerous biological processes such as breathing, body heat generation, different activities, for example, running, twirling, dancing, etc (Sezer & Koç, 2021). In the field of new-age electronic devices miniaturization, portability, durability, multitasking, high computational capability, and flexibility have become primary considerations (Jain & Mishra, 2015; Sik et al., 2019; Ali et al., 2019; H. Liu et al., 2018), Piezoelectricity, has emerged as the field of study that has shown extensive commitment in energy scavenging led by certain smart materials.

1.1 Smart Materials

When exposed to external stimuli like temperature, stress, magnetic and electric fields, pH, or other chemical effects, "smart materials" can mutate some or all of their properties (configuration, color, structure, electrical properties, etc.) and perhaps even their functions. The substances possess adaptive capabilities and pertain to a class of modern materials that could detect specific impulses from the surroundings and manifest their roles in the response. These can also be named as "intelligent," "adaptive," or even "responsive" as they can sense the external effects and summon the responses at the actuators (Salim & Abdulrazig 2020; Ghareeb & Farhat 2018; Qader et al., 2019). Piezoelectric materials, shape memory polymers, magnetostrictive materials are some examples of such Smart materials.

1.2 Piezoelectricity and Piezoelectric Materials

Piezoelectricity is the capacity of a particular material (piezoelectric materials) to garner a voltage when subjected to mechanical force. The contrary of the effect is also pertinent, i.e., an electrical voltage all over the sample causes stress within the specimen.

The word "piezo" comes from the Greek word "piezein," which means 'to squeeze' or 'to press" (S. M. Salim and Abdulrazig 2020). Piezoelectric materials mainly include ceramics, crystals, and organic materials. Some of the piezoelectric substances are Quartz, $BaTiO_3$, Lead Zirconate Titanate PZT:($Pb[Zr(x)Ti(1-x)]O_3$), Polyvinylidene fluoride (PVDF), etc. Piezoceramics prospered as the most versatile material in the family of savvy/smart materials. Ease in the availability of commercialisations is the main factor in piezoelectric materials' wide popularity. In addition, piezoelectric devices can work as sensors, actuators, and transducers. Piezo-sensors convert Mechanical stress ® EE (i.e. Electrical energy). Thus it is also referred to as "generators."

Actuators that employ the converse piezoelectric effect to transform EE ® ME (i.e. Mechanical Energy) are thus referred to as "motors," whereas all these effects are present in the same device in the case of transducers (Nuffer and Bein 2006). They are durable, flexible, structurally, and functionally compatible without engineering devices. Moreover, they don't get affected by external environmental factors like humidity, and thus, it has become the most valuable ambient energy harvesting technology (Y. Sun et al.

2019; Shirvanimoghaddam and Shirvanimoghaddam 2019; Mahmud et al. 2018). The materials have a vast field of applications, including micro-electro-mechanical systems (MEMS), optoelectronic devices, wireless electronics, microfluidic drive technology, wave energy conversion technology, microphones, etc. The promising application areas of piezoelectric smart materials include structural health monitoring (SHM), biomedical engineering, transportation, and aviation. Using piezoelectric generators, one can outfit internal biological tasks (heartbeat, blood flow, etc.) and anthropogenic energy produced from various human activities on the shoes and wrists to power the electronic devices. These substances can likewise be utilized in ecological monitoring (Mahmud et al. 2018; Zheng et al. 2017; Zhu et al. 2013; Bhalla et al. 2018; Gomez et al., n.d.).

As it is well established that Piezoelectric materials can generate power through excitations and strain, the effect is also manifested in the nanoscale size. They undoubtedly are going to play a significant role in constructing future smart cities with a healthier environment and self-powered monitoring systems. The latest research & development of such nano-scale substances are broadening the range of applicability. This chapter provides an in-depth review of the application and commercialization of such smart materials and the remarkable progress of nanotechnology and applications, accompanied by the capitalization of these technologies.

2. HISTORY AND BACKGROUND

2.1 Discovery (1880)

In 1880, an article (Noor et al. 2020) was reported by two brothers from France, P. Curie and J. Curie. It had been shown for the first time that electric charges are generated when a mechanical strain is forced in a few explicit materials. The article showed some experimentation on the crystals of quartz, tourmaline, Rochelle salt. Thus, the picture of piezoelectricity first came across the scientific world. Nonetheless, The Curie siblings didn't anticipate that the gems would likewise show the opposite effect also. A year later, Gabriel Lippman (Hippel Von 1950) concluded mathematically from the Fundamental Thermodynamics Principles that the opposite piezoelectric impact can likewise be seen. In 1882, the Curie brothers affirmed Lippman's hypothesis tentatively and proceeded with their works for the mathematical proof of the opposite impact in those translucent piezoelectric structures (Shirane, Sawaguchi, and Takagi 1951).

2.2 Laboratory Interest (1882-916)

Eventually, more experimental work and mathematical effort have begun exploring and defining the crystals' piezoelectric phenomenon. In 1910, the most impactful outcome was seen when Woldemar Voigt published "Textbook on Crystal Physics", (Iurgiutiu 2002). He listed twenty natural piezoelectric crystal classes and a rigorous thermodynamic approach of crystal materials employing proper tensorial analysis defined all 18 possible macroscopic piezoelectric constants. But it remained as a laboratory interest till 1916, although it created a significant impact in the early radioactive research.

2.3 First Age Applications & World War I (1917-1940)

Practical utilization of piezoelectric phenomenon for the first time emerged in 1917 when World War I was nearing its end. Paul Langevin, a Ph.D. candidate of P. Curie, developed the first ultrasonic submarine detector employing piezoelectric crystals in France. It was composed of thin quartz crystal transducers. The spacing between the object and the sensor can be measured by transmitting a high-frequency wave out from the transducer and assessing the interval between signals traveling to a submerged object and echoing it back to the device. That piezoelectric material's resonant frequency was around 50 Hz. In the United Kingdom, scientists Rutherford and Robert Boyle revealed that they were building a quartz piezoelectric device for underwater submarine detection.

The invention of Langevin stimulated the researchers to work in piezoelectricity and to continue the development of those devices further. Piezoelectricity became very impactful in varied areas. Those developments from 1920 to 1940 include ultrasonic transducers, ultrasonic time domain reflectometers, microphones, pick-ups, accelerometers, ceramic phonograph cartridges, signal filters, flaw detection inside metals, frequency stabilizers, and many more.

2.4 World War II & Afterwar: Second Age Applications

The newly discovered ferroelectrics group synthesized from ceramic substances which would perform well on piezoelectric parameters compared to quartz is a distinct advantage. In 1946, Arthur von Hippel reported the first ferroelectric material, $BaTiO_3$ (Abadi, Darlis, & Suraatmadja, 2018). Following the discovery of ferroelectric materials, numerous efforts were made to seek innovative applications and a competitive landscape for piezoelectric devices (Naresh et al. 2018; Xuezheng Jiang et al. 2014). Nonetheless, the U.S. (i.e. United States) efforts were inadequate due to their intellectual property rights & regulations, limitations, and the confidentiality preferences established during the expansion of such matters amidst the war situation. Based on the sheer nature of the field's wartime roots and the desire to protect employable intellectual rights, the growth of piezoelectric substances & frameworks in the US was limited to the firms involved in the innovation.

Given the fact that new materials were created first—quartz crystals were the primary profitably grown piezoelectric material—researchers actively sought out more substantial resources. Despite material improvements and the maturation of manufacturing techniques, the U.S. economy grew at a slower rate compared to country Japan. Because of the deficiency of present-day advancements, the piezoelectric business in the U.S has found it somewhat precarious to prosper. Even so, over the next two decades, the Japanese teamed up in an inclusive space and made real strides in developing a unique movement of industrial workplaces. A few of the implementations observed during that period included piezoelectric igniters, air ultrasonic transducers, audio tone transducers, phonograph cartridges, ultrasonic reverse parking assists in automobiles, smoke alarms, surface acoustic wave filters, tone generators, etc. (Ahmad & Abdul, 2019; Gao et al., 2018).

Frederick Lack created a crystal insensitive to a temperature, which would be operational in various conditions. Also, Issac Koga (Salim & Abdulrazig 2020) created a temperature-stable crystal in Japan. Japanese piezoelectric inventions primarily included piezoceramic filters for radios & TVs, buzzers, audio transducers, piezoelectric igniters for natural gas and butane appliances, and another similar device (Arshadi & Cobbold, 2017; Budenkov et al., 2009; Zahn, 2004). Accordingly, an arising research field with far more prominent potential, bounty more applications were noticed, and serious promoting and commercialization started.

3. GENERATION OF ELECTRICITY USING PIEZOELECTRIC MATERIAL

Energy demand has increased rapidly, and in the recent decade, this surge appears to be accelerating at an alarming rate. Because of the current shutdown of large industrial zones in China, several localities are experiencing frequent power outages, disrupting work hours thereby affecting the entire supply chain. Even countries such as Australia, the United Kingdom, and India are facing an energy crisis due to a sudden rise in demand following the COVID-19 second wave. Consequently, petroleum oil prices are also at an all-time high, impeding the global economy's recovery (Uchino & Uchino, 2015).

Amidst these power and energy supply deficiencies, the age of power generation through Piezoelectric materials is investigated.

3.1 Piezoelectric Effect

Piezoelectricity is explained by the ability of certain materials to generate an electric charge in response to applied mechanical stress, with the opposite happening when the used electric field changes. When subjected to the influence of an electric field, they, too, change dimension. A change in the electric field causes strain, which provokes the inverse piezoelectric effect (Garimella, Sastry, & Mohiuddin, 2015).

3.2 Mechanism

A piezoelectric crystal contains negative and positive charges that are symmetrically arranged, resulting in an electrically neutral crystal. These charges form dipoles one on top of the other and align to form a shape known as "Weiss domains," which are randomly oriented but can be aligned at high temperatures and electric fields, a process known as poling. Flexible piezoelectric materials are considered for use in high-power machines due to their capability to withstand enormous mechanical strain. Enhancing the effective coupling and thus the effectiveness of dipoles is considered the most effective method of improving these materials' ability to handle large strains. Because of these attributes, the material can be used as a power harvesting medium. The most well-known use among the common advantages of piezoelectric material is an electric lighter (Hong et al., 2017; Mahmud et al., 2018).

3.3 Piezoelectric Tiles

These are used to transform kinetic energy or the energy harnessed by the body and convert, to electricity; a brief explanation as to how these entire system functions are furnished below:

Figure 1. Generation of electricity with the help of piezo sensors (Naresh et al. 2018) (Reprint with permission)

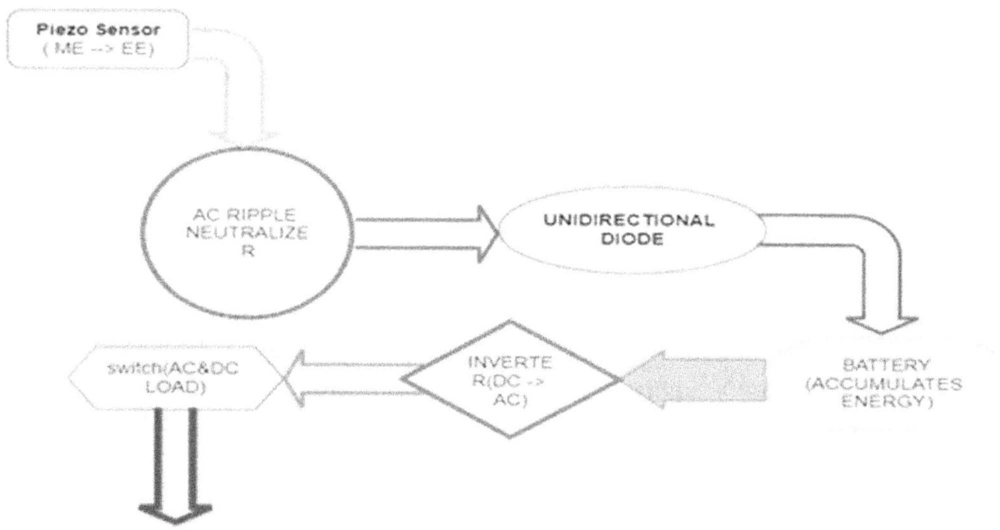

3.4 Piezoelectric Road and Highways

Piezoelectric materials, such as ceramic or crystal, can garner a significant portion of electricity in the mere existence of mechanical pressure, which can also be productive for generating sufficient energy to power up large structures if their traffic or passenger crowd is adequate. Since the mechanical strain is converted to electrical energy and stored in batteries which are again used to generate mechanical stress and so on. As a result, a self-powered device is created. Regardless of whether it is a rail line station or a street, the traffic issue can be utilized in favor where piezoelectric material frameworks are being laid to scourge energy requirements (Lippmann, 2010). Let us explain it's mechanism through an example of a car:

As the car passes, its load and the consequent deformation due to the passing automobile cause mechanical stress to be generated which in turn garner electrical energy. It should be noted that the roadway collector may be pre-compacted, allowing energy to be generated even when pressure is replayed. Numerous aspects influence the effectiveness of piezoelectric roads— vehicle speed, weight, traffic flow potential, etc. . Electricity production tends to increase as speed increases. In essence, the greater the constraint applied, the more significant the disruption of the piezoelectric materials, the greater the energy outcome (Yu et al., 2020)

3.5 Piezoelectric Railway Tracks

Railways carry a lot of weight, which causes enormous mechanical stress and pressure. A piezoelectric material rail-line is provided just 2-3 kilometers before the railway station, allowing enough energy to be generated and stored in the nearby railway station for later use.

Figure 2. Piezoelectric railway track (Gao et al. 2018) (Reprint with permission)

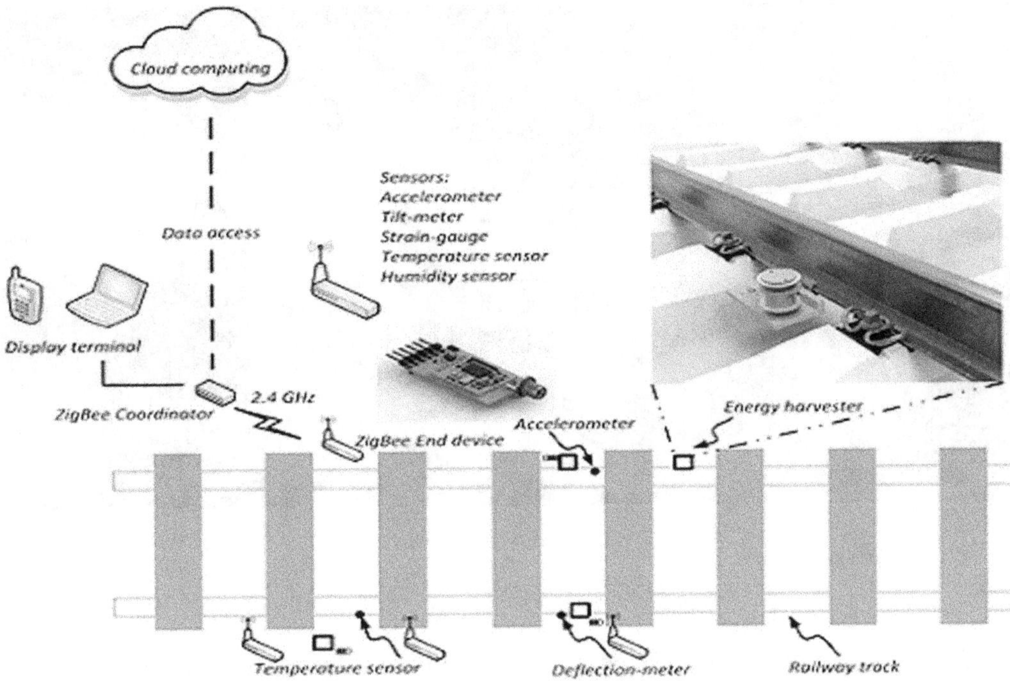

Figure 2 demonstrates a railroad track framework with piezoelectric strip and stack vitality interceptors, a piezoelectric patch-type energy harvester is installed at the rail's base. A linking machine would also install a piezoelectric stackable energy harvester over the bottom of a steel rail (Ramanathan et al., 2020). A piezoelectric stack gadget and a magnetic base connect the piezoelectric stack and fence (demonstrated in figure 3). Block Diagram of integrating the energy from multiple sources and supplies is presented in figure 4. The envisaged piezoelectric stack machine would include a displacement transmission rod, a deformation spring, a force transmission unit, a piezoelectric stack, an entire metal shell, screw bolts, and a wire hole. The compression spring converts the transverse track dispersion of the rail into a force transmitted to the piezoelectric stack. Nevertheless, each type of harvester would have its energy scourging framework.

Figure 3. Energy harvesting in railway using piezoelectric stack (Gao et al. 2018) (Reprint with permission)

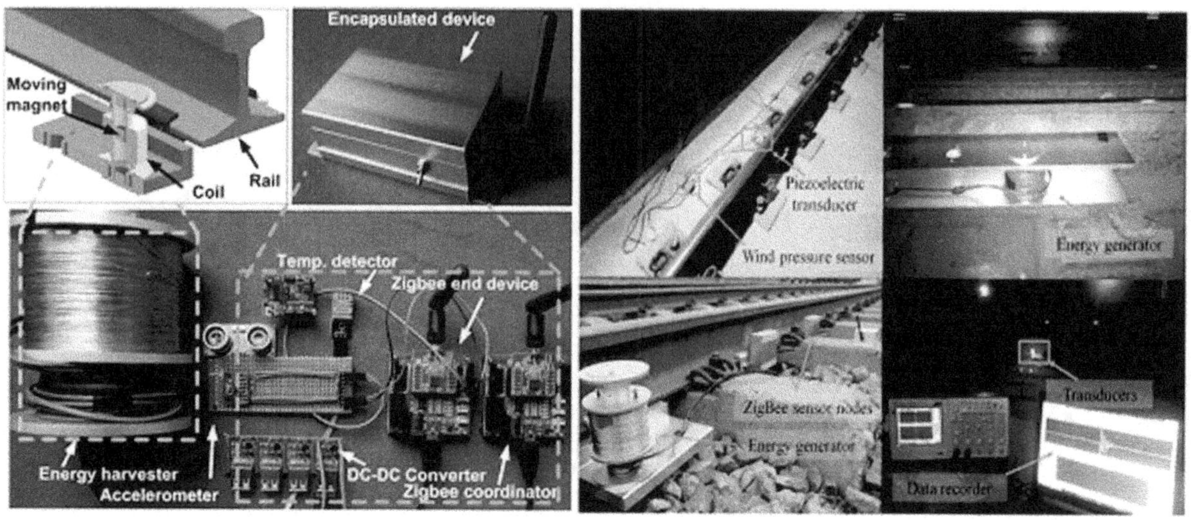

Figure 4. Block Diagram of integrating the energy from multiple sources and supplies (Laskar 2017)

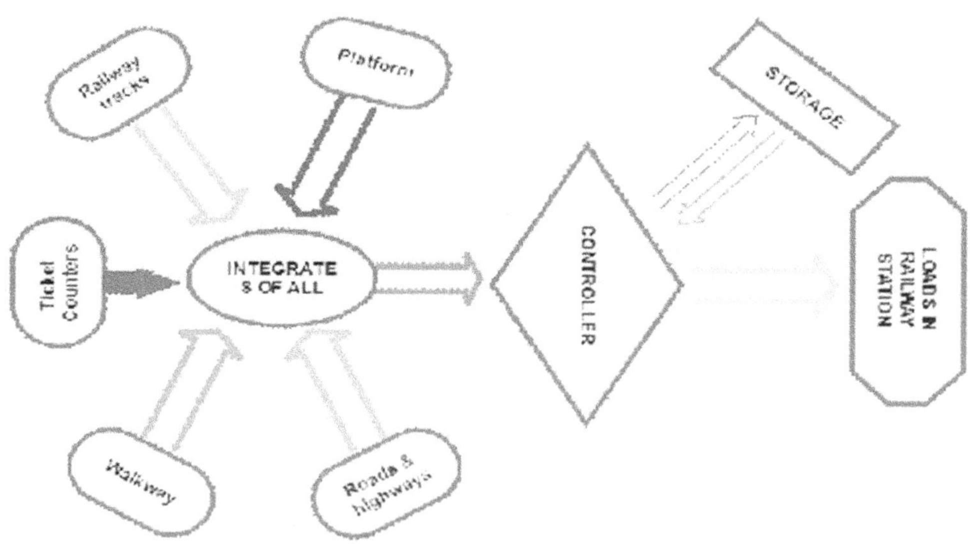

4. ACTUATORS: "THE POTENTIAL DARK HORSE IN OUR ERA"

The commercialization of piezoelectric actuators in various industries such as IT, robotics, biomedical, energy, ecological engineering places, depicted in figure 8, places them in a unique position of market leadership, enhancing future research and development. This section attempts to describe their applications in various industries (Uchino 2008)

Figure 5. Piezoelectric actuator developments are motivated typically by three areas: IT, Bio and Eco-engineering (Uchino, 2008)

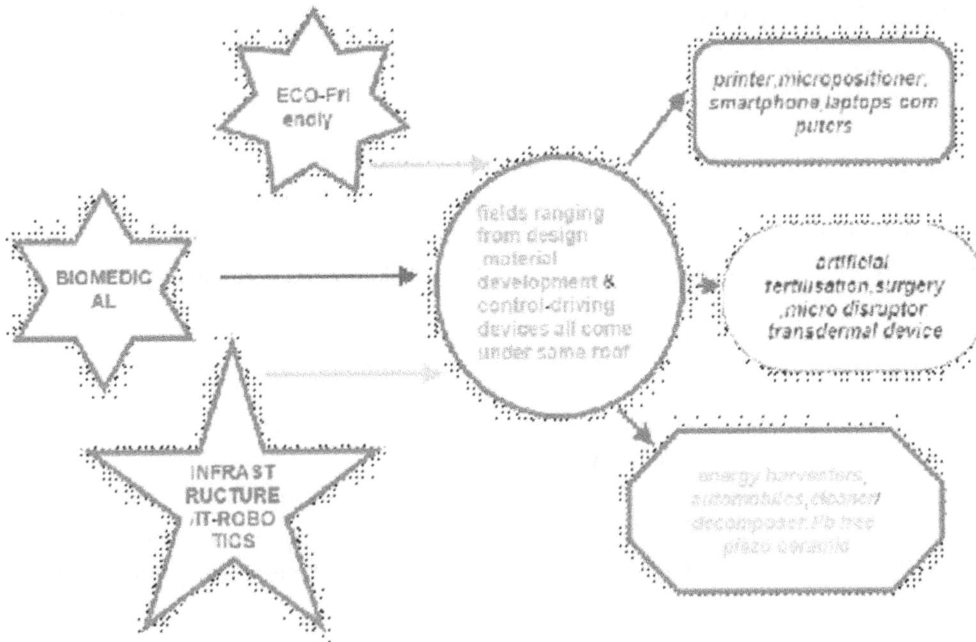

The first fabrication of piezoelectric actuators in large-scale production was in the form of a dot-matrix printer by NEC, fascinated by Penn State University's discovery of the co-fired multilayer actuator (MLA).

4.1 IT Tech Applications

Piezoelectric materials are a significant component of cellular phones. Even in laptops and computers, one can witness their dominance in the form of piezo pumps and piezoelectric transformers (Zhou et al., 2011). The simplicity and cost-effective solutions have also been an integral part of their production involvement in these devices on an industrial scale, thus adding to their dominance in the market.

4.1.1 Cellular Phone

Camera clicks, which have become "the new normal" in this generation, are another application of piezoelectric sensors and materials. As per 2006, 80 percent of smartphones had autofocus and even zooming, which utilized the piezoelectric sensors (J. Li et al., 2015).

4.1.2 Laptop-Computer

The piezoelectric pump has enabled significant advancements in laptop and computer cooling modules, as well as further miniaturization without sacrificing efficiency. A liquid crystal clear display that works over a very thin panel is an example of the above-mentioned sensors and is highly efficient with no electromagnetic noise alike transformers or fluorescent backlight etc. The piezo transformers have stood up to all of these requirements, and more research is being conducted for further development. The piezo-input transformers, the yield kiosks are made on a piezo machine, and the input/output voltage is changed using wave power transfer (H. Q. Li et al., 2004; Minary-Jolandan & Yu, 2013).

4.1.3 Printers

Epson's commercialization of the co-firing technique of Lead Zirconate Titanate with Zirconium dioxide (ZrO2) substances in piezoelectric inkjet printers led to substantial improvements in the vibrational stability for various inks during the piezoelectric chamber's (Xiaoning Jiang et al. 2014), patented pull–push-pull dynamic action. This showed a significant reduction in manufacturing costs as well.

4.2 Robotics

Among the significant applications are in the robotics domain which includes vibration damping, micro vehicles, shape control, micro positioners.

The most common application of piezoelectric actuators is positioners—Piezoelectric materials are extremely suitable for positioning due to their small size, high accuracy, and resolution, quick response, noiselessness, and a lack of electromagnetic radiation compared to the electromagnetic driving systems (T. X. Wu, 2014).

Many companies are commercialized by micro vehicles and microrobots made up of electromagnets and multilayer (ML) actuators. The concept currently involves "inchworm" movements over a steel surface having a mobility resolution of 30 nm. For entrapping nanoparticles and fibers, micro-instruments such as forceps, pincers, and drillers are readily accessible in tandem with the robot base (Strassberger & Waller, 2000; T. X. Wu & Thompson, 2002; Remington, 2014).

The evolving aircraft wings or the "Smart Wings'' used by the United States Air Force in their fighter aircraft incorporate accentuating mechanics, which controls the wing shape and morphing (Bein & Hanselka, 2014).

Likewise, ML actuators can be utilized to restrict the vibration of the body, and the technology can be implemented where a lightweight, electromagnetic noise-free, smooth framework which is responsive is required.

4.3 Biomedical Applications

4.3.1 Microsurgery

Countless plastic surgeons employ microsurgery to carry out specific methodologies such as free tissue transfer (relocating tissue from one part of the body to the other), reattachment of shattered sections (replantation), and composite tissue transplantation. General surgery, optometry, orthopedic surgery, gynecological surgery, neurological surgery, plastic surgery, pediatric surgery, and many other processes presently use microsurgery techniques that rely on piezoelectric instrumentation (Shin & Hong, 2020).

4.3.2 Micro Medical

Piezoelectrics are used in various applications, such as total analysis systems (TAS), compact DNA analyzing devices, electronic refrigeration systems, and portable medication delivery equipment, especially for insulin dispersion. Micro-pumps are customarily crafted and engineered to satisfy a subset of features, the significance of which varies based on the requirements; consequently, variations among models are frequent. For instance, in the specialty of drug delivery systems (DDS), it is well understood that the potency of specific drugs is necessarily correlated in how they're distributed. Micro-pumps crafted for drug delivery can and should pump precise volumes of fluid with strong controllability and match numerous essential aspects such as small diameter, low power consumption, and biocompatibility (Ali et al., 2019; Nisar et al., 2008).

4.3.3 Drug Delivery

Drug delivery frameworks are advances intended to transport therapeutic agents at a designated place in a controlled manner. Transdermal drug delivery can benefit from high-power ultrasonic delivery (Shung, Cannata, & Zhou 2007; Shawgo et al., 2002). Structure of PZT/ MEMS Si-device, blood tester as depicted in figure 8 below

Figure 6. A blood tester, Structure of a PZT/silicon MEMS device, (Uchino and Uchino 2015) (Reprint with permission)

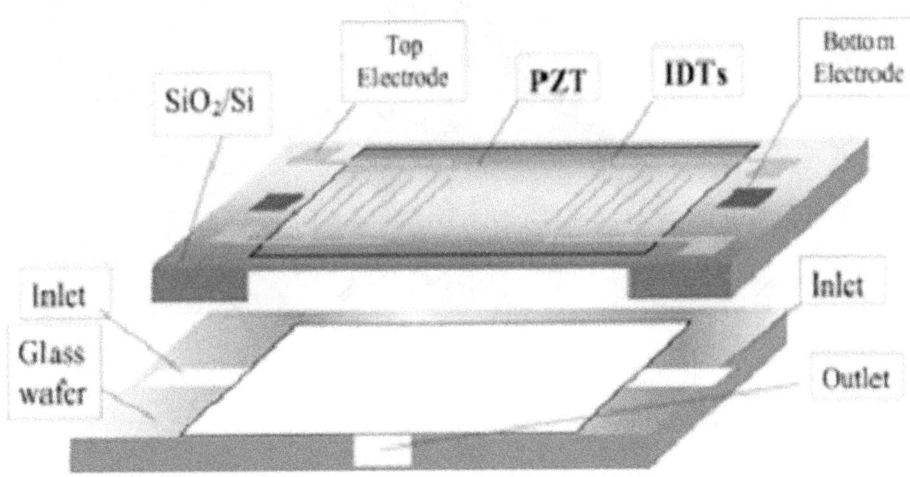

It has also been used in artificial fertilization systems and ultrasonic distillation. Piezoelectrics are valuable in keeping up with ecological balance and have monstrous potential in the next few decades to address energy challenges that are still unfolding in this era. The continuous advancements in lead-free electronics, ultrasonic cleaners, noise detection, and cancellation have considerably improved its further development and modeling.

4.4 Reduction of Toxic Gases

Highly pressurized fuel and fast injection regulation are developed to increase diesel engine efficiency. However, conventional diesel engines release toxic fumes such as sulfur oxides and nitrogen oxides. With the help of piezoelectric innovative diesel injection valves with PZT-based piezoelectric multilayer (ML) actuators to address this issue. Figure 9 depicts a standard rail-type diesel injection valve with an ML piezo-actuator for high-pressure fuel and quick injection control.

These devices have exhibited reliable endurance at elevated temperatures (150 °C) over a longer duration (10 years). The PZT piezoelectric actuator is critical for increasing combustion efficiencies and minimizing toxic fumes (Ali et al. 2019; S. Wu et al. 2021; Rödel et al. 2015; Shin and Hong 2020).

Figure 7. A typical Diesel injection valve (Uchino and Uchino 2015) (Reprint with permission)

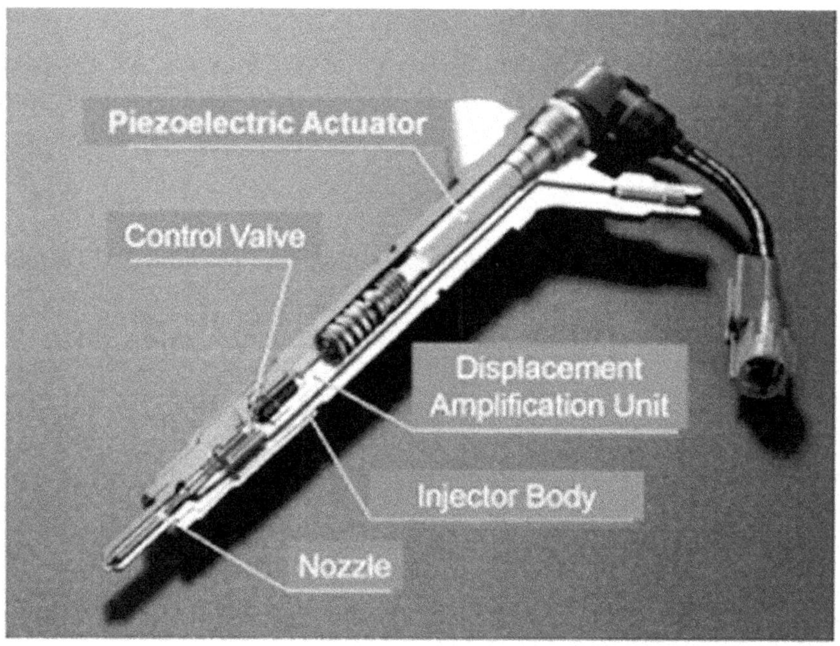

5. PIEZOELECTRIC SENSORS

Measurement is of prime importance in characterizing any physical phenomenon. A sensor is a device that responds to different physical stimuli like light, sound, heat, and magnetism and transfers a resulting signal for measurement.

Although they can be used for instant processes, they have certain properties that make them superior to other materials (Liang et al. 2018). They are much sensitive, can work at a wide range of temperatures, and are insensitive to the electric and magnetic fields and electromagnetic radiation in high frequency; they also act as a generator, i.e., they convert the electrical signal to mechanical stresses .

Piezoceramics, thin-film piezo elements and crystal materials are mainly used to develop a sensor. Piezoceramics (e.g. PZT) have high sensitivity but are not stable. On the other hand, single crystals like gallium orthophosphate ($GaPO_4$), quartz (SiO_2), langasite ($La_3Ga_5SiO_{14}$), aluminum nitride (AlN) are less sensitive but are highly stable. Also, some organic materials in teeth and bones show the properties (Garimella, Sastry, and Mohiuddin 2015) .Piezoelectric sensors are used in many fields of application like in medical, industrial, aero crafts, instrumentations, etc.

5.1 Engine Knock Sensors

Are the essential components used in automobiles. Manufacturers face problems related to the control of engine parameters like detonation and pre-ignition. They both are abnormal combustion in the engine consequently damaging the same. The Engine knock sensors can detect the detonation and preignition problem (Kulkarni, Khusru, & Aiyappa 2018).

5.2 Pressure Sensors

Pressure is one of the most vital units useful for operation in IC engines. Piezoelectric pressure detectors outperform general electromechanical pressure sensors because of their feedback and signal transformation. Piezoelectric pressure sensors depicted in figure 12, also detect acceleration. As a result, distinctive pressure sensors are intended to reduce acceleration understanding in applications of pressure sensors in vibrating or non-static contexts (Sasaki & Iguchi, 2000; Ricote & Hungrı, 2007; Chang et al., 2012).

Figure 8. A typical piezoelectric pressure sensor (Xu and Ma 2019). (Reprint with permission)

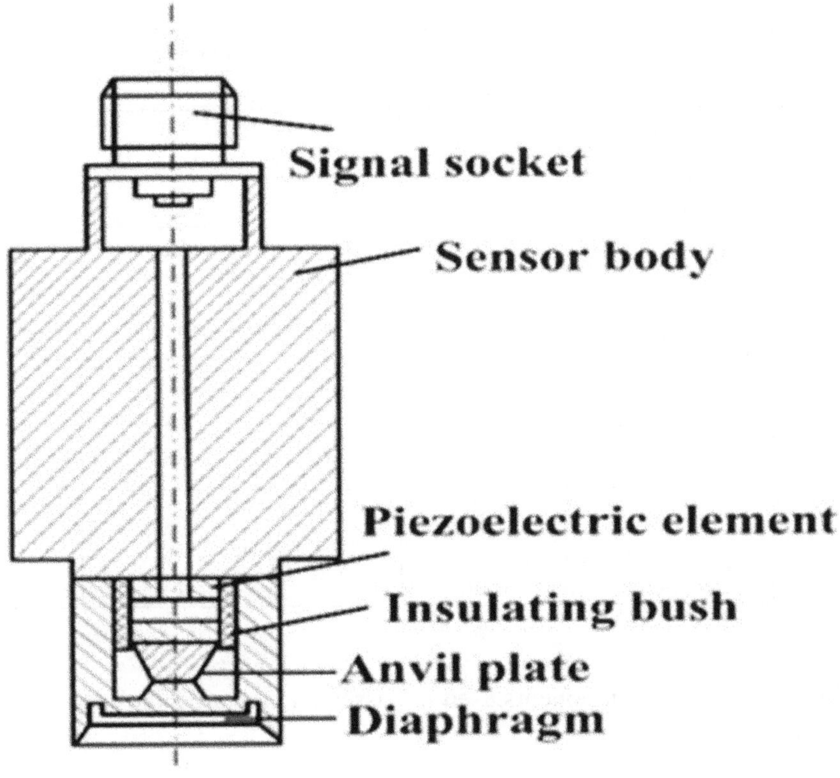

5.3 Ultrasound Imaging

Piezoelectric transducers are used in ultrasound techniques. Ultrasound sensors measure the distance of desired objects by transmitting ultrasonic sound waves and responding as electrical signals (Lu et al., 2015; Westerveld et al., 2021; Wissmeyer et al., 2018; Dangi et al., 2019; Acosta, 2020). It is a widely used instrument for determining the inner condition of materials, and it has been widely used in medical settings as well as laboratories. Piezoelectric sensors are also used in process control, quality control, quality assurance, and sonar systems in the industries (Wang et al., 2020; Fei, Chiu, & Qian, 2016; Acosta, 2020).

6. LIFELINE OF THE TRANSPORTATION

Piezoelectric material has captured the commercial market in all divisions possible over the last few decades, whether it is used as an actuator or a sensor. They discovered numerous potential uses, notably in the transportation industry. Some of the breakthroughs like fuel injection systems, energy harvesters, active noise cancellation devices have earned piezoelectric materials as a potential competitor in the recent world market. Its ability to withstand harsh environmental conditions makes it an outstanding candidate. Table 1 reflects the major impacts and layouts that contenders use in the automotive industry (Jhalani et al., n.d.; Hu & Lin, 2015; Piezoelectric, 2017; Sezer & Koç, 2021).

Table 1. Effect direction of various PZT devices (Nuffer and Bein 2006)

Design	Direct piezoeffect		Indirect piezoeffect	Both effects
	Sensor	Generator	Actuator	transducer
Bulk Material (d_{33}- effect)	Accelerometers Knock sensors Pressure/Force sensor	High voltage spark igniters	-	Ultrasonic sonar devices Distance meters
Multilayer (d_{33}- effect)	-	Energy harvesting	Active vibration reduction Nano- Positioning	-
Patch (monolithic) (d_{31}- effect)	Dynamic strain sensor	Energy harvesting	Active vibration reduction	-
Patch (fibre composites) (d_{31} and d_{33}- effect)	Dynamic strain sensors	Energy harvesting	Active vibration reduction	-
Bi-/ trimorph (d_{31} + bimorp effect)	-	-	Textile machine Fans	-
Special designs	-	Cymbal transducers for Energy harvesting	Ultrasonic Motors	Transformers

Some of the major piezoelectric components used in passenger cars are as follows:

6.1 Knock Sensors

Knock sensors are installed just next to the engine to detect anomalous burnout. The vibrating framework is squished in between piezoelectric material and a seismic mass, enabling vibration factions to join the piezo component. The piezo element transforms the pulses to electric charge in proportion to the force applied. The substance must be thermally stable to overcome temperature fluctuations. Besides that, the piezoelectric coefficient of the substance has to be nearly temperature-independent and steady throughout its lifetime (Ladd, Olsen, & Beshouri, 2017; Kulkarni, Khusru, & Aiyappa 2018; Pradesh et al., 2020).

6.2 Fuel Injection Systems

Its ability to work in extreme weather conditions, quick response time, capacity to perform any intended portfolio of valve displacement, and potential to eliminate energy consumption by more than 20% make it superior to traditional magnetic frameworks, laying the foundation for piezoelectric stack actuator systems to deliver more.

6.3 Distance Sensors

One of the several important applications of ultrasonic transducers in vehicles is parking assist. The gadget scatters ultrasonic waves, which are reflected by the obstruction, and are converted to an electrical signal by a similar transducer, which was earlier used as an ultrasonic sensor (Vitola, Pozo, & Tibaduiza, 2017).

7. CONTINUED APPLICATIONS AND RECENT DEVELOPMENTS

7.1 Health Monitoring of Railway Systems

Various researches are being conducted using piezoelectric materials as detecting elements for health monitoring and damage detection. of rail wheel wear. The objective is to discover variability in the overall performance of the wheel due to surface changes occurring because of friction. Theoretical work over this behavior of train wheels has incentivized us to use this approach. Piezoelectric sensors are installed in specific wheel areas to transfer surface deformation into electronic signals (Agarwal, 2018; Deepika, Kaur, & Kumari, 2019; Yan, Jin, & Sun, 2019; B. Liu et al., 2019).

In transducers, ultrasound is produced by applying a voltage to piezoelectric precious stones. The piezoelectric strategy necessitates using a liquid between the test and the fabric to allow ultrasound from the precious stone to emigrate into the material. The coupling liquid can be oil, gel, or glue, as long as it is appropriate for the material to be evaluated and remains on the fabric's surface during the review. A piezoelectric sensor is a combination of an eddy current sensor and an ultrasonic sensor, but it has some limitations due to temperature and the use of a complaint, making it unsuitable. Therefore to overcome this limitation, more research is required (Alemi, Corman, & Lodewijks, 2016; Zhang et al. 2018). A comprehensive comparison of different types of sensors that are commonly used for Non-Destructive Testing and Evaluation of rail wheel cracks is presented in table 2 below.

Table 2. Comparison of different types of Sensors (Ramanathan et al., 2020)

Sensors	Surface cracks	Internal cracks	External disturbances	Temperature	Couplant	Expensive
Ultrasonic sensor	Detects	Detects	Detects	N/A	Detects	N/A
Acoustic Emission Sensor	Detects	N/A	Detects	N/A	Detects	N/A
Hall Effect Sensor	Detects	N/A	N/A	Detects	N/A	N/A
Eddy Current Sensor	Detects	N/A	N/A	N/A	N/A	Detects
Piezo electric sensor	Detects	Detects	N/A	Detects	Detects	N/A
EMAT sensor	Detects	Detects	N/A	N/A	N/A	N/A

7.2 Active Vibration Reduction

One of the primary sources of sound pollution reduction is the reduction in sound emitted by vehicles. The power engine, particularly the oil pan, is the primary source of it. The reduction will not result in a lowering in sound pollution but also in reducing the wastage of energy, thus leading towards more cost-effective and sustainable usage of fuel and vehicles. Piezoelectric devices, particularly piezoelectric actuators, are being modeled and extensively tested to carry out this ambitious task economically and feasibly. The potential of piezoelectrics for AVC (i.e. Active Vibration Control) is extensive, as evidenced by numerous studies and projects (Omidi, Mahmoodi, and Jr., 2016; Shivashankar & Gopalakrishnan, 2020; Aridogan & Basdogan, 2015; Herold & Mayer, 2016).

7.3 High Temperature Piezoelectric

Because the oil temperature raises oil pan temperatures far beyond 100°C, the evolution of high-temperature piezoelectric actuator configurations was indeed primarily driven by the practical implications of an active damping concept for the same. The figure depicts piezoceramic wafers with uniformly metalized surfaces that operate in the lateral d31- mode. The piezoceramic is sandwiched between sequestering and contact layers. The contacting cover is made of a flexible conductive material that takes the shape & dimensions of the piezoceramic wafer. The flexibility of the contacting material guarantees high durability and dependability (Martin & Blumfield, 2015; Zu, Wu, & Wang, 2016).

7.4 Miniaturized Electronics

The power amplification framework seems to be another significant obstacle to AVR – systems for automobiles. It must be electrically compliant with the onboard power supply, should save space, be light, and vibration resilient. A miniaturized power amplifier for piezo was developed and investigated under the supervision of the University of Twente, as shown in the figure below. The weight, robustness, and self-heating during the operation were significantly improved by using an Al encasement & stuffing the enclosure of the amplifiers with a polyurethane resin. Piezo film research is advancing rapidly, with a fascinating range of applications using film actuators and film sensors (Guo and Li 2018; M. Salim, Salim, and Chandran 2018).

7.5 Smart Skin

The surface energy of a piezoelectric film can be sensed and micro-deflected. These capabilities could be valuable in effective surfaces for noise cancellation, extending laminar to turbulent flow boundaries, and other specific systems. Extensive work is being done in all of the above related applications at the university and government laboratories (Zaszczy, Gradys, & Sajkiewicz, 2020; Lee et al., 2020).

7.6 Recent Trends in Aircraft Morphing

Smart materials, piezo-actuators, and sensors have hugely transformed the aero crafts, robotics, biomedical, and automobile industries. Nowadays, smart materials are being used in various fields, thereby increasing the research trends of these technologies (J. Sun et al. 2016; Dangi et al., 2019; Jain & Mishra, 2015).

Disruption in the commercial market can be forced upon research and its application, as evidenced by the papers and their frequency. The number of papers relating to smart materials has increased dramatically over the last three decades, from a few to several thousand. Significant changes have also occurred in sectors such as the aerospace industry and domains aircraft morphing. (Bashir & Rajendran, 2019).

Smart materials have become increasingly impactful, in the aerospace industry. These technologies are used to manufacture various vehicles and have significantly transformed aviation by increasing potency by decreasing the overall weight and increasing fuel efficiency. Thus, smart technologies have become crucial in aircraft morphing predominantly using different composites (Bashir et al., 2018; Jodin et al., 2017; Herrera, 2018).

8. A GLANCE THROUGH COMMERCIAL EYES

A material capable of converting electrical energy to mechanical energy and converse. The above feature provides the potential for commercial production of PZT material, which is used in the development of many sectors such as the building and construction industry, the medical industry, the aerospace industry, and so on. All of the above has given piezoelectrics an excellent worldwide platform and market (Shibata et al., 2018).

8.1 Medical Sector

Due to various appealing characteristics, innovative piezoelectric materials are receiving recognition in the medical industry. Because of the excellent biocompatibility and biodegradability of piezoelectric intelligent materials, there has been massive progress in using piezoelectric materials in biomedical devices. The burgeoning use of piezoelectric intelligent materials in the healthcare industry to design medical equipment and better understand the piezoelectric sensor is cruising the market growth. Medical devices constantly being researched include actuators and sensors, energy harvesters, and active neural scaffolds for tissue engineering. Like sensors and actuators used in medical devices, Piezoelectric smart materials can convert flow rate and pressure to garner or ingest energy. Piezoelectric devices are primarily used for imaging in medicine. They would be used in a non-medical way to check fetuses. It helps detect creaks (Chorsi et al., 2018).

8.2 Aerospace Industry

The material also impacts the aircraft's weight, service life, function, and strength. As a result, the development of novel materials is typically regarded as an advancement in the aviation sector. In terms of piezoelectric material applications, there are two key features: shape control and vibration control. In aspects of shape-changing, it refers to the modification of an aerodynamic part. Traditionally, the control surface of an aircraft is still governed implicitly and tends to lack versatility. Nevertheless, the piezoelectric actuator can be a ground-breaking control system methodology; it significantly enhances effectiveness and maneuverability due to its flexible, convenient, and thin actuator (Chorsi et al., 2018).

8.3 Building and Construction Industry

The ability of smart piezo materials to transform mechanical disturbances ® electrical power enables the use of renewable and sustainable energy in dwellings through power generation and autonomous smart sensing. Elucidate the process of piezoelectric properties of various structural systems, as well as the most recent deployments and innovations in the building and construction sector, is essential for the successful use of piezoelectric smart materials for sustainable building production besides that the significant advancements for enhancing crosslinking piezoelectricity and utilizing the material for various construction operations are expected to result in more renewable and sustainable building systems. As a result of these advantages, there is a rise in prices for piezoelectric smart materials in the building and construction sector (Chen et al., 2019).

9. MERITS AND DEMERITS

9.1 Comparison of Piezoelectric with Electromagnetic Sensors

a. More suitable for miniaturization—According to the findings of a 1992 market research for eighty industries in Japan, micro motors in the range 5-8 mm were in high demand for official sectors and factory automation apparatus. However, producing conventional EM motors with sufficient energy efficiency is difficult. Because the piezo device has a higher stored energy density than an EM type, micro motors can be 1/10 lesser in volume and weight.
b. No EM noise — Because magnetic shielding isn't required, we can keep the design as small as possible.
c. Non-flammable—In addition, they are less prone to overloading or short circuits at the output terminal.
d. Greater efficiency (Uchino, 2008)

In addition, piezoelectric devices claim some more advantages, which are described as follows:

- Flexible - The piezoelectric materials can be made into various shapes and sizes due to their high flexibility, therefore resulting in quick popularity in the commercial market.
- Piezoelectric materials can use human activities to generate electricity from mechanical strain/ force. So, they do not need any external power source.
- Piezoelectric materials can be used at a wide range of temperatures, making them superior to other materials.
- Piezoelectric materials comparatively have a higher frequency response.
- Miniaturised, Portable & Compatible – These materials are incredibly fit and compatible with almost every engineering device due to their small dimensions.

But they also suffer from the following limitations or drawbacks:

- Piezo materials, despite being self-generating, create relatively modest electric charges, necessitating the use of a high-impedance wire to connect them to an electrical interface.
- Piezo materials can't be used for static processes. They are only suitable for dynamic measurements.
- The piezo materials produce small electric charges, i.e., the output is not very high. An external circuit will be necessary to optimize their utilization.
- Some piezo-crystals are water-soluble. A portion of these materials also gets dissolved in a high humid atmosphere.

10. FUTURE PROSPECTS

The development of SMs (smart materials) for applications such as autonomous wireless sensor networks, self-tuned vibration energy harvesting devices, seismological systems, and so on is critical. These astute materials can be used to create and implement intelligent structural components. Smart materials are sensory cue materials that incorporate various vibration control elements like piezo-shape memory alloys and electro, and magneto-rheological fluid. Smart materials are like biological systems in explicit ways. For example, piezoelectric hydrophones, which are comparable to fish's ears for detecting vibrations, piezoelectric with electro-mechanical recombination, shape-memory materials bearing the potential to relive the original form, and electro-rheological fluids with mutable viscosity strength, etc. Such power initiated a massive research interest, enabling to consider and use new imaginative innovation into smooth, various utilitarian packs to develop breakthroughs. The review discussed in this book chapter presents the above-mentioned multi-responsive materials, in-depth synopsis of some of the innovative materials, art components, and reshaping the domain of intelligent materials research.

11. CONCLUSIVE REMARKS

Within the last few decades, piezoelectric materials have progressed significantly in getting worldwide business markets across all areas and are the foundation material for actuators and sensors.

Commercialization of any item or material is dependable over different elements, including market reach, scaling (the product's market potential to arrive at buyers effectively and extensively), and the material/product's capacity to be upgraded further. Any item we see today was once only a thought that was improved, modernized, and capitalized, making it ready to enter the market.

All through this chapter, it has been discussed about various segments of piezoelectric materials with the help of certain examples and explanations alongside their created items that are now on the lookout and have been redesigned, and some in the improvement stage. One of the critical factors behind developing various new piezoelectric materials and crystals like lead-free piezoelectrics, single crystals, high-temperature piezoelectrics, and nano piezo-materials is the constant enhancement cycle of different & unique piezoelectric materials with enhanced electromechanical, biocompatible, and thermal properties. Most of the upgradation in frequency and power generation of the piezoelectric devices is due to the emersion of nonlinear and broadband harvesters in place of conventional linear, beam-based piezoelectric transducers have indeed been frequently employed.

The development of utilization-based tools to extract energy from the flowing fluid, through a human body, from animals, vehicles/ their motion is possible with the use of the piezoelectric devices, and are being used extensively and intelligently with further enhancements. Medical instruments and their developments rely heavily on piezoelectric devices, and the use of these materials has piqued the interest of researchers looking for more advanced wearable medical gadgets.

Several of the points introduced above are as yet going through different examinations and tasks. Moreover, there are vast opportunities in this domain for further improvements, upgrades, also new products that will soon occupy a place in the industry shortly. As a result, the evolution and hunt continue for further developments in every way possible.

ACKNOWLEDGMENT

This work was supported by Department of Science and Technology [DST], Ministry of Science and Technology, Government of India funded project under the MI IC#5' Conversion of Sunlight to Storable fuels' issued by DBT-DST Joint Funding Opportunity, Central Government of India through Mission Innovation Programme DST[DST/TMD[EWO]/IC5-2018/06] [S. Roy]. We also like to acknowledge Science and Engineering Research Board [SERB] grants funded by Department of Science and Technology [DST] Central, Government of India through Teachers Associateship for Research Excellence [TAR/2018/000195] [S. Roy] for providing financial support.

REFERENCES

Abadi, P. B., Darlis, D., & Suraatmadja, M. S. (2018). Green Energy Harvesting from Human Footsteps. *MATEC Web of Conferences, 11015*, 6–9. 10.1051/matecconf/201819711015

Acosta, S. (2020). Solvability for Photoacoustic Imaging with Idealized Piezoelectric Sensors. *IEEE Transactions on Ultrasonics, Ferroelectrics, and Frequency Control, 3010*(c), 1–10. doi:10.1109/TUFFC.2020.3005037 PMID:32746192

Agarwal, V. (2018). Smart Sensors for Structural Health Monitoring- Overview, Challenges and Advantages. *Sensors & Transducers, 221*(3), 1–8.

Ahmad, S., & Abdul, M. (2018). Energy Harvesting from Pavements and Roadways : A Comprehensive Review of Technologies, Materials, and Challenges. *International Journal of Energy Research*, *43*(August), 1974–2015. doi:10.1002/er.4350

Alemi, A., Corman, F., & Lodewijks, G. (2016). Condition Monitoring Approaches for the Detection of Railway Wheel Defects. *Journal of Rail and Rapid Transit*, *0*(0), 1–21. doi:10.1177/0954409716656218

Ali, F., Raza, W., Li, X., Gul, H., & Ki-hyun, K. (2019). Piezoelectric Energy Harvesters for Biomedical Applications. *Nano Energy*, *57*, 879–902. doi:10.1016/j.nanoen.2019.01.012

Aridogan, U., & Basdogan, I. (2015). A Review of Active Vibration and Noise Suppression of Plate-like Structures with Piezoelectric Transducers. *Journal of Intelligent Material Systems and Structures*, *26*(12), 1455–1476. doi:10.1177/1045389X15585896

Arshadi, R., & Cobbold, R. (2017). A Pioneer in the Development of Modern Ultrasound: Robert William Boyle (1883-1955). *Ultrasound in Medicine & Biology*. doi:10.1016/j.ultrasmedbio.2006.07.030 PMID:17189042

Bashir, M., & Rajendran, P. (2019). Recent Trends in Piezoelectric Smart Materials and Its Actuators for Morphing Recent Trends in Piezoelectric Smart Materials and Its Actuators for Morphing Aircraft Development. *International Review of Mechanical Engineering*, *13*(August), 117. doi:10.15866/ireme.v13i2.15538

Bashir, M., Rajendran, P., Sharma, C., & Smrutiranjan, D. (2018). ScienceDirect Investigation of Smart Material Actuators & Aerodynamic Optimization of Morphing Wing. *Materials Today: Proceedings*, *5*(10), 21069–21075. doi:10.1016/j.matpr.2018.06.501

Bein, T. & Hanselka, H. (2014). The Integrated Project ' Intelligent Materials for Active Noise Reduction '– An Overview and First Results The Integrated Project. *The Integrated Project, Intelligent Materialsd for Active Noise Reduction.* Vibration & Harshness Congress, Graz, *2-3*.

Bhalla, S., Moharana, S., Talakokula, V., & Kaur, N. (2018). Piezoelectric Materials: Applications in SHM, Energy Harvesting & Biomechanics. *Features Book Reviews*, *43*(February), 148–149.

Budenkov, G. A., Ivannikov, V. P., Kabakova, A. V., & Strizhak, V. A. (2009). Ultrasonic Quality Control of Press-Fit Joints. *Acoustic Methods*, *45*(8), 575–581. doi:10.1134/S1061830909080105

Chang, Y., Stephen F Poterala, Z. Y., Trolier-mckinstry, S., & Gary, L. (2009). Messing, Yunfei Chang, Stephen F Poterala, Zupei Yang, and Susan Trolier-mckinstry. 2012. "Electromechanical Coupling over a Broad Temperature Range Electromechanical Coupling over a Broad Temperature Range. *Applied Physics Letters*, *232905*, 2007–2010. doi:10.1063/1.3271682

Chen, J., Qiu, Q., Han, Y., & Lau, D. (2018, September). Piezoelectric Materials for Sustainable Building Structures : Fundamentals and Applications. *Renewable & Sustainable Energy Reviews*, *101*, 14–25. doi:10.1016/j.rser.2018.09.038

Chorsi, M. T, Curry, E., Chorsi, H., Das, R., Baroody, J., Purohit, P., Ilies, H., & Nguyen, T. D. (2018). Piezoelectric Biomaterials for Sensors and Actuators. *Advanced Materials*, *1802084*, 1–15. doi:10.1002/adma.201802084 PMID:30294947

Dangi, A., Cheng, C., Agrawal, S., Datta, G. R., Benoit, R., Pratap, R., Trolier-Mckinstry, S., & Sri-Rajasekhar, K. (2019). A Photoacoustic Imaging Device Using Piezoelectric Micromachined Ultrasound Transducers (PMUTs). *IEEE Transactions on Ultrasonics, Ferroelectrics, and Frequency Control*. IEEE. . doi:10.1109/TUFFC.2019.2956463

Deepika, G J, Kaur, G., & Kumari, N. (2019). Railway Accident Monitoring. *International Journal on Recent and Innovation Trends in Computing and Communication*, *5*(5), 704–706.

Evangeliou, N., Tsampiras, P., & Tzes, A. (2018). Development of a Tendon-Driven Robotic Tool Targeting Visual-Servoing Minimally-Invasive Surgical Operations. *26th Mediterranean Conference on Control and Automation*. IEEE. 10.1109/MED.2018.8442724

Fei, C., Chiu, C. T., & Qian, X. (2016). Author ' s Accepted Manuscript. *Nano Energy*. doi:10.1016/j.nanoen.2016.06.048

Ferrari, A., & Pizzo, P. (2016). Fully Predictive Common Rail Fuel Injection Apparatus Model and Its Application to Global System Dynamics Analyses. *International Journal of Engine Research*. doi:10.1177/1468087416653246

Gao, M., Li, Y., Lu, J., Wang, Y., Wang, P., & Wang, L. (2018). Condition Monitoring of Urban Rail Transit by Local Energy Harvesting. *International Journal of Distributed Sensor Networks*, *14*(11). doi:10.1177/1550147718814469

Garimella, R. C., Sastry, V. R., & Mohiuddin, M. S. (2015). Piezo- Gen- An Approach to Generate Electricity from Vibrations. *Procedia Earth and Planetary Science*, *11*, 445–456. doi:10.1016/j.proeps.2015.06.044

Ghareeb, N., & Farhat, M. (2018). *Nanotechnology & Applications Smart Materials and Structures : State of the Art and Applications*. Nanotechnology & Applications.

Gomez, A., Gich, M., Puig, T., & Obradors, X. (n.d.). Piezo-Generated Charge Mapping Revealed through Direct Piezoelectric Force Microscopy. *Nature Communications*, 1–10. doi:10.103841467-017-01361-2 PMID:29062016

Guo, J., & Li, X. (2018). Surface Effects on an Electrically Permeable Elliptical Nano-Hole or Nano-Crack in Piezoelectric Materials under Anti-Plane Shear. *Acta Mechanica*, *229*(10), 4251–4266. doi:10.100700707-018-2232-1

Herold, S., & Mayer, D. (2016). Adaptive Piezoelectric Absorber for Active Vibration Control. *Actuators*, *5*(1), 7. doi:10.3390/act5010007

Herrera, G. (2018). Lead-Free Piezoelectric Ceramics as Candidates on the Development of Morphing Winglets in Aircrafts. *Journal of Marine Science and Engineering*, *8*(1), 32–42. doi:10.17265/2161-6213/2018.1-2.005

Hippel Von, A. (1950). Ferroelectricity, Domain Structure, and Phase Transitions of Barium Titanate. *Reviews of Modern Physics*, *22*(3), 22. doi:10.1103/RevModPhys.22.221

Hong, L., Qu, Y., Dhupia, J. S., Sheng, S., Tan, Y., & Zhou, Z. (2017). *A Novel Vibration-Based Fault Diagnostic Algorithm for Gearboxes Under Speed Fluctuations Without Rotational Speed Measurement*. Elsevier. doi:10.1016/j.ymssp.2017.02.024

Hu, Y., & Lin, Z. (2015). Recent Progress in Piezoelectric Nanogenerators as a Sustainable Power Source in Self-Powered Systems and Active Sensors. *Nano Energy*, *14*, 1–12. doi:10.1016/j.nanoen.2014.11.038

Imran, P., & Uzun, M. (2011). The Requirement for Piezoelectric Smart Material for Current and Future Applications. *Journal of Engineering and Natural Sciences*, (216), 395–411.

Iurgiutiu, V., & Ictor, G. (2002). Electro-Mechanical Impedance Method for Crack Detection in Thin Plates. *Journal of Intelligent Material Systems and Structures*, *12*(October). doi:10.1106/HHXL-TLM8-8CM5-6OAK

Jain, S., & Mishra, R. (2015). *A Review on Piezoelectric Material as a Source of Generating Electricity and Its Possibility to Fabricate Device*. Int. J. Systems, Control and Communications.

Jhalani, V. A., Zhou, J. J., Park, J., Dreyer, C., & Bernardi, M. (n.d.) Piezoelectric Electron-Phonon Interaction from Ab Initio Dynamical Quadrupoles: Impact on Charge Transport. In *Wurtzite GaN*, 1–6.

Jiang, X., Kim, K., Zhang, S., Johnson, J., & Salazar, G. (2014). High-Temperature Piezoelectric Sensing. *Sensors (Basel)*, *14*(1), 144–169. doi:10.3390140100144 PMID:24361928

Jiang, X., Li, Y., Li, J., Wang, J., & Yao, J. (2014, January). Piezoelectric Energy Harvesting from Traffic- Induced Pavement Vibrations. *Journal of Renewable and Sustainable Energy*, *043110*(4), 043110. Advance online publication. doi:10.1063/1.4891169

Jodin, G., Scheller, J., Duhayon, E., Rouchon, J. F., & Braza, M. (2017). Implementation of a Hybrid Electro-Active Actuated Morphing Wing in Wind Tunnel. *Diffusion and Defect Data, Solid State Data. Part B, Solid State Phenomena*, *260*, 85–91. . doi:10.4028/www.scientific.net/SSP.260.85

Kulkarni, H., Zohaib A Khusru, K., & Shravan Aiyappa, K. (2018). ScienceDirect Application of Piezoelectric Technology in Automotive Systems. *Materials Today: Proceedings*, *5*(10), 21299–21304. doi:10.1016/j.matpr.2018.06.532

Ladd, J. W., Olsen, D. B., & Beshouri, G. (2017). Evaluation of Operating Parameters and Fuel Composition on Knock in Large Bore Two-Stroke Pipeline Engines. *Fuel*, *202*, 165–174. doi:10.1016/j.fuel.2017.04.038

Laskar, A. R. (2017). Piezoelectricity : An Energy Source for Future Railway Stations. *SEU Journal of Science and Engineering*, *11*(2), 13–18.

Lee, H., Park, H., Park, K., & Yi, H. (2020). Application of Adaptive Wave Cancellation Underwater to a Piezoelectric-Material-Based Multilayer Sensor. *Sensors (Basel)*, *3*, 1–12. doi:10.339020010134 PMID:31878227

Li, H. Q., Roberts, D. C., Steyn, J. L., Turner, K. T., Yaglioglu, O., Hagood, N. W., Spearing, S. M., & Schmidt, M. A. (2004). Fabrication of a High Frequency Piezoelectric Microvalve. *Sensors and Actuators*, *111*(1), 51–56. doi:10.1016/j.sna.2003.10.013

Li, J., Zhou, X., Zhao, H., Shao, M., Fan, Z., & Liu, H. (2015). Design and Experimental Tests of a Dual-Servo Piezoelectric Nanopositioning Stage for Rotary Motion. *The Review of Scientific Instruments, 045002*(4), 1–7. doi:10.1063/1.4918295 PMID:25933888

Liang, Z., Yan, C., Rtimi, S., & Bandara, J. (2018). Piezoelectric Materials for Catalytic/Photocatalytic Removal of Pollutants: Recent Advances and Outlook. *Applied Catalysis B: Environmental*. doi:10.1016/j.apcatb.2018.09.028

Lippmann, M G. (2010). Magazine and Journal of Science On the Principle of the Conservation of Electricity. *Taylor and Francis, 5982* (May). . doi:10.1080/14786448108627081

Liu, B., Jia, L., Qin, Y., Liu, Z., Diao, L., & An, M. (2019). *Proceedings of the 4th International Conference on Electrical and Information Technologies for Rail Transportation (EITRT) 2019*. Springer.

Liu, H., Zhong, J., Lee, C., & Seung-wuk, L. (2018). Mechanisms, and Applications APPLIED PHYSICS REVIEWS A Comprehensive Review on Piezoelectric Energy Harvesting Technology : Materials, Mechanisms, and Applications. *Applied Physics Reviews, 041306*(4), 041306. doi:10.1063/1.5074184

Lu, Y., Hao-yen, T., & Fung, S., Boser, B., & Horsley, D. (2015). Pulse-Echo Ultrasound Imaging Using Transmit Beam-Forming. *Journal of Microelectromechanical Systems*, 1–9.

Luo, T., Jiang, S., Moro, A., Wang, C., Zhou, L., & Luo, F. (2017, December). Measurement and Validation of Hole-to-Hole Fuel Injection Rate from a Diesel Injector. *Flow Measurement and Instrumentation, 61*, 66–78. doi:10.1016/j.flowmeasinst.2018.03.014

Mahmud, M. A. (2018). Recent Advances in Nanogenerator-Driven Self-Powered Implantable Biomedical Devices. *Advanced Energy Materials, 1701210*(2), 1–25. doi:10.1002/aenm.201701210

Martin, T., Stevenson, D. G., & Cowin A Blumfield, P. I. (2015). Piezoelectric Materials for High Temperature Transducers and Actuators. *Journal of Materials Science Materials in Electronics, 26*(12), 9256–9267. doi:10.100710854-015-3629-4

Minary-jolandan, M., & Yu, M. (2010). 2013. “Shear Piezoelectricity in Bone at the Nanoscale Shear Piezoelectricity in Bone at the Nanoscale. *Applied Physics Letters, 153127*(15), 2008–2011. doi:10.1063/1.3503965

Naresh, K, A Balaji, M Rambabu, & G Nagaraju. (2018). IRJET- Practical Oriented Foot Step Electric Power Generation By Using Piezo Material And Microcontroller In Campus Practical Oriented Foot Step Electric Power Generation. *International Research Journal of Engineering and Technology.*

Nisar, A., Afzulpurkar, N., Mahaisavariya, B., & Tuantranont, A. (2008). MEMS-Based Micropumps in Drug Delivery and Biomedical Applications. *Sensors and Actuators. B, Chemical, 130*(2), 917–942. doi:10.1016/j.snb.2007.10.064

Noor, N. A., Rashid, M., Mustafa, G. M., Khan, M. I., Mahmood, A., & Ramay, S. M. (2020). Study of Pressure Induced Physical Properties of ZnZrO 3 Perovskite Using Density Functional Theory. *Chemical Physics Letters, 753*(January), 137601. doi:10.1016/j.cplett.2020.137601

Nuffer, J., & Bein, T. (2006). Applications of Piezoelectric Materials in Transportation Industry. *Global Symposium on Innovative Solutions for the Advancement of the Transport Industry.*

Omidi, E., Mahmoodi, S. N., & Shepard, W. S. Jr. (2016). Mechatronics Multi Positive Feedback Control Method for Active Vibration Suppression in Flexible Structures. *Mechatronics*, *33*, 23–33. doi:10.1016/j.mechatronics.2015.12.003

Piezoelectric, C. (2017). Author's Accepted Manuscript. *Nano Energy*. doi:10.1016/j.nanoen.2017.05.029

PradeshM.UpadhyayB.TongyaB.GuptaC.PradeshM. (2020). Study of alternative fuels and their knock behavior. https://ssrn.com/abstract=3372828

Qader, I. N., Kok, M., Dagdelen, F., & Aydogdu, Y. (2019). A Review of Smart Materials: Researches and Applications. *ECJSE*, (3), 755–788.

Ramanathan, K., Dhanraj, J., Sekaran, C., & Manohar, L. (2020). A Comprehensive Review on Rail Wheel Crack Inspection System. *JMCMS*. . doi:10.26782/jmcms.2020.03.00014

Remington, P. J. (2014). Wheel/Rail Rolling Noise, II: Validation of the Theory. Acoustical Society of America, 1824–32.

Ricote, J., & Hungrı, T. (2007). High-Sensitivity Piezoelectric, Low-Tolerance-Factor Perovskites by Mechanosynthesis. *Chemistry of Materials*, (8), 4982–4990.

Rödel, J., Webber, K. G., Dittmer, R., Jo, W., Kimura, M., & Damjanovic, D. (2015). Feature Article Transferring Lead-Free Piezoelectric Ceramics into Application. *Journal of the European Ceramic Society*, *35*(6), 1659–1681. doi:10.1016/j.jeurceramsoc.2014.12.013

Safaei, M., Sodano, H. A., & Anton, S. R. (2019). A Review of Energy Harvesting Using Piezoelectric Materials : State-of-the-Art a Decade Later (2008 – 2018). *Smart Materials and Structures*, *28*(11), 113001. doi:10.1088/1361-665X/ab36e4

Salikhov, R. F., Makushev, Y. P., Musagitova, G. N., Volkova, L. U., & Suleymanov, R. S. (2019). Diagnosis of Fuel Equipment of Diesel Engines in Oil-and-Gas Machinery and Facilities Diagnosis of Fuel Equipment of Diesel Engines in Oil-and- Gas Machinery and Facilities. *AIP Conference Proceedings*, *050009*(August), 050009. doi:10.1063/1.5122152

Salim, M. & Abdulrazig, O. (2020). Using Smart-Piezoelectric Materials to Generate Electricity. *Bright Star Journal for Scientist Research*, *1*(x), 0–1.

Salim, M., Salim, D., Chandran, D., Aljibori, H. S., & Kherbeet, A. S. (2018). Review of Nano Piezoelectric Devices in Biomedicine Applications. *Journal of Intelligent Material Systems and Structures*, *29*(10), 2105–2121. doi:10.1177/1045389X17754272

Sasaki, H., & Iguchi, K. (2000). Driving Method of Piezoelectric Transformer and Driving Circuit for the Same. *US Patent 6,133,672*, no. 19.

Sezer, N., & Koç, M. (2020, November). Nano Energy A Comprehensive Review on the State-of-the-Art of Piezoelectric Energy Harvesting. *Nano Energy*, *80*, 105567. doi:10.1016/j.nanoen.2020.105567

Shawgo, R. S., Richards Grayson, A. C., Li, Y., & Cima, M. J. (2002). BioMEMS for Drug Delivery. *Current Opinion in Solid State and Materials Science*, *6*(4), 329–334. doi:10.1016/S1359-0286(02)00032-3

Shibata, K., Wang, R., Tou, T., & Koruza, J. (2018). Applications of Lead-Free Piezoelectric Materials. *Advanced Materials*, *612–16*(8), 612–616. doi:10.1557/mrs.2018.180

Shin, D. & Hong, S. (2020). Recent Advances in Organic Piezoelectric Biomaterials for Energy and Biomedical Applications. *Nanomaterials, 10* (123), 1–15.

Shirane, G., Sawaguchi, E., & Takagi, Y. (1951). Dielectric Properties of Lead Zirconate. *Physical Review*, *84*(3), 476–481. doi:10.1103/PhysRev.84.476

Shirvanimoghaddam, M., Shirvanimoghaddam, K., Abolhasani, M. M., Farhangi, M., Zahiri Barsari, V., Liu, H., Dohler, M., & Naebe, M. (2019). Towards a Green and Self-Powered Internet of Things Using Piezoelectric Energy Harvesting. *IEEE Access: Practical Innovations, Open Solutions*, *7*, 94533–94556. doi:10.1109/ACCESS.2019.2928523

Shivashankar, P., & Gopalakrishnan, S. (2020). Review on the Use of Piezoelectric Materials for Active Vibration, Noise, and Flow Control. *Smart Materials and Structures*, *29*(5), 053001. doi:10.1088/1361-665X/ab7541

Shung, K. K., Cannata, J. M., & Zhou, Q. F. (2007). Piezoelectric Materials for High Frequency Medical Imaging Applications : A Review. *Journal of Electroceramics*, *19*(1), 139–145. doi:10.100710832-007-9044-3

Sik, S., Seo, H., Kawahara, M., Glinsek, S., Lee, J., Kim, Y., Kyu, C., & Angus, I. (2018, October). Nano Energy Flexible Vibrational Energy Harvesting Devices Using Strain-Engineered Perovskite Piezoelectric Thin Films. *Nano Energy*, *55*, 182–192. doi:10.1016/j.nanoen.2018.10.068

Strassberger, M., & Waller, H. (2000). Active Noise Reduction by Structural Control Using Piezo-Electric Actuators. *Mechatronics*, *10*(8), 851–868. doi:10.1016/S0957-4158(99)00096-3

Sun, J., Guan, Q., Liu, Y., & Leng, J. (2016). Morphing Aircraft Based on Smart Materials and Structures : A State-of-the-Art Review. *Journal of Intelligent Material Systems and Structures*, *27*(2), 2289–2312. Advance online publication. doi:10.1177/1045389X16629569

Sun, Y., Chen, J., Li, X., Lu, Y., Zhang, S., & Cheng, Z. (2019). Nano Energy Flexible Piezoelectric Energy Harvester / Sensor with High Voltage Output over Wide Temperature Range. *Nano Energy*, *61*(March), 337–345. doi:10.1016/j.nanoen.2019.04.055

Uchino, K. (2008). Piezoelectric Actuators 2006. *Journal of Electroceramics*, *20*(3-4), 301–311. doi:10.100710832-007-9196-1

Uchino, K., & Uchino, K. (2015). Glory of Piezoelectric Perovskites. *Science and Technology of Advanced Materials*, *16*(4), 46001. doi:10.1088/1468-6996/16/4/046001 PMID:27877827

Vitola, J., Pozo, F., & Tibaduiza, D. (2017). Distributed Piezoelectric Sensor System for Damage. *Sensors*, 1–26. . doi:10.3390/s17061252

Wang, J., Zheng, Z., Chan, J., & John, T. W. (2020, August 24). Yeow. 2020. "Capacitive Micromachined Ultrasound Transducers for Intravascular Ultrasound Imaging. *Microsystems & Nanoengineering*, *6*(1), 73. doi:10.103841378-020-0181-z PMID:34567683

Westerveld, W. J., Shnaiderman, R., Ntziachristos, V., Rottenberg, X., Severi, S., & Rochus, V. (2021). Sensitive, Small, Broadband and Scalable Optomechanical Ultrasound Sensor in Silicon Photonics. *Nature Photonics*, *15*(May), 341–345. doi:10.103841566-021-00776-0

Wissmeyer, G., Miguel A Pleitez, A. R., & Ntziachristos, V. (2018). Looking at Sound : Optoacoustics with All-Optical Ultrasound Detection. *Light, Science & Applications*, *7*(1), 53. doi:10.103841377-018-0036-7 PMID:30839640

Wu, S., Yu, X., Zhang, J., Zhang, Y., Zhu, Y., & Zhu, M. (2021). Construction of BiOCl / CuBi 2 O 4 S-Scheme Heterojunction with Oxygen Vacancy for Enhanced Photocatalytic Diclofenac Degradation and Nitric Oxide Removal. *Chemical Engineering Journal*, *411*(January), 128555. doi:10.1016/j.cej.2021.128555

Wu, T. X., & Thompson, D. J. (2000). 2002. "The Vibration Behavior of Railway Track at High Frequencies under Multiple Preloads and Wheel Interactions The Vibration Behavior of Railway Track at High Frequencies under Multiple Preloads and Wheel Interactions. *Acoustical Society of America*, *1046*(3), 1046. doi:10.1121/1.1288408 PMID:11008807

Wu, T. X., & Thompson, D. J. (2014). The Vibration Behavior of Railway Track at High Frequencies under Multiple Preloads and Wheel Interactions. *Acoustical Society of America*, *108*(3), 1046–1053. doi:10.1121/1.1288408 PMID:11008807

Xu, F., & Ma, T. (2019). Modeling and Studying Acceleration-Induced Effects of Piezoelectric Pressure Sensors Using System Identification Theory. *Sensors (Basel)*, *19*(5), 1052. Advance online publication. doi:10.339019051052 PMID:30832233

Yan, J., Jin, H., Sun, H., & Qing. (2019). Active Monitoring of Fatigue Crack in the Weld Zone of Bogie Frames Using Ultrasonic Guided Waves. *Sensors (Basel)*, *19*(15), 19. doi:10.339019153372 PMID:31370343

Yu, H., Liu, Y., Tian, X., Zhang, S., & Liu, J. (2020). Sensors and Actuators A : Physical A Precise Rotary Positioner Driven by Piezoelectric Bimorphs : Design, Analysis and Experimental Evaluation. *Sensors and Actuators. A, Physical*, *313*, 112197. doi:10.1016/j.sna.2020.112197

Zahn, M. (2004). Arthur Robert von Hippel (1898–2003). *News & Views*, *3*(March), 2004.

Zaszczy, A., Gradys, A., & Sajkiewicz, P. (2020). Progress in the Applications of Smart Piezoelectric Materials for Medical Devices. *Polymers*, 1–19. PMID:33266424

Zhang, X., Qin, D., Chen, L., Liang, W., & Xu, J. (2018). Research on Recognition and Location of Rail Damage Acoustic Emission Signal Based on Dual Sensors. *7th Asia- Pacific Workshop on Structural Health Monitoring*, 1–6.

Zheng, Q., Shi, B., Li, Z., & Lin Wang, Z. (2017). Recent Progress on Piezoelectric and Triboelectric Energy Harvesters in Biomedical Systems. *Advanced Science News*, 1–23. . doi:10.1002/advs.201700029

Zhou, Q., Lau, S., Wu, D., & Kirk Shung, K. (2011). Progress in Materials Science Piezoelectric Films for High Frequency Ultrasonic Transducers in Biomedical Applications. *Progress in Materials Science*, *56*(2), 139–174. doi:10.1016/j.pmatsci.2010.09.001 PMID:21720451

Zhu, G., Lin, Z., Jing, Q., Bai, P., Pan, C., Yang, Y., Zhou, Y., & Wang, Z. L. (2013). Toward Large-Scale Energy Harvesting by a Nanoparticle-Enhanced Triboelectric Nanogenerator. *Nano Letters*, *13*(2), 847–853. doi:10.1021/nl4001053 PMID:23360548

Zu, H., Wu, H., & Wang, Q. (2016). High-Temperature Piezoelectric Crystals for Acoustic Wave Sensor Applications. *IEEE Transactions on Ultrasonics, Ferroelectrics, and Frequency Control*, *15261*(c), 1–55. doi:10.1109/TUFFC.2016.2527599 PMID:26886982

Chapter 12
Structural Health Monitoring by Electrical Resistance Change Method:
Using Nanoparticles

Mehmet Guclu
https://orcid.org/0000-0001-6611-1203
Istanbul University-Cerrahpaşa, Turkey

Fatih Turan
Eskisehir Technical University, Turkey

ABSTRACT

Structural health monitoring (SHM) is widely used to examine the structural integrity of a structure to improve safety considerations, minimize maintenance cost, and avoid sudden breakdowns that might occur under various loadings under service conditions. Fiber reinforced polymer (FRP) composites are widely used in aerospace, automotive, marine, energy, infrastructure, armor, and biomedical applications due to their high specific stiffness and strength, high degree of dimensional and thermal stability, and good resistance to corrosion. Thus, SHM of the damages developed in FRPs during service conditions is very crucial. However, traditionally used sensors for SHM in composites provide limited sensitivity and might negatively affect the structural integrity. Therefore, a self-sensing approach by evaluating the electrical resistance change (ERC) of FRPs has been successfully used by many researchers. The purpose of this chapter is to describe the ERC techniques using various nanoparticles such as graphene, carbon nanotube, expanded graphite, and carbon black.

DOI: 10.4018/978-1-6684-7358-0.ch012

INTRODUCTION

Fiber reinforced polymer (FRP) composites are widely used in aerospace, automotive, marine, energy, infrastructure, armor and biomedical applications due to their high specific stiffness and strength and excellent fatigue characteristics as compared to traditional materials such as steel and aluminum. In traditional materials, the crack initiates at the surface and progress through the thickness. Crack progression in the metals can be easily observed by visual inspection. However, damages in composite materials such as fiber breakage, matrix cracking, fiber-matrix debonding, and delamination between plies, usually occur under the surface during service condition and is not visible with the naked eye. These damages might reduce the mechanical performance of the composite material causing sudden catastrophic failure in FPRs. Thus, structural health monitoring (SHM) of FPRs is crucial to minimize the safety concerns and maintenance costs. Traditional SHM methods such as ultrasonic inspection, laser vibration, acoustic emission, X-Ray, eddy current, and thermal wave maps have been widely used for over 20 years to monitor the integrity of FRPs. These methods are working out of service conditions. However, critical damages in FRPs occur under operating conditions. Therefore, in-situ health monitoring techniques such as vibration analysis, piezoresistive strain gauges, fiber optic, and piezoelectric sensors have been used to continuously detect damage mechanisms in composite structures. However, these methods require too many sensors and might negatively affect the structural integrity of composites. The concept of damage detection by electrical resistance change (ERC) method, in which composite structures also act as a sensor, eliminates these disadvantages. The composite structures are required to possess sufficient electrical conductivity for successful application of ERC method, especially ones with non-conductive glass fibers reinforcement. In the following, the electrical conductivity of an insulating material by adding nanoparticles to composite materials and the in-situ damage progression of composites by SHM method is reviewed.

BACKGROUND

Structural Health Monitoring Methods

Traditional non-destructive testing methods such as ultrasonic, X-Ray have been used for many years for the detection of damages in composite materials. However, these methods are difficult to use due to complex materials, intensive and long labor, high costs and damage detection in certain areas. In addition, maintenance costs are high due to the materials and equipment used (Cai et al., 2012).

In the classical SHM method, different types of sensors are integrated into the structures to detect changes such as strain, temperature, vibration, and stress. For this, resistance change sensors, fiber optic sensors, piezoelectric sensors, eddy current sensors and microelectromechanical system (MEMS) sensors are used. The use of the classical SHM methods in composite materials is by active and passive ways, depending on the use of actuators (Cuc et al., 2007). In passive SHM, it is formed as a result of processing and interpreting the signals obtained by simply listening to the irregularities caused by various situations without any intervention (Figure 1- a). In the active SHM method, the data is created by the actuators and obtained as a result of the interpretation of the data by means of sensors (Figures 1-b).

Figure 1.

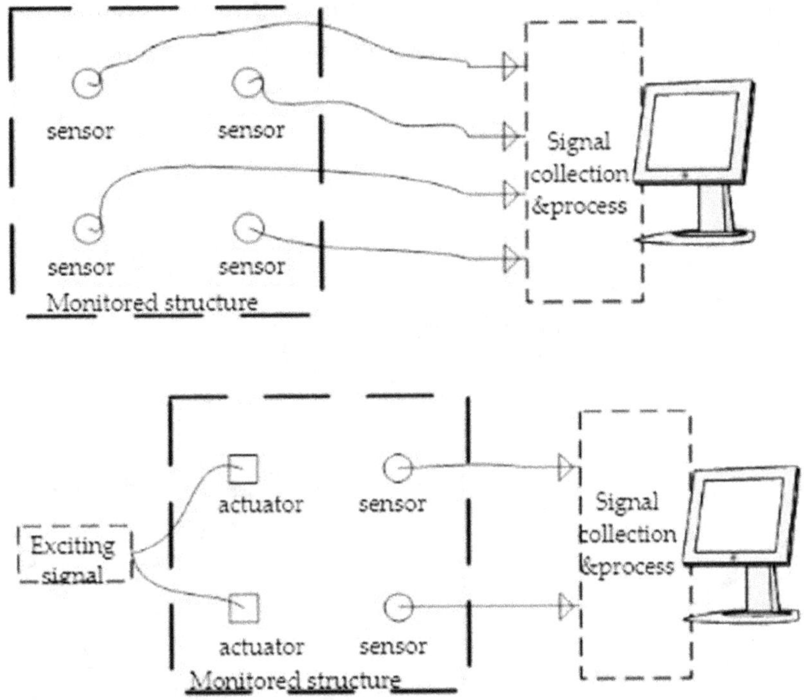

With the SHM method, accurate and reliable damage determination of the whole material can be made instantly and cheaper with the right designs. In the following, it has been investigated that the electrical conductivity of an insulating material by adding nanoparticles to glass fiber reinforced epoxy matrix materials and the damage progression of an insulating material by SHM method, in the whole part, with instant data. The electrical resistance changes of the network formed as a result of the addition of different types of nanoparticles (such as carbon nanotube, graphene, carbon black) into the epoxy matrix was investigated by structural health monitoring.

Electrical Resistance Change (ERC) Method

When the ERC method was first applied, damage detection in carbon fiber reinforced polymer composites was succeeded by utilizing the electrical conductivity of carbon fibers (J.-M. Park et al., 2007). However, the use of carbon fibers as sensors makes it possible to monitor the damage as a result of conductive carbon fiber breakages. The fracture of carbon fibers occurs at the last stage of the strength of the composite structure (J. M. Park et al., 2008) and therefore it is not possible to monitor the damages in matrix phase and fiber-matrix interphase by conductive carbon fibers. The required electrical conductivity for ERC method can be obtained by dispersing conductive fillers into matrix phase above the percolation threshold. The percolation threshold is defined as the minimum filler amount to form electrically conductive network. The percolation threshold may range from 0.01 wt% to 10 wt% depending on the various parameters such as filler size, filler geometry, and dispersion level. The conductivity is also governed by the interparticle distance which is known as tunneling effect. The electrons can bounce between nano particles when the distance between nano particles is sufficient for electrons to bounce between nano particles. The required distance, which is known as tunneling distance, varies between 1.7 nm (Mohiuddin & Hoa, 2013) and 5 nm (Chaurasia & Seidel, 2014) depending on the nano structure and polymer type. Figure 2 shows the schematic view of conductive network within a matrix phase.

Figure 2.

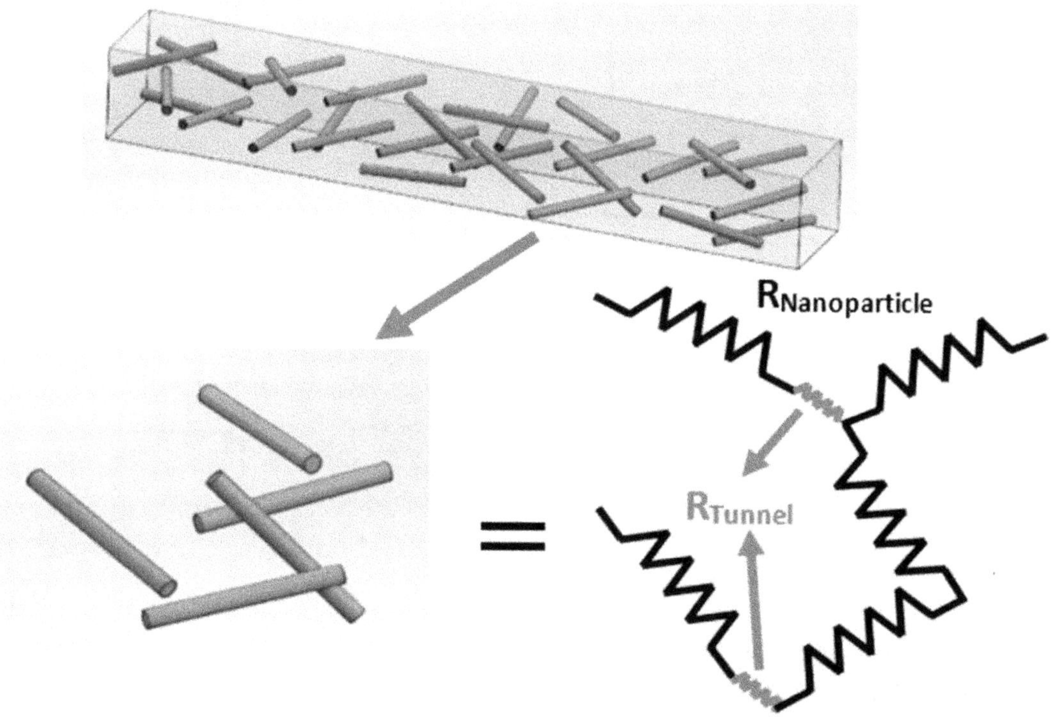

Below the percolation threshold, the composite material behaves like a capacitive material and conductivity is achieved by the tunneling effect between nanoparticles. Above the percolation threshold, the polymer exhibits a resistive property because the nanoparticles are in contact with each other throughout the entire volume. In a capacitive material, the conductivity varies depending on the frequency of the applied voltage, while in a resistive material, the conductivity is independent of the frequency of the applied voltage. For the successful application of ERC method, the structure should be resistive material. It is known that the electrical conductivity of the matrix increases with increasing nanoparticle content. However, it is not reasonable to use much higher number of nanoparticles than the percolation threshold for the following reasons: i) Increase rate in the conductivity is not exceptional after percolation threshold. ii) Introducing high number of nanoparticles might increase the possibility of forming nanoparticle aggregates which act as stress concentrations weaking the mechanical property of the composite structure. On the other hand, a noisy electrical signal during resistance measurement may cause instrumentation problems below the percolation threshold. Therefore, the percolation threshold is an important parameter for the successful application of ERC method.

The modeling of the nanoparticle added composite material as an electrical circuit is given in Figure 3. Figure 3a shows the situation where percolation behavior is provided (resistance behavior) and Figure 3b shows the situation where percolation is not fully achieved (capacitive behavior).

Figure 3.

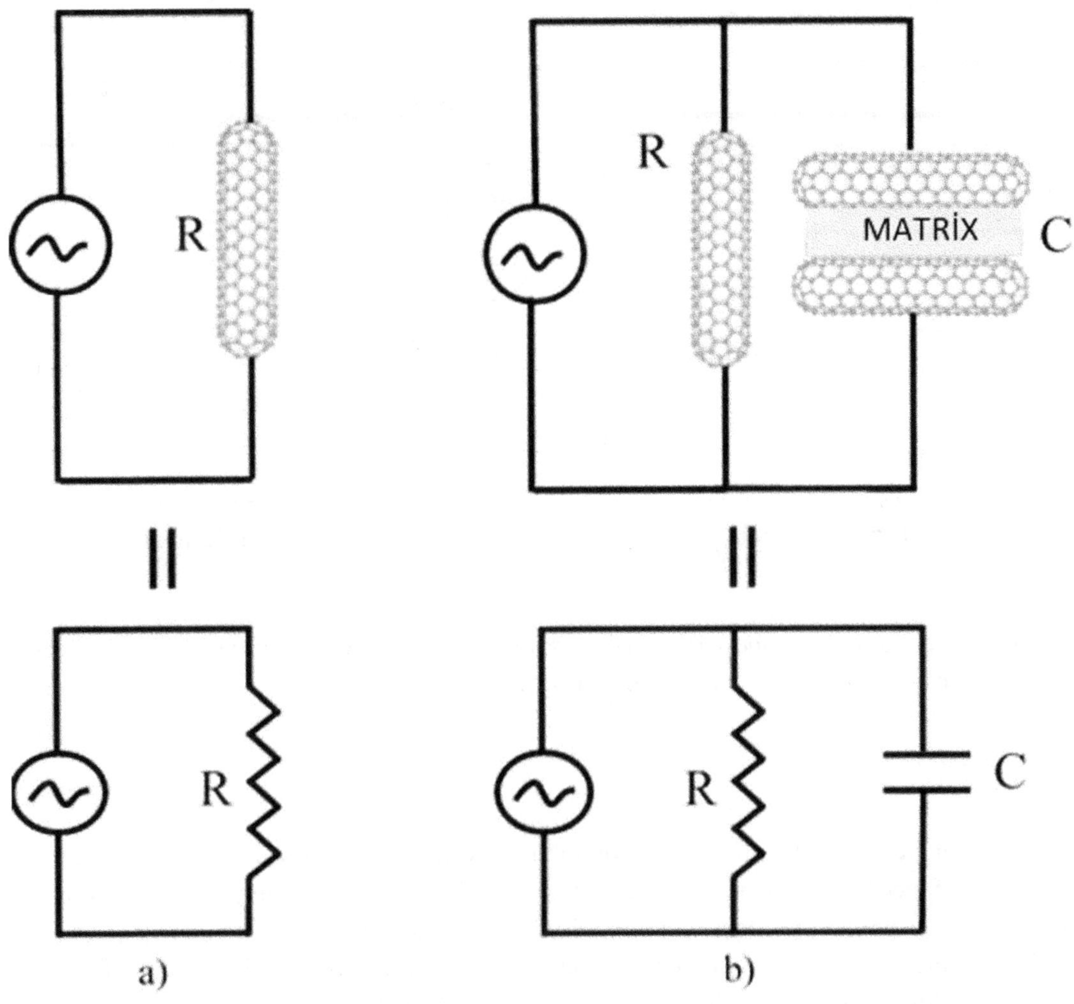

Figure 4 also shows the graphical representation of capacitive and resistive materials with corresponding CNT dispersion states. In Figure 4, the conductivity increases with increasing frequency for the materials, a and b indicating that they are capacitive materials. Consequently, the percolation threshold of nanoparticles to form conductive network is not achieved in a and b materials. On the other hand, the conductivity exhibits no dependence on the frequency indicating that the percolation threshold is achieved throughout the entire structure for the resistive materials c, d, and e.

Figure 4.

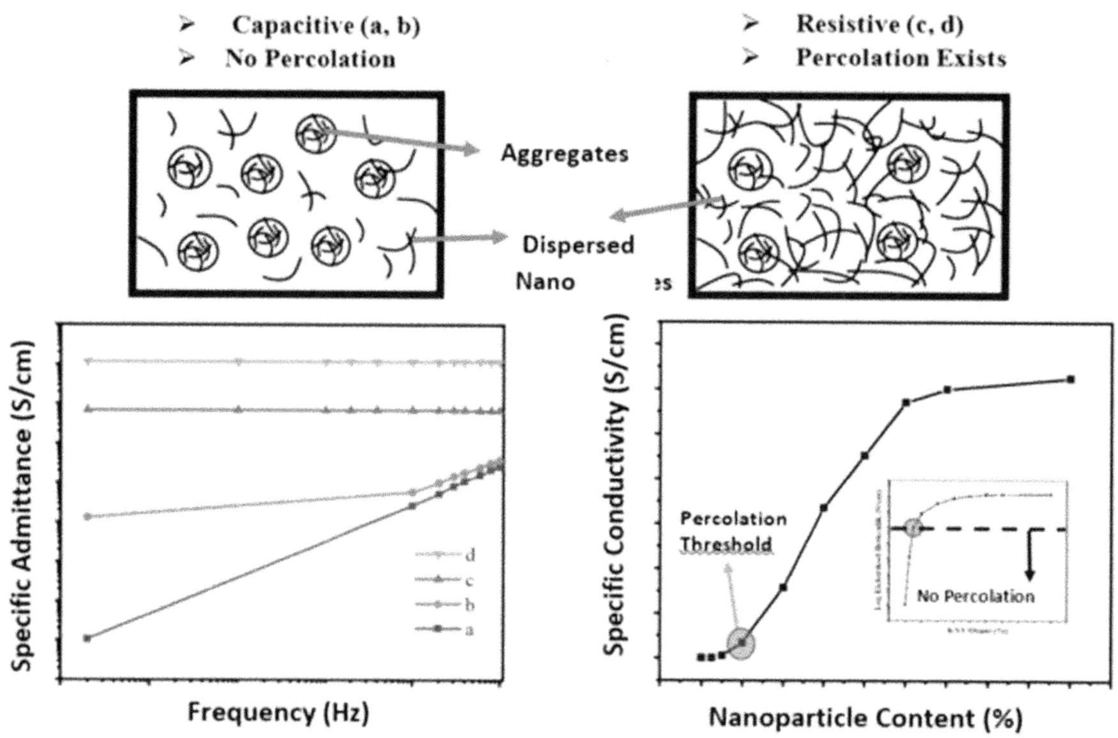

The conductivity values of nanoparticle-filled samples under alternating current depending on the frequency (f) are expressed with a complex equation 1.1:

$$\sigma^*(w) = \sigma' + j\sigma'' \tag{1.1}$$

In this equation, $\sigma¢$ and $\sigma¢¢$ represent the conductivity value of the resistive CNTs and the capacitive part, respectively. Since CNTs can be modeled as resistors, there is no frequency-dependent conductivity (Equation 1.2), whereas in capacitors, conductivity can vary depending on frequency (Equation 1.3). The specific conductivity at any frequency is obtained after taking the absolute value of the complex conductivity and dimensioning it with the sample geometry (Equation 1.4).

$$\sigma' = \frac{1}{R} \tag{1.2}$$

$$\sigma'' = 2\pi fC \tag{1.3}$$

$$\sigma(w) = |\sigma^*| \frac{l}{A} = \sqrt{\sqrt{(\sigma')^2 + (\sigma")^2}} \qquad (1.4)$$

According to Ohm's Law, electrical resistance (R, ohm) is the ratio of the voltage across a conductor to the current flowing through it (Eq 1.5). In this equation, E (volt) and I (amper) represent the applied voltage and the measured current, respectively.

$$R = \frac{E}{I} \qquad (1.5)$$

In a conductive material, R depends on the specific resistivity (ρ), conductor length (l) and cross - sectional area (A) (Equation 1.6).

$$R = \frac{\rho l}{A} \qquad (1.6)$$

The inverse of the specific resistivity gives the specific conductivity of the material (Equation 1.7);

$$R = \frac{1}{\rho} \qquad (1.7)$$

The unit of specific conductivity is S/m (Siemens/meter).

Once the composite becomes electrically conductive, change in the electrical resistance by the application of loading can be used as an effective tool for self-sensing of damage monitoring. The contact points between conductive fillers decrease upon the application of load resulting increase in the electrical resistance of the composite structure as shown in Figure 5. Beside the reduction in contact points between nano particles, the increase in the tunneling distance due to applied strain may also cause increase in the resistance of polymer nanocomposites.

Figure 5.

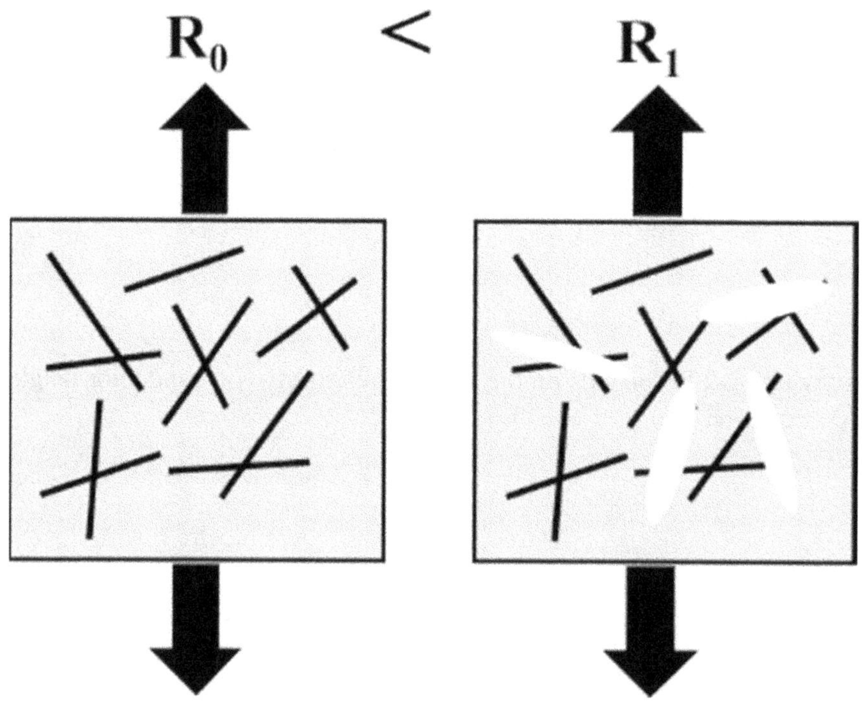

Dispersion Methods

Nanoparticles are required to well dispersed in order to achieve percolating network in a matrix phase for desired functionality (Thostenson & Chou, 2006). There are many dispersion methods of nanoparticles in thermoset based matrix such as high shear mechanical mixer (Bryning et al., 2005; Sandler et al., 1999), ultrasonication (Lau et al., 2005; Miyagawa & Drzal, 2004), and calendaring (Gojny et al., 2004). As a general procedure for all methods, nanoparticles are first added to the matrix material (Figure 6a) and mixed by hand to wet it using a stick (Figure 6b). Then, the dispersion process is followed by using a dispersion method (Figure 6c). Finally, the hardener is placed and the curing process is carried out (Figure 6d).

Figure 6.

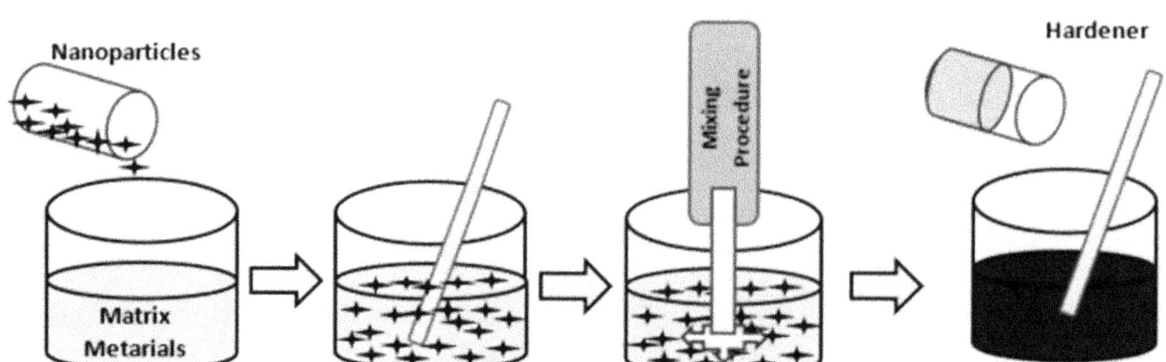

The mechanical mixer provides the mixture of nanoparticle and matrix material as a result of the rotation of the rod attached to the end of a high-speed motor as shown in Figure 7a. Size and shape of the propeller, mixing speed, and mixing time are the main parameters that affect the dispersion quality of the nanoparticles in viscous resin medium. Although this method has a practical use, it is difficult to disentangle the agglomerates of highly reactive CNTs within a highly viscous polymer matrix.

Ultrasonication is one of the most commonly used methods for dispersing micro- and nano-sized particles in a solution. Ultrasonication is performed using an ultrasonic bath or an ultrasonic probe. The schematic of the ultrasonic probe method is shown in Figure 7b. In ultrasonication, ultrasound waves are applied to the mixture at a frequency of up to 20000 cycles per second. As the frequency amount increases, the dispersion rate of nanoparticles also improves. In ultrasonication, sound waves propagate through the liquid by creating high pressure and low pressure cycles. In the low-pressure cycle, small gas bubbles are formed in the liquid. These gas bubbles form cavitation by bursting in the high-pressure cycle. Thus, as a result of acoustic cavitation, the pellets in the solution are disintegrated. The parameters that affect the dispersion in ultrasonication are power, time and probe surface area (Shojaeiarani et al., 2020; Zhao et al., 2013) . In the ultrasonication method, the mixture must be of low viscosity for effective dispersion of nanoparticles. For this reason, in high viscosity nanoparticle/matrix mixtures, the viscosity is reduced by diluting it with a solvent compatible with the polymer. Long-term mixing improves the dispersion of nanoparticles in the viscous medium. However, long-term, high-power application to the mixture increases the temperature and damages the structure of CNTs (Lu et al., 1996). For this reason, it is necessary to provide the mixture with optimum parameters so that CNTs do not lose their properties.

In the calendaring process, mixing is performed by allowing the matrix material with nanoparticles to pass between the shafts between 3 harmonious working spindles as shown in Figure 7c. While the material is passing through, a large shear force is applied, and the dispersion process takes place. The distances between the shafts are in micro dimensions and the speed of each shaft is gradually increasing.

Figure 7.

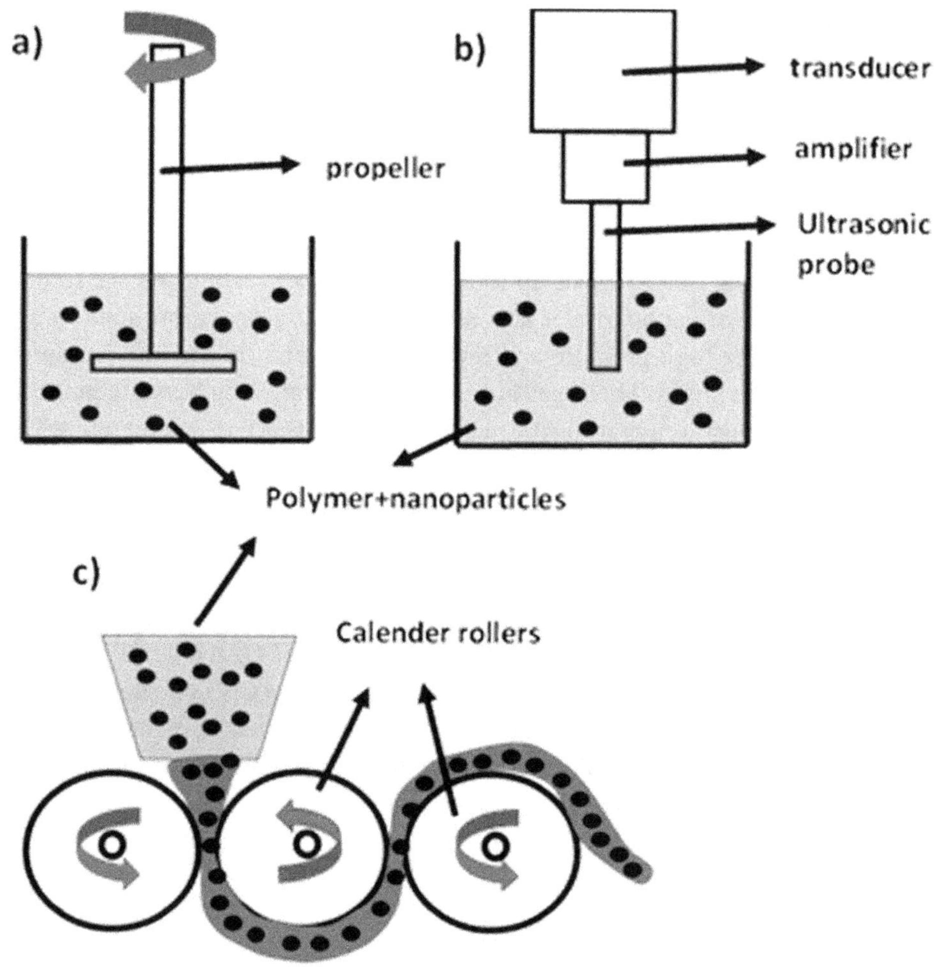

The mixing process is very important to take advantage of the electromechanical properties of nanoparticles. The mixing procedure varies depending on the mixing process used, the mixing time, the weight ratio of nanoparticles, the substance and amount used for the proportional distribution of the mixture, the heating-cooling procedure during mixing, and the speed of the mixing process. While increasing the mixing time increases the dispersion of nanoparticles, it causes a decrease in the length/diameter ratio of the nanoparticles, resulting in a decrease in their electrical and mechanical properties. In addition, if the number of nanoparticles is higher than the threshold value, it will cause aggregation of the nanoparticles in the matrix material, resulting in a decrease in their mechanical properties. The weight of nanoparticles must have a minimum value at which electrical conductivity is achieved. This value is called the percolation value. In order to obtain all these values, it is necessary to test the experiments at different rates and evaluate the results.

Piezoresistivity

In general, carbon based fillers such as carbon nanotubes (CNTs), carbon black (CB), graphene sheets, and graphite nanoplates (GNPs), and expanded graphite (EG) are used to monitor strains and damages in polymeric composites due to their high inherent electrical conductivity. Among them, CNTs, CNFs, GNPs, and graphene sheets possess excellent electrical conductivity and large aspect ratios. Therefore, they are more commonly used for SHM applications in composites. The nanoparticle modification not only offers enhancement in the electrical conductivity for ERC application but also improvement in the mechanical performance of polymer composites.

The ERC method works based on the piezoresistivity of a material. Mechanical stimulus such as strain and stress results with the change in the electrical resistance (ΔR) in a piezoresistve material. Commercial strain gages made of semiconducting foils work based on piezoresitive principle. The degree of piezoresistivity of a material is quantified by a parameter known as gage factor (GF). GF is calculated using equation 1.8.

$$GF = \frac{\left(\dfrac{\Delta R}{R_0}\right)}{\varepsilon} \tag{1.8}$$

where, R_0 and ε represent the initial resistance and the applied strain, respectively. GF of traditional metallic strain gages may range between 2.1 and 2.3. In nano composites, piezoresistivity depends on filler material, polymer material, the dispersion state of filler, and the amount of filler material within matrix phase. Although the inherent piezoresistivity of nano fillers are quite high (Coetzee et al., 2020), the piezoresisitive behavior of nanoparticle modified polymer composites is much lower than the pure nano fillers. This is mainly due to the fact that the applied strain is not fully transferred to the filler material itself. Numerous works have been published to understand the piezoresisstive effect of nanocomposites including tunneling, nanoparticle interaction, nanoparticle morphology, and loading conditions (Celzard et al., 1996; Ren et al., 2015; Sun et al., 2009).

The aspect ratio (length/diameter) is one the most important parameters that affect the percolation threshold and piezoresistive behavior of nanocomposites (Bauhofer & Kovacs, 2009). The larger the aspect ratio, the lower the percolation threshold and higher the contact points between nano particles within polymer phase.

It should be noted that piezoresistivitiy is the changes in the electrical resistance under applied strain. However, the changes in the electrical resistance due to damages such as matrix cracking, delamination, and fiber fractures is not considered piezoresistivity. Therefore, this aspect of nanocomposites can be used as structural health monitoring technique.

SHM OF NANOCOMPOSITES USING ERC METHOD BY DIFFERENT NANOPARTICLES

SHM Using Carbon Nanotubes (CNT)

CNTs were first studied by Sumio et al. (Iijima, 1991). After this observation, Ebbesen and Ajayan (Ebbesen & Ajayan, 1992) conducted studies on large-scale CNT synthesis. These developments have been a turning point for future studies in terms of synthesis and characterization of CNTs. Among the conductive fillers, carbon nanotubes (CNTs) have been extensively used nano fillers in conductive composite structures thanks to their extraordinary mechanical, electrical and thermal properties. Theoretical and experimental studies show that CNTs have a modulus of elasticity of around 1 –1.8 TPa and a tensile strength in the range of 10 –63 GPa (Ruoff et al., 2003). CNTs obtain these superior mechanical properties thanks to their sp2 atomic structure. CNTs are cylindrical graphene structures formed by the bonding of carbon atoms to each other in a hexagonal structure and this structure periodically repeating itself. As a result of this periodicity, each carbon atom is bonded to three adjacent carbon atoms (Anantram & Léonard, 2006). This covalent bond, also called sigma(σ) bond, is a strong chemical bond and plays an important role in the superior mechanical performance of CNTs (Figure 8).

Figure 8.

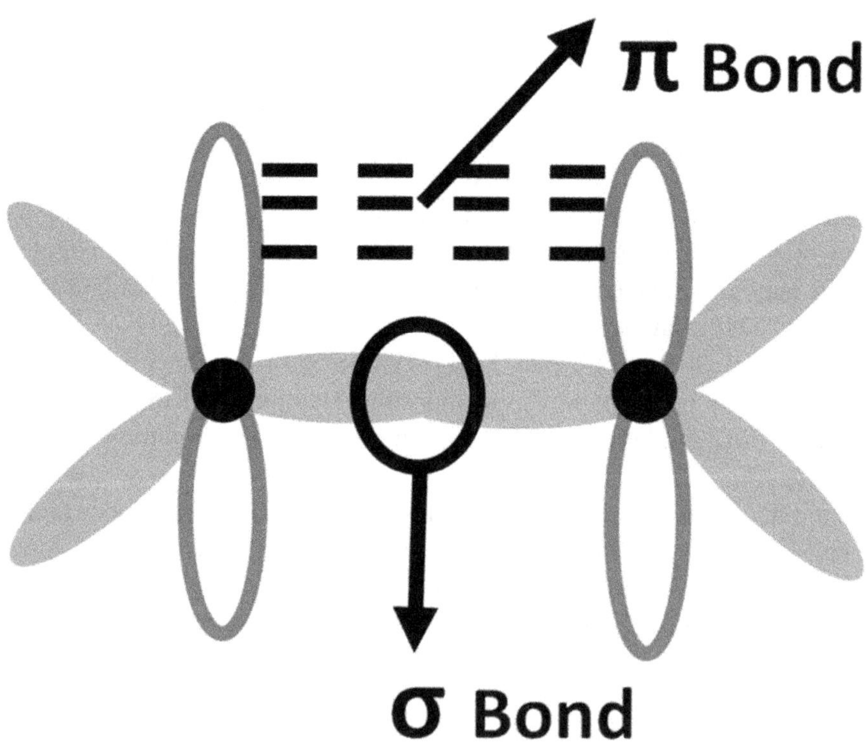

In addition to their superior mechanical properties, CNTs also have excellent electrical and thermal properties. π orbital, which is perpendicular to the graphene plane and therefore to the nanotube surface, creates a high electron density π network, forming the electronic property of CNTs (Gojny et al., 2004b). There are four electrons in the outermost shell of each carbon atom. While three of them are used to form covalent bonds with other carbon atoms, a single electron remaining in the π orbital provides electrical conductivity. Since CNTs have an extremely large aspect ratio, surface area, and superior inherent electrical conductivity, the percolation threshold can be reached at extremely small CNT concentrations, lower than 0.1 wt% .

CNTs, as in the figure 8, are formed by wrapping a graphene plane consisting of a single row of carbon atoms. In this structure, the chiral vector is the winding vector and is expressed as:

$$Ch = na1 + \mathrm{m}a2 \quad (1.9)$$

In Figure 9 and equation 1.9, vectors $a1$ and $a2$ are the unit vectors of the hexagonal lattice structure in the graphene plane and θ is the angle of the chiral vector with respect to the X-axis. The integers n and m show how CNTs are bent (Dresselhaus et al., 1995). Studies (Chen et al., 2004a; Mintmire et al., 1992) show that CNTs exhibit metallic or semiconductor behavior depending on their (n, m) coefficients.

Figure 9.

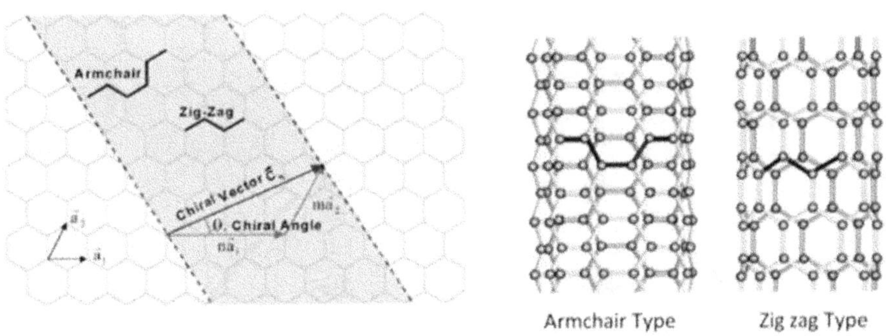

While All Seat type (n=m and θ=30°) CNTs show metallic properties, zig zag type (m=0 and θ=0°) and chiral type (0<θ<30°) CNTs, if (n-m)/3 ratio is an integer, it shows metallic properties.In all other cases, CNTs exhibit a semiconducting behavior (Thostenson et al., 2001).

In the literature, by using the electromechanical properties of CNTs, the homogeneous distribution of the polymer matrix composite material is ensured, and the deformation and damage mechanisms that occur using the piezoresistive sensor feature have been controlled (Houshyar et al., 2005; B. Kim et al., 2003).

Since CNTs were first introduced into FRPs as sensors for SHM purposes by Fiedler et al. in 2004 numerous works related to the use of CNTs as sensor within FRPs for ERC application have been published (Gao et al., 2009b; Paleari et al., 2021b; Thostenson & Chou, 2008c, 2008d). In general, these studies show that in-situ monitoring of different types of damage mechanisms under various loading conditions such as static, fatigue, and impact loads is possible by ERC method. Moreover, strength and stiffness degradation due to fatigue loading can be determined using ERC method.

Alexopoulos et al., 2010 used CNT embedded glass fibers for SHM of the composite material. The purpose was to enhance both the mechanical behavior and electrical response of the composite specimens.

It was found that CNT embedded fiber did not degrade the mechanical properties. Direct correlation between mechanical loading and the electrical resistance change was successfully established.

Thostenson & Chou, 2008 carried out research to investigate the onset of local damage in fastened joints, such as delamination and cracking by ERC method using carbon nanotubes as in-situ sensors. Figure 10a shows the schematic of the experimental setup used in this study. Figure 10b and 10c show the load-displacement and electrical resistance change response of a single lap joint and double lap joint, respectively. The electrical resistance linearly increases with increasing applied load at the initial state. Then the linearity in the electrical resistance change curve changes to nonlinear response as a result of damage accumulation at higher loads. Finally, sudden sharp increase in the resistance curve is seen at the peak load that indicates the failue of the joint.

Figure 10.

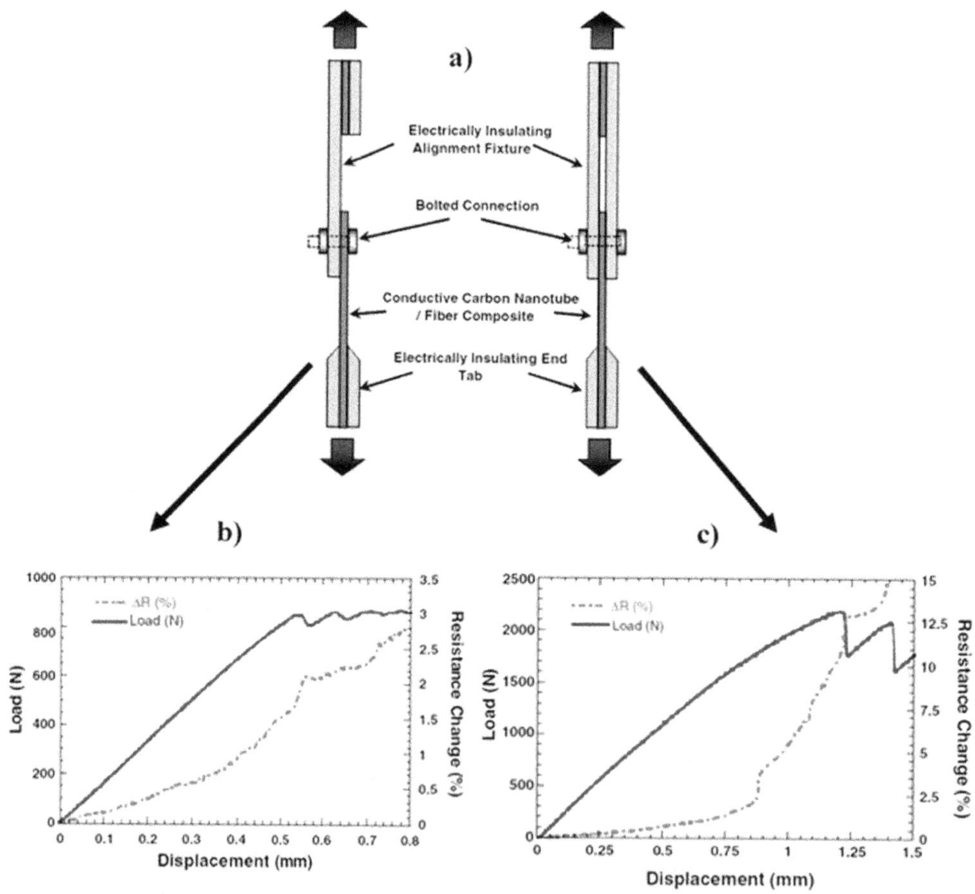

Thostenson & Chou, 2008 utilized conductive networks of carbon nanotubes as in situ sensors for detecting damage accumulation during cyclic loading of advanced fiber composites. The in-situ resistance measurement during fatigue loading was combined with the load-deflection data to gain insight toward the evolution and accumulation of damage. At first two cycles where the applied strain is low, the electrical resistance change curve shows a reversible response as well as it follows the strain as shown in Figure 11a and b. However, when the strain level is increased at the following cycles, the electrical resistance change curve shows irreversible response indicating that the some damages such as matrix cracking and fiber breaks occurs within the composite.

Figure 11.

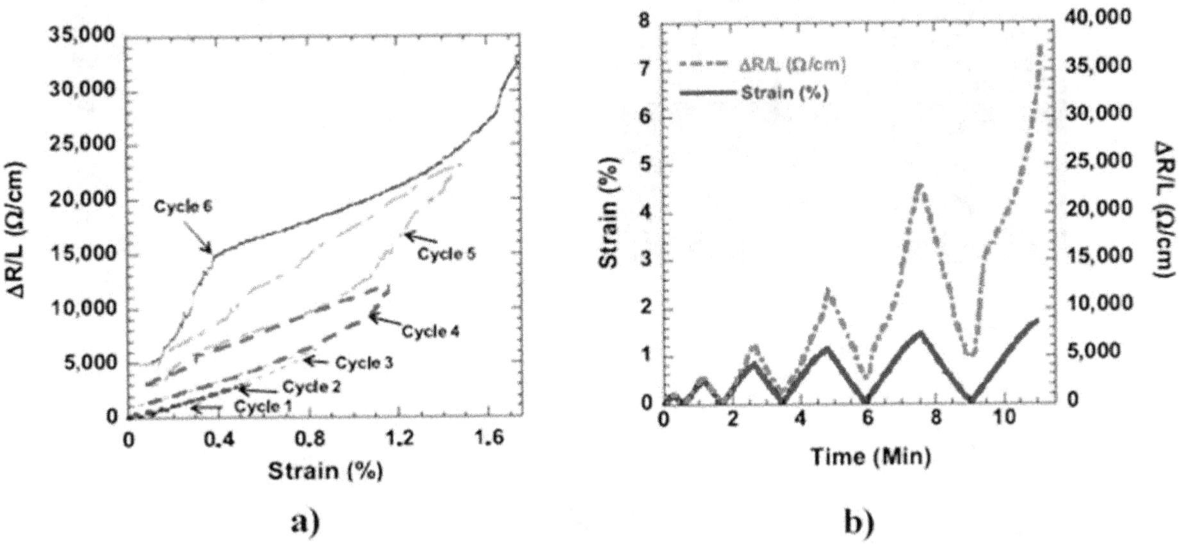

(Gao et al., 2009) combined the electrical response of composite structures with the acoustic emission data to sense the damage initiation and evolution in laminated composites. (Paleari et al., 2021) used carbon black (6 wt.%) and multiwalled carbon nanotubes (0.75 and 1 wt.%) to develop self -monitoring plies to be used in glass fiber reinforced epoxy-based composites in order to conduct structural health monitoring under three-point bending tests as well as ball drop impact tests. The results showed that CB based monitoring system exhibits higher sensitivity with the gauge factor of 6.1. Viets et al., 2014 carried a study to determine damage mapping in glass fiber reinforced polymer composites under impact load using in-plane and through-thickness electrical resistance measurements. CB and MWCNT were used as filler to obtain conductivity in the composite structures. Electrical measurement results were compared with ultrasonic C-scans of the damage area as shown in Figure 12. The results showed a good correlation in the positions of the impact area. Also, different type of nanoparticle modifications of the composite-laminates resulted in a significant variation of the sensitivity of the damage mapping method.

Figure 12.

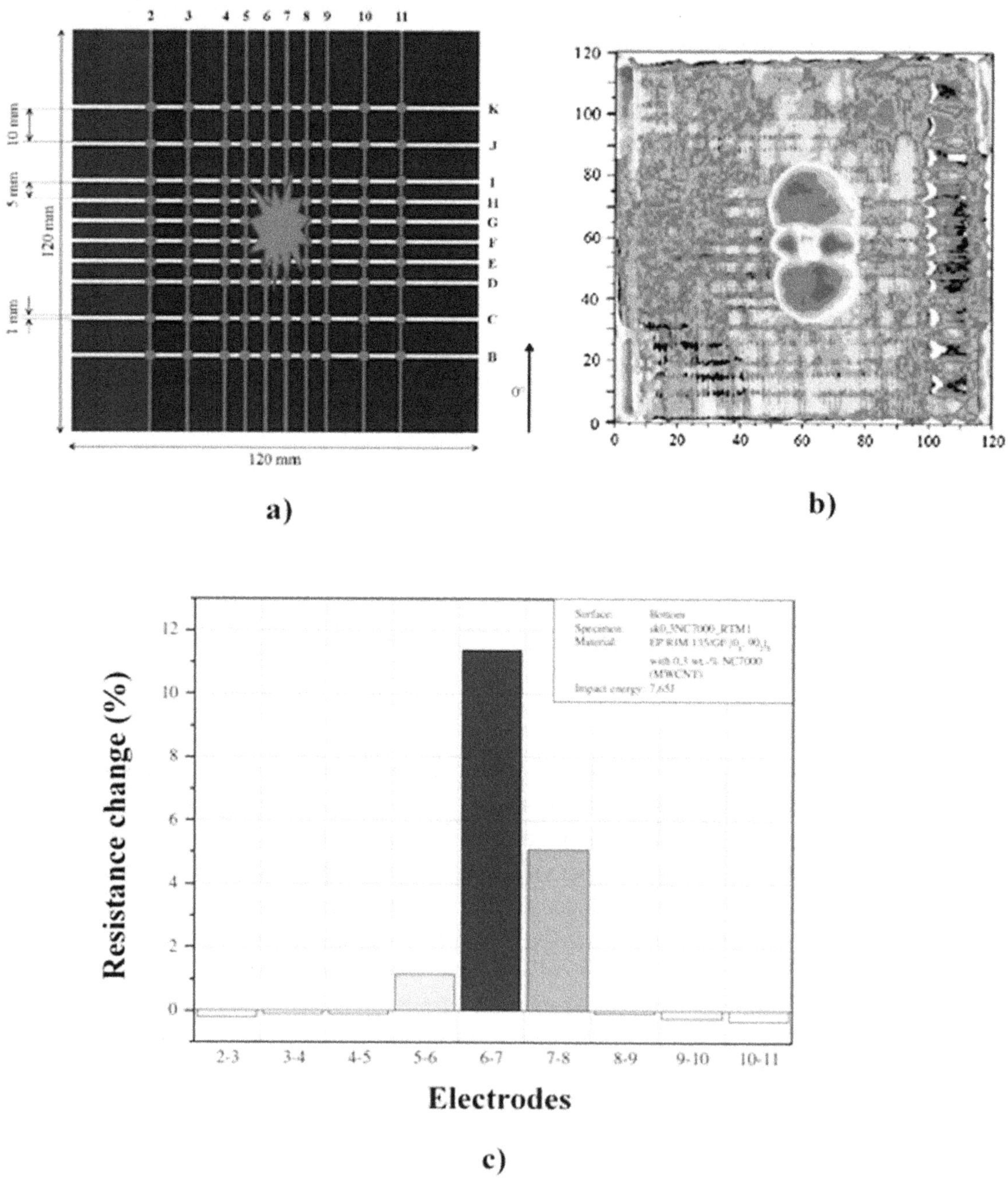

Liu et al., 2012 investigated the feasibility of ERC method on single-lap joints subjected to tensile loading using CNTs. The results showed that conductive bonded material is sensitive to crack propagation within the matrix material. S. Kim et al., 2018 evaluated the oxidation effect on the interfacial, electrical, and mechanical properties of polyurethane (PU)/MWCNT aircraft topcoat layers. It was found that electrical resistance increased with increasing oxidation time.

Since the dispersion of CNTs within polymer at high concentrations is difficult, a thin sheet made from well controlled dispersion of CNTs called buckypaper has been developed and used as sensing material in composites. Kravchenko et al., 2018 used conductive interlaminar interfaces made of CNT based buckypaper films in glass fiber reinforced composites for SHM under fatigue loading in 3-point bending configuration using ERC methods. The concept of using conductive interface showed to be an effective tool to assess the structural integrity of composites.

SHM Using Graphite (EG) Based Nanoparticles

Nano graphite platelets (NGPs) are stack of monolayer carbon atom flat-structure graphene (Wang et al., 2010). They are 2D carbon nanomaterials with nano scale thickness. Since the electrical resistivity of NGPs is only about 10–6 Ω cm which is lower than that of copper and silver, NGPs are excellent conductive filler to be used as sensor material for successful SHM application of composites using ERC method.

Expanded graphite (EG) is one of the multifunctional carbon-based nanoparticles. EG is a material with low density, high surface area, high thermal and electrical conductivity (Yasmin et al., 2006). Many studies have been conducted on the use of EGs as fillers in polymer composites for various functional applications. These studies include heat resistant resistors, sealing gaskets, and flame resistant composites (Mazela et al., 2020; Zheng & Wong, 2003). The structure of natural graphite consists of a large number of graphene layers connected by van der Waals bonds (Chen et al., 2004). The production of EG is briefly as follows: First of all, natural graphite is subjected to chemical oxidation with the ultrasonication process with H2SO4 and HNO3 acids, and H2SO4 and HNO3 molecules are placed between the graphene sheets. This process is called graphite intercalation. Intercalated graphite is exposed to high temperature (600 °C) in a furnace. H2SO4 and HNO3 intercalation molecules trapped between the graphene layers due to the effect of high temperature decompose, exerting force on the graphene sheets and allowing the sheets to expand throughout the thickness (Figure 13). This expansion can be between 80-100 times. As a result of expansion, accordion-shaped, high surface area, porous and low density EG is obtained. The distance between graphene layers in EG material can vary from 2 nm to 10 μm.

Figure 13.

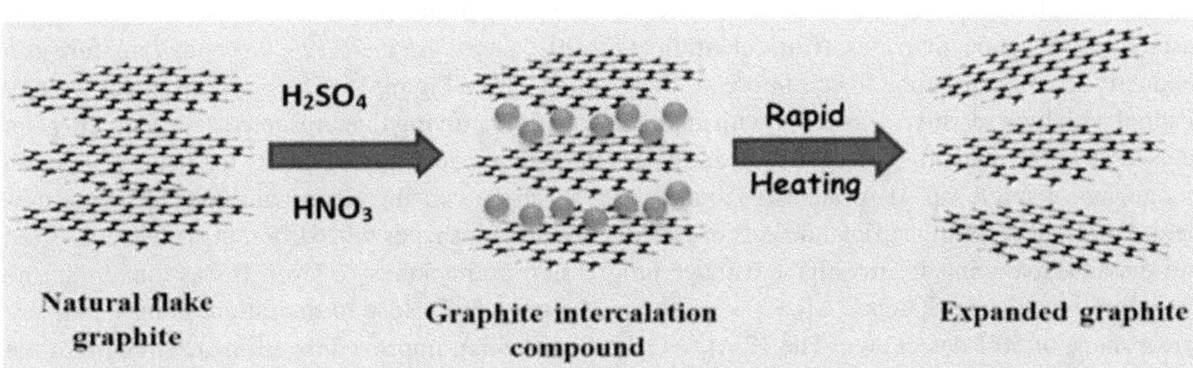

(Spinelli et al., 2020) studied the self-sensing behavior of epoxy nanocomposites using CNT and EG particles. Figure 14 shows the electrical resistance change response of 2 wt. % MWCNT and 2 wt.

% EG under static loading condition. The gage factor of 3.9 and 39.7 were determined for composites filled with CNT (1D) and EG (2D) particles, respectively. The higher sensitivity of 2D systems than 1D system is due to the fact that there is a strong reduction in the conducting pathways of 2D system. On the other hand, the densely packed conducting pathways in the conducting network in 1D system is slightly reduced under deformation.

Figure 14.

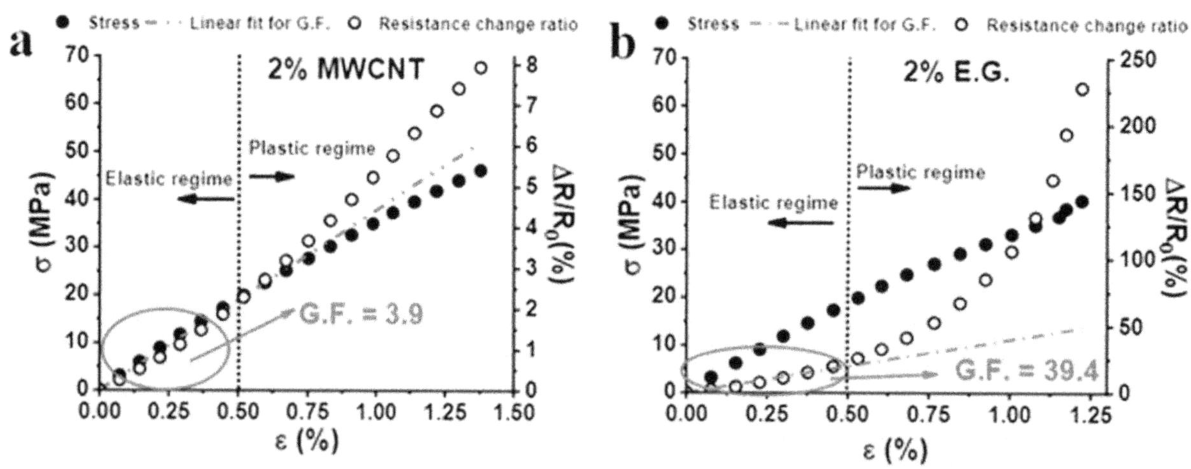

SHM Using Graphene (G) Nanoparticles

Graphene thanks to its high aspect ratio, excellent electrical conductivity, and high mechanical properties is suitable for many functional applications of polymer nanocomposites in automotive, aerospace, electronics, and packaging.

Eswaraiah et al., 2011 developed lightweight and flexible strain sensor using functionalized graphene (f-G) modified polyvinylidene fluoride (PVDF) by solvent casting technique. The study revealed that 2 wt% f-G–PVDF composite films exhibited better strain sensing performance than that of its counterparts based on CNT nanocomposites. (Chiacchiarelli et al., 2013) used graphene (2% wt) epoxy mixture as a conductive coating in carbon fiber reinforced composite as seen in Figure 15. The suggested methodology resulted with high resistive sensitivity with gage factor of 56.7 proving that presented sensor coating can be used as SHM method. (Sam-Daliri et al., 2020 studied the capability of GNPs for structural health monitoring of single lap Al-Al adhesive joints under shear stress using electrical impedance measurement. Different types of artificial defects of circular and square shapes with different sizes are inserted into the adhesive joints to investigate damage progression in the adhesive layer. It was found that the maximum impedance changes (ΔR/R) were obtained almost 60% close to the failure point for square defect shape of 50% defect area. The 12 wt.% GNP modification improved the ultimate strength of the adhesive layer about 74%.

Figure 15.

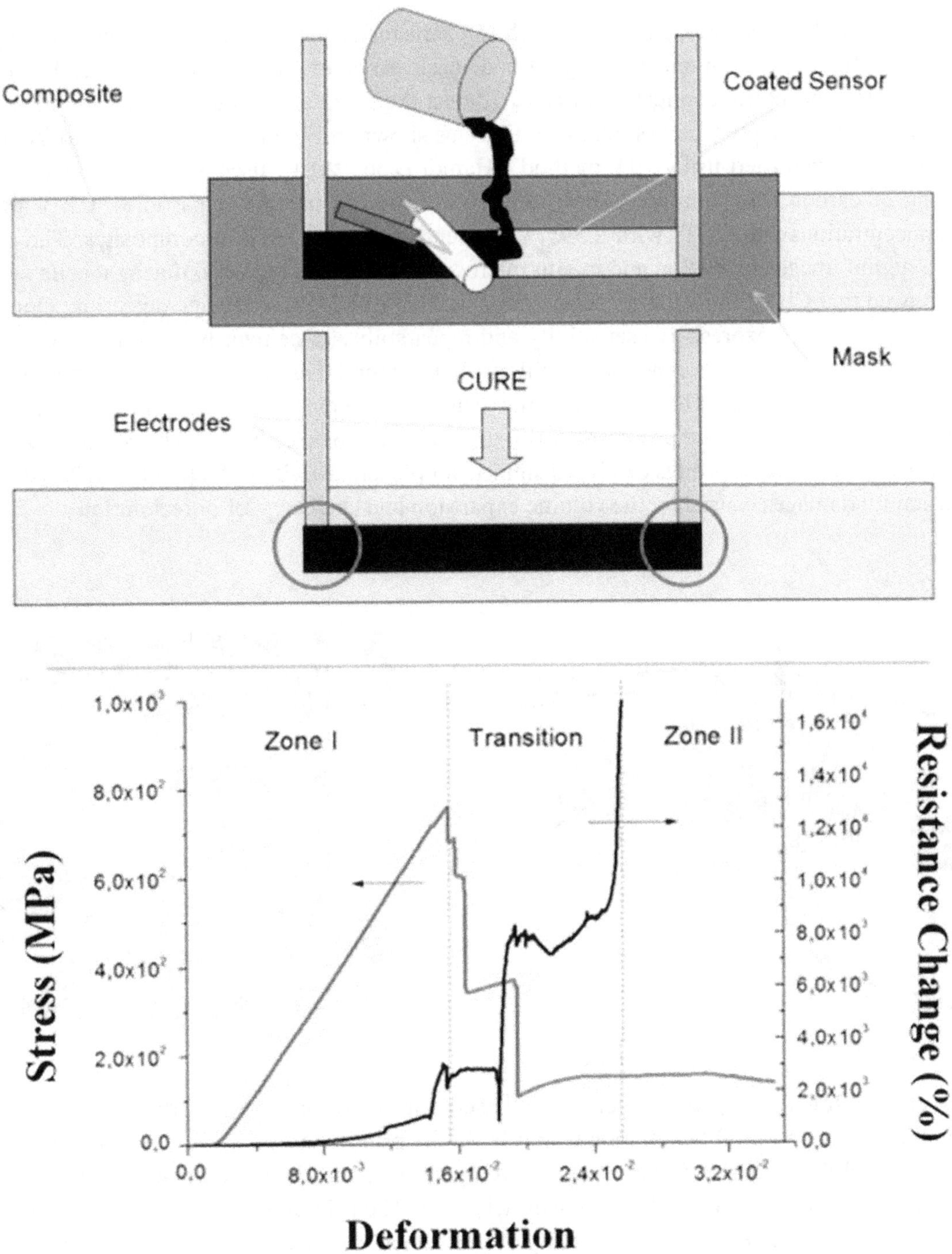
Composite
Coated Sensor
Mask
CURE
Electrodes
Zone I
Transition
Zone II
Stress (MPa)
Resistance Change (%)
Deformation

SHM Using Carbon Black(CB) Nanoparticles

(Tallman et al., 2015) has carried out a research to characterize the electrical impedance tomography (EIT) of glass fiber/epoxy laminates using carbon black nanoparticles as sensor material. The study demonstrated the successful application of EIT to detect the hole damages in the fiber reinforced composites filled with conductive carbon black particles as shown in Figure 16. Also the damage location was successfully identified using EIT method. (Hamdi et al., 2020) used ERC method for damage monitoring of carbon fiber reinforced Polyamide 6 composites using CB particles. CB with 0 and 8 wt% concentrations and CNTs with 2.5 wt% concentration were used in composites. The acoustic emission, digital image correlation and in-situ microscopy were also recorded during tensile tests. The results showed that CB exhibits higher sensitivity than that of CNTs to damage detection. Dong et al., 2019 investigated the piezoresistive sensitivity and repeatability of cement-based sensors with carbon black (CB) under various environmental conditions. It was found that the water content of cementitious composites caused significant fluctuations on the resistivity and piezoresistivity, and the optimal water content for cement-based sensor possesses was found to be approximately 8%. The results also showed that the water saturated composites exhibited significant piezoresistivity reduction by 30.7%, due to the microstructural damages caused by the volume expansion and shrinkage of pore solution.

Figure 16.

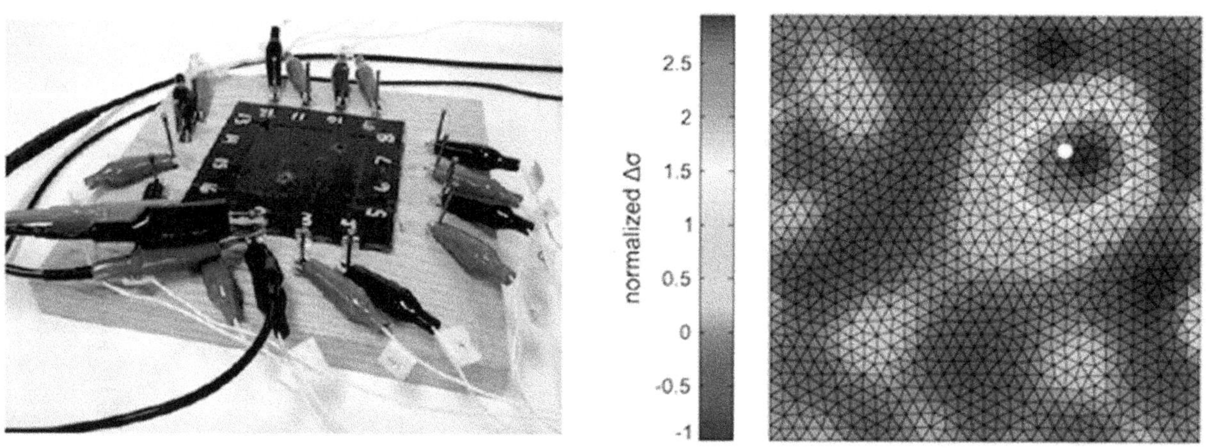

(Liao et al., 2021) developed nanocomposite-based sensor ink made from carbon black and polyvinyl pyrrolidone for structural health monitoring by inkjet printing approach. Thanks to having strong structural adaptability and high flexibility the developed sensor can be directly deposited onto the surface of carious substrates to form dense sensor networks. The printable sensor is lightweight and can be deployed at a large scale. The sensing ability of this sensor was successfully verified experimentally from quasi-static up to 500 kHz. (Ke et al., 2016) used hybrid filler systems consisting of CNTs, at the concentrations of 0.5–1 wt % and CB at the concentrations of 0.5–4 wt % to tune both electrical conductivity and piezoresistive sensitivity of melt mixed poly(vinylidene fluoride) PVDF nanocomposites for SHM applications. It was found that CNTs filled PVDF composites have higher electrical conductivity than that of CB filled PVDF composites at the same filler content. This is due to the fact that CNTs have larger aspect ratio than CBs. However, piezoresitivity sensitivity is larger when used CBs. Therefore, tunable electrical conductivity and strain sensitivity was achieved by combining the advantages of CNTs and CB via the adjustable conductive network structure. The results also showed that lower CNT/CB ratio results in higher strain sensitivity.

Foteinidis & Paipetis, 2021) reported 2D and 3D impedance mapping tomography of high velocity impacted glass fiber epoxy composites using hybrid modification of CNT and CB particles. 0.5% wt MWCNTs and 2% wt CB were dispersed in the epoxy resin to exceed the percolation threshold for successful application of ERC method. Yin et al., 2017) reported a highly durable ternary conductive nanocomposite made of polydimethylsiloxane (PDMS), CB and MWCNTs to fabricate stretchable strain sensors to be used in a flexible aircraft. The results showed that the nanocomposite with homogeneous hybrid filler of 1.75 wt % CB and 3 wt % MWCNTs exhibits a highly strain sensitive characteristics of good linearity. Monteiro et al., 2017 investigated the sensing capability of piezoresistive cement-based composites loaded with CB particles for real time service monitoring of civil infrastructures. Composites filled with CB particles up to 10% by weight were mechanically tested under cyclic uniaxial compression and the response of resistance change was recorded. A reversible piezoresistive behavior and a quasi-linear relation between the fractional change in resistivity and the compressive strain was found. It was concluded that CB/cement composites can be used as effective tool to monitor compressive strain in civil infrastructures.

CONCLUSION

The concept of the structural health monitoring of FRPs by ERC method was introduced. Examples of ERC methods using CNTs were reviewed. The electrical and mechanical properties of CNTs were introduced. The percolation threshold and the drawbacks of using much nanoparticles were also discussed.

The application ERC method to monitor damage level in large scale composites such as airplane wings and wind turbine blades is crucial for the enhancement of safety and minimize the maintenance cost. However, one of the most important drawback of the successful application of ERC method on the large scale products is the difficulty of uniform dispersion of nanoparticles within matrix phase. Once the practical way of uniform dispersion process is succeeded for large volume mixtures, this method would find a way to be used in a daily life more commonly.

REFERENCES

Alexopoulos, N. D., Bartholome, C., Poulin, P., & Marioli-Riga, Z. (2010). Structural health monitoring of glass fiber reinforced composites using embedded carbon nanotube (CNT) fibers. *Composites Science and Technology*, *70*(2), 260–271. doi:10.1016/j.compscitech.2009.10.017

Anantram, M. P., & Léonard, F. (2006). Physics of carbon nanotube electronic devices. *Reports on Progress in Physics*, *69*(3), 507–561. doi:10.1088/0034-4885/69/3/R01

Bauhofer, W., & Kovacs, J. Z. (2009). A review and analysis of electrical percolation in carbon nanotube polymer composites. *Composites Science and Technology*, *69*(10), 1486–1498. doi:10.1016/j.compscitech.2008.06.018

Bryning, M. B., Islam, M. F., Kikkawa, J. M., & Yodh, A. G. (2005). Very low conductivity threshold in bulk isotropic single-walled carbon nanotube-epoxy composites. *Advanced Materials, 17*(9), 1186–1191. doi:10.1002/adma.200401649

Cai, J., Qiu, L., Yuan, S., Shi, L., Liu, P., & Liang, D. (2012). Structural Health Monitoring for Composite Materials. In *Composites and Their Applications*. InTech. doi:10.5772/48215

Celzard, A., McRae, E., Deleuze, C., Dufort, M., Furdin, G., & Marêché, J. F. (1996). Critical concentration in percolating systems containing a high-aspect-ratio filler. *Physical Review. B, 53*(10), 6209–6214. doi:10.1103/PhysRevB.53.6209 PMID:9982020

Chaurasia, A. K., & Seidel, G. D. (2014). Computational micromechanics analysis of electron-hopping-induced conductive paths and associated macroscale piezoresistive response in carbon nanotube–polymer nanocomposites. *Journal of Intelligent Material Systems and Structures, 25*(17), 2141–2164. doi:10.1177/1045389X13517314

Chen, G., Weng, W., Wu, D., Wu, C., Lu, J., Wang, P., & Chen, X. (2004a). Preparation and characterization of graphite nanosheets from ultrasonic powdering technique. *Carbon, 42*(4), 753–759. doi:10.1016/j.carbon.2003.12.074

Chen, G., Weng, W., Wu, D., Wu, C., Lu, J., Wang, P., & Chen, X. (2004b). Preparation and characterization of graphite nanosheets from ultrasonic powdering technique. *Carbon, 42*(4), 753–759. doi:10.1016/j.carbon.2003.12.074

Chiacchiarelli, L. M., Rallini, M., Monti, M., Puglia, D., Kenny, J. M., & Torre, L. (2013). The role of irreversible and reversible phenomena in the piezoresistive behavior of graphene epoxy nanocomposites applied to structural health monitoring. *Composites Science and Technology, 80*, 73–79. doi:10.1016/j.compscitech.2013.03.009

Coetzee, D., Venkataraman, M., Militky, J., & Petru, M. (2020). Influence of Nanoparticles on Thermal and Electrical Conductivity of Composites. *Polymers, 12*(4), 742. doi:10.3390/polym12040742 PMID:32230802

Cuc, A., Giurgiutiu, V., Joshi, S., & Tidwell, Z. (2007). Structural health monitoring with piezoelectric wafer active sensors for space applications. *AIAA Journal, 45*(12), 2838–2850. doi:10.2514/1.26141

Dong, W., Li, W., Lu, N., Qu, F., Vessalas, K., & Sheng, D. (2019). Piezoresistive behaviours of cement-based sensor with carbon black subjected to various temperature and water content. *Composites. Part B, Engineering, 178*, 107488. doi:10.1016/j.compositesb.2019.107488

Dresselhaus, M. S., Dresselhaus, G., & Saito, R. (1995). Physics of carbon nanotubes. *Carbon, 33*(7), 883–891. doi:10.1016/0008-6223(95)00017-8

Ebbesen, T. W., & Ajayan, P. M. (1992). Large-scale synthesis of carbon nanotubes. *Nature, 358*(6383), 220–222. doi:10.1038/358220a0

Eswaraiah, V., Balasubramaniam, K., & Ramaprabhu, S. (2011). Functionalized graphene reinforced thermoplastic nanocomposites as strain sensors in structural health monitoring. *Journal of Materials Chemistry, 21*(34), 12626. doi:10.1039/c1jm12302e

Foteinidis, G., & Paipetis, A. S. (2021). A Novel Composite with Structural Health Monitoring Functionality via 2D and 3D Impedance Mapping Topography. *Applied Sciences (Basel, Switzerland)*, *11*(4), 1647. doi:10.3390/app11041647

Gao, L., Thostenson, E. T., Zhang, Z., & Chou, T. W. (2009a). Coupled carbon nanotube network and acoustic emission monitoring for sensing of damage development in composites. *Carbon*, *47*(5), 1381–1388. doi:10.1016/j.carbon.2009.01.030

Gao, L., Thostenson, E. T., Zhang, Z., & Chou, T.-W. (2009b). Coupled carbon nanotube network and acoustic emission monitoring for sensing of damage development in composites. *Carbon*, *47*(5), 1381–1388. doi:10.1016/j.carbon.2009.01.030

Gojny, F. H., Wichmann, M. H. G., Köpke, U., Fiedler, B., & Schulte, K. (2004a). Carbon nanotube-reinforced epoxy-composites: Enhanced stiffness and fracture toughness at low nanotube content. *Composites Science and Technology*, *64*(15 SPEC. ISS.), 2363–2371. doi:10.1016/j.compscitech.2004.04.002

Gojny, F. H., Wichmann, M. H. G., Köpke, U., Fiedler, B., & Schulte, K. (2004b). Carbon nanotube-reinforced epoxy-composites: Enhanced stiffness and fracture toughness at low nanotube content. *Composites Science and Technology*, *64*(15), 2363–2371. doi:10.1016/j.compscitech.2004.04.002

Hamdi, K., Aboura, Z., Harizi, W., & Khellil, K. (2020). Structural health monitoring of carbon fiber reinforced matrix by the resistance variation method. *Journal of Composite Materials*, *54*(25), 3919–3930. doi:10.1177/0021998320921476

Houshyar, S., Shanks, R. A., & Hodzic, A. (2005). The effect of fiber concentration on mechanical and thermal properties of fiber-reinforced polypropylene composites. *Journal of Applied Polymer Science*, *96*(6), 2260–2272. doi:10.1002/app.20874

Iijima, S. (1991). Helical microtubules of graphitic carbon. *Nature*, *354*(6348), 56–58. doi:10.1038/354056a0

Ke, K., Pötschke, P., Wiegand, N., Krause, B., & Voit, B. (2016). Tuning the Network Structure in Poly(vinylidene fluoride)/Carbon Nanotube Nanocomposites Using Carbon Black: Toward Improvements of Conductivity and Piezoresistive Sensitivity. *ACS Applied Materials & Interfaces*, *8*(22), 14190–14199. doi:10.1021/acsami.6b03451 PMID:27171017

Kim, B., Lee, J., & Yu, I. (2003). Electrical properties of single-wall carbon nanotube and epoxy composites. *Journal of Applied Physics*, *94*(10), 6724–6728. doi:10.1063/1.1622772

Kim, S., Zare, Y., Garmabi, H., & Rhee, K. Y. (2018). Variations of tunneling properties in poly (lactic acid) (PLA)/poly (ethylene oxide) (PEO)/carbon nanotubes (CNT) nanocomposites during hydrolytic degradation. *Sensors and Actuators. A, Physical*, *274*, 28–36. doi:10.1016/j.sna.2018.03.004

Kravchenko, O. G., Pedrazzoli, D., Bonab, V. S., & Manas-Zloczower, I. (2018). Conductive interlaminar interfaces for structural health monitoring in composite laminates under fatigue loading. *Materials & Design*, *160*, 1217–1225. doi:10.1016/j.matdes.2018.10.045

Lau, K. T., Lu, M., Lam, C. K., Cheung, H. Y., Sheng, F. L., & Li, H. L. (2005). Thermal and mechanical properties of single-walled carbon nanotube bundle-reinforced epoxy nanocomposites: The role of solvent for nanotube dispersion. *Composites Science and Technology, 65*(5 SPEC. ISS.), 719–725. doi:10.1016/j.compscitech.2004.10.005

Liao, Y., Zhou, P., Pan, D., Zhou, L., & Su, Z. (2021). An ultra-thin printable nanocomposite sensor network for structural health monitoring. *Structural Health Monitoring, 20*(3), 894–903. doi:10.1177/1475921719859338

Liu, Y., Rajadas, A., & Chattopadhyay, A. (2012). A biomimetic structural health monitoring approach using carbon nanotubes. *JOM, 64*(7), 802–807. doi:10.100711837-012-0357-6

Lu, K. L., Lago, R. M., Chen, Y. K., Green, M. L. H., Harris, P. J. F., & Tsang, S. C. (1996). Mechanical damage of carbon nanotubes by ultrasound. In Carbon, 34(6). doi:10.1016/0008-6223(96)89470-X

Mazela, B., Batista, A., & Grześkowiak, W. (2020). Expandable Graphite as a Fire Retardant for Cellulosic Materials—A Review. *Forests, 11*(7), 755. doi:10.3390/f11070755

Mintmire, J. W., Dunlap, B. I., & White, C. T. (1992). Are fullerene tubules metallic? *Physical Review Letters, 68*(5), 631–634. doi:10.1103/PhysRevLett.68.631 PMID:10045950

Miyagawa, H., & Drzal, L. T. (2004). Thermo-physical and impact properties of epoxy nanocomposites reinforced by single-wall carbon nanotubes. *Polymer, 45*(15), 5163–5170. doi:10.1016/j.polymer.2004.05.036

Mohiuddin, M., & Hoa, S. V. (2013). Estimation of contact resistance and its effect on electrical conductivity of CNT/PEEK composites. *Composites Science and Technology, 79*, 42–48. doi:10.1016/j.compscitech.2013.02.004

Monteiro, A. O., Cachim, P. B., & Costa, P. M. F. J. (2017). Self-sensing piezoresistive cement composite loaded with carbon black particles. *Cement and Concrete Composites, 81*, 59–65. doi:10.1016/j.cemconcomp.2017.04.009

Paleari, L., Bragaglia, M., Fabbrocino, F., & Nanni, F. (2021a). Structural monitoring of glass fiber/epoxy laminates by means of carbon nanotubes and carbon black self-monitoring plies. *Nanomaterials (Basel, Switzerland), 11*(6), 1543. doi:10.3390/nano11061543 PMID:34208091

Paleari, L., Bragaglia, M., Fabbrocino, F., & Nanni, F. (2021b). Structural Monitoring of Glass Fiber/Epoxy Laminates by Means of Carbon Nanotubes and Carbon Black Self-Monitoring Plies. *Nanomaterials (Basel, Switzerland), 11*(6), 1543. doi:10.3390/nano11061543 PMID:34208091

Park, J.-M., Kim, D.-S., Kim, S.-J., Kim, P.-G., Yoon, D.-J., & DeVries, K. L. (2007). Inherent sensing and interfacial evaluation of carbon nanofiber and nanotube/epoxy composites using electrical resistance measurement and micromechanical technique. *Composites. Part B, Engineering, 38*(7–8), 847–861. doi:10.1016/j.compositesb.2006.12.004

Park, J. M., Kim, P. G., Jang, J. H., Wang, Z., Kim, J. W., il Lee, W., Park, J. G., & Lawrence DeVries, K. (2008). Self-sensing and dispersive evaluation of single carbon fiber/carbon nanotube (CNT)-epoxy composites using electro-micromechanical technique and nondestructive acoustic emission. *Composites. Part B, Engineering, 39*(7–8), 1170–1182. doi:10.1016/j.compositesb.2008.03.004

Ren, X., Chaurasia, A. K., Oliva-Avilés, A. I., Ku-Herrera, J. J., Seidel, G. D., & Avilés, F. (2015). Modeling of mesoscale dispersion effect on the piezoresistivity of carbon nanotube-polymer nanocomposites via 3D computational multiscale micromechanics methods. *Smart Materials and Structures, 24*(6), 065031. doi:10.1088/0964-1726/24/6/065031

Ruoff, R. S., Qian, D., & Liu, W. K. (2003). Mechanical properties of carbon nanotubes: Theoretical predictions and experimental measurements. *Comptes Rendus Physique, 4*(9), 993–1008. doi:10.1016/j.crhy.2003.08.001

Sam-Daliri, O., Farahani, M., Faller, L.-M., & Zangl, H. (2020). Structural health monitoring of defective single lap adhesive joints using graphene nanoplatelets. *Journal of Manufacturing Processes, 55*, 119–130. doi:10.1016/j.jmapro.2020.03.063

Sandler, J., Shaffer, M. S. P., Prasse, T., Bauhofer, W., Schulte, K., & Windle, A. H. (1999). Development of a dispersion process for carbon nanotubes in an epoxy matrix and the resulting electrical properties. *Polymer, 40*(21), 5967–5971. doi:10.1016/S0032-3861(99)00166-4

Shojaeiarani, J., Bajwa, D., & Holt, G. (2020). Sonication amplitude and processing time influence the cellulose nanocrystals morphology and dispersion. *Nanocomposites, 6*(1), 41–46. doi:10.1080/20550324.2019.1710974

Spinelli, G., Lamberti, P., Tucci, V., Guadagno, L., & Vertuccio, L. (2020). Damage Monitoring of Structural Resins Loaded with Carbon Fillers: Experimental and Theoretical Study. *Nanomaterials (Basel, Switzerland), 10*(3), 434. doi:10.3390/nano10030434 PMID:32121350

Sun, Y., Bao, H.-D., Guo, Z.-X., & Yu, J. (2009). Modeling of the Electrical Percolation of Mixed Carbon Fillers in Polymer-Based Composites. *Macromolecules, 42*(1), 459–463. doi:10.1021/ma8023188

Tallman, T. N., Gungor, S., Wang, K. W., & Bakis, C. E. (2015). Damage detection via electrical impedance tomography in glass fiber/epoxy laminates with carbon black filler. *Structural Health Monitoring, 14*(1), 100–109. doi:10.1177/1475921714554142

Thostenson, E. T., & Chou, T. W. (2006). Processing-structure-multi-functional property relationship in carbon nanotube/epoxy composites. *Carbon, 44*(14), 3022–3029. doi:10.1016/j.carbon.2006.05.014

Thostenson, E. T., & Chou, T. W. (2008a). Carbon nanotube-based health monitoring of mechanically fastened composite joints. *Composites Science and Technology, 68*(12), 2557–2561. doi:10.1016/j.compscitech.2008.05.016

Thostenson, E. T., & Chou, T. W. (2008b). Real-time in situ sensing of damage evolution in advanced fiber composites using carbon nanotube networks. *Nanotechnology, 19*(21), 215713. doi:10.1088/0957-4484/19/21/215713 PMID:21730592

Thostenson, E. T., Ren, Z., & Chou, T.-W. (2001). Advances in the science and technology of carbon nanotubes and their composites: A review. *Composites Science and Technology, 61*(13), 1899–1912. doi:10.1016/S0266-3538(01)00094-X

Turan, F., Guclu, M., Gurkan, K., Durmus, A., & Taskin, Y. (2022). The effect of carbon nanotubes loading and processing parameters on the electrical, mechanical, and viscoelastic properties of epoxy-based composites. *Journal of the Brazilian Society of Mechanical Sciences and Engineering*, *44*(3), 93. doi:10.100740430-022-03393-2

Viets, C., Kaysser, S., & Schulte, K. (2014). Damage mapping of GFRP via electrical resistance measurements using nanocomposite epoxy matrix systems. *Composites. Part B, Engineering*, *65*, 80–88. doi:10.1016/j.compositesb.2013.09.049

Wang, S. J., Geng, Y., Zheng, Q., & Kim, J.-K. (2010). Fabrication of highly conducting and transparent graphene films. *Carbon*, *48*(6), 1815–1823. doi:10.1016/j.carbon.2010.01.027

Yasmin, A., Luo, J.-J., & Daniel, I. M. (2006). Processing of expanded graphite reinforced polymer nanocomposites. *Composites Science and Technology*, *66*(9), 1182–1189. doi:10.1016/j.compscitech.2005.10.014

Yin, F., Ye, D., Zhu, C., Qiu, L., & Huang, Y. (2017). Stretchable, Highly Durable Ternary Nanocomposite Strain Sensor for Structural Health Monitoring of Flexible Aircraft. *Sensors (Basel)*, *17*(11), 2677. doi:10.339017112677 PMID:29156620

Zhao, Z., Yang, Z., Hu, Y., Li, J., & Fan, X. (2013). Multiple functionalization of multi-walled carbon nanotubes with carboxyl and amino groups. *Applied Surface Science*, *276*, 476–481. doi:10.1016/j.apsusc.2013.03.119

Zheng, W., & Wong, S.-C. (2003). Electrical conductivity and dielectric properties of PMMA/expanded graphite composites. *Composites Science and Technology*, *63*(2), 225–235. doi:10.1016/S0266-3538(02)00201-4

Chapter 13
Synthesis and Characterization of Carbon-Based Quantum Dot From Rice, Sugar, and Aloe-Vera

Sunil Kumar Pradhan
https://orcid.org/0000-0003-1615-6168
School of Electronics Engineering, Vellore Institute of Technology, Chennai, India

Sumit Kashyap
Department of Electronics and Communication Engineering, National Institute of Technology, Andhra Pradesh, India

Akula Maheswari Durga
Department of Electronics and Communication Engineering, National Institute of Technology, Andhra Pradesh, India

Jostin Vara
Department of Electronics and Communication Engineering, National Institute of Technology, Andhra Pradesh, India

ABSTRACT

Carbon quantum dots (CQDs), a novel class of zero-dimensional carbon nanostructures with diameters below 10 nm, have recently come into existence. The authors have synthesized and analyzed carbon-based quantum dots from aloe-vera, rice, and sugar. The characterization of these quantum dots is done via UV-visible spectrophotometer. The quantum dots that are made from chemicals are toxic and are harmful to both humans and animals. However, the carbon quantum dots (CQD) which they have synthesized in this research work are bio-compatible, less toxic, and have the property of high aqueous dispersibility. CQDs are being used in various fields of applications such as bio-imaging, medical diagnosis, bio-sensing, chemical sensing, photo catalysis, and photovoltaic devices. This study helps the scientific community to understand the nature of carbon quantum dots and gives a brief idea of suitable materials that create CQD.

DOI: 10.4018/978-1-6684-7358-0.ch013

INTRODUCTION

Now-a-days, the usage of chemicals has increased to a great extent. For instance, the biomedical industry makes extensive use of toxic semiconductor QDs including cadmium selenide (CdSe), cadmium sulphide (CdS), silver sulphide (Ag_2S), and PbSe (lead selenide) (Hetsch et al., 2013; Prabaharan & Mano, 2006; Sun et al., 2013; Yang et al., 2019). These chemicals are toxic and are not eco-friendly as they may cause harm to people if they can be used frequently. QD degradation with release of free cadmium and generation of reactive oxygen species makes them toxic. Noble metal nanoparticles, namely silver and gold nanoparticles with alginate residues, are used because of their improved effectiveness in medication delivery and wound healing applications (Brus, 1984; Brus, 1986; Kramer & Sargent, 2011; Sun et al., 2006). The cytotoxicity of such noble metal nanoparticles is not entirely absent. The level of pollution has been increasing (water, air, soil pollution) due to the increase in the usage of chemicals. The health of humans and animals are also at risk. Some of the species have become endangered. In the realm of biomedical application, the nearly non-toxic, luminous CQDs are likewise regarded as the ideal substitute for hazardous semiconductor QDs like cadmium selenide (CdSe), cadmium sulphide (CdS), silver sulphide (Ag_2S), and lead selenide (PbSe) (Ekimov & Onushchenko, 1981; Kim, Jo, Yoon et al, 2019; Tang et al., 2012; Zhang et al., 2019). These functional nanoparticles are widely used as a promising candidate in bio-imaging, sensors, drug delivery, ink, photo catalysis, and optoelectronics because of their distinctive properties, including tunable photoluminescence, photo stability, low cytotoxicity, robust chemical inertness, and excellent biocompatibility, etc. (Kim, Han, Yang et al, 2019; Moraitis et al., 2019; Russo et al., 2007; Rybak, 2006).

QDs are semiconducting materials with diameter range of 2 to 10 nanometer and of 10 to 15 atoms. Due to the unique dimensions of their energy band gaps, they display quantum mechanical activity (Agarwal et al., 2015; Antoci et al., 2007; Yang et al., 2014; Yu et al., 2014). The electrons in quantum dots are confined to a single place in space. Since electrons are believed to be localized at a location and have no freedom in any direction, a change in any direction will change the properties. Quantum confinements, a rare phenomenon, play a part in the optical features of quantum dots. A QD may excite an electron from the valence band to the conduction band, leaving a positive hole in its place, when a photon with sufficient energy (equal to or larger than the band gap) impacts it. The generation of an electron-hole pair, often known as an "exciton," is a typical event in semiconducting materials. When an exciton is compressed inside a QD, confinement energy is created because the average exciton size (exciton Bohr radius) is less than the QD's size. The band gap of the QD, the quantum confinement energy, and the bound exciton energy can all be used to describe the total energy of a fluorescing photon. The wavelength of the light emitted by the quantum dots depends upon the size of the quantum dots. The quantum dots of larger size emit light of higher wavelength and smaller frequency (Tian et al., 2009; Wang et al., 2011; Wei et al., 2013; Yu et al., 2015). The quantum dots of smaller size emit light of smaller wavelength and higher frequency. The color produced from quantum dots of smaller size is nearer to blue and the color produced from quantum dots of larger size is nearer to red and the medium sized particles give color nearer to yellow. Generally, the core of the quantum dots is crystalline metalloid and is quantum confined to 1 to 5nm size. This core consists of metal complexes like semiconductors, magnetic transition metals and noble metals. Indium phosphate (InP), galium arsenide (GaAs), gallium nitride (GaN), and indium arsenate (InAs) metalloid cores are QD semiconductors from the III-V class. Zinc sulphide (ZnS), zinc-selenium (ZnSe), and cadmium-selenium (CdSe) cores are QD semiconductors from the II to IV class (Baker & Baker, 2010; Bhattacharya et al., 2012; Chen & Yan, 2011; Chen et al., 2004). Quantum dot semiconductors' level of toxicity is influenced by a number of variables, including QD size, charge, concentration, outer coating bioactivity, oxidative, photolytic, and mechanical stability. After oxidation or photolytic degradation of core coatings, the toxicity of quantum dot semiconductors is determined. Two frequently utilized QD core metalloid complex components are cadmium and selenium, which are poisonous to humans and the environment and are known to produce acute and chronic toxicities in vertebrates. Cd is a potential carcinogen (agent that causes cancer) and lives between 15 and 20 years in people. It is systematically disseminated to all tissues and has the potential to cross the blood-brain barrier and the placenta, with the liver and kidneys being its primary target organs for toxicity. From Kesterson Reservoir in California and Belews Lake in North Carolina, where the surrounding environment is severely impacted, the hazardous potential of Se pollution is well established. This is due to the increase in concentration of Se in environment. The risk from quantum dot semiconductors to human lie and environment is very high and need to be evaluated. The rise in the use of Quantum dot (QD)products in environment and society also increases the potential toxicity. It's necessary to control the use of these quantum dot semiconductors not only in humans but also in industries. The exposure to these chemicals causes various diseases in humans. Carbon quantum dots are the kind of nanoparticles with size less than 10 nm and made of carbon. CQDs were initially discovered in the year 2004 during a purification process of single walled carbon nanotubes (De & Karak, 2013; Del Valle, 2004; Khademhosseini & Langer, 2007; Nie et al., 2014). There was a lot of research done to know the properties and applications of CQD.CQDs have high stability, low toxicity, better conductivity compared to QDs and are environmentally friendly. CQDs are more focused and conducted wide number of experiments for its strong and tunable fluorescence emission properties, which are more useful in biomedicine, optoelectronics applications and are biocompatible and has best solubility in water. This helps CQD to serve as proton conducting nanoparticles (Fowley et al., 2012; Jeong et al., 2009; Sevilla & Fuertes, 2009; Welsher et al., 2009). CQDs can be synthesized in two methods known as Bottom-Up and Top-Down approach. The Top-Down method points out the breaking of larger carbon bond connections like graphite into CQDs using laser and electrochemical techniques and in Bottom-Up method, synthesizing CQDs from small procedures like carbohydrates, polymer silica. Applications of CQDs in biomedical is high due to its biological and environmentally friendly decomposition. Also, in order to sustain with the nor- mal QD, high Q yields are to be achieved. There are various applications of QCDs for example in bio-imaging, sensing, drug delivery, catalysis, optoelectronics etc. Here, we have synthesized and analyzed carbon-based quantum dots from aloe-vera, rice and sugar. The characterization of these quantum dots is done via UV-Visible spectrophotometer.

MATERIALS AND METHODS

In this research we have used Aloe vera, sugar, rice, ethanol, distilled water, beakers, test tube and aluminium foil as necessary items to synthesize our carbon-based quantum dots. The equipment's which are deployed in this work are induction stove and ultra-sonicator. The synthesized CQDs are finally characterized via UV-Visible spectrophotometer. A strong piece of lab equipment with an ultrasonic electric generator is called an ultra sonicator. Piezoelectric crystals, which directly respond to electricity by producing a mechanical vibration, are used by the transducer to convert the electrical signal. The vibration is preserved and amplified by the sonicator before reaching the probe. The probe quickly goes up and down while moving in rhythm with the vibration to convey it to the solution. Based on the characteristics of the solution, the sonicator operator can adjust the amplitude. Although a large tip can reach more of the solution, a small tip creates a stronger reaction. Sound waves are used in sonication to stir up the particles in a solution. To disintegrate materials, it transforms an electrical signal into a physical vibration. These disturbances have the ability to mix solutions, hasten the dissolving of a solid into a liquid, such as the solubility of sugar into water, and remove dissolved gas from liquids. Sonication is used in DNA testing to disassemble molecules and disrupt cells, releasing proteins for analysis. Spectrophotometry, often known as UV-Vis Spectroscopy, is a quantitative method for calculating how much a chemical substance absorbs light (Loh et al., 2010; Mochalin & Gogotsi, 2009; Nie et al., 2014; Shen et al., 2011). This is accomplished by comparing the amount of light that passes through a sample to the amount of light that goes through a reference sample or a blank. Glass, liquids, solids, thin films, and other sample types can all be employed with this method. The energy content of the atoms or molecules increases as light is absorbed by matter. When ultraviolet radiations are absorbed, this causes the electrons' ground state to be excited and move up into a higher energy state. By absorbing energy in the form of UV light, molecules with n- or π-electrons can excite these electrons to higher anti-bonding molecular orbitals. Longer wavelengths of light may be absorbed by electrons the more readily excited they are. The methodology involves the carbonization process, ultrasonication and then analysis of samples using UV-Visible spectrophotometer (Lei et al., 2011; Pan, Zhang, Li, & Wu, 2010; Zhao et al., 2011). It provides broad definitions and discussions of the topic and incorporate views of others (literature review) into the discussion to support, refute, or demonstrate our position on the topic.

Figure 1. Process of carbonization of aloe-vera

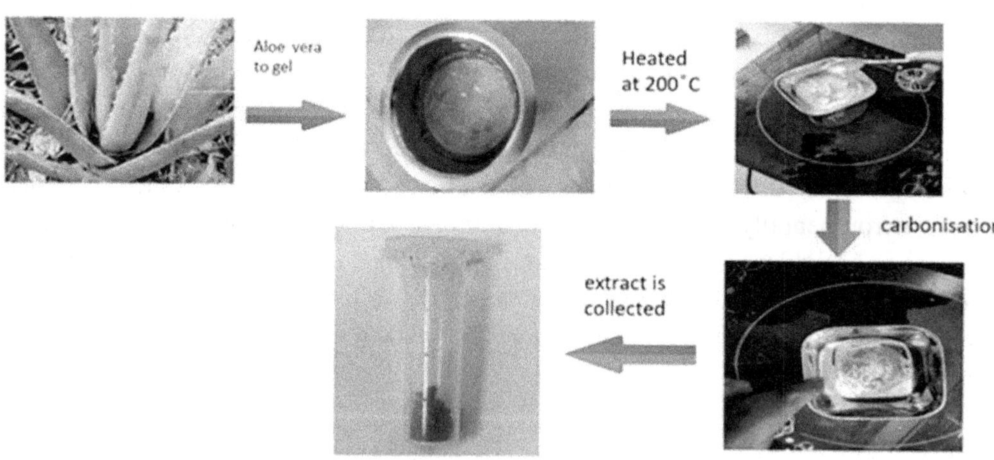

In this case Samples are taken aloe vera, rice, sugar. Aloe vera (scientific name: Aloe barbadensis Mill.) leaves are gathered and cleaned with double-distilled water for additional use. Aloe vera is made into aloe vera gel and then we have performed carbonization of aloe vera gel and collected the sample. The process of carbonization of aloe-vera is shown in the Figure 1. Carbonization is performed on rice and sugar also, which is shown in the Figure 2.

Figure 2. Process of carbonization of sugar and rice respectively

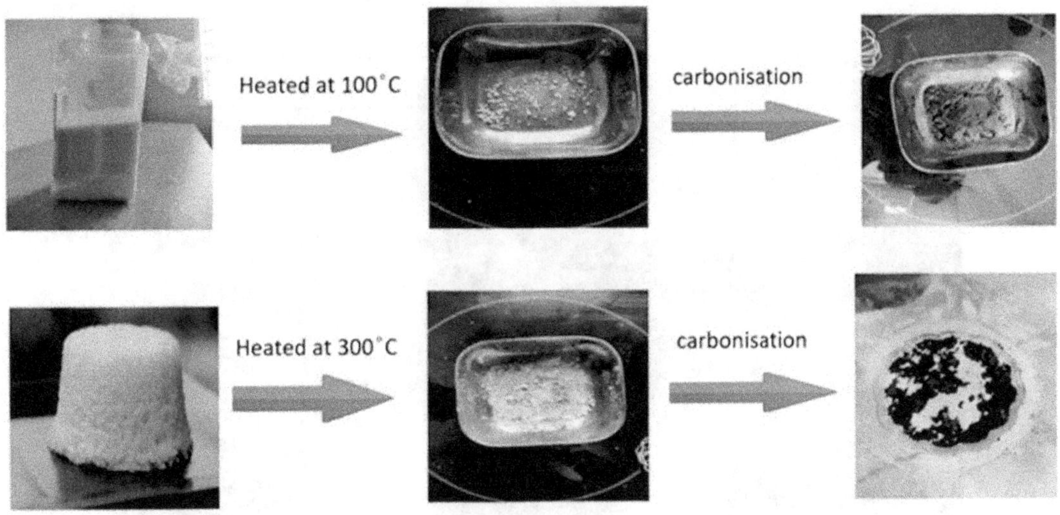

The extract is added in 10ml ethanol and the sonication is performed on (aloe vera + ethanol) mixture. The process of ultra-sonication to obtain CQD solution is shown in the Figure 3.

Figure 3. process of ultra-sonication to obtain CQD solution

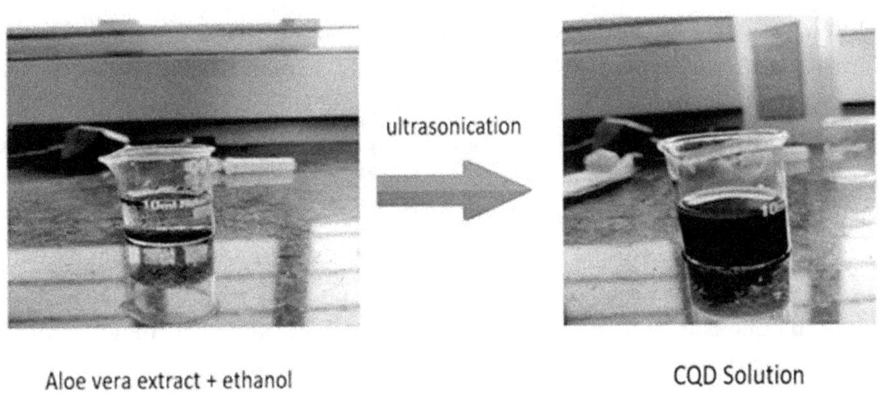

The sugar sample (1), rice sample (2) and the aloe vera sample (3) af ter ultrasonication are shown in the Figure 4.

Figure 4. CQD solutions of sugar, rice and aloe vera respectively

RESULTS AND DISCUSSION

The CQD solutions of sugar, rice and aloe-vera respectively are analyzed via UV- Visible spectrophotometer. The characterization is shown in the Figure 5. As it can be analyzed, the aqueous dispersibility of aloe vera is more than the aqueous dispersibility of rice and sugar. The aqueous dispersibility of rice is higher than that of sugar. When UV light is passed through the samples, they exhibit different behavior (Eda et al., 2010; Gupta et al., 2011; Li et al., 2011; Pan, Zhang, Li, Wu et al, 2010). From the graphs, we observe that the absorption is more for sugar and less for aloe vera. So, aloe vera can emit more energy when UV light is passed through it. It is also can be interpreted from Figure 5 that, the highest absorbance peak for sugar is at 272 nm, at 279nm for rice and at 290 nm for aloe-vera. It implies the formation of quantum dots as it signifies the existence of peaks of the absorbance is in the UV region.

Figure 5. Absorbance w.r.t wavelength for aloe vera, rice and sugar

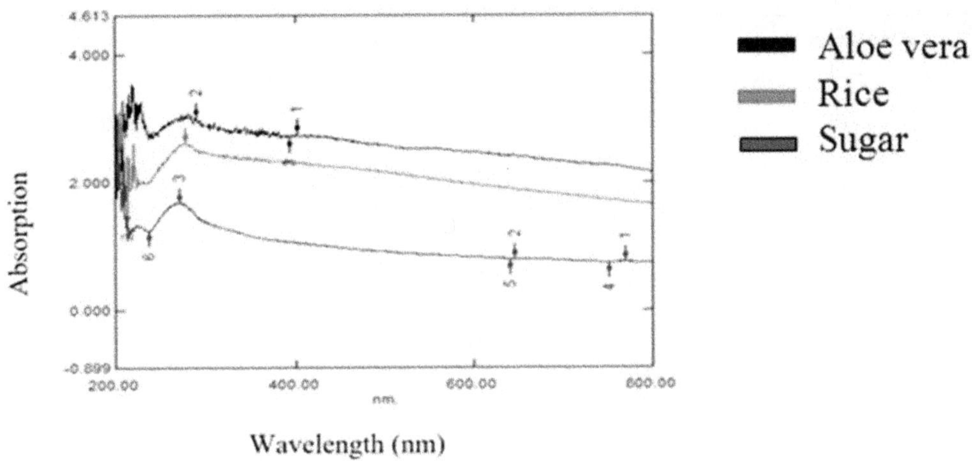

Figure 6. Spectra of aloe vera showing Absorbance w.r.t wavelength

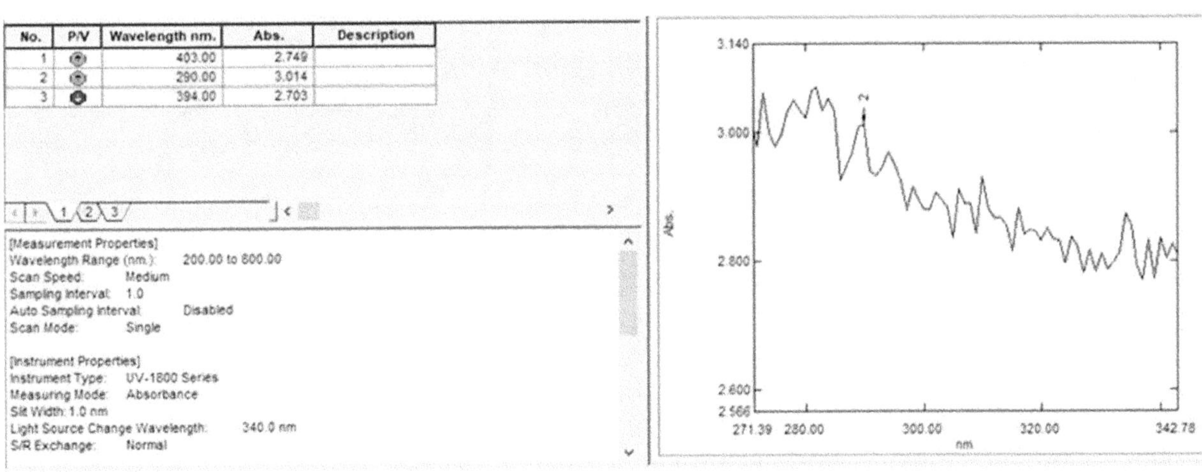

Figure 7. Spectra of rice showing Absorbance w.r.t wavelength

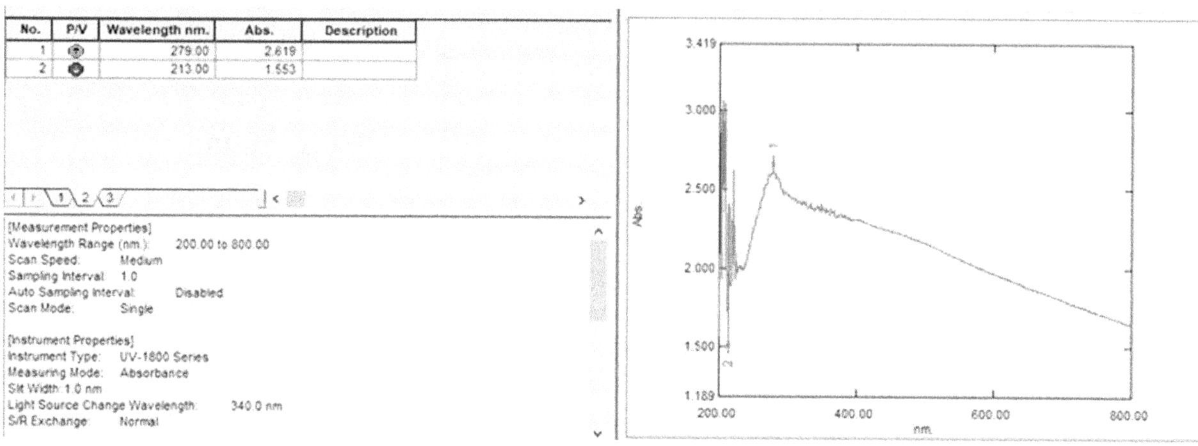

Figure 8. Spectra of sugar showing Absorbance w.r.t wavelength

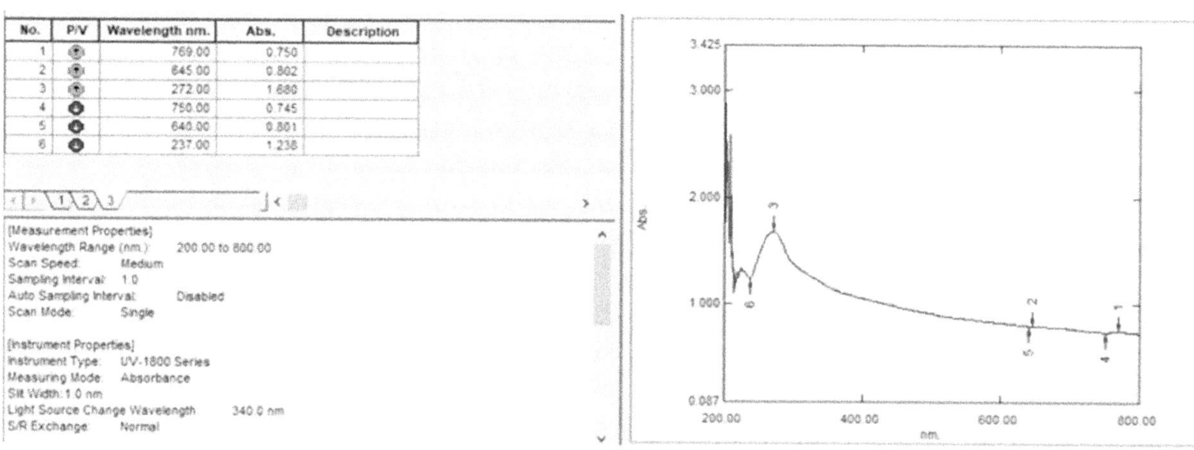

Figures 6, 7 and 8 depict the individual spectra of aloe vera, rice and sugar showing Absorbance (in Y-axis) w.r.t wavelength (in X-axis) measured from UV-Visible spectrophotometer. In our study, optical energy band gap by "TAUC method" is also investigated (Liu et al., 2011; Yan et al., 2010). The graph is plotted for analyzing optical energy band gap for the quantum dots. The quantity $h\nu$ (the photon energy) on the abscissa. The quantity $(\alpha h\nu)^2$ on the ordinate where α is the absorption coefficient of the material, ν is the frequency, and h is the plank's constant. A tangent is drawn to the graph plotted and intersection point on the X-axis is the energy band gap of the carbon quantum dots. From the analysis it is found that the band gap of Aloe Vera Quantum Dot = 2.39 eV, band gap of Rice Quantum Dot = 2.43 eV, and the band gap for sugar quantum dot is 2.77 eV. The analysis and demonstration can be seen in the Figures 9, 10 and 11 respectively.

Figure 9. energy band gap of quantum dots of aloe vera

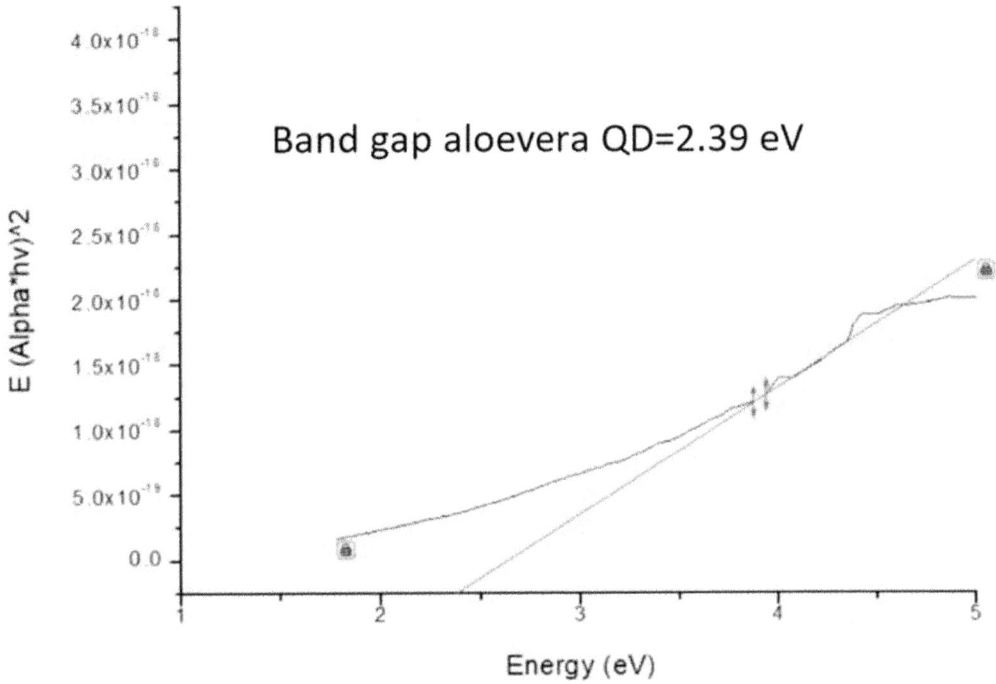

Figure 10. energy band gap of quantum dots of rice

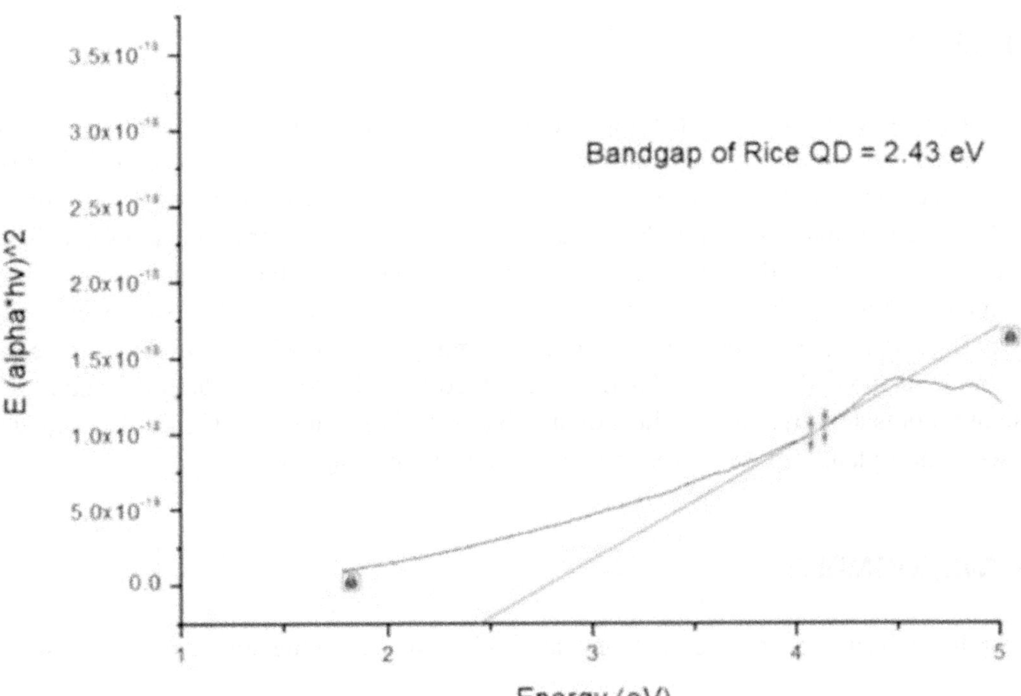

Figure 11. energy band gap of quantum dots of sugar

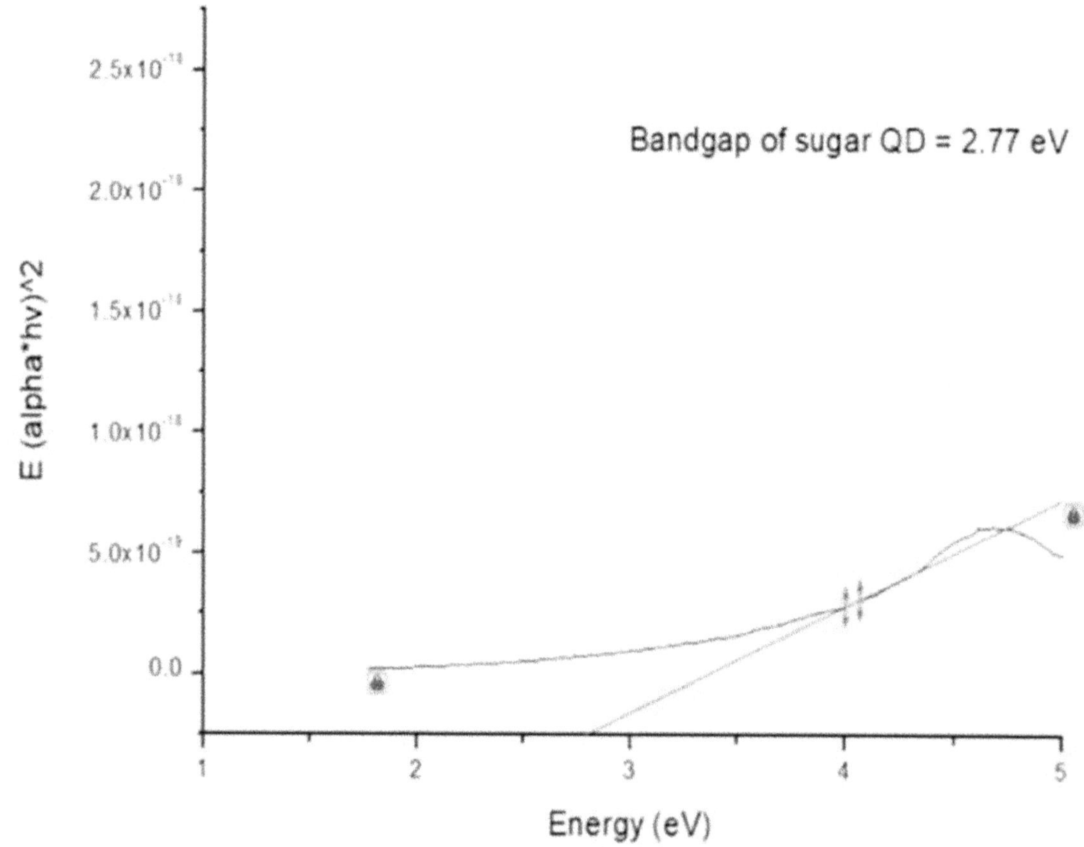

CONCLUSION

In this research work, we have synthesized carbon quantum dots which exist naturally. The synthesis procedure is adapted with an objective of non-toxic effect and non- hazardous consideration of sample preparation. The quantum dots which have been synthesized are expected to be below 10nm. Here, we have synthesized and analyzed carbon-based quantum dots from aloe-vera, rice and sugar. The characterization of these quantum dots is done via UV-Visible spectrophotometer. The carbon quantum dots (CQD) which we have synthesized in this research work are bio-compatible, less toxicity and has the property of high aqueous dispersibility. The energy band gap of these quantum dots is also calculated through TAUC method. This study helps the research community to understand the nature of carbon-based quantum dots and gives a brief idea of suitable materials that create CQD. These quantum dots can be used in non-planar image sensor and various bio-medical applications.

ACKNOWLEDGMENT

This research received no specific grant from any funding agency in the public, commercial, or not-for-profit sectors.

REFERENCES

Agarwal, T., Narayana, S. G., Pal, K., Pramanik, K., Giri, S., & Banerjee, I. (2015). Calcium alginate-carboxymethyl cellulose beads for colon-targeted drug delivery. *International Journal of Biological Macromolecules*, *75*, 409–417. doi:10.1016/j.ijbiomac.2014.12.052 PMID:25680962

Antoci, V. Jr, Adams, C. S., Hickok, N. J., Shapiro, I. M., & Parvizi, J. (2007). Antibiotics for local delivery systems cause skeletal cell toxicity in vitro. *Clinical Orthopaedics and Related Research*, *462*, 200–206. doi:10.1097/BLO.0b013e31811ff866 PMID:17572634

Baker, S. N., & Baker, G. A. (2010). Luminescent carbon nanodots: Emergent nanolights. *Angewandte Chemie International Edition*, *49*(38), 6726–6744. doi:10.1002/anie.200906623 PMID:20687055

Bhattacharya, M., Malinen, M. M., Lauren, P., Lou, Y. R., Kuisma, S. W., Kanninen, L., & Laukkanen, A. (2012). Nanofibrillar cellulose hydrogel promotes three-dimensional liver cell culture. *Journal of Controlled Release*, *164*(3), 291–298. doi:10.1016/j.jconrel.2012.06.039 PMID:22776290

Brus, L. (1984). Electron-electron and electron-hole interactions in small semiconductor crystallites: The size dependence of the lowest excited electronic state. *The Journal of Chemical Physics*, *80*(9), 4403–4409. doi:10.1063/1.447218

Brus, L. (1986). Electronic wave functions in semiconductor clusters: Experiment and theory. *Journal of Physical Chemistry*, *90*(12), 2555–2560. doi:10.1021/j100403a003

Chen, J. L., & Yan, X. P. (2011). Ionic strength and pH reversible response of visible and near-infrared ñuorescence of graphene oxide nanosheets for monitoring the extracellular p^H. *Chemical Communications*, *47*(11), 3135–3137. doi:10.1039/c0cc03999c PMID:21270983

Chen, S. C., Wu, Y. C., Mi, F. L., Lin, Y. H., Yu, L. C., & Sung, H. W. (2004). A novel p^H- sensitive hydrogel composed of N, O-carboxymethyl chitosan and alginate cross-linked by genipin for protein drug delivery. *Journal of Controlled Release*, *96*(2), 285–300. doi:10.1016/j.jconrel.2004.02.002 PMID:15081219

De, B., & Karak, N. (2013). A green and facile approach for the synthesis of water soluble ñuorescent carbon dots from banana juice. *RSC Advances*, *3*(22), 8286–8290. doi:10.1039/c3ra00088e

Del Valle, E. M. (2004). Cyclodextrins and their uses: A review. *Process Biochemistry (Barking, London, England)*, *39*(9), 1033–1046. doi:10.1016/S0032-9592(03)00258-9

Eda, G., Lin, Y. Y., Mattevi, C., Yamaguchi, H., Chen, H. A., Chen, I. S., Chen, C.-W., & Chhowalla, M. (2010). Blue photoluminescence from chemically derived graphene oxide. *Advanced Materials*, *22*(4), 505–509. doi:10.1002/adma.200901996 PMID:20217743

Ekimov, A. I., & Onushchenko, A. A. (1981). Quantum size effect in three-dimensional microscopic semiconductor crystals. *JETP Letters*, *34*, 346–349.

Fowley, C., McCaughan, B., Devlin, A., Yildiz, I., Raymo, F. M., & Callan, J. F. (2012). Highly luminescent biocompatible carbon quantum dots by encapsulation with an amphiphilic polymer. *Chemical Communications*, *48*(75), 9361–9363. doi:10.1039/c2cc34962k PMID:22892652

Gupta, V., Chaudhary, N., Srivastava, R., Sharma, G. D., Bhardwaj, R., & Chand, S. (2011). Luminscent graphene quantum dots for organic photovoltaic devices. *Journal of the American Chemical Society*, *133*(26), 9960–9963. doi:10.1021/ja2036749 PMID:21650464

Hetsch, F., Zhao, N., Kershaw, S. V., & Rogach, A. L. (2013). Quantum dot field effect transistor. *Materials Today*, *16*(9), 312–325. doi:10.1016/j.mattod.2013.08.011

Jeong, J., Cho, M., Lim, Y. T., Song, N. W., & Chung, B. H. (2009). Synthesis and characterization of a photoluminescent nanoparticle based on fullerene-silica hybridization. *Angewandte Chemie International Edition*, *48*(29), 5296–5299. doi:10.1002/anie.200901750 PMID:19533701

Khademhosseini, A., & Langer, R. (2007). Micro engineered hydrogels for tissue engineering. *Biomaterials*, *28*(34), 5087–5092. doi:10.1016/j.biomaterials.2007.07.021 PMID:17707502

Kim, H. J., Jo, J. H., Yoon, S. Y., Jo, D. Y., Kim, H. S., Park, B., & Yang, H. (2019). Emission enhancement of Cu-doped InP quantum dots through double shelling scheme. *Materials (Basel)*, *12*(14), 2267. doi:10.3390/ma12142267 PMID:31311083

Kim, M., Han, C. Y., Yang, H., & Park, B. (2019). Band to band tunneling at the zinc oxide (ZnO) and lead selenide (PbSe) quantum dot contact; interfacial charge transfer at ZnO/PbSe/ZnO probe device. *Materials (Basel)*, *12*(14), 2289. doi:10.3390/ma12142289 PMID:31319559

Kramer, I. J., & Sargent, E. H. (2011). Colloidal quantum dot photovoltaics: A path forward. *ACS Nano*, *5*(11), 8506–8514. doi:10.1021/nn203438u PMID:21967723

Lei, W. W., Portehault, D., Dimova, R., & Antonietti, M. (2011). Boron carbon nitride nanostructures from salt melts: Tunable water-soluble phosphors. *Journal of the American Chemical Society*, *133*(18), 7121–7127. doi:10.1021/ja200838c PMID:21506566

Li, Y., Hu, Y., Zhao, Y., Shi, G. Q., Deng, L., Hou, Y. B., & Qu, L. (2011). An electrochemical avenue to green-luminescent graphene quantum dots as potential electron-acceptors for photovoltaics. *Advanced Materials*, *23*(6), 776–780. doi:10.1002/adma.201003819 PMID:21287641

Liu, R. L., Wu, D. Q., Feng, X. L., & Mullen, K. (2011). Bottom-up fabrication of photoluminescent graphene quantum dots with uniform morphology. *Journal of the American Chemical Society*, *133*(39), 15221–15223. doi:10.1021/ja204953k PMID:21894989

Loh, K. P., Bao, Q., Eda, G., & Chhowalla, M. (2010). Graphene oxide as a chemically tunable platform for optical applications. *Nature Chemistry*, *2*(12), 1015–1024. doi:10.1038/nchem.907 PMID:21107364

Mochalin, V. N., & Gogotsi, Y. (2009). Wet chemistry route to hydrophobic blue fluorescent nanodiamond. *Journal of the American Chemical Society*, *131*(13), 4594–4595. doi:10.1021/ja9004514 PMID:19290627

Moraitis, P., van Leeuwen, G., & van Sark, W. (2019). Visual appearance of nanocrystal-based luminescent solar concentrators. *Materials (Basel)*, *12*(6), 885. doi:10.3390/ma12060885 PMID:30884811

Nie, H., Li, M., Li, Q., Liang, S., Tan, Y., Sheng, L., Shi, W., & Zhang, S. X. (2014). Carbon dots with continuously tunable full-color emission and their application in radiometric pH sensing. *Chemistry of Materials*, *26*(10), 3104–3112. doi:10.1021/cm5003669

Pan, D. Y., Zhang, J. C., Li, Z., Wu, C., Yan, X. M., & Wu, M. H. (2010). Observation of p^H- solvent-, spin-, and excitation-dependent blue photoluminescence from carbon nanoparticles. *Chemical Communications*, *46*(21), 3681–3683. doi:10.1039/c000114g PMID:20396809

Pan, D. Y., Zhang, J. C., Li, Z., & Wu, M. H. (2010). Hydrothermal route for cutting graphene sheets into blue-luminescent graphene quantum dots. *Advanced Materials*, *22*(6), 734–738. doi:10.1002/adma.200902825 PMID:20217780

Prabaharan, M., & Mano, J. F. (2006). Stimuli-responsive hydrogels based on polysaccharides incorporated with thermo-responsive polymers as novel biomaterials. *Macromolecular Bioscience*, *6*(12), 991–1008. doi:10.1002/mabi.200600164 PMID:17128423

Russo, R., Malinconico, M., & Santagata, G. (2007). Effect of cross-linking with calcium ions on the physical properties of alginate ðlms. *Biomacromolecules*, *8*(10), 3193–3197. doi:10.1021/bm700565h PMID:17803277

Rybak, M. J. (2006). The pharmacokinetic and pharmacodynamic properties of vancomycin. *Clinical Infectious Diseases*, *42*(Suppl.1), S35–S39. doi:10.1086/491712 PMID:16323118

Sevilla, M., & Fuertes, A. B. (2009). Chemical and structural properties of carbonaceous products obtained by hydrothermal carbonization of saccharides. *Chemistry (Weinheim an der Bergstrasse, Germany)*, *15*(16), 4195–4203. doi:10.1002/chem.200802097 PMID:19248078

Shen, J. H., Zhu, Y. H., Chen, C., Yang, X. L., & Li, C. Z. (2011). Facile preparation and up conversion luminescence of graphene quantum dots. *Chemical Communications*, *47*(9), 2580–2582. doi:10.1039/C0CC04812G PMID:21173992

Sun, D., Ban, R., Zhang, P. H., Wu, G. H., Zhang, J. R., & Zhu, J. J. (2013). Hair ðber as a precursor for synthesizing of sulfur-and nitrogen-co-doped carbon dots with tunable luminescence properties. *Carbon*, *64*, 424–434. doi:10.1016/j.carbon.2013.07.095

Sun, Y. P., Zhou, B., Lin, Y., Wang, W., Fernando, K. S., Pathak, P., Meziani, M. J., Harruff, B. A., Wang, X., Wang, H., Luo, P. G., Yang, H., Kose, M. E., Chen, B., Veca, L. M., & Xie, S.-Y. (2006). Quantum-sized carbon dots for bright and colorful photoluminescence. *Journal of the American Chemical Society*, *128*(24), 7756–7757. doi:10.1021/ja062677d PMID:16771487

Tang, L., Ji, R., Cao, X., Lin, J., Jiang, H., Li, X., Teng, K. S., Luk, C. M., Zeng, S., Hao, J., & Lau, S. P. (2012). Deep ultraviolet photoluminescence of water-soluble self-passivated graphene quantum dots. *ACS Nano*, *6*(6), 5102–5110. doi:10.1021/nn300760g PMID:22559247

Tian, L., Ghosh, D., Chen, W., Pradhan, S., Chang, X., & Chen, S. (2009). Nanosized carbon particles from natural gas soot. *Chemistry of Materials*, *21*(13), 2803–2809. doi:10.1021/cm900709w

Wang, X., Qu, K., Xu, B., Ren, J., & Qu, X. (2011). Microwave assisted one-step green synthesis of cell-permeable multicolor photoluminescent carbon dots without surface passivation reagents. *Journal of Materials Chemistry*, *21*(8), 2445–2450. doi:10.1039/c0jm02963g

Wei, J., Shen, J., Zhang, X., Guo, S., Pan, J., Hou, X., Zhang, H., Wang, L., & Feng, B. (2013). Simple one-step synthesis of water-soluble ñuorescent carbon dots derived from paper ash. *RSC Advances*, *3*(32), 13119–13122. doi:10.1039/c3ra41751d

Welsher, K., Liu, Z., Sherlock, S. P., Robinson, J. T., Chen, Z., Daranciang, D., & Dai, H. (2009). A route to brightly fluorescent carbon nanotubes for near-infrared imaging in mice. *Nature Nanotechnology*, *4*(11), 773–780. doi:10.1038/nnano.2009.294 PMID:19893526

Yan, X., Cui, X., & Li, L. S. (2010). Synthesis of large, stable colloidal graphene quantum dots with tunable size. *Journal of the American Chemical Society*, *132*(17), 5944–5945. doi:10.1021/ja1009376 PMID:20377260

Yang, X., Zhuo, Y., Zhu, S., Luo, Y., Feng, Y., & Dou, Y. (2014). Novel and green synthesis of high-ñuorescent carbon dots originated from honey for sensing and imaging. *Biosensors & Bioelectronics*, *60*, 292–298. doi:10.1016/j.bios.2014.04.046 PMID:24832204

Yang, Z., Gao, M., Wu, W., Yang, X., Sun, X. W., Zhang, J., Wang, H.-C., Liu, R.-S., Han, C.-Y., Yang, H., & Li, W. (2019). Recent advances in quantum dot-based light-emitting devices: Challenges and possible solutions. *Materials Today*, *24*, 69–93. doi:10.1016/j.mattod.2018.09.002

Yu, H., Zhao, Y., Zhou, C., Shang, L., Peng, Y., Cao, Y., Wu, L. Z., Tung, C. H., & Zhang, T. (2014). Carbon quantum dots/TiO_2 composites for efficient photocatalytic hydrogen evolution. *Journal of Materials Chemistry. A, Materials for Energy and Sustainability*, *2*(10), 3344–3351. doi:10.1039/c3ta14108j

Yu, J., Song, N., Zhang, Y. K., Zhong, S. X., Wang, A. J., & Chen, J. (2015). Green preparation of carbon dots by Jinhua bergamot for sensitive and selective ñuorescent detection of Hg^{2+} and Fe^{3+}. *Sensors and Actuators. B, Chemical*, *214*, 29–35. doi:10.1016/j.snb.2015.03.006

Zhang, X., Li, Q., Yan, S., Lei, W., Chen, J., & Qasim, K. (2019). A novel phototransistor device with dual active layers composited $CsPbBr_3$ and ZnO quantum dots. *Materials (Basel)*, *12*(8), 1215. doi:10.3390/ma12081215 PMID:31013902

Zhao, H. X., Liu, L. Q., Liu, Z. D., Wang, Y., Zhao, X. J., & Huang, C. Z. (2011). Highly selective detection of phosphate in very complicated matrixes with an off–on fluorescent probe of europium adjusted carbon dots. *Chemical Communications*, *47*(9), 2604–2606. doi:10.1039/c0cc04399k PMID:21234476

KEY TERMS AND DEFINITIONS

Absorption: A physical or chemical phenomena or process known as absorption occurs when atoms, molecules, or ions enter a liquid or solid material's bulk phase.

Carbonization: By use of destructive distillation, carbonization refers to the conversion of organic materials like dead plants and animal remains into carbon.

Energy Band Gap: A band gap, also known as an energy gap in solid-state physics, is an energy range in a solid where no electronic states can exist.

Nano-Materials: Nano-materials, in general, refer to substances with a unit size between 1 and 100 nm.

Quantum Dot: A quantum dot is a semiconductor nanometer-sized particle that typically has a core-shell structure.

Ultrasonication: In order to extract different substances from plants, microalgae, and seaweeds, sonication is the process of using sound energy to stir up particles in a sample. Since ultrasonic frequencies are typically employed, the procedure is also referred to as "ultrasonication" or "ultra-sonication."

UV-Visible Spectrophotometer: In the ultraviolet and the entire, neighboring visible parts of the electromagnetic spectrum, absorption spectroscopy or reflectance spectroscopy is referred to as UV-visible spectrophotometer.

Compilation of References

Abadi, P. B., Darlis, D., & Suraatmadja, M. S. (2018*). Green Energy Harvesting from Human Footsteps. MATEC Web of Conferences, 11015, 6–9. 10.10*51/matecconf/201819711015

Abarrategui, C. L., Espi, V. F., Alvarez, M. B. L., Pereira, C. D., & Garay, H. (2016). The intrinsic antimicrobial activity of citric acid-coated manganese ferrite nanoparticles is enhanced after conjugation with the antifung*al peptide Cm-p5. Internati*onal Journal of Nanomedicine, 11, 3849–3857. doi:10.2147/IJN.S107561 PMID:27563243

Abbas, A., Yasamin, G., Ahmad, G., Bahram, H., Samira, D., Mohamm*adreza, N., & Hashem, S. (2015). The Effect of Charge* at the Surface of Silver Nanoparticles on Antimicrobial Activity against Gram-Positive and Gram-Negative Bacteria: A Preliminary Study. Journal of Nanomaterials, 1–8.

Abbas, M. I., El-Khatib, A. M., Dib, M. F., Mustafa, H. *E., Sayyed, M. I., & Elsafi, M. (*2022). The influence of Bi2O3 nanoparticle content on the γ-ray interaction parameters of silicon rubber. Polymers, 14(5), 1048. doi:10.3390/polym14051048 PMID:35267871

Ab*del-Karim, R., Reda, Y., & Abdel-Fa*ttah, A. (2020). Nanostructured materials-based nanosensors. Journal of the Electrochemical Society, 167(3), 037554. doi:10.1149/1945-7111/ab67aa

Abdelbasir, S. M., McCourt, K. M., Lee, *C. M., & Vanegas, D. C. (2020). Wa*ste-Derived Nanoparticles: Synthesis Approaches, Environmental Applications, and Sustainability Considerations. Frontiers in Chemistry, 8, 782. doi:10.3389/fchem.2020.00782 PMID:*33110911*

*Abdolahzadeh, T., Morshedia*n, J., & Ahmadi, S. (2022). Preparation and characterization of nano WO3/Bi2O3/GO and BaSO4/GO dispersed HDPE composites for X-ray shielding application. Polyolefins Journ*al, 9.*

Abou-Hadid, A. F., Usama, El-Behairy, A., Elmalih, M. M., Amdeha, ·E., Ahmed, El Naggar, M. A., Taha, M. H., & Hussein, E. M.(2022). Production of efficient carbon fiber *from different solid waste residuals fo*r adsorption of hazardous metals from wastewater samples. Biomass Conversion and Biorefinery, 1, 1–16. doi:10.1007/S13399-022-03097-6

Abraham, K. (2020). How comparabl*e are sodium-*ion *b*atteries to lithium-ion counterparts? ACS Energy Letters, 5(11), 3544–3547.

Abualnaja, K. M., ElAassar, M. R., Ghareeb, R. Y., Ibrahim, A. A., & Abdelsalam, N. R. (2021). Development of photo-induced Ag0/TiO2 nano*composite coating* fo*r* photocatalysis, self-cleaning and antimicrobial polyester fabric. Journal of Materials Research and Technology, 15, 1513–1523. doi:10.1016/j.jmrt.2021.08.127

Abunahel, B. M., Ramli, R. M., Quffa, K. M., & Azman, N. Z. N. (2018). Effect of nanofibrous porosity on the X-ray attenuation properties *of electrospun n-Bi2O3/epoxy–polyvinyl alcohol (PV*A) *n*anofiber mats. Applied Physics. A, Materials Science & Processing, 124(8), 1–9. doi:10.100700339-018-1975-9

Acosta, S. (2020). Solvability for Photoacoustic Imaging with Idealized Piezoelectric Sensors. IEEE Transactions on Ultrasonics, Ferroelectrics, and Frequency Control, 3010(c), 1–10. doi:10.1109/TUFFC.2020.3005037 PMID:32746192

Adachi, K., Ohta, K., & Mizuno, T. (1994). Photocatalytic reduction of carbon dioxide to hydrocarbon using copper-loaded titanium dioxide. Solar Energy, 53(2), 187–190. doi:10.1016/0038-092X(94)90480-4

Adam, J. D., Davis, L. E., Dionne, G. F., Schloemann, E. F., & Stitzer, S. N. (2002). Ferrite devices and materials. IEEE Transactions on Microwave Theory and Techniques, 50(3), 721–737. doi:10.1109/22.989957

Adebayo-Tayo, B., Salaam, A., & Ajibade, A. (2019). Green synthesis of silver nanoparticle using Oscillatoria sp. extract, its antibacterial, antibiofilm potential and cytotoxicity activity. Heliyon, 5(10), e02502. doi:10.1016/j.heliyon.2019.e02502 PMID:31667375

Agarwal, H., Venkat Kumar, S., & Rajeshkumar, S. (2017). A review on green synthesis of zinc oxide nanoparticles – An eco-friendly approach. Resource-Efficient Technologies, 3(4), 406–413. doi:10.1016/j.reffit.2017.03.002

Agarwal, T., Narayana, S. G., Pal, K., Pramanik, K., Giri, S., & Banerjee, I. (2015). Calcium alginate- carboxymethyl cellulose beads for colon-targeted drug delivery. International Journal of Biological Macromolecules, 75, 409–417. doi:10.1016/j.ijbiomac.2014.12.052 PMID:25680962

Agarwal, V. (2018). Smart Sensors for Structural Health Monitoring- Overview, Challenges and Advantages. Sensors & Transducers, 221(3), 1–8.

Ahlawat, J., & Narayan, M. (2020). Introduction to Active, Smart, and Intelligent Nanomaterials for Biomedical Application. Intelligent Nanomaterials for Drug Delivery Applications, 1–16. doi:10.1016/b978-0-12-817830-0.00001-1

Ahmad, S., & Abdul, M. (2018). Energy Harvesting from Pavements and Roadways : A Comprehensive Review of Technologies, Materials, and Challenges. International Journal of Energy Research, 43(August), 1974–2015. doi:10.1002/er.4350

Ahmed, B., Shah, G. B., Malik, A. H., & Rizwan, M. (2020). Gamma-ray shielding characteristics of flexible silicone tungsten composites. Applied Radiation and Isotopes, 155, 108901. doi:10.1016/j.apradiso.2019.108901 PMID:31655354

Ahmed, T., Shahid, M., Noman, M., Bilal Khan Niazi, M., Zubair, M., Almatroudi, A., Khurshid, M., Tariq, F., Mumtaz, R., & Li, B. (2020). Bioprospecting a native silver-resistant Bacillus safensis strain for green synthesis and subsequent antibacterial and anticancer activities of silver nanoparticles. Journal of Advanced Research, 24, 475–483. doi:10.1016/j.jare.2020.05.011 PMID:32566283

Ahn, E. Y., Jin, H., & Park, Y. (2019). Assessing the antioxidant, cytotoxic, apoptotic and wound healing properties of silver nanoparticles green-synthesized by plant extracts. Materials Science and Engineering C, 101, 204–216. doi:10.1016/j.msec.2019.03.095 PMID:31029313

Ahn, R. Y. S., Han, M. H., & Kim, C. O. (1996). Synthesis of yttrium iron garnet precursor particles by homogeneous precipitation. Journal of Materials Science, 31(16), 4233–4240. doi:10.1007/BF00356444

Aichele, T., Lorenz, A., Hergt, R., & Görnert, P. (2003). Garnet layers prepared by liquid phase epitaxy for microwave and magneto-optical applications – a review. Crystal Research and Technology, 38(78), 575–587. doi:10.1002/crat.200310071

Akgenc, B., Sarikurt, S., Yagmurcukardes, M., & Ersan, F. (2021). Aluminum and lithium sulfur batteries: A review of recent progress and future directions. Journal of Physics Condensed Matter, 33(25), 253002.

Akgönüllü, S., Yavuz, H., & Denizli, A. (2020). SPR nanosensor based on molecularly imprinted polymer film with gold nanoparticles for sensitive detection of aflatoxin B1. Talanta, 219, 121219. doi:10.1016/j.talanta.2020.121219 PMID:32887120

Akhmetzhan, A., Abeu, N., Longinos, *S. N., Tashenov, A., M*yrzakhmetova, N., Amangeldi, N., & Toktarbay, Z. (2021). Synthesis and Heavy-Metal Sorption Studies of N, N-Dimethylacrylamide-Based Hydrogels. Polymer*s, 13(18), 3084.*

Akhtar, M. N., Yahya, N., & Nasir, N. (2011). Novel EM Antenna based on Y3Fe5O12 magnetic feeders for improved MVO. IEEE. doi: 10.1109/SIECPC.2011.5876902

Akpan, E. I., Shen, X., Wetzel, B., & Friedrich, K. (2019). Design and synthesis of polymer nanocomposites. Polymer Composites with Functionalized Nanoparticles. Elsevier.

Al-Ansari, M. M., Al-Dahmash, N. D., & Ranjitsingh, A. *J. A. (2021). Synthesis* of *s*ilver nanoparticles using gum Arabic: Evaluation of its inhibitory action on Streptococcus mutans causing dental caries and endocarditis. Journal of Infection and Public Hea*lth, 14(3), 324–330. doi:10.*1016/j.jiph.2020.12.016 PMID:33618277

Al-Balushi, M. A., Ahmed, N. M., Zyoud, S. H., Mohammed Ali, M. K., Akhdar, H., Aldaghri, O. A., & Ibnaouf, K. H. *(2022). Ionization r*adiation shielding effectiveness of lead acetate, lead nitrate, and bism*uth nitrate-doped zinc* oxide nanorods thin films: A comparative evaluation. Materials (Basel), 15(1), 3. doi:10.3390/ma15010003 PMID:35009*152*

Al-Turjman, F. (2020). Int*e*lligence and security in big 5G-oriented IoNT: An overview. Future Generation Computer Systems, 102, 357–368. doi:10.1016/j.future.2019.08.009

Al Aboody, M. S., & Mickymaray, S. (2020). Anti-fungal efficacy and mechanisms of flavonoids. Antibiotics (Basel, *Switzerl*and), 9(45), 1–43. PMID:31991883

Alahmad, A., Feldhoff, A., Bigall, N. C., Rusch, P., Scheper, T., & Walter, J. G. (2021). Hypericum perforatum L.-mediated green synthesis of *silver nanoparticles exhi*bit*i*ng antioxidant and anticancer activities. Nanomaterials (Basel, Switzerland), 11(487), 1–25. doi:10.3390/nano11020487 PMID:33673018

Alarifi, A., Deraz, N. M., & Shaban, S. (2009). Structural, morphological and magnetic properties of NiFe2O4 nano-particles. Journal of Alloys and Compounds, 486(1-2), 501–506. doi:1*0.1016/j.jallcom.2009.06.192*

*Alemi, A., Cor*man, F., & Lodewijks, G. (2016). Condition Monitoring Approaches for the Detection of Railway Wheel Defects. Journal of Rail and Rapid Transit, 0(0), 1–21. doi:10.1177/0954409716656218

*Alexopoulos, N. D., Bartholome, C., Pou*lin, P., & Marioli-Riga, Z. (2010). Structural health monitoring of glass fiber reinforced composites using embedded carbon nanotube (CNT) fibers. Composites Science and Technology, 70(2), 260–271. doi:10.1016/j.compscitech.2009.10.017

Ali, A., Rahman, K. S., Ali, L., Akh*taruzzam*an, M., Sopian, K., Radiman, S., & Amina, N. (2017). A computational study on the energy bandgap engineering in performance enhancement of CdTe thin film solar cells. Results in Physics, 7, 1066–1072. doi:10.1016/j.rinp.2017.0*2.032*

Ali, F., Raza, W., Li, X., Gul, H., & Ki-hyun, K. (2019). Piezoelectric Energy Harvesters for Biomedical Applications. Nano Energy, 57, 879–902. doi:10.1016/j.nanoen.2019*.01.012*

Allam, E. A., El-Sharkawy, R. M., El-Taher, A., Shaaban, E. R., Massoud, E. E. S., & Mahmoud, M. E. (2022). Enhancement and optimization of gamma radiation shielding by doped $n_a$no HgO into nanoscale bentonite. Nuclear Engineering and Tech*nology, 54(6), 2253–2261. doi:10.1016/j.net.2021.12.026*

*Allawadhi, P., S*ingh, V., Khurana, A., Khurana, I., Allwadhi, S., Kumar, P., Banothu, A. K., Thalugula, S., Barani, P. J., Naik, R. R., & Bharani, K. K. (2021). Silver nanoparticle based multifunctional approach for combating COVID-19. Sensors International, *2(May), 100101. doi*:10.1016/j.sintl.2021.100101 PMID:34766057

Almasoud, N., Alhaik, H., Almutairi, M., Houjak, A., Hazazi, K., Alhayek, F., Aljanoubi, S., Alkhaibari, A., Alghamdi, A., Soliman, D. A., Alomar, T. S., & Awad, M. A. (2021). *Green nanotechnology synthesized* silver nanoparticles: Characterization and testing its antibacterial activity. Green Processing and Synthesis, 10(1), 510–528. doi:10.1515/gps-2021-0048

Alshahri, S., Alsuhybani, M., Alosime, E., Almurayshid, M., Alrwais, A., & Alotaibi, S. (2021). LDPE/bismuth oxide nanocomposite: Preparation, characterization and application in X-ray shielding. Polymers, 13(18), 3081. doi:10.3390/polym13183081 PMID:34577982

Ambade, R. B., Veerasubramani, G. K., Ambade, S. B., Christy, M., Eom, W., Shin, H., ... Han, T. H. (2021). Photonic split-second induced mesoporous TiO2-Graphene architectures for efficient sodium-ion batteries. *Carbon, 178*, 332–344.

Amdeha, E. (2021). Recovery of Nanomaterials from Agricultural and Industrial Wastes for Water Treatment Applications. In A. S. H. Makhlouf & G. A. M. Ali (Eds.), Topics in Mining, Metallurgy and Materials Engineering (pp. 385–417). *Springer.* doi:10.1007/978-3-030-68031-2_14

Amdeha, E., Mohamed, R. S., & Dhmees, A. S. (2021). Sonochemical assisted preparation of ZnS–ZnO/MCM-41 based on blast furnace slag and electric arc furnace dust for Cr (VI) photoreduction. Ceramics International, 47(16), 23014–23027. doi:10.1016/j.ceramint.2021.05.015

Amir, M., Baykal, A., Guner, S., Sertkol, M., Sozeri, H., & Toprak, M. (2015). *Synthesis* and Characterization of CoxZn1−xAlFeO4 Nanoparticles. Journal of Inorganic and Organometallic Polymers and Materials, 25(4), 747–754. doi:10.100710904-014-0153-6

Amiri, S., & Shokrollahi, H. (2013). Magnetic and structural properties of RE doped Co-ferrite (RE = Nd, Eu, Gd) nanoparticles synthesized by co-precipitation. Journal of Magnetism and Magnetic Materials, 345, 18–23. doi:10.1016/j.jmmm.2013.05.030

Anantram, M. P., & Léonard, F. (2006). Physics of carbon nanotube electronic devices. Reports on Progress in Physics, 69(3), 507–561. doi:10.1088/0034-4885/69/3/R01

Anjum, S., Ishaque, S., Fatima, H., Farooq, W., Hano, C., Abbasi, B. H., & Anjum, I. (2021). *Emerging Applications of Nanotechnology* in Healthcare Systems: Grand Challenges and Perspectives; MDPI. Pharmaceuticals, 14(8), 707. doi:10.3390/ph14080707 PMID:34451803

Antoci, V. Jr, Adams, C. S., Hickok, N. J., Shapiro, I. M., & Parvizi, J. (2007). *Antibiotics* for local delivery systems cause skeletal cell toxicity in vitro. Clinical Orthopaedics and Related Research, 462, 200–206. doi:10.1097/BLO.0b013e31811ff866 PMID:17572634

Antony, V. (2021). A review on synthesis, characterization and potential biological applications of superpara*magnetic iron oxide nanoparticles*. Current Research in Green and Sustainable Chemistry, 4, 100042. doi:10.1016/j.crgsc.2020.100042

Anupama, M. K., Srinatha, N., Matteppanavar, S., Angadi, B., Sahoo, B., & Rudraswamy, B. (2018). Effect of Zn substitution on the structural and magnetic properties of nanocrystalline *NiFe2O4 ferrites. Ceramics International, 44*(5), 4946–4954. doi:10.1016/j.ceramint.2017.12.087

Arbain, R., Othman, M., & Palaniandy, S. (2011). Preparation of iron oxide nanoparticles by mechanical milling. Minerals Engineering, 24(1), 1–9. doi:10.1016/j.mineng.2010.08.025

Aridogan, U., & Basdogan, I. (2015). *A Review of* Active Vibration and Noise Suppression of Plate-like Structures with Piezoelectric Transducers. Journal of Intelligent Material Systems and Structures, 26(12), 1455–1476. doi:10.1177/1045389X15585896

Arif, M., Muhmood, T., Zhang, M., Amjad Majeed, M., Honglin, Y., Liu, X., & Wang, X. (2022). Highly visible-light active, eco-friendly artificial enzyme and 3D Bi4Ti3O12 biomimetic nanocomposite for efficient photocatalytic tetracycline hydro*chloride degradation a*nd Cr(VI) reduction. Chemical Engineering Journal, 434, 134491. doi:10.1016/j.cej.2021.134491

Arjmandi, R., Hassan, A., Majeed, K., & Zakaria, Z. (2015). Rice husk filled polymer composites. International Journal of Poly*mer Science, 2015.*

Arshadi, R., & Cobbold, R. (2017). A Pioneer in the Development of Modern Ultrasound: Robert William Boyle (1883-1955). Ultrasound in Medicine & Biology. doi:10.1016/j.ultrasmedbio.2006.07.030 PMID:17189042

Arshak, K., Nicholson, M., Arshak, A., Moore, E. G., Gaidan, I., & Cunniffe, C. (*2006). Examining the effects of polymer* b*in*der in Fe2O3/ZnO thick film sensors on the response to propanol vapour. In the 29th International Spring Seminar on Electronics Technology, (pp. 489-494). UPM.

Arshak, K., Twomey, K., & *Egan, D. (2002). A* Ceramic Thick Film Humidity Sensor Based on MnZn Ferrite. Sensors (Basel), 2(2), 50–61. doi:10.339020200050

Arulmurugana, R., Jeyadevan, B., Vaidyanathan, G., & Sendhilnathan, S. (2005). Effect of zinc substitution on Co-Zn and Mn-Zn ferrite nanoparticles *prepared by co-precipitati*on. Journal of Magnetism and Magnetic Materials, 288, 470–477. *doi:10.1016/j.jmmm.2004.09.138*

*Aryan, R., Ruby, & Mehata, M. S. (2021). Green synthesi*s of silver nanoparticles using Kalanchoe pinnata leaves (life plant) and their antibacterial and photocatal*ytic act*iv*i*ties. Chemical Physics Letters, 778, 138760–138770. doi:10.1016/j.cplett.2021.138760

Ashour, A. H., El-Batal, A. I., Maksoud, M. I. A. A., El-Sayyad, G. S., Labib, S., Abdeltwab, E., & El-Okr, M. M. (2018). Antimicrob*ial activity of metal-substitut*ed *c*obalt ferrite nanoparticles synthesized by sol-gel technique. Particuology, 40, 141–151. doi:10.1016/j.partic.2017.12.001

Askaruly, K., Azat, S., Sartova, Z., Yeleuov, M., Kerimkulova, A.*, & Bekseitova, K. (20*20*).* Obtaining And Characterization Of Amorphous Silica From Rice Husk. Journal of Chemical Technology & Metallurgy, 55(1).

Assar, S. T., Abosheiasha, H. F., & El Nimr, M. K. (2014). Comparison study 1 of the magnet*ic permeability and dc conductivity of Co-N*i*-Li* ferrite nanoparticles and their bulk counterparts. Journal of Magnetism and Magnetic Materials, 354, 1–6. doi:10.1016/j.jmmm.2013.10.029

Aswathi, V. P.*, Meera, S., Ann Maria, C. G., & Nidhin,* M. *(*2022). Green synthesis of nanoparticles from biodegradable waste extracts and their applications: a critical review; Nanotechnology for Environmental Engineering. Springer.

Atif, M., Asghar, M. W., N*adeem, M., Khalid,* W., Ali, Z., & Badshah, S. (2018). Synthesis and investigation of structural, magnetic and dielectric properties of zinc substituted cobalt ferrites. Journal of Physics and Chemistry *of Solids, 123,* 36–42. doi:10.1016/j.jpcs.2018.07.010

Autthawong, T., Chimupala, Y., Haruta, M., Kurata, H., Kiyomura, T., Yu, A.-s., ... Sarakonsri, T. (2020). Ultrafast-charging and long cycle-life anode materials of TiO 2-bronze/nitrogen-doped graphene nanocomposites for high-performance l*ithium-ion batteries. R*SC Advances, 10(71), 43811–43824.

Avdyli, M., & Kastrati, E. (2020). Aspects of the use of nanotechnology and nanomedicine in medicine. International Journal of Engineer$_{ing}$ an$_{d\ In}$no$_{vati}$ve $_{Tec}$h$_{n}$ology, 10(*3), 10–12.*

Awwad, A. M., & Salem, M., N., M.*Aqa*rbeh, M., & M.Abdulaziz, F. (2020). Green synthesis, characterization of silver sulfide nanoparticles and antibacterial activity evaluation. Chemistry International, 6(1), 42–48.

Ayatzhan, A., *Tashenov, A., Nurgeldi, A., Zhanar, O., Zhe*xenbek, T., Kaldibek, A., & Nuraje, N. (2021). P (DADMAAC-co-DMAA): Synthesis, thermal stability, and kinetics. Polymers for *Advanced Technologies, 32(7), 2669–2675.*

Azadmanjiri, J., Salehani, H. K., Barati, M. R., & Farzan, F. (2007). Preparation and electromagnetic properties of Ni1-xCuxFe2O4 nanoparticle ferrites by sol-gel auto-co^mbustion method. Materials Letters, 61(1), 84–87. doi:10.1016/j.matlet.2006.04.011

Azarbani, F., *& Shiravand, S. (2020*). Green synthesis of silver nanoparticles by Ferulago macrocarpa flowers extract and their antibacterial, antifungal and toxic effects. Green Chemistry Letters and Reviews, 13(1), 41–49. doi:10.1080/17518253.2020.1726504

Azat, S., *Arkhangelsky, E., Pa*pathanasiou, T., Zorpas, A. A., Abirov, A., & Inglezakis, V. J. (2020). Synthesis of bio-sourced silica–Ag nanocomposites and amalgamation reaction with mercury in aqueous solutions. Comptes Rendus. Chimie, 23(1), 77–92.

Azat, S., Busquets, R., Pavlenko, V. V., *Kerimkulova, A. R., W*hitby, R. L., & Mansurov, Z. A. (2014). Applications of activated carbon sorbents based on greek walnut. []. Trans Tech Publications Ltd.]. Applied Mechanics and Materials, 467, 49–51.

Azat, S., Korobei*nyk, A. V., Moustakas, K., & Inglezakis,* V. *J.* (2019). Sustainable production of pure silica from rice husk waste in Kaz*akhstan. Journal of C*leaner Production, 217, 352–359. doi:10.1016/j.jclepro.2019.01.142

Azat, S., Sartova, Z., Bekseitova, K., & Askaruly, K. (2019). Extraction of high-purity silica from rice husk via hydrochloric acid leaching treatment. Turkish Journ*al of Chemistry, 43(5)*, 1258–1269. doi:10.3906/kim-1903-53

AZoNano. (2013). Yttrium Iron Oxide / Yttrium Ferrite (Y3Fe*5O12) Nanoparticles – Properties, Applications. AZoNa*no.com.

Badran, H., Yahia, I. S., Hamdy, M. S., & Awwad, N. S. (2017). Lithium-doped hydroxyapatite nano-composites: Synthesis, characterization, gamma attenuation coefficient and dielectric properties. Radiation Physics and Chemistry, 130, 85–91. doi:10.1016/j.radphysc*hem.2016.08.001*

*Bagherzadeh, S. B., Kaz*emei*n*i, M., & Mahmoodi, N. M. (2021). Preparation of novel and highly active magnetic ternary structures (metal-organic framework/cobalt ferrite/graphene oxide) for effective visible-light-driven photocatalytic and photo-F*enton-like degradation of orga*nic contaminants. Journal of Colloid and Interface Science, 602, 73–94. doi:10.1016/j.jcis.2021.05.181 PMID:34118607

Bai, X., Zhang, H., Lin, J., & Zhang, G. (2021). UV-ozone contributions towards facile self-assembly and high performa*nce of silicon-carbon fibe*r materials as lithium-ion battery anodes. Journal of Colloid *and Interface Science, 598, 339–347. doi:10.1016/j.jcis.2021.04.044*

Baig, N., Kammakakam, I., Falath, W., & Kammakakam, I. (2021). Nanomaterials: A review of synthesis methods, properties, recent progress, and challenges. Materials Advances, 2(6), 1821–*1871. doi:10.1039/D0MA00807A*

*Baik, N. S., Sak*ai, G., Miura, N., & Yamazoe, N. (2000). Hydrothermally treated sol solution of tin oxide for thin-film gas sensor. Sensors and Actuators. B, Chemical, 63(1-2), 74–79. doi:10.1016/S0925-4005(99)00513-4

Baimenov*, A., Berillo, D., Azat, S., Nur*gozhin, T., & Inglezakis, V. (2020). Removal of Cd2+ from water by use of super-macroporous cryogels and comparison to commercial adsorbents. Polymers, 12(10), 2405.

Baker, S. N., & Baker, G. A. (2010). Luminescent carbon nanodots: Emergent *nanolights. Angewandte* Chemie International Edition, 49(38), 6726–6744. doi:10.1002/anie.200906623 PMID:20687055

Bakker, K., Apelqvist, J., & Schaper, N. C. (2012). *Practial guidelines on the management and prev*ention of the diabetic foot 2011. Diabetes/Metabolism Research and Reviews, 28(30), 225–231. doi:10.1002/dmrr.2253 PMID:22271742

Bamzai, K., & Bhat, M. (2014). Electrical and magnetic properties of some rare earth orthoferrites (RFeO3 where R = Y, Ho, Er) systems. Integrated Ferroelectrics, 158(1), 108–122. doi:10.1080/10584587.2014.957591

Barbillon, G. (2019*). Fabrication and SER*S performances of metal/Si and metal/ZnO nanosensors: A review. Coatings, 9(2), 86. doi:10.3390/coatings9020086

Barnard, A. S. (2012). Modelling of the *reactivity and stability of carbon nanotubes un*der environmentally relevant conditions. Physical Chemistry Chemical Physics, 14(29), 10080–10093. doi:10.1039/c2cp40862g PMID:22576146

Bashir, M., & Rajendran, *P. (2019). Recent Trends in Piezoelectric Smart Materials and* Its Actuators for Morphing Recent Trends in Piezoelectric Smart Materials and Its Actuators for Morphing Aircraft Development. International Review of Mechanical Engineering, 13(August), 117. doi:10.15866/ireme.v13i2.15538

Bashir, M., Rajendran, P., *Sharma, C., & Smrutiran*jan, D. (2018). ScienceDirect Investigation of Smart Material Actuators & Aerodynamic Optimization of Morphing Wing. Materials Today: Proceedings, 5(10), 21069–21075. doi:10.1016/j.matpr.2018.06.501

Bastrukov, S., K*hoo, J. Y., Lukiyanchuk, B., & Mo*lodtsova, I. (2012). Micromagnetic theory of spin relaxation and ferromagnetic resonance in multilayered metallic films, 2012. ar*Xiv:1210.2609*

*Batool, S., Id*rees, M., Kong, J., Zhang, J., Kong, S., & Dong, M. (2020). Assessment of the electrochemical behaviour of silicon@ carbon nanocomposite anode for lithium-ion batteries. Academic Press.

Bauhofer, W., & Kovacs, *J. Z. (2009*). A review and analysis of electrical percolation in carbon nanotube polymer composites. Composites Science and Technology, 69(10), 1486–1498. *doi:10.1016/j.compscit*ech.2008.06.018

Behzad, F., Naghib, S. M., kouhbanani, M. A. J., Tabatabaei, S. N., Zare, Y., & Rhee, K. Y. (2021). An o*verview of the plant-mediated green synthesis of noble metal nanoparticles f*or antibacterial applications. Journal of Industrial and Engineering Chemistry, 94, 92–104. doi:10.1016/j.jiec.2020.12.*005*

Bein, T. & Hanselka, H. (2014). The Integrated Project ' Intelligent Materials for A*ctive Noise Reduction '– An Overview and First Results The* Integrated Project. The Integrated Project, Intelligent Materials*d for Active Noise Red*uction. Vibration & Harshness Congress, Graz, 2-3.

Belov, L. A., Smolskiy, S. M., & Kochemasov, V. N. (2012). Handbook of RF, Micr*owave, and Millimeter-wave Components. A*rtech House.

Belrhazi, H., Hafidi, M. Y. E., & Hafidi, M. E. (2019). Permanent magnets elaboration from BaFe12O19 hexaferrite material: Simulation and prototype. Research and Developments in Material Science, 11, 1–5.

Bhalla, S., Moharana, S., Talakokula, V., & Kaur, N. (2018). *Piezoelectric Materials:* Appl*ications in SHM, Energy Harvesting & Biomechanics. Features Book Reviews, 43(February), 148–149.

Bhandare, S. V., Kumar, R., Anupama, A. V., Choudhary, H. K., Jali, V. M., & Sahoo, B. (2017). Annealing te*mperature-dependent structural and magnetic pro*perties of MnFe2O4 nanoparticles grown by sol-gel auto-combustion method. Journal of Magnetism and Magnetic Materials, 433, 29–34. doi:10.1016/j.jmmm.2017.02.040

Bharadwaj, K. K., Rabha, B., Pati, S., Choudhury, B. K., Sarkar, T., Gogoi, S. K., Kakati, N., Baishya, D., Kari, Z. *A., & Edinur, H. A. (2021). Green synthesis of sil*ver nanoparticles using diospyros malabarica fruit extract and assessments of their antimicrobial, anticancer and catalytic reduction *of 4-nitrophe*nol (4-np). Nanomaterials (Basel, Switzerland), 11(8), 1999–2043. doi:10.3390/nano11081999 PMID:34443829

Bharamagoudar, R. C., Patil, A. S., Mathad, S. N., Kumbar, V. M., & Kankanawadi, L. B. (2018). Magnetic and antibacterial studies of Nanoferrites prepared by self propagating high-temperature synthesis route. Acta Chemica IASI, 26(2), 249–262. doi:10.2478/achi-2018-0016

Bharti, M. K., Chalia, S., *Thakur, P., Sridhara,* S. N., Thakur, A., & Sharma, P. B. (2021). Nanoferrites heterogeneous catalysts for biodiesel production from soybean and canola oil: A review. Environmental Chemistry *Letters, 19(5), 3727–3746.* doi:10.100710311-021-01247-2 PMID:33967660

Bhat, A. S. (2014). Nanobots: The future of Medicine. International Journal of Engineer*ing and Management Scien*ces, 5(1), 44–49.

Bhattacharya, M., Malinen, M. M., Lauren, P., Lou, Y. R., Kuisma, S. W., Kanninen, L., & Laukkanen, A. (2012). Nanofibrillar cellu*lose hydrogel promotes* three-dimensional liver cell culture. Journal of Controlled Release, 164(3), 291–298. doi:10.1016/j.jconrel.2012.06.039 PMID:22776290

Bhukal, S., Shi*vali, & Singhal, S. (2014).* Magnetically separable copper substituted cobalt-zinc nano-ferrite photocatalyst with enhanced photocatalytic activity. Material*s Science in Semiconductor Processing, 26, 467–476.* doi:10.1016/j.mssp.2014.05.023

Bian, Y., Su, L., Yu, Z., Lv, Z., Chen, H., Zhou, Y., . . . Cao, H. (*2020). Graphite/copper phthalocyanine composite cathode fo*r overcharge protection and gas evolution suppression in aluminum-ion batteries at room temperature. Academic Press.

Bindhu, M. R., Umadevi, M., Esmail, G. A., Al-Dhabi, N. A., & Arasu, *M. V. (2020). Gre*en synthesis and characterization of silver nanoparticles from Moringa oleif*era flower and assessment of antimicrobial and sensing* properties. Journal of Photochemistry and Photobiology. B, Biology, 205, 111836. doi:10.1016/j.jphotobiol.2020.111836 PMID:32172135

Bock, J. A. N., & Su G. J. (1970). Interpre*tation of the inf*rared spectra of fused silica. Journal of the American Ceramic Society, 53(2), 69–73. doi:10.1111/j.1151-2916.1970.tb12012.x

Bodade, A. B., Bodade, A. B., Wankhade, H. G., Chaudhari, D. G., & Kothari, D. C. (2012). Conduction mechanism and gas sensing properties of CoFe2O4 n*anocomposite thic*k *fi*lms for H2S gas. Talenta, 89, 183–188. doi:10.1016/j.talanta.2011.12.013 PMID:22284478

Bonnet, M., Delapalme, A., Fuess, H., & Thomas, M. (1975). Re*finement of the structure of y*ttrium iron garnet (YIG). A case of severe extinction and absorption. Acta Crystallographica. Section B, Structural Crystallography and Crystal Chemistry, 31(9), 2233–2240. doi:10.1107/S0567740875007315

Botcha, S., & Prattipati, S. D. (2020). Callus Extract Mediated Green Synthesis *of Silver Nanoparticles, Their Characte*rization and Cytotoxicity Evaluation Against MDA-MB-231 and PC-3 Cells. BioNanoScience, 10(1), 11–22. doi:10.100712668-019-00683-3

Bradley, F. N. (1980). Materials for magnetic functions, 247-290. Hayden book company, Inc.

Bragg, W. H. (1915). The structure of the spinel group of crystals. Philosophical M*agazine, 30(176), 305–31*5.

Braginsky, V.B. (1987), Experimental observation of fundamental microwave absorption in high-quality dielectric crystal$_s$. Physics Letters A, 120(6), 300-305.

Brar, S. K., Verma, M., Tyagi, R. D., & *Surampal*l*i,* R. Y. (2010). Engineered nanoparticles in wastewater and wastewater sludge-Evidence and impacts. Waste Management (New York, N.Y.), 30(3), 504–520. doi:10.1016/j.wasman.2009.10.012 PMID:19926463

Brindley, G. W., & Zussman, J. (1959). In*fra-red absorption dat*a *fo*r serpentine minerals.

Britt, J., & Ferekides, C. (1993). Thin Film CdS/CdTe solar cell with 15.8% efficiency. Applied Physics Letters, 62(22), 2851–2852. doi:10.1063/1.109629

Brus, L. (1984). Electron-electron and electron-hole interactions in small semiconductor crystallites: The size dependence of the lowest excited electronic state. The Journal of Chemical Physics, 80(9), 4403–4409. doi:10.1063/1.447218

Brus, L. (1986). Electronic wave functions in semiconductor clusters: Experiment and theory. Journal of Physical Chemistry, 90(12), 2555–2560. doi:10.1021/j100403a003

Bryning, M. B., Islam, M. F., Kikkawa, J. M., & Yodh, A. G. (2005). Very low conductivity threshold in bulk isotropic single-walled carbon nanotube-epoxy composites. Advanced Materials, 17(9), 1186–1191. doi:10.1002/adma.200401649

Budenkov, G. A., Ivannikov, V. P., Kabakova, A. V., & Strizhak, V. A. (2009). Ultrasonic Quality Control of Press-Fit Joints. Acoustic Methods, 45(8), 575–581. doi:10.1134/S1061830909080105

Buggenhout, J., Brijs, K., Celus, I., & Delcour, J. A. (2013). The breakage susceptibility of raw and parboiled rice: A review. Journal of Food Engineering, 117(3), 304–315.

Burst, J. M., Duenow, J. N., Albin, D. S., Colegrove, E., Reese, M. O., Aguiar, J. A., Jiang, C. S., Patel, M. K., Al-Jassim, M. M., Kuciauskas, D., Swain, S., Ablekim, T., Lynn, K. G., & Metzger, W. K. (2016). CdTe solar cells with open-circuit voltage breaking the 1 V barrier. Nature Energy, 1(3), 16015. doi:10.1038/nenergy.2016.15

Buscaglia, V., Caracciolo, F., Bottino, C., Leoni, M., & Nanni, P. (1997). Reaction diffusion in the Y2O3-Fe2O3 system. Acta Materialia, 45(3), 1213–1224. doi:10.1016/S1359-6454(96)00246-7

Caglar, M., Zor, M., Ilican, S., & Caglar, Y. (2006). Effect of indium incorporation on the optical properties of spray pyrolyzed Cd0.22Zn0.78S thin films. Czechoslovak Journal of Physics, 56(3), 277–287. doi:10.100710582-006-0088-4

Cai, C., Wang, R., Liu, S., Yan, X., Zhang, L., Wang, M., Tong, Q., & Jiao, T. (2020). Synthesis of self-assembled phytic acid-MXene nanocomposites via a facile hydrothermal approach with elevated dye adsorption capacities. Colloids and Surfaces. A, Physicochemical and Engineering Aspects, 589, 124468. doi:10.1016/j.colsurfa.2020.124468

Cai, J., Qiu, L., Yuan, S., Shi, L., Liu, P., & Liang, D. (2012). Structural Health Monitoring for Composite Materials. In Composites and Their Applications. InTech. doi:10.5772/48215

Cai, T., Wang, L., Liu, Y., Zhang, S., Dong, W., Chen, H., Yi, X., Yuan, J., Xia, X., Liu, C., & Luo, S. (2018). Ag3PO4/Ti3C2 MXene interface materials as a Schottky catalyst with enhanced photocatalytic activities and anti-photocorrosion performance. Applied Catalysis B: Environmental, 239, 545–554. doi:10.1016/j.apcatb.2018.08.053

Caltun, O. F., Rao, G. S. N., Rao, K. H. B., Rao, P., Kim, C. G., Kim, C. O., Dumitru, I., Lupu, N., & Chiriac, H. (2007). High magnetostrictive cobalt ferrite for sensor applications. Sensor Letters, 5(1), 45–47. doi:10.11661.2007.027

Candless, B., Hegedus, A. S. S, & Sites, J.R.. (2003). Handbook of Photovoltaic Science and Engineering, 617-662. John Wiley and Sons.

Cao, S. W., Zhu, Y. J., Cheng, G. F., & Huang, Y. H. (2009). ZnFe2O4 nanoparticles: Microwave-hydrothermal ionic liquid synthesis and photocatalytic property over phenol. Journal of Hazardous Materials, 171(1–3), 431–435. doi:10.1016/j.jhazmat.2009.06.019 PMID:19564076

Carregal-Romero, S., Caballero-Diaz, E., Beqa, L., Abdelmonem, A. M., Ochs, M., Huhn, D., Suau, B. S., Valcarcel, M., & Parak, W. J. (2013). Muliplexed sensing and imaging with colloidal nano- and microparticles. Annual Review of Analytical Chemistry (Palo Alto, Calif.), 6(1), 53–81. doi:10.1146/annurev-anchem-062012-092621 PMID:23451718

Carta, D., C*asula, M. F., Falqui, A., Lo*che*,* D., Mountjoy, G., Sangregorio, C., & Corrias, A. (2009). A structural and magnetic investigation of the inversion degree in ferrite nanocrystals MFe2O4 (M = Mn, Co, Ni). The Journal of Physical Chemistry C, 113(20), 8606–8615. doi:10.*1021/jp901077c*

Castillo-Henríquez, L., Alfaro-Aguilar, K., Ugalde-álvarez, J., Vega-Fernández, L., de Oca-Vásquez, G. M., & Vega-Baudrit, J. R. (2020). Green synthesis of gold and silver nanoparticles from plant extracts and their possible applications *as antimicrobial agents in the agr*icul*tural area. Nanomaterials, 10(1763), 1763–1797.

Cecchini, R., Gajjela, R. S., Martella, C., Wiemer, C., Lamperti, A., Nasi, L., Lazzarini, L., Nobili, L. G., & Longo, M. (2019). High-De*nsity S*b2*Te3* Nanopillars Arrays by Templated, Bottom-Up MOCVD Growth. Small, 15(37), 1901743. doi:10.1002mll.201901743 PMID:31222940

Celinski, Z., Urquhart, K. B., & Heinrich, B. (1997), Using ferromagnetic resonance to measure the magnetic moments of ultrathin films. Jou*rnal of* Mag*n*etism and Magnetic Materials, 166(1–2), 6-26.

Celzard, A., McRae, E., De*leuze, C., Dufort, M., Furdin, G., & Marêché, J. F. (1996). Critical concentration in percolating s*ystems containing a high-aspect-ratio filler. Physical Review. B, 53(10), 6209–6214. doi:10.1103/PhysRevB.53.*6209 PMID:9982020*

Cerdan, K., Ouyang, W., Colmenares, J. C., Muñoz-Batista, M. J., Luque, R., & Balu, A. M. (2019). Facile mechanochemical modification of g-C3N4 for selective photo-oxidation of benzyl alcohol. Chemical Engineering Science, 194, 78–84. doi:10.1016/j.ces.20*18.04.001*

Chang, M., Morgan, G., Bed*i*er, F., Chieng, A., Gomez, P., Raminani, S., & Wang, Y. (2020). Recent Advances in Nanosensors Built with Pre-Pulled Glass Nanopipettes and Their Applications in Chemical and Biological Sensing. Journal of the Electrochemical Society, 167(3), 037533. doi:10.1149/1945-7111/*ab64be PMID:34326553*

*Chang, Y., Stephe*n *F* Poterala, Z. Y., Trolier-mckinstry, S., & Gary, L. (2009). Messing, Yunfei Chang, Stephen F Poterala, Zupei Yang, and Susan Trolier-mckinstry. 2012. "Electromechanic$_{a}$l Coupling over a B*road Temperature Range Electrom*ech*an*ical Coupling over a Broad Temperature Range. Applied Physics Letters, 232905, 2007–2010. doi:10.1063/1.3271682

Chatterjee, S., Saha, S., Barman, S. R., Khan, I., Pao, Y.-P., Lee, S., Choi, D., & Lin, Z.-H. (2020). Enhanced sensing performance of triboelectric nanosensors by solid-liquid contact electrification. Nano Energy, 77, 1*05093. doi:10.1016/j.* nanoen.2020.105093

Chauhan, K. R., Burgess, I. J., Chang, G. S., & Mukhopadhyay, I. (2014). Preparation of CdTe thin film by electrodeposition in butyl methyl imidazoli*um bath at 80 C. Journal of Elect*roanalytical Chemistry (Lausanne, Switzerland), 713, 70–76. doi:10.1016/j.jelechem.2013.11.032

Chaurasia, A. K., & Seidel, G. D. (2014). Computational micromechanics analysis of electron-hopping-induced conductive paths and assoc*iated macroscale piezoresistive r*esp*o*nse in carbon nanotube–polymer nanocomposites. Journal of Intelligent Material Systems and Structures, 25(17), 2141–2164. doi:10.1177/1045389X13517314

Chen, G., Weng, W., Wu, D., Wu, C., Lu, J., Wang, P., & Chen, X. (2004a). Preparation and characterization of gr*aphite nanosheets* from ultrasonic powdering technique. Carbon, 42(4), 753–759. doi:10.1016/j.carbon.2003.12.074

Chen, J., Qiu, Q., Han, Y., & Lau, D. (2018, September). Piezoelectric M*aterials fo*r *Su*stainable Building Structures : Fundamentals and Applications. Renewable & Sustainable Energy Reviews, 101, 14–25. doi:10.1016/j.rser.2018.09.038

Chen, J. L., & Yan, X. P. (2011). Ionic strength and pH reversible response of visible and near-infrared ﬁuo*rescence of graphene oxide nanoshe*ets for monitoring the extracellular pH. Chemical Communications, 47(11), 3135–3137. doi:10.1039/c0cc03999c PMID:21270983

Chen, K., Chou, W., Liu, L., Cui, Y., Xue, P., & Jia, M. (2019). Electrochemical sensors fabricated by electrospinning technology: An overview. Sensors (Basel), 19(17), 3676. doi:10.339019173676 PMID:31450877

Chen, K.-H., Namkoong, M. J., Goel, V., Yang, C., Kazemiabnavi, S., Mortuza, S., ... Sakamoto, J. (2020). Efficient fast-charging of lithium-ion batteries enabled by laser-patterned three-dimensional graphite anode architectures. Journal of Power Sources, 471, 228475.

Chen, Q., Wu, J., Ou, X., Huang, B., Almutlaq, J., Zhumekenov, A. A., Guan, X., Han, S., Liang, L., Yi, Z., Li, J., Xie, X., Wang, Y., Li, Y., Fan, D., Teh, D. B. L., All, A. H., Mohammed, O. F., Bakr, O. M., & Liu, X. (2018). All-inorganic perovskite nanocrystal scintillators. Nature, 561(7721), 88–93. doi:10.103841586-018-0451-1 PMID:30150772

Chen, S. C., Wu, Y. C., Mi, F. L., Lin, Y. H., Yu, L. C., & Sung, H. W. (2004). A novel pH- sensitive hydrogel composed of N, O-carboxymethyl chitosan and alginate cross-linked by genipin for protein drug delivery. Journal of Controlled Release, 96(2), 285–300. doi:10.1016/j.jconrel.2004.02.002 PMID:15081219

Cheng, C., Zhang, R., Wang, J., Zhang, Y., Wen, C., Tan, Y., & Yang, M. (2020). An ultrasensitive and selective fluorescent nanosensor based on porphyrinic metal–organic framework nanoparticles for Cu 2+ detection. Analyst (London), 145(3), 797–804. doi:10.1039/C9AN02231G PMID:31894762

Cheng, Y., Wang, Z., Chang, L., Wang, S., Sun, Q., Yi, Z., & Wang, L. (2020). Sulfur-mediated interface engineering enables fast SnS nanosheet anodes for advanced lithium/sodium-ion batteries. ACS Applied Materials & Interfaces, 12(23), 25786–25797.

Cheong, S. W., & Mostovoy, M. (2007). Multiferroics: A magnetic twist for ferroelectricity. Nature Materials, 6(1), 13–20. doi:10.1038/nmat1804 PMID:17199121

Cherepanov, V., Kolokolov, I., & L'Vov, V. (1993). The Saga of YIG: Spectra, Thermodynamics, Interaction and Relaxation of Magnons in a Complex Magnet. Physics Reports, 229(3), 84–144.

Chi, K., Li, Q., Meng, X., Liu, L., & Fu, W. (2017). Morphological control and characterization of CdTe microstructure arrays synthesized by one- step electrodeposition. Materials Letters, 194, 78–80. doi:10.1016/j.matlet.2017.02.023

Chiacchiarelli, L. M., Rallini, M., Monti, M., Puglia, D., Kenny, J. M., & Torre, L. (2013). The role of irreversible and reversible phenomena in the piezoresistive behavior of graphene epoxy nanocomposites applied to structural health monitoring. Composites Science and Technology, 80, 73–79. doi:10.1016/j.compscitech.2013.03.009

Choi, J. R. (2020). Development of point-of-care biosensors for COVID-19. Frontiers in Chemistry, 8, 517. doi:10.3389/fchem.2020.00517 PMID:32574316

Chorsi, M. T, Curry, E., Chorsi, H., Das, R., Baroody, J., Purohit, P., Ilies, H., & Nguyen, T. D. (2018). Piezoelectric Biomaterials for Sensors and Actuators. Advanced Materials, 1802084, 1–15. doi:10.1002/adma.201802084 PMID:30294947

Choti, C., Sharma, V., Tarachand, Deshpandey, U., Okram, G. S. (2017). Stable Dispersibility study of Cadmium Telluride Nanoparticles in Deionized water. SF Nano Res Let., 1.

Chou, R. L., Lin, M. S., & Chou, K. S (1986). Characteristic of CdTe Grown on Silicon by Low Pressure metal Organic Chemical Vapour Deposition. Appl. Phys. Leets., 48(8), 523–525. doi:10.1063/1.96494

Choudhary, D., Patil, T., Raichurkar, P., & Daberao, A. (2018). Review on Nanotechnology & Its Application in Coating Industry. International Journal of Scientific and Engineering Research, 09(4).

Chowdhury, S., Mishra, R., Saha, P., & Kushwaha, P. (2011). Adsorption thermodynamics, kinetics and isosteric heat of adsorption of malachite green onto chemically modified rice husk. Desalination, 265(1-3), 159–168.

Chu, Y., Zheng, X., & Fan, J. (2022). Preparation of sodium and boron co-doped graphitic carbon nitride for the enhanced production of H2O2 via two-electron oxygen reduction and the degradation of 2,4-DCP via photocatalytic *oxidation coupled with Fenton* oxidation. Chemical Engineering Journal, 431, 134020. doi:10.1016/j.cej.2021.134020

Coetzee, D., Venkataraman, M., Militky, J., & Petru, M. *(2020). Influence of Nanoparticles on Th*ermal and Electrical Conductivity of Composites. Polymers, 12(4), 742. doi:10.3390/polym12040742 PMID:32230802

Coots, J., Gandhi, *V., Onakoya, T., Chen, X., &* Larriba-Andaluz, C. (2020). A parallelized tool to calculate the electrical mobility of charged aerosol nanoparticles and ions in the gas phase. Journal of Aerosol Science, 147, 105570. doi:10.1016/j.jaerosci.2020.105570

Corbin, J. C., Moallemi, A., Liu, F., *Gagné, S., Olfert, J. S., Smallwood, G. J., & Lobo, P. (2020*). Closure between particulate matter concentrations measured ex situ by thermal–optical analysis and in situ by the CPMA–elec*trometer reference* mass system. Aerosol Science and Technology, 54(11), 1293–1309. doi:10.1080/02786826.2020.1788710

Cornish, K., & Li, Z. (2017). Guayule medical radiation attenuation glove. Rubber & Plastics News, Technical No*te, October, 2, 20-25.*

Corregidor, V., Saucedo, E., Fornaro, L., Castaño, J. L., & Dieguez, E. (2002). Defects in CdTe polycrystalline films grown by physical vapour deposition. Materials Science and Engineering B, 91–92, 525–528. doi:10.1016/S0921-5107(01)01063-7

Cost*a, A. C. F. M., Morelli,* M. *R.*, & Kiminami, R. H. G. A. (2007). Microstructure and magnetic properties of Ni1-xZnx-Fe2O4 synthesized by combustion reaction. Journal of Materials Science, 42(3), 779–783. doi:10.100710853-006-1440-6

Cottet, L., Almeida, C. A., Naidek, N., Viantea, M. F., Lopesa, M. C., & Deba*cher, N. A. (2014*). Adsorption characteristics of montmorillonite clay modified with iron oxide with respect to methylene blue in aqueous media. Applied Clay Science, 95, 25–31. doi:10.1016/j.clay.2014.03.023

Craig, B., *Schoetz, T., Cruden, A., & de Leon, C. P. (2020*). Review of current progress in non-aqueous aluminium batteries. Renewable & Sustainable Energy Reviews, 133, 110100.

Cuc, A., Giurgiutiu, V., Joshi, S., & Tidwell, Z. (2007). Structural health *monitoring with piezoelectric wafer active sens*ors for space applications. AIAA Journal, 45(12), 2838–2850. doi:10.2514/1.26141

Cui, Z., He, S. A., Liu, *Q., Guan, G., Zhang, W., Xu, C., ... Zou, R. (2020). Graphene-Like Carbon Film Wrapped Tin (II) Sulfide Nanosheet Arrays on Porous Carbon Fibers with Enh*anced Electrochemical Kinetics as High-Performance Li and Na Ion Battery Anodes. Advancement of Science, 7(18), 1903045.

Cunningham, J. R. Jr, & Anderson, E. E. (1960). Samarium substitution in yttrium iron *garnet. Journal of Applied Physics, 31(5),* S45–S46. doi:10.1063/1.1984598

Dahiya, S., & Duhan, S. (2018). Metal–ferrite nanocomposites for targeted drug delivery. Applications of Nanocomposite Materials in Drug Delivery.

Dairy, A. R. A., Al-Hmoud, L. A., & Khatatbet, H. A. (2019). Magnetic and structural properties of barium hexaferrite nanopar*ticles doped wit*h titanium. Symmetry, 11(6), 732. doi:10.3390ym11060732

Dalt, S. D., Takimi, A. S., Sousa, V. C., & Bergmann, C. P. (2009*). Magnetic and structural characterization of nanostructured* Mg*Fe*2O4 synthesized by combustion reaction. Particulate Science and Technology, 27(6), 519–527. doi:10.1080/02726350903328662

Dangi, A., Cheng, C., Agrawal, S., Datta, G. R., Benoit, R., Pratap, *R., Trolier-Mckinstry, S.*, & Sri-Rajasekhar, K. (2019). A Photoacoustic Imaging Device Using Piezoelectric Micromachined Ultrasound Transducers (PMUTs). IEEE Transactions on Ultrasonics, Ferroelectrics, and *Frequency Control. IEEE. . doi:10.1109/*TUFFC.2019.2956463

Das, B., Sharma, M., Baruah, M. J., Mounash, B. P., Karunakar, G. V., & Bania, K. K. (2020). Gold nanoparticle supported on mesoporous vanadium oxide for photo-oxidation of 2-naphtho*l with hydrogen* per*ox*ide and aerobic oxidation of benzyl alcohols. Journal of Environmental Chemical Engineering, 8(5), 104268. doi:10.1016/j.jece.2020.104268

Das, P., Dutta, T., Manna, S., Loganathan, S., & Basak, P. (2022). Facile green *synthesis of non-genotoxic, non-he*mol*y*tic organometallic silver nanoparticles using extract of crushed, wasted, and spent Humulus lupulus (hops): Characterization, anti-bacterial, and anti-cancer studies. Environmental Research, 204(PA), 111962.

Daulbayev, C., Sultanov, F., Korobeinyk, A. V., Y*eleuov, M., Azat, S.,* Ba*k*bolat, B., & Mansurov, Z. (2021). Bio-waste-derived few-layered graphene/SrTiO3/PAN as efficient photocatalytic system for water splitting. Applied Surface Science, 549, 149176.

De, B., & Karak, N. (2013). A green a*nd facile approach for the synt*he*s*is of water soluble ñuorescent carbon dots from banana juice. RSC Advances, 3(22), 8286–8290. doi:10.1039/c3ra00088e

Deepika, G J, Kaur, G., & Kumari, N. (20*19). Railway Accident Monitor*ing. International Journal on Recent and Innovation Trends in Computing and Communication, 5(5), 704–706.

Deivanayaki, S., Jayamurugan, P., Mariappan, R., & Ponnuswamy, V. (2010). Optical and structural *characterization of CdTe th*in *f*ilms by chemical bath deposition technique. Chalcogenide Letters, 7, 159–163.

Delacotte, C., Whitehead, G. F. S., Pitcher, M. J., Robertson, C. M., Sharp, P. M., Dyer, M. S., Alaria, J., Claridge, J. B., Darling, G. R., Allan, D. R., W*inter, G., & Rosseinsky, M. J. (2018).* Structure determination and crystal chemistry of large repeat mixed-layer hexaferrites. International Union of Crystallography Journal, 5(6), 681–698. doi:10.1107/S2052252518011351 PMID:30443353

Del Valle, E. M. (2004). Cyclodextrins and their uses: A review. Process Biochemistry (Barking, London, England), 39(9), 1033–1046. doi:10.101*6/S0032-9592(03)00258-9*

*Demir, M. E., C*he*hade*, G., Dincer, I., Yuzer, B., & Selcuk, H. (2019). Synergistic effects of advanced oxidization reactions in a combination of TiO2 photocatalysis for hydrogen production and wastewater treatment applications. I*nternational Journ*al of Hydrogen Energy, 44(43), 23856–23867. doi:10.1016/j.ijhydene.2019.07.110

Deng, J., Zhao, S., Liu, Y., Liu, C., & Sun, J. (2020). Nanosensors for diagnosis of infectious *diseases. ACS Applied Bio Materia*ls, 4(5), 3863–3879. doi:10.1021/acsabm.0c01247 PMID:35006812

Devi, G. S., Hyodo, T., Shimizu, Y., & Egashira, M. (2002). Synthesis of mesoporous TiO2-based powders and their gas-sensing properties. Sensors and Actuators. B, Chemical, *87(1),* 122–129. doi:10.1016/S0925-4005(02)00228-9

Dewangan, P. K., Khan, F., Shrivas, K., & Sahu, V. (2019). Determination of uranium in environmental sample by nanosensor graphene quantum dots. Journal of Radi*oanalytical and Nuclear Chemistry, 320(3), 7*57–763. doi:10.100710967-019-06512-x

Dharmadasa, I. M., Herrenden-Harker, W. G., & Williams, R. H. (1986). Metals on cadmium telluride: Schottky barriers and interface reactio*ns. Applied Physics Letters, 48(26), 1802–18*04. doi:10.1063/1.96792

Dhillon, S., Hernández, G., Wagner, N. P., Svensson, A. M., & Brandell, D. (2021). Modelling capacity fade in silicon-graphite composite electrodes for lithium-ion *batteries. Electrochimic*a *Act*a, 377, 138067.

Ding, M. (1992). Rice husk silicon and its applications. Inorg. Chem. Ind, 24(6), 36–42.

Ding, X., Liang, D., & Zhao, H. (2021). Enhanc*ed elect*rochemical Performance Promoted by Tin in silica Anode materials for Stable and High-capacity Lithium-ion Batteries. Academic Press.

Diso, D. G., Fauzi, F., Echendu, O. *K., Olusola, O. I., & Dharmadasa,* I. *M.* (2016). Optimisation of CdTe electrodeposition voltage for development of CdS/CdTe solar cells. Journal of Materials Science, 27, 12464–12472.

Divya, M., Natarajan, S., Lee, Y.-S., & Aravindan, V. (2020). Achieving high-energy dual carbon Li-ion capacitors with unique *low-and high-temperature performance from spe*nt *L*i-ion batteries. Journal of Materials Chemistry. A, Materials for Energy and Sustainability, 8(9), 4950–4959.

Divyanshi, T. (2018). Nanotechnology Market By Type (Nanosensor and Nanodevice) and Application (Electr*onics, Energy, Chemical Manu*facturing, Aerospace & Defense, Healthcare, and Others). Global Opportunity Analysis and Industry Forecast, 2021-2030; Report Cod*e: A04928. Allied Market Research. https://www.alliedmarketresearch.com/nanotechnology-market*

Dong, W., Li, W., Lu, N., Qu, F., Vessalas, K., & Sheng, D. (2019). Piezoresistive behaviours of cement-based sensor with carbon black subjected to various t*emperature and water content. Comp*os*i*tes. Part B, Engineering, 178, 107488. doi:10.1016/j.compositesb.2019.107488

Dong, X., Zheng, X., Deng, Y., Wang, L., Hong, H., & Ju, Z. (2020). SiO 2/N-doped graphene aerogel composite anode for lithium-ion batteries. Journal of Materials Science, 55, 13023–13035.

*Dong, Y., Chang, S. Q., Zhang, H. X., Ren, C., Ka*ng, B., Dai, M. Z., & Dai, Y. D. (2012). Effects of WO3 particle size in WO3/epoxy resin radiation shielding material. Chinese Physics Letters, 29(10), 108102. doi:10.1088/0256-307X/29/10/108102

Donga, S., *& Chanda, S. (2021). Facile green synthesis of silver nanoparticles using* Mangifera indica seed aqueous extract *and* its antimicrobial, antioxidant and cytotoxic potential (3-in-1 system). Artificial Cells, Nanomed$_{ic}$i$_{ne}$, and Biotechnology, 49(1), 292–302. doi:10.1080/*21691401.2021.1899193 PMID:33733973*

Dorj, U.-O., Lee, M., Choi, J.-y., Lee, Y.-K., & Jeong, G. (2017). The Intelligent Healthcare Data Management System Using Nanosensors. Journal of Sensors, 2017, *2017. doi:10.1155/20*17/7483075

Dresselhaus, M. S., Dresselhaus, G., & Saito, R. (1995). Physics of carbon nanotubes. Carbon, 33(7), 883–891. doi:10.1016/0008-6223(95)00017-8

Du, M., Zeng, G., Ye, C., Jin, H., Huan$_g$, J., Sun, D., Li, Q., Chen, B., & Li, X. (2020). Solvent-*free photo-thermocatalytic oxidation of benzyl a*lcohol on Pd/TiO2 (B) nanowires. Molecular Catalysis, 483, 110771. doi:10.1016/j.mcat.2020.110771

Dutta, T., Ghosh, N. N., Das, M., Adhikary, R., Mandal, V., & Chattopadhyay, A. P. (2020). Green synthesis of antibacterial and antifungal silver nanoparticles using Citrus limetta peel extract: Experimental and theoretical studies. Journal of Environmental Chemical E*ngineering, 8(4), 104019. doi:10.1*016/j.jece.2020.104019

d'Espinose de la Caillerie, J. B., Kermarec, M., & Clause, O. (1995). 29Si NMR observation of an amorphous magnesium silicate formed during impregnation of silica with Mg (II) in aqueous solution. Journal of Physical Chemistry, 99(47), 17273–172*81. doi:10.1021/j*100047a036

Ebbesen, T. W., & Ajayan, P. M. (1992). Large-scale synthesis of carbon nanotubes. Nature, 358(6383), 220–222. doi:10.1038/358220a0

Echendu, O. K., Okeoma, K. B., Oriku, C. I., & Dharmadasa, I. M. (2016). Electrochemical Depos*ition of CdTe Semiconductor Thi*n *Fi*lms for Solar Cell Application Using Two-Elecytrode and Three Electrode Configuration: A Comparative Study (Vol. 2*016). Advance in Materials Science and Engineering.*

Echiegu, E. A. (2016). Nanotechnology as a Tool for Enhanced Renewable Energy Application in Developing Countries (Editorial). Journal of Fundamentals of Renewable Energy J and Applicati*ons, 6(6). doi:10.4172/2090-4541.1000e113*

Eda, G., Lin, Y. Y., Mattevi, C., Yamaguchi, H., Chen, H. A., Chen, I. S., Chen, C.-W., & Chhowalla, M. (2010). Blue photolum*inescence from chemically derived graphene oxide. Advanced Materials, 22(4), 505–509. doi:10.1002/adma.200901996 PMID:20217743*

*Ekimov, A. I., & Onush*chenko, A. A. (1981). Quantum size effect in three-dimensional microscopic semiconductor crystals. JETP Letters, 34, 346–349.

El-Bassuony, A. A. H. (2020). Influence of High Annealing Temperature on Structural, Magnetic and Antimicrobial Activity *of Silver Chromite Nanoparticles for Biomedical Applicati*ons. Journal of Inorganic and Organometallic Polymers and Materials, 30(5), 1821–1828. doi:10.100710904-019-01306-w

El-Fouani, A. H., Aamouche, A., Mohseni, *F., Amaral, J. S., Tobaldi, D. M., &* Pu*ll*ar, R. C. (2019). Effect of surfactants on the optical and magnetic properties of cobalt zinc ferrite Co0.5Zn0.5Fe2O4. Journal of Alloys and Compounds, 774, 1250–1259. doi:10.1016/j.jallcom.2018.09 3 9 3

El-Khatib, A. M., Abbas, M. I, Elzah*er, M.* A., Badawi, M. S., Alabsy, M. T., Alharshan, G. A., & Aloraini, D. A. (2019). Gamma attenuation coefficients of nano cadmium oxide/high density polyethylene composites. Scientific Reports, 9(1), 1–11. doi:10.103841598-019-52220-7 PMID:31690761

El-Khawaga, *A. M., Farr*ag, *A.* A., Elsayed, M. A., El-Sayyad, G. S., & El-Batal, A. I. (2021). *Promising Antimicrobial and Azo* Dye Removal Activities of Citric Acid-Functionalized Magnesium Ferrite Nanoparticles. Journal of Cluster S*cience, 33(1), 197–213. do*i:10.100710876-020-01944-y

El-Salamony, R. A., Amdeha, E., El Shafey, A. M., & Al Sabagh, A. M. (2021). Preparation and characterisation of Ce-doped SiO2 nano-materi*als as effective photo-catalyst u*nd*er* visible light. International Journal of Environmental Analytical Chemistry, 1–16. doi:10.1080/03067319.2020.1865328

El-Sheekh, M. M., Shabaan, M. T., Hassan, L., & Morsi, H. H. (2020). Antiviral activity of algae biosynthesized silver and *gold nanoparticles again*st Herps Simplex (HSV-1) virus in vitro using cell-line culture technique. International Journal of Environmental Health Research, 1–12. PMID:32627584

Eren, E. (2009). Investigation of a basic dye removal fr*om aqueous solutio*n *on*to chemically modified Unye bentonite. Journal of Hazardous Materials, 166(1), 88–93. doi:10.1016/j.jhazmat.2008.11.011 PMID:19111978

Eskandarinezhad, S., Khosravi, R. K. *R., Amarzadeh,* M., Mondal, P., & Fiho, F. J. C. M. (2021). Application of different Nanocatalysts in industrial effluent treatment: A review. Journal of Xomposites and Compounds, 2(5). Advanc*e online publication. doi:1*0.52547/jcc.3.1.5

Eswaraiah, V., Balasubramaniam, K., & Ramaprabhu, S. (2011). Functionalized graphene reinforced thermoplastic nanocomposites as strain sensors in structural health monitoring. Journal of Materials Chemistry, 21(34), 12626. doi:10.1039/c1jm12302e

Evangeliou, N., Tsampiras, P., & T*zes, A. (2018*). Development of a Tendon-Driven Robotic Tool Targeting Visual-Servoing Minimally-Invasive Surgical Operations. 26th Mediterranean Conference on Control and Automation. IEEE. 10.1109/MED.2018.8442724

Faegh, E., Ng, B., Hayman, D., & Mustain, W. E. (2021). Practical assessment of the performance of aluminium battery technologies. Nature Energy, 6(1), 21–29.

Fainstein, A., Winkler, E., Butera, A., & Tallon, J. (1999). Magnetic interactions and magnon gap in the ferromagnetic superconductor RuSr2*GdCu2O8. Physical Review B: Condensed Matter and Materials Physics, 60(*18), R12597–R12600. doi:10.1103/PhysRevB.60.R12597

Fakher, M., Rahmanian, S., & Hosseini-Hashemi, S. (2019). On the carbon nanotube mass nanosensor by integral form of *nonlocal elasticity. International* Journal of Mechanical Sciences, 150, 445–457. doi:10.1016/j.ijmecsci.2018.10.051

Fan, H., Zhao, X., Yang, J., Shan, X., Yang, L., Zhang, Y., $_L$i, X., & Gao, M. (2012). ZnO-graphene composite for photocatalytic degradation of methylene blue dye. Catalysis Communications, 29, 29–*34. doi:10.1016/j.catcom.2012.09.013*

*Fa*n, X., Wang, Z., Cai, T., Yang, Y., Wu, H., Cao, S., ... Zhang, W. (2021). An integrated highly stable anode enabled by carbon nanotube-reinforced all-carbon binder for enhanced performance in lithium-ion battery. Carbon, 182, 749–*757. doi:10.1*016/j.carbon.2021.06.065

Fang, B., Xing, Z., Guo, M., Qiu, Y., Cui, Y., Li, Z., Wang, Y., Chen, P., & Zhou, W. (2021). Phosphorus-doping CdS@ NiFe layered double hydroxide as Z-Scheme heterojunction for enhanced photocatalytic and photo-fe$_n$t$_o$n degradation performance. Separation and Purification Technology, 274, 119066. doi:10.1016/j.seppur.20*21.119066*

*Fang, L., Sun, Q., Du*an, Y.-H., Zhai, J., Wang, D., & Wang, J. X. (2020). Preparation of transparent BaSO4 nanodispersions by high-gravity reactive precipitation combined with surface modification for transparent X-ray shielding nanocomposite films. Frontiers of Chemical Science and Engineering, 15(4), *902–912. doi:10.100711705-020-1985-y*

FAO. (2018). FAO Rice Market Monitor. FAO. https://www.fao.org/economic/est/publications/rice-publications/rice-market-monitor-rmm/en/

Fard, L. S., Peighambardoust, N. S., Jang, H. W., Dehghan, A., Saligheh, N. N. K., Iranpour, M., & Rajabi, M. I. (2020). The rechargeable aluminu$_m$-$_i$on battery with dif*ferent composite cathodes: A review. Jou*rnal of Composites and Compounds, 2(4), 138–146.

Faria, M. C. S., Rosemberg, R. S., Bomfeti, C. A., Monteiro, D. S., Barbosa, F., Oliveira, L. C. A., Rodriguez, M., Pereira, M. C., & Rodrigues, J. L. (2014). Arsenic removal from contaminated water by ultrafine d-FeOOH adsorbents. Chemical Engineering Journal, 237, 47–54. doi:*10.1016/j.cej.201*3.10.006

Farmani, A., Soroosh, M., Mozaffari, M. H., & Daghooghi, T. (2020). Optical nanosensors for cancer and virus detections Nanosensors for Smart Cities. Elsevier.

Fei, C., Chiu, C. T., & Qian, X. (2016). Author ' s Accep*ted Man*u*s*cript. Nano Energy. doi:10.1016/j.nanoen.2016.06.048

Feng, W., Maça, R. R., & Etacheri, V. (2020). High-energy-density sodium-ion hybrid capacitors enabled by interface-engineered hierarchical TiO2 nanosheet anodes. ACS Appli*ed Materials & Inte*r*f*aces, 12(4), 4443–4453.

Feng, X., Yu, Z., Long, R., Li, X., Shao, L., Zeng, H., Zeng, G., & Zuo, Y. (2020). Self-assembling 2D/2D (MXene/LDH) materials achieve ultra-high adsorption of heavy metals Ni$_2$+ through terminal group modification. Separation an*d Purification Technology, 253, 1*17525. doi:10.1016/j.seppur.2020.117525

Fernandes, R. J. C., Maghalhaes, C. A. B., Amorim, C. O., Amaral, V. S., Almeida, B. G., Castanheira, E. M. S., & Coutinho, P. J. G. (2019). Magnetic nanoparticles of zinc/calcium ferrite decorated with silver for photode*gradation of dyes. Materials (Basel),* 12(*21*), 3582. doi:10.3390/ma12213582 PMID:31683585

Ferrari, A., & Pizzo, P. (2016). Fully Predictive Common Rail Fuel Injection Apparatus Model and Its Application to Global System Dynamics Analyses. International Journal of Engine Research. doi:10.1177/1468087416653246

Fetisov, Y. K., & Srinivasan, G. (2005a). Ferrite piezoelectric microwave pha*se shifter: Studies on* el*ectric* field tunability. Electronics Letters, 41(19), 1066. doi:10.1049/el:20051653

Fetisov, Y. K., & Srinivasan, G. (2005b). Electrically tunable ferrite-ferroelectric microwave delay lines. Applied Physics Letters, 87(10), 103502. doi:10.1063/1.2037860

Fiebig. (2005). Revival of the magnetoelectric effect. Journal of Physics. D, Applied Physics, 38(8), R123–R152. doi:10.1088/0022-3727/38/8/R01

Figueroa, J. D., Fout, T., Plasynski, S., McIlvried, H., & Srivastava, R. D. (2008). Advances in CO2 capture technology-The U.S. Department of Energy's Carbon Sequestration Program. International Journal of Greenhouse Gas Control, 2(1), 9–20. doi:10.1016/S1750-5836(07)00094-1

Firby, C. J. & Eleezzabi, A. Y. (2017). Nonlinear optical modulation in a plasmonic Bi:YIG Mach-Zehnder Inferometer. SPIE Proceedings, Integrated Optics: Devices, Materials and Technologies XXI.

Firby, C. J., & Eleezzabi, A. Y. (2016). Design of integrated YIG-based isolators and high speed modulators. SPIE Proceedings, 9746.

Flores, R. G., Andersen, S. L. F., Maia, L. K. K., Jose, H. J., & Moreira, R. (2012). Recovery of iron oxides from acid mine drainage and their application as adsorbent or catalyst. Journal of Environmental Management, 111, 53–60. doi:10.1016/j.jenvman.2012.06.017 PMID:22820746

Fokina, E. L., Budim, N. I., Kochnev, V. G., & Chernik, G. G. (2004). Planetary mills of periodic and continuous action. Journal of Materials Science, 39(16-17), 5217–5221. doi:10.1023/B:JMSC.0000039213.44891.7d

Fontainha, C. C., Baptista-Neto, A. T., & Faria, L. O. (2016). Polymer-based nanocomposites of P (VDF-TrFE)/Bi. Journal of Materials Science, 4.

Forestier, H. (1928). Magnetic Transformations of the Sesquioxide of Iron, of Its Solid Solution and of Its Ferromagnetic Combinations (In Fr.). Annales de Chimie. (Paris), 9, 316–401.

Foteinidis, G., & Paipetis, A. S. (2021). A Novel Composite with Structural Health Monitoring Functionality via 2D and 3D Impedance Mapping Topography. Applied Sciences (Basel, Switzerland), 11(4), 1647. doi:10.3390/app11041647

Fowley, C., McCaughan, B., Devlin, A., Yildiz, I., Raymo, F. M., & Callan, J. F. (2012). Highly luminescent biocompatible carbon quantum dots by encapsulation with an amphiphilic polymer. Chemical Communications, 48(75), 9361–9363. doi:10.1039/c2cc34962k PMID:22892652

Fredrick, T. (2022). 5 Most Advanced Countries in Nanotechnology. Insider Monkey. https://www.insidermonkey.com/blog/5-most-advanced-countries-in-nanotechnology-1098326/5/

Frunza, L., Diamandescu, L., Zgura, I., Frunza, S., Ganea, C. P., Negrila, C. C., Enculescu, M., & Birzu, M. (2018). Photocatalytic activity of wool fabrics deposited at low temperature with ZnO or TiO2 nanoparticles: Methylene blue degradation as a test reaction. Catalysis Today, 306, 251–259. doi:10.1016/j.cattod.2017.02.044

Fu, Y., & Ma, Q. (2020). Recent developments in electrochemiluminescence nanosensors for cancer diagnosis applications. Nanoscale, 12(26), 13879–13898. doi:10.1039/D0NR02844D PMID:32578649

Galashev, A. Y., Ivanichkina, K. A., Katin, K. P., & Maslov, M. M. (2020). Computer test of a modified silicene/graphite anode for lithium-ion batteries. ACS Omega, 5(22), 13207–13218.

Ganesan, V., Hariram, M., Vivekanandhan, S., & Muthuramkumar, S. (2020). Periconium sp.(endophytic fungi) extract mediated sol-gel synthesis of ZnO nanoparticles for antimicrobial and antioxidant applications. Materials Science in Semiconductor Processing, 105, 104739. doi:10.1016/j.mssp.2019.104739

Gao, L., Thostenson, E. T., Zhang, Z., & Chou, T. W. (2009a). Coupled carbon nanotube network and acoustic emission monitoring for sensing of damage development in composites. Carbon, 47(5), 1381–1388. doi:10.1016/j.carbon.2009.01.030

Gao, M., Li, Y., Lu, J., Wang, Y., Wang, P., & Wang, L. (2018). Condition Monitoring of Urban Rail Transit by Local Energy Harvesting. International Journal of Distributed Sensor Networks, 14(11). doi:10.1177/1550147718814469

Gao, Y., Yan, S., He, Y., Fan, Y., Zhang, L., Ma, J., Hou, R., Chen, L., & Chen, J. (2021). A photo-Fenton self-cleaning membrane based on NH2-MIL-88B (Fe) and graphene oxide to improve dye removal performance. Journal of Membrane Science, 626, 119192. doi:10.1016/j.memsci.2021.119192

Gaponik, N. P., Talapin, D. V., Rogach, A. L., & Eychumuller, A. (2000). Electrochemical synthesis of CdTe Cd-nanocrystal/polypyrrole composites for optoelectronic applications. Journal of Materials Chemistry, 10(9), 2163–2166. doi:10.1039/b001958p

Garadkar, K. M., Pawar, S. J., Hankare, P. P., & Patil, A. A. (2010). Effect of annealing on chemically deposited polycrystalline CdTe thin films. Journal of Alloys and Compounds, 491(1-2), 77–80. doi:10.1016/j.jallcom.2009.10.146

Garay-Rodríguez, L. F., Torres-Martínez, L. M., & Moctezuma, E. (2019). Photocatalytic performance of K2Ti6O13 whiskers to H2 evolution and CO2 photo-reduction. Journal of Energy Chemistry, 37, 18–28. doi:10.1016/j.jechem.2018.11.014

Garcia-Muñoz, P., Fresno, F., de la Peña O'Shea, V. A., & Keller, N. (2020). Ferrite Materials for Photoassisted Environmental and Solar Fuels Applications. In M. Muñoz-Batista, A. Navarrete Muñoz, & R. Luque (Eds.), Heterogeneous Photocatalysis. Topics in Current Chemistry Collections (pp. 107–162). Springer. doi:10.1007/978-3-030-49492-6_4

Garg, A., Sharma, G. S., Goyal, A. K., Ghosh, G., Si, S. C., & Rath, G. (2020). Recent advances in topical carriers of anti-fungal agents. Heliyon, 6(8, e04663), 1–12. doi:10.1016/j.heliyon.2020.e04663 PMID:32904164

Garibo, D., Borbón-Nuñez, H. A., de León, J. N. D., García Mendoza, E., Estrada, I., Toledano-Magaña, Y., Tiznado, H., Ovalle-Marroquin, M., Soto-Ramos, A. G., Blanco, A., Rodríguez, J. A., Romo, O. A., Chávez-Almazán, L. A., & Susarrey-Arce, A. (2020). Green synthesis of silver nanoparticles using Lysiloma acapulcensis exhibit high-antimicrobial activity. Scientific Reports, 10(1), 1–11. doi:10.103841598-020-69606-7 PMID:32732959

Garimella, R. C., Sastry, V. R., & Mohiuddin, M. S. (2015). Piezo- Gen- An Approach to Generate Electricity from Vibrations. Procedia Earth and Planetary Science, 11, 445–456. doi:10.1016/j.proeps.2015.06.044

Gautam, R., & Chattopadhyaya, M. (2016). Nanomaterials for Wastewater Remediation. Academic Press. doi:10.1201/9781315368108

Gawas, U. B., Verenkar, V. M. S., Meena, S. S., & Bhatt, P. J. (2017). Influence of Mn Substitution on Mossbauer and Magnetic Properties of Ni-Zn Ferrite Nanoparticles. Journal of Superconductivity and Novel Magnetism, 30(11), 3241–3246. doi:10.100710948-017-4149-7

Geller, S., & Gilleo, M. A. (1957). The crystal structure and ferrimagnetism of yttrium-iron garnet, Y3Fe2(FeO4)3. Journal of Physics and Chemistry of Solids, 3(1-2), 30–36. doi:10.1016/0022-3697(57)90044-6

Ghareeb, N., & Farhat, M. (2018). Nanotechnology & Applications Smart Materials and Structures : State of the Art and Applications. Nanotechnology & Applications.

Ghazanfar, U., Siddiqi, S. A., & Abbas, G. (2005). Structural analysis of the Mn-Zn ferrites using XRD technique. Journal of Material Science and Engineering: B, 118(1), 84-86.

Ghidan, A. Y., & Al Antary, T. M. (2019). Applications of Nanotechnology in Agriculture. IntechOpen.

Ghojavand, S., Madani, M., & Karimi, J. (2020). Green synthesis, characterization and antifungal activity of silver nanoparticles using stems and flowers of Felty Germ*ander. Journal of Inorganic a*nd *O*rganometallic Polymers and Materials, 30(8), 1–11. doi:10.100710904-020-01449-1

Gilleo, M. A., & Geller, S. (1958). Magnetic and crystallographic properties of substituted yttrium-iron garnet, 3Y2*O3 xM2O3* (*5*-x)Fe2O3. Physical Review, 110(1), 73–78. doi:10.1103/PhysRev.110.73

Gingasu, D., Mindru, I., Patron, L., Moreno, J. M. C., Mocioiu, O. C., Preda, S., Stanica, N., $_{N}i_{t}$a, S., Dobre, N., Popa, M., Gradiste*anu, G., & Chifiriuc, M. C. (2016). Gr*een synthesis methods of CoFe2O4 and Ag-CoFe2O4 nanoparticles using Hibiscus extracts and their antimicrobial potential. Journal of Nanomaterials, 2016, 2106756. doi:10.1155/2016/2106756

Gojny, F. H., Wichmann, M. H. G., Köpke, U., Fiedler, B., & Schulte, K. (2004a). *Carbon nanotube-reinforced epoxy-composites: Enhanced stiffness and f*racture toughness at low nanotube content. Composites Science and Technology, 64(15 SPEC. ISS.), 2363–2371. doi:10.1016/j.compscitech.2004.04.002

Gomathi, A. C., Rajarathinam, S. R. X., Sadiq, A. M., Rajeshkumar, S., Rajarathinam, S. R. X., & Sadiq, A. M. (2019). Anticancer activity of silver nanoparticles synthesized using a*queous fruit shell extract* of Tamarindus indica on MCF-7 human breast cancer cell line. Journal of Drug Delivery Science and Technology, 1–28.

Gomez, A., Gich, M., Puig, T., & Obradors, X. (n.d.). Piezo-Generated Charge Mapping Revealed through Direct Piezo-electric Force Microscopy. Nature *Communications, 1–10. doi:10.103841467-017-0136*1-2 PMID:29062016

Gondek, J. J., Koprowski, J., & Wojcicki, M. A. (1984).. . Active and Passive Electronic Components, 11, 147.

Gong, S., & Cheng, W. (2017). One-dimensional nanomaterials for soft electronics. Advanced Electronic Materials, 3(3), 1600314. do*i:10.1002/aelm.20160031*4

*Go*nzalez, J. J., Mazario, E., Menendez, N., Marcos, J. S., Bonilla, A. M., & Herrasti, *P. (2016). Comparison of ferrite nanoparticles obtained electrochemically for catal*ytical reduction of hydrogen peroxide. Journal of Solid State Electro-chemistry, 20(4), 1191–1198. doi:10.100710008-015-2938-0

Gopakumar, D. A., Pai, A. R., Pasquini, D., Shao-Yuan, L., Abdul *Khalil, H. P. S., & T*homas, S. (2018). Nanomaterials-State of Art, New Challenges, and Opportunities. Nanoscale Materials in Water Purification, 1–24. doi:10.1016/B978-0-12-813926-4.00001-X

Gore, S. K., & Jadhav, S. S. (2020). Basics of ferrites. In R. S. Mane & V. V. Jadhav (Eds.), Spinel Ferrite Nanostructures for *Energy Storage Devices (pp. 1–11). Elsevier.* doi:10.1016/B978-0-12-819237-5.00001-8

Goulon, Rogalev, Wilhelm, Goujon, Yaresko, Brouder, & Ben Youssef. (2012). Site-selective couplings in x-ray-detected magnetic resonance spectra of rare-earth-substituted yttrium iron garnets. New Journal of Physics, 14(6), 063001.

Gozuak, F., Kose*oglu, Y., Baykal, A., & Kavas, H. (2009). S*ynthesis and characterization of CoxZn1-xFe2O4 magnetic nanoparticles via a PEG-assisted route. Journal of Magnetism and Magnetic Materials, 321(14), 2170–2177. d*oi:10.1016/j.jmmm.2009.01.*008

Graboski, A. M., Martinazzo, J., Ballen, S. C., Steffens, J., & Steffens, C. (2020). Nanosensors for water quality control Nanotechnology in t_{h}e Beverage Industry. Elsevier.

Graham, E., Goodi*ng, D., Gruszko, J., Grant, C., Na*ranjo, B., & Winslow, L. (2019). Light yield of perovskite nanocrystal-doped liquid scintillator. Journal of Instrumentation: An IOP and SISSA Journal, 14(11), P11024. doi:10.1088/1748-0221/14/11/P1*1024*

Grattan, K. T. V., & Meggitt, B. T. (Eds.). (1999). Optical Fiber Sensor Technology: Volume 3: Applications and Systems. Springer Science & Business Media.

Green, M. A., & Bremner, S. P. (2016). Energy conversion approaches and mat*erials for high-efficie*ncy photovoltaics. Nature Materials, 16(1), 23–34. doi:10.1038/nmat4676 PMID:27994249

Gu, S., & Zhu, A. (2020). Graphene nanosheets loaded Fe3O4 nanoparticles as a promising anode material for lithium ion batteries. *Journal of Alloys a*nd *C*ompounds, 813, 152160. doi:10.1016/j.jallcom.2019.15*2160*

*Gu, S., Zhang, D., Luo, S., & Yang, H. (2019). Construction of direct Z-Scheme photocatalyst by Mg1.2Ti1.8O5 and g-C3N4 nan*osheets toward photocatalytic H2 production and disinfection. International *Journal of Photo*energy, 2019, 1–9. doi:10.1155/2019/6307858

Guan, X., Xu, B., & Gong, J. (2020). Hierarchically architected polydopamine modified BaTiO3@ P (VDF-TrFE) nanocomposite fiber mats for flexible *piezoelectric nanogenerator*s *an*d self-powered sensors. Nano Energy, 70, 104516. doi:10.1016/j.nanoen.2020.104516

Gul, A. R., Shaheen, F., Rafique, R., Bal, J., Waseem, S., & Park, T. J. (2021). Grass-mediated biogenic synthesis of silver *nanoparticles and their drug delivery evaluation: A biocompatible anti-canc*er therapy. Chemical Engineering Journal, 407, 1–54. doi:10.1016/j.cej.2020.127202

Gul, I. H., Abbasi, A. Z., Amin, F., Anis-ur-Rehman, M., & Maqsood, A. (2007). Structural, magnetic and electrical properti*es of Co1-xZnxFe2O4 synthesized by co-precipitation method. Journal of Magnetism* and Magnetic Materials, 311(2), 494–499. doi:10.1016/j.jmmm.2006.08.005

Guo, J., & Li, X. (2018). Surface Effects on an Electrically Permeable Elliptical Nano-Hole or Nano-Crack in Piezoelectric Materials under Anti-Plane Shear. Acta Mechanica, 229(10), 4251–4266. doi:10.100700707-01*8-2232-1*

*Guo, L., Shen, X., Meng, X., & Feng, Y. (20*10). Effect of Sm3+ ions doping on structure and magnetic properties of nanocrystalline NiFe2O4 fibers. Journal of Alloys and Compounds, 490(1-2), 301–306. doi:10.1016/j.jallcom.2009.09.182

Guo, P.-H., Cheng, H.-M., Chen, J.-S., Wu, Y.-S., Yang, C.*-C., Zeng, F.-Y., ... Chen, P.-T. (20*21). Novel Triaxial Raman Scanning Platform for Evaluating Integrity of Graphite Electrodes in Li-Ion Batteries. IEEE Access: Practical Innovations, Open Solutions, 9, 81895–81901.

Guo, Q., *Shao, H., Zhang, K., Chen, G., K*ong, W., Feng, X., ... Dong, C. (2021). CoP Nanoparticles Intertwined with Graphene Nanosheets as a Superior Anode for $_{H}$alf/Full Sodium-Ion Batteries. ChemElectroChem, 8(11), 2022–2027.

Guo, X. Z., Ravi, B. G., Devi, P. S., Hanson, J. C., Margolies, J., Gambino, R. J., Parise, J. B., & Sampath, S. (2005, August). Synthesis of yttrium iron garnet (YIG) by citrate–nitrate gel combustion and prec*ursor plasma spray* processes. Journal of Magnetism and Magnetic Materials, 295(2), 145–154. doi:10.1016/j.jmmm.2005.01.007

Gupta, N., Goyal, S., *& Pawar,* S. (2020). Green synthesis of silver nanoparticles based on moringa oleifera leaf extract and its application in antimicrobial activity. International Journal of Advanced Science and Technology, 29(8 Special I$_{s}$sue), 1045–1052.

Gupta, N., Verma, A., Kashyap, S. C., & Dube, D. C. (2005). Dielectric behavior of spin-deposited nanocrystalline nickel-zinc ferrite thin films processed by citrate-route. Solid State Communications, 134(10), 689–694. doi:10.1016/j.ssc.2005.02.037

Gupta, V., Chaudhary, N., Srivastava, R., Sharm*a, G. D., Bhardwaj, R., & Chand, S. (2011). Lum*inscent graphene quantum dots for organic photovoltaic devices. Journal of the American Chemical Society, 133(26), 9960–9963. doi*:10.1021/ ja*2036749 PMID:21650464

Guptaa, P. (2006). High efficiency ultra-thin sputtered CdTe solar cells.Solar Energy Mat*erials &amp. Solar Cells, 90(15), 2263–2271. doi:10.1016/j.solmat.2006.02.029*

Habibullah, G., Viktorova, J., & Ruml, T. (2021). Current Strategies for Noble Metal Nanoparticle Synthesis. Nanoscale Research Letters, 16(1), 47. doi:10.118611671-021-03480-8 PMID:33721118

Habisreutinger, S. N., Schmidt-Mende, L., & Stolarczyk, J. K. (2013). Photocatalytic reduction of CO2 on TiO2 and other semiconductors. Angewandte Chemie International Edition, 52(29), 7372–7408. doi:10.1002/anie.201207199 PMID:23765842

Haggag, E. G., Elshamy, A. M., Rabeh, M. A., Gabr, N. M., Salem, M., Youssif, K. A., Samir, A., Bin Muhsinah, A., Alsayari, A., & Abdelmohsen, U. R. (2019). Antiviral potential of green synthesized silver nanoparticles of lampranthus coccineus and malephora lutea. International Journal of Nanomedicine, 14, 6217–6229. doi:10.2147/IJN.S214171 PMID:31496682

Hahn, O. (1936). Applied Radiochemistry. Cornell University Press.

Hajalilou, A., Kamari, H. M., & Shameli, K. (2017). Dielectric and electrical characteristics of mechanically synthesized Ni-Zn ferrite nanoparticles. Journal of Alloys and Compounds, 708, 813–826. doi:10.1016/j.jallcom.2017.03.030

Halo, T. L., McMahon, K. M., Angeloni, N. L., Xu, Y., Wang, W., Chinen, A. B., Malin, D., Strekalova, E., Cryns, V. L., Cheng, C., Mirkin, C. A., & Thaxton, C. S. (2014, December 02). NanoFlares for the detection, isolation, and culture of live tumor cells from human blood. Proceedings of the National Academy of Sciences of the United States of America, 111(48), 17104–17109. doi:10.1073/pnas.1418637111

Hamdi, K., Aboura, Z., Harizi, W., & Khellil, K. (2020). Structural health monitoring of carbon fiber reinforced matrix by the resistance variation method. Journal of Composite Materials, 54(25), 3919–3930. doi:10.1177/0021998320921476

Hammad, M., Hardt, S., Mues, B., Salamon, S., Landers, J., Slabu, I., Wende, H., Schulz, C., & Wiggers, H. (2020). Gas-phase synthesis of iron oxide nanoparticles for improved magnetic hyperthermia performance. Journal of Alloys and Compounds, 824, 153814. doi:10.1016/j.jallcom.2020.153814

Han, M., Lv, Z., Hou, L., Zhou, S., Cao, H., Chen, H., ... Cai, M. (2020). Graphitic multi-walled carbon nanotube cathodes for rechargeable Al-ion batteries with well-defined discharge plateaus. Journal of Power Sources, 451, 227769.

Han, M., Vestal, C. R., & Zhang, Z. J. (2004). Quantum couplings and magnetic properties of CoCrxFe2-xO4 (0<x<1) spinel ferrite nanoparticles synthesized with reverse micelle method. The Journal of Physical Chemistry B, 108(2), 583–587. doi:10.1021/jp035966m

Han, N., & Ho, J. C. (2014). One-dimensional nanomaterials for energy applications. Nanocrystalline Materials, 75-120.

Hariharan, D., Thangamuniyandi, P., Christy, A. J., Vasantharaja, R., Selvakumar, P., Sagadevan, S., & Nehru, L. (2020). Enhanced photocatalysis and anticancer activity of green hydrothermal synthesized Ag@ TiO2 nanoparticles. Journal of Photochemistry and Photobiology. B, Biology, 202, 111636. doi:10.1016/j.jphotobiol.2019.111636 PMID:31739259

Harris, V. G., Geiler, A., Chen, Y., Yoon, S. D., Wu, M., Yang, A., Chen, Z., He, P., Parim, P. V., Zuo, X., Patton, C. E., Abe, M., Acher, O., & Vittoria, C. (2009). Recent advances in processing and applications of microwave ferrites. Journal of Magnetism and Magnetic Materials, 321(14), 2035–2047. doi:10.1016/j.jmmm.2009.01.004

Harvey, D. (2000). Modern Analytical Chemistry. McGraw-Hill.

Hasanzadeh, M., Shadjou, N., & de la Guardia, M. (2015). Iron, and iron-oxide magnetic nanoparticles as signal-amplification elements in electrochemical biosensing. Trends in Analytical Chemistry, 72, 1–9. doi:10.1016/j.trac.2015.03.016

Hashem, E. M., Hamza, M. A., El-Shazly, A. N., Abd El-Rahman, S. A., El-Tanany, E. M., Mohamed, R. T., & Allam, N. K. *(2021). Novel Z-Scheme/Type-II CdS@ZnO/g-C*3N4 ternary nanocomposites for the durable photodegradation of organics: Kinetic and mechanistic insights. Chemosphere, 277, 128730. doi:10.1016/j.ch$_e$mosphere.2020.128730 PMID:33189399

Hashemi, Z., Mohammadyan, M., Naderi, S., Fakhar, M., Biparva, P., Akhtari, J., & Ebrahimzadeh, M. A. (2021). *Green synthesis of silver nanoparticles us*ing *F*erula persica extract (Fp-NPs): Characterization, antibacterial, antileishmanial, and in vitro anticancer activities. Materials Today Communications, 27(102264), 1–10. doi:10.1016/j.mtcomm.2021.102264

Hash*emian, S., & Foroghimoqh*adam, A. (2014). Effect of copper doping on CoTiO3 ilmenite type nanoparticles for removal of congo red from aqueous solution. Chemical Engineering Journal, 235, 299–306. doi:10.1016/j.cej.2013.08.089

Hashemian, S., & Salimi, M. (2010). Nano composite a potentia*l low* co*st* adsorbent for removal of cyanine acid. Chemical Engineering Journal, 188, 57–63. doi:10.1016/j.cej.2012.02.008

Hathiya, L. J., Baraliya, J. D., Das, A., Sen, D., Gismelseed, A. M., Yousif, A. A., & Joshi, H. H. (2022). Anomalous magnetic behaviour at nano-scale of Mn2+-substituted magnesio-ferrite synthesized by auto-combustion technique. Indian Journal of Physics, 96(8), 2323–2335. *doi:10.100712648-021-02169-z*

*Hat*hiya, L. J., Baraliya, J. D., Jethva, H. O., & Joshi, H. H. (2022). Structural, dielectric and ac electrical properties of nano-structured Mn0.6Co0.2Ni0.2Fe2O4 ternary ferrite synthesized by citrate-gel auto-combustion technique. Indian Journal of Pu*re and Applied Physics, 60, 59–71.*

Hathiya, L. J., Tanna, A. R., Baraliya, J. D., & Joshi, H. H. (2020). Study of ternary nano-structured fe*rrite Mn0.6Co0.2Ni0.2Fe2O4 for its magnetic and chemical dye degradation propert*ies. In Proceedings of the 3rd International Conference on Condensed Matter and Applie*d Physics (*ICC-2019), AIP Conference Proceedings, 2220, 020155. AIP.

Haw, C., Chiu, W., Abdul Rahman, S., Khiew, P., Radiman, S., Abdul Shukor, R., Hamid, M. A. A., & Ghazali, N. (2016). The design of new magn*etic-photocatalyst nanocomposites* (C*o*Fe2O4-TiO2) as smart nanomaterials for recyclable-photocatalysis applications. New Journal of Chemistry, 40(2), 1124–1136. doi:10.1039/C5NJ02496J

Hawar, S. N., Al-Shmgani, H. S., Al-Kubaisi, Z. [A]., Sulaiman, G. M., Dewir, Y. H., & Ri*kisahedew, J. J. (2022). Green Synthe*sis *of* Silver Nanoparticles from Alhagi graecorum Leaf Extract and Evaluation of Their Cytotoxicity and Antifungal Activity. Journal of Nanomaterials, 2022, 1–8. doi:10.1155/2022/1058119

Haye, E., Capon, F., Barrat, S., Boulet, P., Andre, R., Carteret, C., & Bruyere, S. (2015). Properties *of rare-earth o*rtho*f*errites perovskite driven by steric hindrance. Journal of Alloys and Compounds, 657, 631–638. doi:10.1016/j.jallcom.2015.10.135

Heju, X., Xin, L., Liu, L., Pang, D., Jiao, Y., Cong, R., & Yu, W. (20*19). Large area MoS2/Si heterojunction-b*ased solar cell through sol-gel method. Materials Letters, 238, 13–16. doi:10.1016/j.matlet.2018.11.051

Hema, E., Manikandan, A., $_G$ayathri, M., Durka, M., Antony, A. S., & Venkatraman, B. R. (2016). Role of Mn2+ - *doping on structural, morphological, optical,* magnetic and catalytic properties spinel ZnFe2O4 nanoparticle. Journal of Nanoscience and Nanotechnology, 16(6), 5929–5943. doi:10.1166/jnn.2016.11037 PMID:27427654

Hemeda, O. M., Eid, M. E. A., Sharshar, T., Ellabany, H. M., & He*naish, A. M. A. (2021). Synthesis o*f *nan*ometer-sized PbZrxTi1-xO3 for gamma-ray attenuation. Journal of Physics and Chemistry of Solids, 148, 109688. doi:10.1016/j.jpcs.2020.109688

Hemlata, M., Meena, P. R., Singh, A. *P., & Tejavath, K. K. (2020). Bi*osynthesis of Silver Nanoparticles Using Cucumis prophetarum Aqueous Leaf Extract and Their Antibacterial and Antiproliferative Activity against Cancer Cell Lines. ACS Omeg*a, 5(10), 5520–5528. doi:10.*1021/acsomega.0c00155 PMID:32201844

Herbin, H. B., Aravind, M., Amalanathan, M., Mary, M. S. M., Lenin, M. M., Parvathiraja, C., Siddiqui, M. R., Wabaidur, S. M., & Islam, M. A. (2022). Synthesis of Silver Nanoparticles Using Syzygium malaccense Fruit Extract and Evaluation of Their Catalytic Activity and Antibacterial Properties. Journal of *Inorganic and Organometallic Polymers and* Materials, 32(3), 1–13. doi:10.100710904-021-02210-y

Herold, S., & Mayer, D. (2016). Adaptive Piezo*electric Absor*ber for Active Vibration Control. Actuators, 5(1), 7. doi:10.3390/act5010007

Herrera, G. (2018). Lead-Free Piezoelectric Ceramics as Candidates on the Development of Morphing Winglets in Aircrafts. Journal of Marine Science and Engineering, 8(1), 32–42. doi:10.17265/2161-6213/2018.1-2.005

Hes$_a$ri, M. Mohammadi, P., Khademi, F., Shackbaei, D., Momtaz, S., M*osaefi, N., Farzaei,* M. H., & Abdollahi, M. (2021). Current advances in the use of Nanophytomedicine Therapies for Human Cardiovascular diseases. International Journal of Nanomedicine, *16, 3292–3315.*

Hetsch, F., Zhao, N., Kershaw, S. V., & Rogach, A. L. (2013). Quantum dot field effect transistor. Materials Today, 16(9), 312–325. doi:10.1016/j.mattod.2013.08.011

Hibst, H. (1982). Hexagonal ferrites from *melts and* aqueous solutions, magnetic recording materials. Angewandte Chemie International Edition, 21(4), 270–282. doi:10.1002/anie.198202701

Hilpert, S. (1909). Correspondence as to Structure and Origin in Magnetic Properties of Ferrite an*d Iron Oxide. (In Ger.). Berichte der Deutsch*en *Ch*emischen Gesellschaft, 42, 2248–2261. doi:10.1002/cber.190904202121

Hilpert, S. (1909). Manufacturing Method of Magnetic Materials Having Small Electric Conductivity for Electric and Magnetic Apparatus. German Pat., *226, 347*, *227*, 787.

Hippel Von, A. (1950). Ferroelectricity, Domain Structure, and Phase Transitions of Barium Titanate. Reviews of Modern Physics, 22(3), 22. doi:10.1103/RevModPhys.22.221

Hjiri, *M. (2020). Highly sensitive NO 2 gas sensor based* on *h*ematite nanoparticles synthesized by sol–gel technique. Journal of Materials Science Materials in Electronics, 31(6), 5025–5031. doi:10.100710854-020-03069-4

Holm, U., Sohlstrom, H., & Brogardh, T. (1984). YIG-Sensor Design for Fiber Optical Magn*etic-Field Measurements. P*roceed*ings - Society of Photo-Optical Instrumentation Engineers, 514, 333–336.

Homg, R. Y., Fu, H. P., Di, G. Q., Zheng, Y., & Wei, D. G. (2008, March 15). Facile route to γ-Fe2O3/SiO2 nanocomposite used as a precursor of magn*etic fluid. J. Magn. Magn. Mat*er. Materials Chemistry and Physics, 108(1), 132–141. doi:10.1016/j.matchemphys.2007.09.014

Hong, L., Qu, Y., Dhupia, J. S., Sheng, S., Tan, Y., & Zhou, Z. (2017). A Novel Vibration-Based *Fault Diagnostic Algorithm for* Gear*b*oxes Under Speed Fluctuations Without Rotational Speed Measurement. Elsevier. doi:10.1016/j.ymssp.2017.02.024

Hongjie, Z., Ji, Z., Yang, B., Zhilun, G. & L$_{on}$gtu, L. (2004) Effect of Bi-sub$_s$titution on the ie*lectric properties of poly*cry*st*alline yttrium iron garnet. Journal of Magnetism and Magnetic Materials, 280(2-3), 208–213. doi:10.1016/j.jmmm.2004.03.014

Hosseini, S., & Amoozadeh, A. (2018). Nano-TiO2-P25-SO3H as a new and robust photo-catalyst: The acceleration effect of selective oxidat*ion of aromatic alcohols to aldehydes under blue LED irradiation. Jou*rnal of Photochemistry and Photobiology A Chemistry, 364, 516–523. doi:10.1016/j.jphotochem.2018.06.035

Hou, Y., Li, M., Gu, Y., Yang, Z., Li, R., & Zhang, Z. (2018). Gamma ray shielding property *of tungsten* powder modified continuous basalt fiber reinforced epoxy matrix composites. Polymer Composites, 39(S4), E2106–E2115. doi:10.1002/pc.24469

Houshyar, S., Shanks, R. A., & Hodzic, A. (2005). The effect of fiber concentration on mechanical and thermal properties of fiber-reinforced polypropylene composites. Journal of Applied Polymer Science, 96(6), 2260–2272. doi:10.1002/app.20874

Houston, D. F. (1972). Rice: Chemistry and *Technology (No. HOU 633*.18 (BR 125.9)). American Association of Cereal Chemists.

Hu, L., Shang, C., Wang, X., & Zhou, G. (2021). Fe 7 Se 8 encapsulated in N-doped carbon nanofibers as a stable anode *material for sodium ion batteries. Nano*scale Advances, 3(1), 231–239.

Hu, Y., & Lin, Z. (2015). Recent Progress in Piezoelectric Nanogenerators as a Sustainable Power Source in Self-Powered Systems and Active Sensors. Nano Energy, 14, 1–12. doi:10.1016/j*.nanoen.2014.11.038*

*Hu, Z., Zhang, H., Wang, H., Zha*ng, F., Li, Q., & Li, H. (2020). Nonaqueous aluminum ion batteries: Recent progress and prospects. ACS Materials Letters, 2(8), 887–904.

Huang, H., Zh$_o$u, F, Lu, *P., Li, X., Das, P., Feng, X., ... Wu, Z.-S.* (2020). Design and construction of few-layer graphene cathode for ultrafast and *high-capacity aluminum-ion batteries. Energy Storage Materials, 27, 396–404.*

Huang, L., Zeng, L., Zhu, J., Sun, L., Yao, L., Deng, L., & Zhang, P. (2021). Oxygen-vacancy-rich TiO2-coated carbon nanofibers for fast sodium storage in high-perfor*mance sodium-ion hybrid capacitors. Journal of Powe*r Sources, 493, 229678.

Huang, Q., Ni, S., Jiao, M., Zh*ong, X., Zhou, G., & Cheng, H. M. (2021). Ali*gned Carbon-Based Electrodes for Fast-Charging Batteries: A Review. Small, 17(48), 2007676.

Huang, Y., Wang, K., Guo, T., Li, J., Wu, X., & Zhang, G. (2020). Construction of 2D/2D Bi2Se3/g-*C3N4 nanocomposite with High interfacial charge separation and* photo-heat conversion efficiency for selective photocatalytic CO2 reduction. Applied Catalysis B: Environmental, 277, 119232. doi:10.1016/j.apcatb.2020.119232

$_H$uo, X, Liu, J, Li, *J., Zhang, B., Zha*ng, Y., Yu, Y., & Kang, F. (2020). Hexagonal composite CuSe@ C as a positive electrode for high-performance aluminum batteries. ACS Applied Energy Materials, 3(11), 11445–11455.

Huq, M. A. (2020). Green synthesis of silver nanoparticles usi$_n$g pseudodugane$_l$l$_a$ eburnea MAHUQ-39 and their antimicrobial mechanisms investigation against *drug resistant human pa*th*ogens*. International Journal of Molecular Sciences, 21(4), 1–14. doi:10.3390/ijms21041510 PMID:32098417

Hussain, M., Raja, N. I., Iqbal, M., & Aslam, S. (2019). Applications of Plant Flavonoids in the Green Synthesis of Colloidal Sil*ver Nanoparticles and Impacts on Huma*n Health. Iranian Journal of Science and Technology. Transaction A, Science, 43(3), 1381–1392. doi:10.100740995-017-0431-6

Hussain, S. Z., Ihrar, M., Hussain, S. B., Oh, W. C., & Ullah, K. (2020). A review on graphene based transition metal oxide composites and its application towards supercapacitor electrodes. SN Applied Sciences, 2, *1–23.*

Hüseynli, S., Çimen, D., Bereli, N., & Denizli, A. (2019). Molecular imprinted based quartz crystal microbalance nanosensors for mercury detection. Global Challenges, 3(3), 1800071. doi:10.*1002/gch2.201800071* PMID:31565367

Ibrahim, I., Ali, I. O., Salama, T. M., Bahgat, A. A., & Mohamed, M. M. (2016). Synthesis of mag*netically recyclable spinel ferrite (MFe2O4,* M = Zn, Co, Mn) nanocrystals engineered by sol gel-hydrothermal technology: High catalytic *performances for nitroarene*s reduction. Applied Catalysis B: Environmental, 181, 389–402. doi:10.1016/j.apcatb.2015.08.005

Idrees, M. (2015). Role of nanotechnology in Medical Sciences: A Review. International Journal of Innovative Drug Discovery, 5(1), 14–24.

Ihokura, K., & Watson, I. (1994). The Stannic Oxide Gas Sensor-Principles and Applications. CRC Press, Boa Raton.

Iijima, S. (1991). Helical microtubules of graphitic carbon. Nature, 354(6348), 56–58. doi:10.1038/354056a0

Ijaz, I., Gilani, E., Nazir, A., & Bukhari, A. (2020). Detail review on chemical, physical and green synthesis, classification, characterizations and applications of nanoparticles. Green Chemistry Letters and Reviews, 13(3), 59–81. doi:10.1080/17518253.2020.1802517

Ilahi, I., Khuda, F., Umar Khayam Sahibzada, M., Alghamdi, S., Ullah, R., Zakiullah, Dablool, A. S., Alam, M., Khan, A., & Ali Khan Khalil, A. (2021). Synthesis of silver nanoparticles using root extract of Duchesnea indica and assessment of its biological activities. Arabian Journal of Chemistry, 14(5), 103110. doi:10.1016/j.arabjc.2021.103110

Imran, P., & Uzun, M. (2011). The Requirement for Piezoelectric Smart Material for Current and Future Applications. Journal of Engineering and Natural Sciences, (216), 395–411.

Inglezakis, V. J., Azat, S., Tauanov, Z., & Mikhalovsky, S. V. (2021). Functionalization of biosourced silica and surface reactions with mercury in aqueous solutions. Chemical Engineering Journal, 423, 129745.

Inoue, T., Fujishima, A., Konishi, S., & Honda, K. (1979). Photoelectrocatalytic reduction of carbon dioxide in aqueous suspensions of semiconductor powders [3]. Nature, 277(5698), 637–638. doi:10.1038/277637a0

Iqbal, K., Halsby, K., Murray, R. D., Carroll, P. V., & Petermann, R. (2019). Glucocorticoid management of adrenal insufficiency in the United Kingdom: Assessment using real-world data. Endocrine Connections, 8(1), 20–31. doi:10.1530/EC-18-0418 PMID:30562160

Iqbal, M. J., & Ashiq, M. N. (2008). Physical and electrical properties of Zr-Cu substituted strontium hexaferrite nanoparticles synthesized by co-precipitation method. Chemical Engineering Journal, 136(2-3), 383–389. doi:10.1016/j.cej.2007.05.046

İrim, Ş. G., Wis, A. A., Keskin, M. A., Baykara, O., Ozkoc, G., Avcı, A., Doğru, M., & Karakoç, M. (2018). Physical, mechanical and neutron shielding properties of h-BN/Gd_2O_3/HDPE ternary nanocomposites. Radiation Physics and Chemistry, 144, 434–443. doi:10.1016/j.radphyschem.2017.10.007

Ishak, W. S. (1998). Magnetostatic wave technology: A review. Proceedings of the IEEE, 76(2), 171–187. doi:10.1109/5.4393

Ishikawa, R., Yamaguchi, T., Ohtaki, Y., Akiyama, R., & Kuroda, S. (2016). Thin film growth of a topological crystal insulator SnTe on the CdTe (111) surface by molecular beam epitaxy. Journal of Crystal Growth, 4531, 124–129. doi:10.1016/j.jcrysgro.2016.08.027

Isshiki, M., & Wang, J. (2017). II-IV Semiconductors for Optoelectronics: CdS, CdSe, CdTe. In S. Kasap & P. Capper (Eds.), Springer Handbook of Electronic and Photonic Materials. Springer Handbooks. Springer. doi:10.1007/978-3-319-48933-9_33

Iurgiutiu, V., & Ictor, G. (2002). Electro-Mechanical Impedance Method for Crack Detection in Thin Plates. Journal of Intelligent Material Systems and Structures, 12(October). doi:10.1106/HHXL-TLM8-8CM5-6OAK

Ivanov, S. (2012). Advanced Functional Materials Science and Technology of Atomic, Molecular, Condensed Matter & Biological Systems. Academic Press.

Ivanovskaya, M., Bogdanov, P., Faglia, G., & Sberveglieri, G. (2000). The features of thin film and ceramic sensors at the detection of CO and NO2. Sensors and Actuators. B, Chemical, 68(1-3), 344–350. doi:10.1016/S0925-4005(00)00455-X

Iwasaki, S. (1984). Perpendicular magnetic recording-evolution and future. IEEE Transactions on Magnetics, 20(5), 657–662. doi:10.1109/TMAG.1984.1063563

Izadkhah, H., Zare, S., Somu, S., Lombardi, F., & Vittoria, C. (2017). Utilizing alternate target deposition to increase the magnetoelectric effect at room temperature in a single phase M-type hexaferrite. *MRS Communications, 7(2), 97*–101. doi:10.1557/mrc.2017.36

Jabeen, S., Qureshi, R., Munazir, M., Maqsood, M., Munir, M., Shah, S. S. H., & Rahim, B. Z. (2021). Application of green synthesized silver nanoparticles in cancer treatment - A critical review. Mate*rials Research Express, 8(9), 092001.* d*oi:1*0.1088/2053-1591/ac1de3

Jacob, J., & Khadar, M. A. (2010). VSM and Mossbauer study of nanostructured hematite. Journal of Magnetism and Magnetic Materials, 322(6), *614–621. doi:10.1016/j.jmmm.2009.1*0.025

Jafari, T., Moharreri, E., Amin, A. S., Miao, R., Song, W., & Suib, S. L. (2016). Photocatalytic water splitting - The untamed dream: A review of *recent advances. Molecules* (B*as*el, Switzerland), 21(7), 900. doi:10.3390/molecules21070900 PMID:27409596

Jahan, I., Erci, F., & Isildak, I. (2019). Microwave-Assisted Green Synthesis of Non-Cytotoxic Silver Nanoparticle*s Using the Aqueous Extract of Rosa san*tana (rose) Petals and Their Antimicrobial Activity. Analytical Letters, 52(12), 1860–1873. doi:10.1080/00032719.2019.1572179

Jain, N., Jain, P., Rajput, D., & Patil, U. K. (2021). Green synthesized plant-based silver nanoparticles: therapeutic prospective for anticancer and antiviral activity. Micro and Nano Systems Letters, *9(1).*

Jain, S., & Mishra, R. (2015). A Review on Piezoelectric Material as a Source of Generating Electricity and Its Possibility to Fabricate Device. Int. J. Systems, Control and Communications.

Jandosov, J., Alavijeh, M., Sultakhan, S., B*aimenov, A., Bernardo, M., Saki*pova, Z., & Berillo, D. (2022). Activated Carbon/Pectin Composite Enterosorbent for Human Protection from Intoxication with Xenobiotics Pb (II) and Sodium Diclofenac. Molecules (Basel, Switzerland), 27(7), 2296.

Jani, K. H., Chhantbar, M. C., & Joshi, H. H. (2008). Study of magnetic ordering in MnAlxCrxFe2−2xO4. Journal of *Magnetism and Magnetic Materials, 320(18), 2208–2214. doi:10.1016/j.jmmm.2008.03.061*

Jaswal, L., & Singh, B., Ferrite materials: A chronological review, Journal of Integrated Science and Technology, 2(2), 69-71.

Jauhar, S., Kaur, J., Goyal, A., & Singhal, S. (2016). Tuning the propert*ies of cobalt Ferrite: A road* tow*a*rds diverse Applications. RSC Advances, 6(100), 97694–97719. doi:10.1039/C6RA21224G

Javaid, M., Haleem, A., Singh, R. P., Rab, S., & Suman, R. (2021). Exploring the potential of nanosensors: A brief overview. Sensors International, 2, 100130. doi:10.1016/j.*sintl.2021.100130*

Jebril, S., Kh*anf*ir Ben Jenana, R., & Dridi, C. (2020). Green synthesis of silver nanoparticles using Melia azedarach leaf extract and their antif*ungal activities: In vitro* and in vivo. Materials Chemistry and Physics, 248, 122898. doi:10.1016/j.matchemphys.2020.122898

Jeisthardt, R. M., & Topic, M. Sites, J. R. (2015). Characteristics of CdTe grown on Si by low pressure met*alorganic chemical vapou*r *dep*osition, 5. IEE.

Jeon, H. Y. (2015). Nanotechnology and formulations of modeling of hydraulic permeability for nonwoven geotextiles. Intechopen. doi:10.5772/61997

Jeong, J., Cho, M., Lim, Y. T., *Song, N. W., & Chung, B. H. (2009).* Synthesis and characterization of a photoluminescent nanoparticle based on fullerene-silica hybridization. Angewandte Chemie International Edition, 48(29), 5296–5299. doi:10.1002/anie.200901750 PMID:19533701

Jeseentharani, V., George, M., Jeyaraj, B., Dayalan, A., *& Nagaraj, K. S. (2013). Synthesis of metal ferrite (*MFe*2O*4, M = Co, Cu, Mg, Ni, Zn) nanoparticles as humidity sensor materials. Journal of Experimental Nanoscience, 8(3), 358–370. doi:10.1080/17458080.2012.690893

Jesudoss, S. K., Vijaya, J. J., Kennedy, L. J., Rajan, P. I., Al-Lohedan, H. A., Ramalingam, R. J., Kaviyarasu, K., & Bououdina, M. (2016). S*tudies on the efficient dual performance of* Mn*1-x*NixFe2O4 spinel nanoparticles in photodegradation and antibacterial activity. Journal of Photochemistry and Photobiology. B, Biology, 165, 121–132. doi:10.1016/j.jphotobiol.2016.10.004 PMID:27776260

Jeyadevan, B., *Chinnasamy, C. N., Shinoda, K., T*ohji, K., & Oka, H. (2003). Mn-Zn ferrite with higher magnetization for temperature sensitive magnetic fluid. Journal of Applied Physics, 93(10), 8450–8452. doi:10.1063/1.1543135

Jhalani, V. A., Zho$_u$, J. J., Park, J., Dreyer, C., & Bernardi, M. (n.d.) Piezoelectric Electron-Phonon Interaction from Ab Ini*tio Dynamic*al *Qu*adrupoles: Impact on Charge Transport. In Wurtzite GaN, 1–6.

Jia, X., Zhou, Y., Zheng, J., Li, Y., Zou, H., & Xie, R. (2016). Cerium doped $_b$arium tantalates: Fabrication, characterization, and investigation of gamma r*adiation attenuation. Journa*l *of* Alloys and Compounds, 688, 679–684. doi:10.1016/j.jallcom.2016.07.188

Jiang, L. L., Yan, C., Yao, Y. X., Cai, W., Huang, J. Q., & Zhang, Q. (2021). Inh*ibiting solvent co-intercalati*on *i*n a graphite anode by a localized high-concentration electrolyte in fast-charging batteries. Angewandte Chemie International Edition, 60(7), 3402–3406.

Jiang, X., Kim, K., Zhang, S., Johnson, J., & Salazar, G. (2014). High-Temperature Piezoelectric Sensing. Sensors (Basel), 14(1), 144–169. doi:10.3390140100144 PM*ID:24361928*

Jiang, X., Li, Y., Li, J., Wang, J., & Yao, J. (2014, January). Piezoelectric Energy Harvesting from Traffic- Induced Pavement Vibrations. Journal of Renewable and Sustainable Energy, 043110(4), 043110. Advance online publication. doi:10.1063/1.4891169

Jin, C., W*ang, K., Oppong-Gyeb*[i,] *A., & Hu, J. (2020). Application of Nanotechnology in Cancer Diagnosis and Therapy - A Mini-Review. Internatio*nal Journal of Medical Sciences, 17(18), 2964–2973. doi:10.7150/ijms.49801

Jin, H., Yu, J., Lin, S., Gao, S., Yang, H., Haick, H., & Liu, Y. (2020). Nanosensor-based flex[ib]le electronic assisted with light fidelity communicating technology for *volatolomics-based teleme*dici*n*e. ACS Nano, 14(11), 15517–15532. doi:10.1021/acsnano.0c06137 PMID:33141556

Jodin, G., Scheller, J., Duhayon, E., Rouchon, J. F., & Braza, M. (2017). Implementation of a Hybrid Electro-Act$_{ive}$ A$_{ctu}$at$_{ed}$ Mo$_r$p$_h$ing Wing in Wind Tunnel. Diffusion and Defect Data, Solid State Data. P*art B, Solid State Phenomena, 260, 85–91. . do*i:10.4028/www.scientific.net/SSP.260.85

Johar, N., Ahmad, I., & Dufresne, A. (2012). Extraction, preparation and characterization of cellulose fibres and nanocrystals from rice husk. Industrial Crops and Products, 37(1), 93–99.

Jorge de Souza, T. A., Rosa S*ouza, L. R., & Franchi,* L. P. (2019). Silver nanoparticles: An integrated view of green synthesis methods, transformation in the environment, and toxicity. Ecotoxicology and Environmental Safety, 171(December 2018), 691–700.

Joseph, J., KerenDeborah, S., Raghavi, R., Mary Shamya, A., & Aruni, W. (2021). Green synth*esis of silver nanoparticl*es using Phyllanthus amarus Seeds and their antibacterial activity assessment. Biomedical and Biotechnology Research Journal, 5(1), 35–38.

Jotania, R. (2014). Crystal structure, magnetic properties and advances *in hexaferrites: A brief review. AIP* Conference Proceedings, 1621, 596–599. doi:10.1063/1.4898528

Jouyban, A., & Rahimpour, E. (2020). Optical sensors based on silver nan0particles for determination of pharmaceuticals: An overview *of advances in th*e *las*t decade. Talanta, 217, 121071. doi:10.1016/j.talanta.2020.121071 PMID:32498884

Kakihana, M. (1996). Invited review "sol–gel" preparation of high temperature superconducting oxides. Journal of Sol-Gel Science and Technology, 6(1), 7–55. doi:10.1007/BF00402588

*Kaliannan, D., Palaninaicker, S., Palaniv*el, V., Mahadeo, M. A., Ravindra, B. N., & Jae-Jin, S. (2019). February). A novel approach to preparation of nano-adsorbent from agricultural wastes (Saccharum officinarum leaves) and its environmental application; Spring*er. Environmental Science and Pollution* Resear*ch* International, 26(6), 5305–5314. doi:10.100711356-018-3734-z

Kambale, R. C., Adhate, N. R., Chougule, B. K., & Kolekar, Y. D. (2010). Magnetic and dielectric properties of mixed spinel Ni-Zn ferrite synthesized by citratenitrate combustion method. Journal of Alloys and C*ompounds,* 491(1-2), 372–377. doi:10.1016/j.jallcom.2009.10.187

Kamble, R., & Mathe, V. (2008). Nanocrystalline nickel ferrite thick film as an efficient gas sensor at room temperature. Sensors and Actuators. B, Chemical, 131(1), 205–209. doi:10.1016/j.snb.2007.11.003

Kapizov, O., Azat, S., Askaruly, K., Zhantikeyev, U., Tauanov, Z., Bergeneva, N. S., & Satayeva, A. *R. (2020). Perspectives of the Silicon Dioxide Production from* Rice Husk in Kazakhstan: An Overview. Eurasian Chemico-Technological Journal, 22(4), 285–293.

Karimi-Maleh, H., Karimi, F., Alizadeh, M., *& Sanati, A.* L. (2020). Electrochemical sensors, a bright future in the fabrication of portable kits in analytical systems. Chemical Record (New York, N.Y.), 20(7), *682–692. doi:10.1002/tcr.201900092 PMID:*31845511

Karthiga, P. (2018). Preparation of silver nanoparticles by Garcinia mangostana stem extract and investigation of the antimicrobial properties. Biotechnology Research and Innovation, 2(1), 30–36. doi:10.1016/j.biori.2017.11.001

Karuppasamy, K., Jothi, V. *R., Nichelson, A., Vikraman, D., Tanveer*, W. H., Kim, H.-S., & Yi, S.-C. (2020). Nanostructured transition metal sulfide/selenide anodes for high-performan*ce sodium-ion batteries. In Nanostructur*ed, Functional, and Flexible Materials for Energy Conversion and Storage Systems (pp. 437-464). Elsevier.

Kasapoglu, N., Birsoz, B., Baykal, A., Koseoglu, Y., & Toprak, M. S. (200*7). Synthesis and magnetic prop*erties of octahedral ferrite NixCo1–xFe2O4 nanocrystals. Central European Journal of Chemistry, 5(2), 570–580.

Kaushik, S., J*angra, G., Kundu, V., Yadav, J. P., & Kaushik,* S. (2020). Anti-viral activity of Zingiber officinale (Ginger) ingredients against the Chikungunya virus. Virusdisease, 31(3), 270–276. doi:10.1007*13337-020-00584-0 PMID:32420412*

Kawano, K., Kotsugi, M., & Ohkochi, T. (2014). Microstructure effect on magnetization and domain structure in Ni0.49Zn0.49Co0.02Fe1.90Ox ferrite. Journal *of Magnetism and Magnetic Materials, 361, 57–60. doi:*10.1016/j.jmmm.2014.02.048

Ke, K., Pötschke, P., Wiegand, N., Krause, B., & Voit, B. (2016). Tuning the Network Structure *in Poly(vinylidene fluoride)/Carbon Nanotube Nanocomposites Using Carbon Black: Toward Improvements of Conductivity and Piezo*resistive Sensitivity. ACS Applied Materials & Interfaces, 8(22), 14190–14199. doi:10.1021/acsami.6b03451 PMID:27171017

Kefayat, U., Bakht Mand, K., Amin Ur, R., & Won Chun, O. (2022). Perspective Chapter: Graphene Based Nanocomposites for Supercapacitor Electrodes. In S. Zoran (Ed.), Updates on Supercapacitors. Rijeka: IntechOpen.

Kefeni, K. K., Mamba, B. B., & Msagati, T. A. M. (2017). Application of spinel ferrite nanoparticles in water and wastewater treatment: A review. Separation and Purification Technology, 188, 399–422. doi:10.1016/j.seppur.2017.07.015

Kerimkulova, A. R., Azat, S., Velasco, L., Mansurov, Z. A., Lodewyckx, P., Tulepov, M. I., & Imangazy, A. M. (2019). Granular rice husk based sorbents for sorption of vapors of organic and inorganic matters. J. Chem. Technol. Metall, 54, 578–584.

Keçili, R., Büyüktiryaki, S., & Hussain, C. M. (2019). Advancement in bioanalytical science through nanotechnology: Past, present and future. Trends in Analytical Chemistry, 110, 259–276. doi:10.1016/j.trac.2018.11.012

Khademhosseini, A., & Langer, R. (2007). Micro engineered hydrogels for tissue engineering. Biomaterials, 28(34), 5087–5092. doi:10.1016/j.biomaterials.2007.07.021 PMID:17707502

Khairy, M., & Gouda, M. E. (2015). Electrical and optical properties of nickel ferrite/polyaniline nanocomposite. Journal of Advanced Research, 6(4), 555–562. doi:10.1016/j.jare.2014.01.009 PMID:26199745

Khalith, S. B. M.; Karuppannan, S. K.; Dowlath, M. J. H.; Raiyan, G. I. D.; Arunachalam, K. D. (2022) Nanostructured Catalytic membranes for water filtration; Nano-Bioremediation: Fundamentals and Applications. Micro and Nano Technologies, 389-412

Khan, I., Saeed, K., & Khan, I. (2019). Nanoparticles: Properties, applications and toxicities. Arabian Journal of Chemistry, 12(7), 908–931. doi:10.1016/j.arabjc.2017.05.011

Khan, M. A., Alam, M. M., Naushad, M., Alothman, Z. A., Kumar, M., & Ahamad, T. (2015). Sol-gel assisted synthesis of porous nano-crystalline CoFe2O4 composite and its application in the removal of brilliant blue-R from aqueous phase: An ecofriendly and economical approach. Chemical Engineering Journal, 279, 416–424. doi:10.1016/j.cej.2015.05.042

Khan, T., Civas, M., Cetinkaya, O., Abbasi, N. A., & Akan, O. B. (2020). Nanosensor networks for smart health care Nanosensors for Smart Cities. Elsevier.

Khan, T., Yasmin, A., & Townley, H. E. (2020). An evaluation of the activity of biologically synthesized silver nanoparticles against bacteria, fungi and mammalian cell lines. Colloids and Surfaces. B, Biointerfaces, 194, 111167. doi:10.1016/j.colsurfb.2020.111156 PMID:32512312

Khan, Y., Qasim Nasar, M., & Numan, M. (2018). Biomimetic Synthesis of Silver Nanoparticles for Breast Cancer Therapeutics and Its Mechanism Bio safety assessments of genetically modified maize View project Green synthesis of Silver Nano particles View project. International Journal of Nanotechnology and Nanomedicine, 3(1), 1–9.

Khanna, P., Kaur, A., & Goyal, D. (2019). Algae-based metallic nanoparticles: Synthesis, characterization and applications. Journal of Microbiological Methods, 163, 105656. doi:10.1016/j.mimet.2019.105656 PMID:31220512

Khanna, V. K. (2019). Nanosensors: physical, chemical, and biological. CRC Press.

Khoo, K. S., Chia, W. Y., Tang, D. Y. Y., Show, P. L., Chew, K. W., & Chen, W.-H. (2020, February 17). Nanomaterials utilization in Biomass for Biofuel and Bioenergy production; MDPI. Energies, 13(4), 892. doi:10.3390/en13040892

Khrypunov, G., Klochko, N., Volkova, N., Lyubov, V. and Li, T. (2011). Single-phase cadmium telluride thin films deposited by electroless electrodeposition. Japanese Journal of Applied Physics, 50, 05FH01- 05FH02.

Kieling, A. G. (2009). Influência da segregação no desempenho de cinzas de casca de arroz como pozolanas e material adsorvente [Influence of segregation on the performance of rice husk ash as pozzolans and absorbant material].

Kim, B., Lee, J., & Yu, I. (2003). Electrical properties of *single-wall carbon nanotube and epoxy* composites. Journal of Applied Physics, 94(10), 6724–6728. doi:10.1063/1.1622772

Kim, H. J., Jo, J. H., Yoon, S. Y., *Jo, D. Y., Kim, H. S., Park, B., & Yang, H. (2019).* Emission enhancement of Cu-doped InP quantum dots through double shelling scheme. Materials (Basel), 12(14*), 2267. doi:10.3390/*ma12142267 PMID:31311083

Kim, J., Raj, M. R., & Lee, G. (2021). *High*-de*fec*t-density graphite for superior-performance aluminum-ion batteries with ultra-fast charging and stable long life. Nano-Micro Letters, 13(1), 171.

Kim, J., Seo, D., Lee, B. C., Seo, Y. S., & Miller, W. H. (2014). Nano-*W dispersed gamma radiation shield*ing materials. Advanced Engineering Materials, 16(9), 1083–1089. doi:10.1002/adem.201400127

Kim, M., Han, C. Y., Yang, H., & Park, B. (2019). Band to band tunneling at the zinc oxide (ZnO) and lead selenide (PbSe) quantum dot contact; interfacial charge transfer at ZnO/PbSe/ZnO probe device. Materials (Basel), 12(14), *2289. doi:10.3390/ma12142289* PM*I*D:31319559

Kim, S., Zare, Y., Garmabi, H., & Rhee, K. Y. (2018). Variations of tunneling properties in poly (lactic acid) (PLA)/poly (ethylene oxide) (PEO)/carbon *nanotubes (CNT) nanocomposites during hyd*rolytic degradation. Sensors and Actuators. A, Physical, 274, 28–36. doi:10.1016/j.sna.2018.03.004

Kim, S. J., Koh, H. J., Ren, C. E., Kwon, O., Maleski, K., Cho, S. Y., Anasori, B., Kim*, C. K., Choi, Y. K., Kim, J., Go*gotsi, Y., & Jung, H. T. (2018). Metallic Ti3C2Tx MXene Gas Sensors with Ultrahigh Signal-to-Noise Ratio. ACS Nano, 12(2), 986–993. doi:10.1021/acsnano.7b07460 PMID:29368519

Kim, S. *Y.*, Kim, C. H., & Yang, C.-M. (2021). Binder-free silicon anodes wrapped in multiple graphene shells for high-performance lithium-ion batteries. Journal of Power Sources, 486, 229350. doi:10.1016/j.jpowsour.2020.229350

Kim, W. *J., Chang, W., Quad*ri, S. B., Wu, H. D., Pond, J. M., Kirchoefer, S. W., Newman, H. S., Chrisey, D. B., & Horwits, J. S. (1999). Electrically and magnetically tunable microwave device using (Ba, Sr) TiO3/Y3Fe5O12 multilayer. Applied Physics. *A, Materials Science & Proc*essing, 71, 1564.

Kim, Y. J., Pyo, S., Kim, S., & Ryu, W.-H. (2021). Group VI metallic pillars for assembly of expanded graphite anodes for high-capacity Na-ion batteries. Carbon, 175, 585–593.

Kimura, T., Takizawa, H., U_he_da, K., Endo, T., & Shimada, M. *(1998). Microwave Synthesis of Yttr*ium Iron Garnet Powder. Journal of the American Ceramic Society, 8(11), 2961–2964. doi:10.1111/j.1151-2916.1998.tb02720.x

Kitagawa, Y., Hiraoka, Y., Honda, T., Ishikura, T., Nakamura, H., & Kimura, T. (2010). Low-field magnetoelectric *effect at room temperatu*re. *Na*ture Materials, 9(10), 797–802. doi:10.1038/nmat2826 PMID:20693994

Klochko, N. P., Volkova, N. D., Kharchenko, M. M., & Kopach, V. R. (2009). Contact electrodeposition o*f CdTe thin films. Functional Materials, 16, 190–191.*

*Kohara, Y., Yamasak*i, Y., Onose, Y., & Tokura, Y. (2010). Excess-electron induced polarization and magnetoelectric effect in yttrium iron garnet. Physical Review. B, 82(10), *104419. doi:10.1103/PhysRevB.82.104419*

Kolahalam, L. A., Viswanath, I. V. K., Diwakar, B. S., Govindh, B., Reddy, V., & Murthy, Y. L. N. (2019). Review on nanomaterials: Synthesis and applications. Materials Today: Proceedings, 18, 2182–2190. d_oi:*10.1016/j.matpr.2019.07.371*

*Kombai*ah, K., Vijaya, J. J., Kennedy, L. J., Bououdina, M., Ramalingam, R. J., & Al-Lohedan, H. A. (2018). Okra extract-assisted green *synthesis of CoFe2O4 nanoparti*cles and their optical, magnetic, and antimicrobial properties. Materials Chemistry and Physics, 204, 410–419. doi:10.1016/j.matchemphys.2017.10.077

Kong, Y., Tang, C., Huang, X., Nanjundan, A. K., Zou, J., Du, A., & Yu, C. (2021). Thermal Reductive Perf*oration of Graphene* Cathode for High-Performance Aluminum-Ion Batteries. Advanced Functional Materials, 31(17), 2010569.

Korenstein, R., Madison, P., & Hallock, P. (1992). Growth of (111) CdTe on GaAs/Si and Si substrates for HgCdTe epitaxy. Journal of V*acuum Science & Technology. B,* Microelectronics and Nanometer Structures: Processing, Measurement, and Phenomena: An Official Journal of the American Vacuum So*ciety, 10(4), 1370. doi:10.1116/1.585870*

*Korot*cenkov, G. (2007). Metal oxides for solid-state gas sensors: What determines our choice? Materials Science and Engineering B, 139(1), 1–23. doi:10.1016/j.mseb.2007.01.044

Korotcenkov, G. (2013). Handbook of Ga*s Sensor Materials: Properties, Ad*vantages and Shortcomings for Applications. New Trends and Technologies (Integrated Analytical Systems) (Vol. 2). Springer. doi:10.1007/978-1-4614-7165-3

Koul, S., & Bhat, B. (1991). Microwave and Millimeter Wave Phase Shifters. Artech House.

Kowase, I., Sato, *T., Yamasawa, K.,* & Miura, Y. (2005). A planar inductor using Mn-Zn ferrite/polyimide composite thick film for low-Voltage and large-current DC-DC converter. IEEE Transactions on Magnetics, 41(10), 223–224. doi:10.1109/TMAG.2005.855164

Kozaw*a, D., Cho, S.-Y., Gong, X., Nguy*en, F. T., Jin, X., Lee, M. A., & Scha*cherl, J. (2020). A fiber optic interface coupled to nanosensors: Applications to protein aggregation and organic* molecule quantification. ACS Nano, 14(8), 10141–10152. doi:10.1021/acsnano.0c03417 PMID:32667777

Kozyrev, A., Ivanov, A., Soldatenkov, O., Tumarkin, A., Ivanova, S., Kayadanov, T., Perkins, J. D., Alleman, J., Ginley, D. S., Sengupta, L., Chiu, L., & Zhang, X. (2003). Millimeter-Wave Lo*aded Line Ferroelectric Phase* Sh*ift*ers. Integrated Ferroelectrics, 55(1), 847–852. doi:10.1080/10584580390259290

Kramer, I. J., & Sar$_g$en$_r$, $_{E. H}$. *(2011). Colloidal quantum dot photovoltaics:* A *pat*h forward. ACS Nano, 5(11), 8506–8514. doi:10.1021/nn203438u PMID:21967723

Kravchenko, O. G., Pedrazzoli, D., Bonab, *V. S., & Manas-Zloczower, I. (2018). Conductiv*e interlaminar interfaces for structural health monitoring in composite laminates under fatigue loading. Materials & Design, 160, 1217–1225. doi:10.*1016/j.matd*es.2018.10.045

Krishnarao, R. V., & Godkhindi, M. M. (1992). Distribution of silica in rice husks and its effect on the formation of silicon carbide. Ceramics Internationa*l, 18(4), 243–249.*

*Kuan*g, P., He, M., Zhu, B., Yu, J., Fan, K., & Jaroniec, M. (2019). 0D/2D NiS2/V-MXene composite for electrocatalytic H2 evolution. Journal of Catalysis, 375, 8–20. doi:10.1016/j.jcat.2019.05.019

Kuang, P., Low, J., Cheng, B., Yu, J., *& Fan, J. (2020). MXene-base*d *pho*tocatalysts. Journal of Materials Science and Technology, 56, 18–44. doi:10.1016/j.jmst.2020.02.037

Kudo, A., & Miseki, Y. (2009). Heterogeneous photocatalyst materials for water splitting. Chemical Society Reviews, 38(1), *253–278. doi:10.1039/B800489G PMID:19088977*

*Kulikowski, J. (1984). Soft magnetic ferrites-de*velopment or stagnation? Journal of Magnetism and Magnetic Materials, 41(1-3), 56–62. doi:10.1016/0304-8853(84)90136-7

Kulkarni, D. R., Malode, S $_{. J}$., Prabhu, K. K., Ayachit, N. H., Kulkarni, R. M., & Shetti, N. P. (20*20). Development of a novel nanosens*or using Ca-doped ZnO for antihistamine drug. Materials Chemistry and Physics, 246, 122791. doi:10.1016/j.matchemphys.2020.122791

Kulkarni, H., Zohaib A Khusru, K., & Shravan Aiyappa, K. (2018). ScienceDirect Application of Piezoelectri$_{c\,T}$ec$_{h}$no$_{l}$o$_{g}$y in Automotive Systems. Materials Today: Proceedings, 5(10), 21299–21*304. doi:10.1016/j.matpr.2018.06.532*

*Kumar, S. R., Go*re, *R.* B., Kulkarni, S. K., & Pandey, R. K. (1991). Rapid Thermal Annealing of Cu-In/Se stacked layers. Semiconductor Science and Technology, 6(9), 940–941. doi:10.1088/0268-1242/6/9/019

Kumar, S. R., Kumar, S., Sharma, S. K., *& Roy, D. (2015). Struc*ture Composition and Optical properties of Non aqueous Deposited ZnCdS Nanocrystalline Film. Materials Today: Proceedings, 2(9), 4563–4568. doi:10.1016/j.matpr.2015.10.071

Kumar, V., & Guleria, P. (2020). Application of DNA-Nano*sensor for E*nvironmental Monitoring: Recent Advances and Perspectives. Current Pollution Reports, 1-21.

Kumar, V., & Swart, H. C. (2020). Luminescence properties of rare-earth doped oxide materials. Spectroscopy of Lanthanide *Doped Oxide Materials.*

*Kumar Panda, M., Ku*mar Dhal, N., Kumar, M., Manjari Mishra, P., & Kumar Behera, R. (2021). Green synthesis of silver nanoparticles and its pot*ential effect o*n *ph*ytopathogens. Materials Today: Proceedings, 35, 233–238. doi:10.1016/j.matpr.2020.05.188

Kuylenstierna, D., Vorobiev, A., Linner, P., & Gevorgian, S. (2006). Composite right/left handed tr*ansmission line phase shifter using ferroe*lectric *v*aractors. IEEE Microwave and Wireless Components Letters, 16(4), 167–169. doi:10.1109/LMWC.2006.872145

Ladd, J. W., Olsen, D. B., & Beshouri, G. (2017). Evaluation of Operating Parameters and Fuel Composition on Knock in Large Bore Two-Strok*e Pipeline Engines. Fuel, 202,* 165*–1*74. doi:10.1016/j.fuel.2017.04.038

Lancaster, M. J., Powell, J., & Porch, A. (1998). Thin-film ferroelectric microwave devices. Superconductor Science and Technology, 11(11), 132*3–1334. doi:10.1088/0953-2048/11/11/021*

La*sk*ar, A. R. (2017). Piezoelectricity : An Energy Source for Future Railway Stations. SEU Journal of Science and Engineering, 11(2), 13–18.

Lassoued, A., Lassoued, M. S., Dkhil, B., Ammar, S., & Gadri, A. (2018). Substituted effect of Al3+ on st*ructural, op*tical, magnetic and photocatalytic activity of Ni ferrites. Journal of Magnetism and Magnetic Materials, 476, 124–133. doi:10.1016/j.jmmm.2018.12.062

Latha, A. V., Ayyappan, M., Kallar, A. R., Kakkadavath, R. V., Vic*tor, S. P., & Selvam, S. (2020). Fluorescence imaging of nitric oxide in li*ving cells using o-phenylenediamine-rhodamine based polymeric nanosensors. Materials Science and Engineering C, 108, 110463. doi:10.1016/j.msec.2019.110463 PMID:31923983

Lau, K. T., Lu, M., Lam, C. K., *Cheung, H. Y., Sheng, F. L.,* & Li, H. L. (2005). Thermal and mechanical properties of single-walled carbon nanotube bundle-reinforced epoxy nanocomposites: The role of solvent for nanotube dispersion. Composites Science and Techn*ology, 65(5 SPEC. ISS.), 719–725. doi:10.10*16/j.compscitech.2004.10.005

Lazarevic, Z. Z., Jovalekic, C., Sekulic, D., Slankamenac, M., Romcevic, M., Milutinovic, A., & Romcevic, N. Z. (2012). Characterization of nanostructured spinel NiFe2O4 obtained by soft mechano*chemical synthesis. Science of Sintering, 44(3*), 331–339. doi:10.2298/SOS1203331L

Lee, H., Park, H., Park, K., & Yi, H. (2020). Application of Adaptive Wave Cancellat*ion Underwater to a Piezo*electric-Material-Based Multilayer Sensor. Sensors (Basel), 3, 1–12. doi:10.339020010134 PMID:31878227

Lee, H. T., Shin, H., Cheon, J., & Weissleder, R. (2015). Recent developments in magnetic diagnostic s*ystems. C*hemical Reviews, 115(19), 10690–10724. doi:10.1021/cr500698d PMID:26258867

Lee, J., Govorov, A. O., Dulka, J., & Kotov, N. A. (2004). Bioconjugates of CdTe Nanowir*es and Au Nanoparticles: Plasmon–Exciton* In*t*eractions, Luminescence Enhancement, and Collective Effects. Nano Letters, 4(12), 2323–2330. doi:10.1021/nl048669h

Lee, J., Hong, Y. K., Bae, S., Jalli, J., Park, J., Abo, G., Donohoe, G., & Choi, B. C. (2011). Integrated Ferrite Film Inductor for Power System-on-Chip (PowerSoC) Smart Phone Applicati*ons. IEEE Transactions on Magnetics, 47(2), 304–307. doi:*10.*1*109/TMAG.2010.2092420

Lee, J. H., Kim, J. W., Levy, M., Kao, A., Noh, S. H., Bozovic, D., & Cheon, J. (2014). Magnetic nanoparticles for ultrafast mechanical control of inner ear hair cells. ACS Nano, 8(7), 6590–6598. doi:10.1021/n*n5020616 PMID:25004005*

Lee, M. A., Wang, S., Jin, X., Bakh, N. A., Nguyen, F. T., Dong, J., & Jones, K. K. (2020). Implantable Nanosensors for Human Steroid Hormone Sensing In Vivo Using a Self-Templating Cor*ona Phase Molecular Recognition.* Advanc*e*d Healthcare Materials, 9(21), 2000429. doi:10.1002/adhm.202000429 PMID:32940022

Lee, S.-M., Kim, J., Moon, J., Jung, K.-N., Kim, J. H., Park, G.-J., ... Lee, J.-W. (2021). A cooperative biphasic MoO x–MoP x promoter enables a fast-charging lit*hium-ion battery. Nature Communications*, *1*2(1), 39.

Lee, S. P., Chen, Y. J., Ho, C. M., Chang, C. P., & Hong, Y. S. (2007). A study on synthesis and characterization of the core-shell materials of Mn1-xZnxFe2O4-polyaniline. *Materials Science and Engineering B,* 143(1-3), 1–6. doi:10.1016/j.mseb.2007.06.029

Lee, Y.-K., Lee, M., Lee, G. W., Eom, K., Lee, M.-H., Lee, S., & Joh, H.-I. (2020). Amorphous MoSx embedded within edges of modified graphite as fast-c*harging anode material for rechargeab*le batteries. Applied Surface Science, 509, 145352. doi:10.1016/j.apsusc.2020.145352

Lei, W. W., Portehault, D., Dimova, R., & Antonietti, M. (2011). Boron carbon nitride nanostructures from salt melts: Tunable water-soluble phosphors. Journal of the American Chemical Society, 133(18), 7121–7127. doi:10.1021/ja200838c PMID:21506566

Lerma, M. S., Zingaro, R. A., & Castillo, S. J. (2001). Optical and structural characterization of CdTe thin films by chemical bath deposition technique, Journal of Oregano met$_a$ll$_{ic.}$ C$_h$e$_m$istry (Weinheim *an der Bergstrasse, Germany), 623, 8*1–*8*6.

Lew, T. T. S., Koman, V. B., Silmore, K. S., Seo, J. S., Gordiichuk, P., Kwak, S.-Y., & Lee, M. A. (2020). Real-time detection of wound-induced H 2 O 2 signalling waves in plants *with optica*l *na*nosensors. Nature Plants, 6(4), 404–415. doi:10.103841477-020-0632-4 PMID:32296141

Lewis, N. S. (2016). Research opportunities to advance solar energy utilization. Science, 351(6271), aad1920. doi:10.*1126cience.aad1920 PMID:26798020*

Li, B., Pan, W., Liu, C., Guo, J., Shen, J., Feng, J., & An, T. (2020). Homogenous magneto-fluorescent nanosensor for tumor-derived exosome isolation and analysis. ACS Sensors, 5(7), 2052–2060. doi:10.1021/acssensors.0c00513 PMID:32594744

Li, D., Wang, N., Wang, F., & Zhao, Q. (2019). Boronate affinity-based surface-imprinted quantum dots as novel fluorescent nanosensors for the rapid and efficient d*etection of rutin. Analytical Methods,* 1*1(2*5), 3212–3220. doi:10.1039/C9AY00787C

Li, F., Huang, Y., Gao, C., & Wu, X. (2021). The enhanced photo-catalytic CO2 reduction performance of g-C3N4 with high selectivity by coupling CoNiSx. Materials Research Bulletin, 144, 111488. doi:10.1016/j.materresbull.2021.111488

Li, H., Sun, B., *Gao, T., Li, H., Ren, Y., & Zhou, G.* (2022). Ti3C2 MXene co-catalyst assembled with mesoporous TiO2 for boosting photocatalytic activity of methyl orange degradation and hydrogen production. Chinese Journal of Catalysis, 43(2), 461–471. doi:10.1016/S1872-2067(21)63915-3

Li, H. Q., Roberts, D. C., Steyn, *J. L., Turner, K. T.,* Yag*l*ioglu, O., Hagood, N. W., Spearing, S. M., & Schmidt, M. A. (2004). Fabrication of a High Frequency Piezoelectric Microvalve. *Sensors and Actuators, 111(1)*, 51–56. doi:10.1016/j.sna.2003.10.013

Li, J., Zhou, X., Zhao, H., Shao, M., Fan, Z., & Liu, H. (2015). Design and Experimental Tests of a Dual-Servo Piezoel*ectric Nanopositioning Stage for* Rotary Motion. The Review of Scientific Instruments, 045002(4), 1–7. doi:10.1063/1.4918295 PMID:25933888

Li, K., Zhang, S., Li, Y., Fan, J., & Lv, K. (2020). MXenes as noble-metal-alternative co catalysts in photocatalysis. Chinese Journal of Catalysis, 42(1), 3–14. doi:10.1016/S1872-2067(20)63630-0

Li, L., Zhang, D., *Deng, J., Gou, Y., Fang, J., Cui,* H., ... Cao, M. (2021). Carbon-based materials for fast charging lithium-ion batteries. Carbon, 183, 721–734.

Li, M., Liu, T., Bi, X., Chen, Z., Amine, K., Zhong, C., *& Lu, J. (2020). Cationic and a*nion*ic* redox in lithium-ion based batteries. Academic Press.

Li, N., Yao, Y., Lv, T., *Chen, Z., Yang, Y., Cao, S., & Chen, T. (202*1). High-performance wire-shaped aluminum ion batteries based on continuous graphene fi*ber cathodes. Journal of Power Sources, 488, 229460.*

Li, R., Gu, Y., Wang, Y., Yang, Z., Li, M., & Zhang, Z. (2017). Effect of particle size on gamma radiation shielding property of gadolinium oxide dispersed epoxy resin matrix composite. Materials Research *Express, 4(3), 035035. doi:10.1088/2053*-*1591*/aa6651

Li, S., Cai, J., Wu, X., Liu, B., Chen, Q., Li, Y., & Zheng, F. (2018). TiO2@Pt@CeO2 nanocomposite as a bifunctional catalyst for enhancing photo-reduction of Cr (VI) and photo-oxidation of benzyl alcohol. Journal of Hazardous Materials, 346, 52–61. doi:10.1016/j.jhazmat.2017.12.001 PMID:29247954

Li, S., Liu, Y.*-M., Zhang, Y.-C., Song, Y., Wang, G.-K., Liu, Y.-X.,* ... Guo, X.-D. (2021). A review of rational design and investigation of binders applied in silicon-based anodes for lithium-ion batteries. Journal of Power Sources, 485, 229331.

Li, W. Y., Xu, L. N., *& Chen, J.* (2005). Co3O4 Nanomaterials in Lithium-Ion Batteries and *Gas Sensors. Advanced Functional Materials, 15(5), 851–857. doi:10.1002/adfm.200400429*

Li, Y., Hu, Y., Zhao, Y., Shi, G. Q., Deng, L., Hou, Y. B., & Qu, L. (2011). An electrochemical avenue to green-luminescent graphene quantum dots as potential electron-acceptors for photovoltaics. Advanced Materials, 23(6), 7*76–780. doi:10.1002/adma.201003*819 PMID:21287641

Li, Y., Wang, W. N., Zhan, Z., Woo, M. H., Wu, C. Y., & Biswas, P. (2010). Photocatalytic reduction of CO2 with H2O on mesoporous silica supported Cu/TiO2 catalysts. Applied Catalysis *B: Environmental,* 100*(*1–2), 386–392. doi:10.1016/j.apcatb.2010.08.015

Li, Z., Li, J., Li, X., & Zhang, W. (2020). Pseudocapacitance effect in Al-C batte*ries with expanded graphite posi*tive electrode at different temperatures. Journal of Power Sources, 467, 228323.

Liang, Z., Yan, C., Rtimi, S., & Bandara, J. (2018). Piezoelectric Materials for Catalytic/Photocatalytic Removal of Pollutants: Recent Advances and Outlook. Applied Catalysis B: Environment*al. doi:*10.*1*016/j.apcatb.2018.09.028

Liao, L., Lu, H. B., Li, J. C., He, H., Wang, D. F., Fu, D. J., Liu, C., & Zhang, W. F. (2007). Size Dependence of Gas Sensitivity of ZnO Nanorods. The Journal of Physical Chemistry C, *111(5), 1900–1903. doi:*10.*1*021/jp065963k

Liao, Y., Zhou, P., Pan, D., Zhou, L., & Su, Z. (2021). An ultra-thin printable nanocomposite sensor network for structural health monitoring. Structural Health Monitoring, 20(3), 894–903. doi:10.1177/1475921719859338

Ligemza, S. (1981). Mossbauer *Study of the 3d5 Shell Deformation* of Fe3+ Ions in MnFe2O4. Physica Status Solidi. B, Basic Research, 105(1), 353–358. doi:10.1002/pssb.2221050141

Lin, C. M., Chuu, D., Xu, J. A., Huang, E., Chou, W. C., Hu, J. Z., & Pei, J. H. *(1998).* H*igh*-pressure phase transitions in Zn 1-x M x Se (M=Cd, Fe, and Mn). J.Phys.Rev.B, 58(1), 16–19. doi:10.1103/PhysRevB.58.16

Lippmann, M G. (2010). Magazine *and Journal of* Science On the Principle of the Conservation of Electricity. Taylor and Francis, 5982 (May). . doi:10.1080/14786448108627081

Liu, B., Jia, L., Qin, Y., Liu, Z., Diao, L., & An, *M. (2019). Proceedings* of the 4th International Conference on Electrical and Information Technologies for Rail Transportation (EITRT) 2019. Springer.

Liu, C., Jin, Z., Cheng, L., Zhao, H., & W*ang, L. (2020). Synthesis of nano*sensors for autonomous warning of damage and self-repairing in polymeric coatings. Nanoscale, 12(5), 3194–3204. doi:10.1039/C9NR09221H PMID:31967166

Liu, G., Abukhadra, M. R., El-Sherbeeny, A. M., M$_0$stafa, A. M., & Elmeligy, M. A. (2020). Insight into the photocatalytic pr*operties of diatomite@Ni/NiO composi*te for effective photo-degradation of malachite green dye and photo-reduction of Cr (VI) under visible light. Journal of Environmental Management, 254, 109799. doi:10.1016/j.jenvman.2019.109799 PMID:31710977

Liu, H., Zhong, J., *Lee, C., & Seung-wuk, L. (2018).* Mechanisms, and Applications APPLIED PHYSICS REVIEWS A Comprehensive Review on Piezoelectric Energy Harvesting Technology : Materials, M*echanisms, and Applications. Applied Physics Reviews, 041306(4), 041306. doi:10.1063/1.5074184*

Liu, I. P., Chen, L. Y., & Lee, Y. (2016). L. Journal of Power Sources, 325, 706–713. doi:10.1016/j.jpowsour.2016.06.095 PMID:12200862

Liu, J., Li, F., Wang, Y., Pan, L., Lin, P., Zhang, B., & Ko, G. (2020). A sensitive and specific nanosensor for monitoring extracellular potassium levels in the brain. Nature Nanotechnology, 15(4), 321–330. doi:10.103841565-020-0634-4 PMID:32042163

Liu, J., Pan, L., Shang, C., Lu, B., Wu, R., Feng, *Y., & Xiong, Z. (2020). A highly sensitiv*e and selective nanosensor for near-infrared potassium imaging. Science Advances, 6(16), eaax9757. doi:10.1126ciadv.aax9757 PMID:32494594

Liu, L., & Corma, A. (2018). Metal Catalysts for Heterogeneous Catalysis: From *Single Atoms to Nanoclusters a*nd *N*anoparticles. Chemical Reviews, 118(10), 4981–5079. doi:10.1021/acs.chemrev.7b00776 PMID:29658707

Liu, R. L., Wu, D. Q., Feng, X. L., & Mullen, K. (2011). Bottom-up fabrication of photoluminescent graphene quantum dots with uniform morphology. Journal of the *American Che*mical Society, 133(39), 15221–15223. doi:10.1021/ja204953k PMID:21894989

Liu, Y., Rajadas, A., & Chattopadhyay, A. (2012). A biomimetic structural health monitoring approach using carbon nanotubes. JOM, 64(7), 802–807. doi:10.100711837-012-0357-6

Liu, *Y. L., Liu, Z. M., Yang*, Y., Yang, H. F., Shen, G. L., & Yu, R. Q. (2005). Simple synthesis of MgFe2O4 nanoparticles as gas sensing materials. Sensors and Actuators. B, Chemical, 107(2), 600–604. doi:10.10*16/j.snb.2004.11.026*

Liu*a*, J., Yua, P., Jina, Q., Chong, Z., Ming, Z., & Harrisb, V. G. (2018). Microwave-accelerated rapid synthesis of high-quality yttrium i$_r$on garnet na*no powders with impro*ved magnetic properties. Mater. Res. Lett., 6(1), 36–40

LiuX.XuH.ZhangM.YuD. G. (2021);Electrospun Medicated Nanofibers for Wou*nd Healing. (•••). MDPI. Membranes (Basel), 11(*10), 770. doi:10.3390/membranes11100770

Loh, K. P., Bao, Q., Eda, G., & Chhowalla, M. (2010). Graphene oxide as a chemically tunab*le platform for optical* ap-*pl*ications. Nature Chemistry, 2(12), 1015–1024. doi:10.1038/nchem.907 PMID:21107364

Lokhande, R. M., Vinayak, V., Mukhamale, *S. V., & Khirade, P. P. (2021). Gamma radiatio*n shielding characteristics of various spinel ferrite nanocrystals: A combined experimental and theoretical investigation. RSC Advances, 11(14), 7925–7937. doi:10.1039/D0RA08372K PMID:35423294

Lo'pez-Ortega, A. L., Lottini, E., *Fernandez, C., & Sangregorio, C. (20*15). Exploring the magnetic properties of cobalt-ferrite nanoparticles for the development of a rare-earth-free permanent magnet. Chemistry of Materials, 27(11), 4048–4056. doi:10.1021/acs.ch*emmater.5b01034*

Lu, A.-H., Sa*l*abas, E. L., & Schüth, F. (2007). Magnetic Nanoparticles: Synthesis, Protection, Functionalization, and Application. Angewandte Chemie International Edition, 46(8), 1222–1244. doi:10.1002/anie.200602866 PMID:17278160

Lu, K. L., *Lago, R. M., Chen, Y. K.,* Gre*e*n, M. L. H., Harris, P. J. F., & Tsang, S. C. (1996). Mechanical damage of carbon nanotubes by ultrasound. In Carbon, 34(6). doi:10.1016/0008-6223(96)89470-X

Lu, N., Jing*, X., Zhang, J., Zhang, P.*, Qiao, Q., & Zhang, Z. (2022). Photo-assisted self-assembly synthesis of all 2D-layered heterojunction photocatalysts with long*-range spatial separation of charge-ca*rriers toward photocatalytic redox reactions. Chemical Engineering Journal, 431, 134001. doi:10.1016/j.cej.2021.134001

Lu, Y., Hao-yen, T., & Fung, S., Boser, B., & Horsley, D. (2015). Pulse-Echo Ultras*ound Imaging Using Transmit* Beam-Forming. Journal of Microelectromechanical Systems, 1–9.

Ludueña, L., Fasce, D., Alvarez, V. A., & Stefani, P. M. (2011). Nanocellulose from rice husk following alkaline treatment to remove silica. BioRes*ources, 6*(2), 1440–1453.

Luk, K.-M., & Leung, K.-W. (2003). Dielectric Resonator Antennas. Research Studies Press Ltd.

Luo, H., Wang, Q.*, Wang, Y., Xu, C., Wang, B., Wang, M., ...* Zhang, Y. (2020). Nano-silicon embedded in MOFs-derived nitrogen-doped carbon/cobalt/carbon nanotubes hybrid composite for enhanced lithium ion storage. Applied Surface Science, 529, 14713*4. doi:10.1016/j.apsusc.2020.147134*

Luo, J., Im, J. H., Mayer, M. T., Schreier, M., Nazeeruddin, M. K., Park, N. G., Tilley, S. D., Fan, H. J., & Grätzel, M. (2014). Water photolysis at 12.3% efficiency via perovskite photovoltaics and Earth-abundant catalysts. Science, 345(6204), 1593–1596. doi:10.1126ci*ence.1258307 PMID:25258076*

*Luo, T., Jia*ng, S., Moro, A., Wang, C., Zhou, L., & Luo, F. (2017, December). Measurement and Validation of Hole-to-Hole Fuel Injection Rate from a Diesel Injector. Flow Measurement and Instrumentation, 61, 66–78. doi:10.1016/j.flowmeasinst.2018.03.014

Luo, X., Han, Y., Chen, X., Tang, W., Yue, T., *& Li, Z. (2020). Carbon dots derived* fluorescent nanosensors as versatile tools for food quality and safety assessment: A review. Trends in Food Science & Technology, 95, 149–161. doi:10.1016/j.tifs.2019.11.017

Lupan, O. I., Chow, L., & Adelung, R. (2019). Nanosensors: Current Status and Perspectives.

Luu, T. H. *T., Duong, D. L.,* Lee*,* T. H., Pham, D. T., Sahoo, R., Han, G., ... Lee, Y. H. (2020). Monodispersed SnS nanoparticles anchored on carbon nanotubes for high-retention sodiu*m-ion batteries.* Jou*r*nal of Materials Chemistry. A, Materials for Energy and Sustainability, 8(16), 7861–7869. doi:10.1039/C9TA13136A

Madan, T., Pandey, S., & Talegaonkar, S. (2015). Nanotechnology: A smart drug delivery tool in modern healthcare. Journal of Chemical and Pharmaceutical Research, 7(6), 257–264.

Madhusanka, S., Sandaruwan, R., Athar, M., Zaib, M., Munasinghe, H. M., Arachchige, B., . . . Gunawardhana, N. (2020). TiO2 Microparticles/Reduced Graphene Oxide Composite as Anode Material for Lithium Ion Battery. Academic Press.

Magneto-Optical Imaging of Superconductors. (2010). Department of Physics, University of Oslo.

Mahdi, M. A., Hassan, Z., Nag, S. S., Hassan, J. J., & Bakhori, S. K. (2012). Structural and optical properties of nanocrystalline CdS thin films prepared using microwave-assisted chemical bath deposition. Thin Solid Films, 520(9), 3477–3484. doi:10.1016/j.tsf.2011.12.059

Maheswari, U. B., Sivakumar, V. M., & Thirumarimurugan, M. (2019). Synthesis, characterization of modified nanoadsorbents and its application in removal of Zn2+ ions from battery effluent. Environmental Chemistry and Ecotoxicology, 1, 2–11. doi:10.1016/j.enceco.2019.05.001

Mahfouz, M. G., Galhoum, A. A., Gomaa, N. A., Abdel-Rehem, S. S., Atia, A. A., Vincent, T., & Guibal, E. (2015). Uranium extraction using magnetic nano-based particles of diethylenetriamine-functionalized chitosan: Equilibrium and kinetic studies. Chemical Engineering Journal, 262, 198–209. doi:10.1016/j.cej.2014.09.061

Mahmood, S. H., Al-Shiab, Q., Bsoul, I., Maswadeh, Y., & Awadallah, A. (2018). Structural and magnetic properties of (Mg,Co)2W hexaferrites. Current Applied Physics, 18(5), 590–598. doi:10.1016/j.cap.2018.02.013

Mahmoodi, N. M. (2013). Magnetic ferrite nanoparticle-alginate composite: Synthesis, characterization and binary system dye removal. Journal of the Taiwan Institute of Chemical Engineers, 44(2), 322–330. doi:10.1016/j.jtice.2012.11.014

Mahmoud, M. E., El-Khatib, A. M., Badawi, M. S., Rashad, A. R., El-Sharkawy, R. M., & Thabet, A. A. (2018a). Fabrication, characterization and gamma rays shielding properties of nano and micro lead oxide-dispersed-high density polyethylene composites. Radiation Physics and Chemistry, 145, 160–173. doi:10.1016/j.radphyschem.2017.10.017

Mahmoud, M. E., El-Khatib, A. M., Badawi, M. S., Rashad, A. R., El-Sharkawy, R. M., & Thabet, A. A. (2018b). Recycled high-density polyethylene plastics added with lead oxide nanoparticles as sustainable radiation shielding materials. Journal of Cleaner Production, 176, 276–287. doi:10.1016/j.jclepro.2017.12.100

Mahmud, M. A. (2018). Recent Advances in Nanogenerator-Driven Self-Powered Implantable Biomedical Devices. Advanced Energy Materials, 1701210(2), 1–25. doi:10.1002/aenm.201701210

Majed, H. M., Khalef, H. Y., Awadh, H. A., Noomi, B. S., Jafar, N. A., & Hadi, K. A. (2021). Evaluation the safety and synergistic effect of NiFe2O4 nanoparticles with antibiotic against Pseudomonas aeruginosa. Iraqi Journal of Veterinary Sciences, 35(1), 71–77. doi:10.33899/ijvs.2020.126298.1294

Majeed, K., Hassan, A., & Bakar, A. A. (2014). Influence of maleic anhydride-grafted polyethylene compatibiliser on the tensile, oxygen barrier and thermal properties of rice husk and nanoclay-filled low-density polyethylene composite films. Journal of Plastic Film & Sheeting, 30(2), 120–140.

Mak, S. Y., & Chen, D. H. (2004). Fast adsorption of methylene blue on polyacrylic acid-bound iron oxide magnetic nanoparticles. Dyes and Pigments, 61(1), 93–98. doi:10.1016/j.dyepig.2003.10.008

Maksoud, M. I. A., El-Sayyad, S. G., Ashour, A. H., El-Batal, A. I., Abd-Elmonem, M. S., Hendawy, H. A. M., Abdel-Khalek, E. K., Labib, S., Abdeltwab, E., & El-Okr, M. M. (2018). Synthesis and characterization of metals-substituted cobalt ferrite (Co(1−x))MxFe2O4; (M = Zn, Cu, Mn; x = 0.05) nanoparticles as antimicrobial agents and sensors for Anagrelide determination in biological samples. Materials Science and Engineering C, 92, 644–656. doi:10.1016/j.msec.2018.07.007 PMID:30184791

Maksoud, M. I. A. A., El-Sayyad, G. S., Ashour, A. H., El-Batal, A. I., Elsayed, M. A., Gobara, M., El-Khawaga, A. M., Abdel-Khalek, E. K., & El-Okr, M. M. (2018). Antibacterial, antibiofilm, and photocatalytic activities of metals- substituted spinel cobalt ferrite nanoparticles Microbial Pathogenesis, 127, 144–158. doi:10.1016/j.micpath.2018.11.045 PMID:30502518

Maksoud, M. I. A. A., El-Sayyad, G. S., Soliman, A. A. L. I., El-Bahnasawy, H. H., & Ashour, A. H. (2020). Influence of Mg2+ substitution on structural, optical, magnetic, and antimicrobial properties of Mn-Zn ferrite nanoparticles. *Journal of Materials Science* Materials in Electronics, 31(3), 2598–2616. doi:10.100710854-019-02799-4

Malini, S., Raj, K., & Madhumathy, S. (2022). El-Hady, Khalid M.; Islam, S.; Dutta, M. (2022) Bioinspired Advances in Nanomaterials for Sustainable Agriculture; p. 1-11; Hindwai. *Journal of Nanomaterials*, 8926133. doi:10 doi:.1155/2022/8926133

Malyshev, V. V., Eryshkin, A. V., Koltypin, E. A., Varfolomeev, A. E., & Vasiliev, A. A. (1994). Gas sensitivity of semiconductor Fe2O3-based thick-film sensors to CH4, H2, *NH3. Sensors and Actuators. B, Chemical, 19(*1-3), 434–436. doi:10.1016/0925-4005(93)01031-X

Mandal, S., Nanavati, S. P., Willock, D. J., & Ananthakrishnan, R. (2022). Band gap engineering of amine functionalized *Ag(I)-based coordination* polymers and their plasmonic Ag0 coupled novel visible light driven photo-redox system for selective oxidation of benzyl alcohol. Applied Catalysis B: Environmental, 303, 120821. doi:10.1016/j.apcatb.2021.120821

Mand Khan, B., Chun Oh, W., Nuengmatch, P., & Ullah, K. (2023). Role of graphene-based nanocomposites as anode material for Lithium-ion batteries. Materials Science and Engineering *B, 287, 116141. doi*:10.1016/j.mseb.2022.116141

Manikandan, V., Kim, J. H., Mirzaei, A., Kim, S. S., Vigneselvan, S., Singh, M., & Chandrasekaran, J. (2018). Effect of temperature on gas sensing properties of lithium (Li) substituted (NiFe2O4) nickel ferrite thin film. Journal of Molecular Structure, 1177, 485–490. doi:10.1016/j.molstruc.2018.09.085

Manj, R. Z. A., Zhang, F., Rehman, W. U., Luo, W., & Yang, J. (2020). Toward understanding the interaction within Silicon-based anodes for stable lithium storage. Academic Press.

Manlika, S., Dungsuwan, W., & Thountom, S. (2018). Synthesis and characterization of Bi0.5(Na1-xKx)0.5TiO3 powders by sol-gel combustion method with glycine fuel. Ceramics International, 44, S168–S171. doi:10.1016/j.ceramint.2018.08.122

Manosalva, N., Tortella, G., Cristina Diez, M., Schalchli, H., Seabra, A. B., Durán, N., & Rubilar, O. (2019). *Green synthesis of silver nano*particles: Effect of synthesis reaction parameters on antimicrobial activity. World Journal of Microbiology & Biotechnology, 35(88), 1–9. doi:10.100711274-019-2664-3 PMID:31134435

Manova, E., Kunev, B., Paneva, D., Mitov, I., Petrov, L., Estournes, C., D'Orlean, C., Rehspringer, J. L., & Kurmoo, M. (2004). Mechano-Synthesis, Characterization, and Magnetic Properties of *Nanoparticles of Cobalt Ferrite, CoFe 2 O 4. Chemistry of Materials, 16(26), 5689–5696. doi:10.1021/cm049189u*

Mansurov, Z. A., Jandosov, J. M., Kerimkulova, A. R., Azat, S., Zhubanova, A. A., Digel, I. E., Savistkaya, I. S., Akimbekov, N. S., & Kistaubaeva, A. S. (2013). Nanostructured carbon materials for biomedical use. Eurasian Chemico-Technological Journal, 15(3), 209–217. doi:10.18321/ectj224

Marinca, T. F., Chicina, I., Isnard, O., & Neamtu, B. V. (2016). Nanocrystalline/nanosized manganese substituted nickel ferrites - Ni1-xMnxFe2O4 obtained by ceramic-mechanical milling route. Ceramics International, 42(4), 4754–4763. *doi:10.1016/j.ceramint.2015.11.155*

Martin, T., Stevenson, D. G., & Cowin A Blumfield, P. I. (2015). Piezoelectric Materials for High Temperature Transducers and Actuators. Journal of Materials Science Materials in Electronics, 26(12), 9256–9267. doi:10.100710854-015-3629-4

Martinez, L. M. P., Torres, S. M., Likodimos, V., Figueiredo, J. L., Faria, J. L., Falaras, P., & Silva, A. M. (2012). Advanced nanostructured photocatalysts based on reduced graphene oxide–TiO2 composites for degradation of diphenhydramine pharmaceutical and methyl orange dye. Applied Catalysis B: Environmental, 123, 241–256. doi:10.1016/j.apcatb.2012.04.045

Mathews, S. A., Babu, D. R., Saravanan, P., & Hayakawa, Y. (2020). Microwave absorption studies of (Ba0.5Sr0.5Fe12O19)1-x/(NiFe2O4)x hard/soft ferrite nanocomposites. Materials Chemistry and Physics, 252, 123063. doi:10.1016/j.matchemphys.2020.123063

Mathur, P., Thakur, A., Lee, J. H., & Singh, M. (2010). Sustained electromagnetic properties of Ni-Zn-Co nanoferrites for the high-frequency applications. Materials Letters, 64(24), 2738–2741. doi:10.1016/j.matlet.2010.08.056

Matsunaga, T., Takeyama, H., Tsujimura, N., & Kamiya, S. (1998). Magnetic Bacteria and Biomineralization. Journal of the Magnetics Society of Japan, 22(S_1_ISFA_97), 434–436. doi:10.3379/jmsjmag.22.S1_434

Matsushita, N., Ichinose, M., Nakagawa, S., & Naoe, M. (1998). Preparation and characteristics of Co-Zn ferrite rigid disks without protective layers for high density recording. IEEE Transactions on Magnetics, 34(4), 1639–1641. doi:10.1109/20.706641

Mat Yusuf, S. N. A., Che Mood, C. N. A., Ahmad, N. H., Sandai, D., Lee, C. K., & Lim, V. (2020). Optimization of biogenic synthesis of silver nanoparticles from flavonoid-rich Clinacanthus nutans leaf and stem aqueous extracts: Biogenic Synthesis of C. nutans AgNPs. Royal Society Open Science, 7(7), 200065. doi:10.1098/rsos.200065 PMID:32874618

Mazela, B., Batista, A., & Grześkowiak, W. (2020). Expandable Graphite as a Fire Retardant for Cellulosic Materials—A Review. Forests, 11(7), 755. doi:10.3390/f11070755

Mazloomi, K., & Gomes, C. (2012). Hydrogen as an energy carrier: Prospects and challenges. Renewable & Sustainable Energy Reviews, 16(5), 3024–3033. doi:10.1016/j.rser.2012.02.028

McColm, I. J., & Clark, N. J. (1988). Forming, Shaping and Working of High Performance Ceramics. Blackie.

McHenry, M. E., & Laughlin, D. E. (2000). Nano-scale materials development for future magnetic applications. Acta Materialia, 48(1), 223–238. doi:10.1016/S1359-6454(99)00296-7

Mehndiratta, P., Jain, A., Srivastava, S., & Gupta, N. (2013). Environment Pollution and Nanotechnology. Environment and Pollution, 2(2), 2013. doi:10.5539/ep.v2n2p49

Meir, R. (2005). Vibrational spectroscopy: A 'vanishing' discipline. Chemical Society Reviews, 34, 734–752. PMID:16100614

Melagiriyappa, E., & Jayanna, H. S. (2009). Structural and magnetic susceptibility studies on samarium substituted magnesium-zinc ferrites. Journal of Alloys and Compounds, 482(1-2), 147–150. doi:10.1016/j.jallcom.2009.03.112

Meshram, A. A., Aashish Moses, K., Baral, S. S., & Sontakke, S. M. (2022). Hydrogen production from water splitting of real-time industry effluent using novel photocatalyst. Advanced Powder Technology, 33(3), 103488. doi:10.1016/j.apt.2022.103488

Mihoc, G., Ianos, R., & Pacurariu, C. (2014). Adsorption of phenol and p-chlorophenol from aqueous solutions by magnetic nanopowder. Water Science and Technology, 69(2), 385–391. doi:10.2166/wst.2013.727 PMID:24473310

Mills, A. A. (2004). The lodestone: History, Physics, and Formation. Annals of Science, 61(3), 273–319. doi:10.1080/00033790310001642812

Minary-jolandan, M., & Yu, M. (2010). 2013. "Shear Piezoelectricity in Bone at the Nanoscale Shear Piezoelectricity in Bone at the Nanoscale. Applied Physics Letters, 153127(15), 2008–2011. doi:10.1063/1.3503965

Mintmire, J. W., Dunlap, B. I., & White, C. T. (1992). Are fullerene tubules metallic? Physical Review Letters, 68(5), 631–634. doi:10.1103/PhysRevLett.68.631 PMID:10045950

Miyagawa, H., & Drzal, L. T. (2004). Thermo-physical and impact properties of epoxy nanocomposites reinforced by single-wall carbon nanotubes. Polymer, 45(15), 5163–5170. doi:10.1016/j.polymer.2004.05.036

Mochalin, V. N., & Gogotsi, Y. (2009). Wet chemistry route to hydrophobic blue fluorescent nanodiamond. Journal of the American Chemical Society, 131(13), 4594–4595. doi:10.1021/ja9004514 PMID:19290627

Mohamed, R. M., Rashad, M. M., Haraz, F. A., & Sigmund, W. (2010). Structure and magnetic properties of nanocrystalline cobalt ferrite powders synthesized using organic acid precursor method. Journal of Magnetism and Magnetic Materials, 322(14), 2058–2064. doi:10.1016/j.jmmm.2010.01.034

Mohammadlou, M., Maghsoudi, H., & Jafarizadeh-Malmiri, H. (2016). A review on green silver nanoparticles based on plants: Synthesis, potential applications and eco-friendly approach. International Food Research Journal, 23(2), 446–463.

Mohanty, A. K., Misra, M., & Drzal, L. T. (2001). Surface modifications of natural fibers and performance of the resulting biocomposites: An overview. Composite Interfaces, 8(5), 313–343.

Moharram, B. M., Nagy, M. E., Shaat, M. K., El Sayed, A. R., Fayiz, M., Dwidar, S. A., & Dorrah, M. E. (2020). Performance of lead and iron oxides nanoparticle materials on shielding properties for γ-rays. Radiation Physics and Chemistry, 173, 108880. doi:10.1016/j.radphyschem.2020.108880

Mohiuddin, M., & Hoa, S. V. (2013). Estimation of contact resistance and its effect on electrical conductivity of CNT/PEEK composites. Composites Science and Technology, 79, 42–48. doi:10.1016/j.compscitech.2013.02.004

Monteiro, A. O., Cachim, P. B., & Costa, P. M. F. J. (2017). Self-sensing piezoresistive cement composite loaded with carbon black particles. Cement and Concrete Composites, 81, 59–65. doi:10.1016/j.cemconcomp.2017.04.009

Moradi, M., Khorasheh, F., & Larimi, A. (2020). Pt nanoparticles decorated Bi-doped TiO2 as an efficient photocatalyst for CO2 photo-reduction into CH4. Solar Energy, 211, 100–110. doi:10.1016/j.solener.2020.09.054

Moradnia, F., Fardood, S. T., Ramazani, A., Osali, S., & Abdolmaleki, I. (2020). Green sol–gel synthesis of CoMnCrO 4 spinel nanoparticles and their photocatalytic application. Micro & Nano Letters, 15(10), 674–677. doi:10.1049/mnl.2020.0189

Moraitis, P., van Leeuwen, G., & van Sark, W. (2019). Visual appearance of nanocrystal-based luminescent solar concentrators. Materials (Basel), 12(6), 885. doi:10.3390/ma12060885 PMID:30884811

Moran, E., Blesa, M. C., Medina, M. E., Tornero, J. D., Menendez, N., & Amador, U. (2002). Nonstoichiometric spinel ferrites obtained from α-NaFeO2 via molten media reactions. Inorganic Chemistry, 41(23), 5961–5967. doi:10.1021/ic011302k PMID:12425621

Moretti, R., & Ottonello, G. (1998). An appraisal of endmember energy and mixing properties of rare earth garnets. Geochimica et Cosmochimica Acta, 62(7), 114_7–1173. doi:10.1016/S0016-7037(98)00033-7

Morisako, A., Naka, T., Ito, K., Takizawa, A., Matsumoto, M., & Hong, Y. K. (2002). Properties of Ba-ferrite/AlN double layered films for perpendicular magnetic recording media. Journal of Magnetism and Magnetic Materials, 242, 304–310. doi:10.1016/S0304-8853(01)01230-6

Moscow, S., Kavinkumar, V., Sriramkumar, M., Jothivenkatachalam, K., Saravanan, P., Rajamohan, N., Vasseghian, Y., & Rajasimman, M. (2022). Impact of Erbium (Er) and Yttrium (Y) doping on BiVO4 crystal structure towards the enhancement of photoelectrochemical water splitting and photocatalytic performance. Chemosphere, 299, 134343. doi:10.1016/j.chemosphere.2022.134343 PMID:35307389

Mostafa, E. M., & Amdeha, E. (2022). Enhanced photocatalytic degradation of malachite green dye by highly stable visible-light-responsive Fe-based tri-composite photocatalysts. Environmental Science and Pollution Research International, 1(46), 1–14. doi:10.100711356-022-20745-6 PMID:35578081

Moulson, A. J., & Herbert, J. M. (2003). Electroceramics: Materials - Properties – Aplications (2nd ed.). John Wiley & Sons Inc. doi:10.1002/0470867965

Mousavi, S. M., Hashemi, S. A., Zarei, M., Amani, A. M., & Babapoor, A. (2018). Nanosensors for chemical and biological and medical applications. Med Chem (Los Angeles), 8(8), 2161-0444.1000515.

Mukasyan, A. S., Epstein, P., & Dinka, P. (2007). Solution combustion synthesis of nanomaterials. Proceedings of the Combustion Institute, 31(2), 1789–1795. doi:10.1016/j.proci.2006.07.052

Mukherjee, K., & Majumder, S. B. (2010). Reducing gas sensing behaviour of nanocrystalline magnesium–zinc ferrite powders. Talanta, 81(4-5), 1826–1832. doi:10.1016/j.talanta.2010.03.042 PMID:20441981

Mukherjee, S., & Patra, C. R. (2017). Biologically synthesized metal nanoparticles: Recent advancement and future perspectives in cancer theranostics. Future Science OA, 3(3), FSO203. doi:10.4155/fsoa-2017-0035 PMID:28884002

Mukhtar, M. W., Irfan, M., Ahmad, I., Ali, I., Akhtar, M. N., Khan, M. A., Abbas, G., Rana, M. U., Ali, A., & Ahmad, M. (2015). Synthesis and properties of Pr-substituted MgZnferrites for core materials and high-frequency applications. Journal of Magnetism and Magnetic Materials, 381, 173–178. doi:10.1016/j.jmmm.2014.12.072

Munawar, A., Ong, Y., Schirhagl, R., Tahir, M. A., Khan, W. S., & Bajwa, S. Z. (2019). Nanosensors for diagnosis with optical, electric and mechanical transducers. RSC Advances, 9(12), 6793–6803. doi:10.1039/C8RA10144B PMID:35518460

Murdock, E. S., Simmons, R. F., & Davidson, R. (1992). Roadmap for 10 Gbit/in/sup 2/ media: Challenges. IEEE Transactions on Magnetics, 28(5), 3078–3083. doi:10.1109/20.179719

Murugesan, A. K., Pannerselvam, B., Javee, A., Rajenderan, M., & Thiyagarajan, D. (2021). Facile green synthesis and characterization of Gloriosa superba L. tuber extract-capped silver nanoparticles (GST-AgNPs) and its potential antibacterial and anticancer activities against A549 human cancer cells. Environmental Nanotechnology, Monitoring & Management, 15, 100460–100471. doi:10.1016/j.enmm.2021.100460

Musa, Azis, R. S., Osman, N. H., Hassan, J., & Zangina, T. (2017). Structural and magnetic properties of yttrium iron garnet (YIG) and yttrium aluminum iron garnet (YAlG) nanoferrite via sol-gel synthesis. Results in Physics, 7, 1135–1142. doi:10.1016/j.rinp.2017.02.038

Muthamma, M. V., Prabhu, S., Bubbly, S. G., & Gudennavar, S. B. (2021). Micro and nano Bi2O3 filled epoxy composites: Thermal, mechanical and γ-ray attenuation properties. Applied Radiation and Isotopes, 174, 109780. doi:10.1016/j.apradiso.2021.109780 PMID:34052516

Muthukumaran, C., Sivakumar, V. M., & Thirumarimurugan, M. (2016). Adsorption isotherms and kinetic studies of crystal violet dye removal from aqueous solution using surfactant modified magnetic nanoadsorbent. Journal of the Taiwan Institute of Chemical Engineers, 63, 354–362. doi:10.1016/j.jtice.2016.03.034

Muzammil, W. K.; Rahman, M. M.; Fazlizan, A.; Ismail, M. A. (2019) Nanotechnology in Renewable Energy: Critical Reviews fir Wind Energy. Nanotechnology: Applications in Energy, Drug and Food, 49-71. doi:10.1007/978-3-319-99602-8_3

Nadargi D., Umar, A., Nadargi, J., Patil, J., Mulla, I., Akbar, S., & Suryavanshi, S. (2022). Spinel Magnesium Ferrite (MgFe2O4): A Glycine-Assisted Colloidal Combustion and Its Potentiality in Gas-Sensing Application. Chemosensors (Basel, Switzerland), 10(9), 361. doi:10.3390/chemosensors10090361

Naguib, M., Kurtoglu, M., Presser, V., Lu, J., Niu, J., Heon, M., Hultman, L., Gogotsi, Y., & Barsoum, M. W. (2011). Two-dimensional nanocrystals produced by exfoliation of Ti3AlC 2. Advanced Materials, 23(37), 4248–4253. doi:10.1002/adma.201102306 PMID:21861270

Naguib, M., Mochalin, V. N., Barsoum, M. W., & Gogotsi, Y. (2014). 25th anniversary article: MXenes: A new family of two-dimensional materials. Advanced Materials, 26(7), 992–1005. doi:10.1002/adma.201304138 PMID:24357390

Nahar, K., Hafezur Rahaman, M., Arifuzzaman Khan, G., Khairul Islam, M., & Md Al-Reza, S. (2021). Green synthesis of silver nanoparticles from Citrus sinensis peel extract and its antibacterial potential. Asian Journal of Green Chemistry, 5, 135–150.

Naik, C. C., Gaonkar, S. K., Furtado, I., & Salker, A. V. (2018). Effect of Cu2+ substitution on structural, magnetic and dielectric properties of cobalt ferrite with its enhanced antimicrobial property. Journal of Materials Science Materials in Electronics, 29(17), 14746–14761. doi:10.100710854-018-9611-1

Naik, M. M., Naik, H. S. B., Nagaraju, G., Vinuth, M., Vinu, K., & Viswanath, R. (2019). Green synthesis of zinc doped cobalt ferrite nanoparticles: Structural, optical photocatalytic and antibacterial studies. Nano Structures & Nano Objects, 19, 100322. doi:10.1016/j.nanoso.2019.100322

Nanogaki, H. (2006). Seed Germination- The Biochemical and Molecular Mechanism. Breeding Science, 56(2), 93–105. doi:10.1270/jsbbs.56.93

Nanografti (n.d.). Use of Nanotechnology in Textile Industry & Its future. Nanografiti. https://nanografi.com/blog/use-of-nanotechnology-in-textile-industry-and-its-future/#:~:text=Nanotechnology%20is%20used%20in%20the,are%20explained%20shortly%20among%20them

Naresh, K, A Balaji, M Rambabu, & G Nagaraju. (2018). IRJET- Practical Oriented Foot Step Electric Power Generation By Using Piezo Material And Microcontroller In Campus Practical Oriented Foot Step Electric Power Generation. International Research Journal of Engineering and Technology.

Naseri, M. G., Halimah, M. K., Dehzangi, A., Kamalianfar, A., Saion, E. B., & Majlis, B. Y. (2014). A Comprehensive overview on the structure and comparison of magnetic properties of nanocrystalline synthesized by a thermal treatment method. Journal of Physics and Chemistry of Solids, 75(3), 315–327. doi:10.1016/j.jpcs.2013.11.004

Nayak, P. K. (2008). Synthesis and characterization of cadmium ferrite. Materials Chemistry and Physics, 112(1), 24–26. doi:10.1016/j.matchemphys.2008.05.018

Nazarenko, O., Yakunin, S., Morad, V., Cherniukh, I., & Kovalenko, M. V. (2017). Single crystals of caesium formamidinium lead halide perovskites: Solution growth and gamma dosimetry. NPG Asia Materials, 9(4), e373–e373. doi:10.1038/am.2017.45

Nazarov, A. V., Ménard, D., Green, J. J., Patton, C. E., Argentina, G. M., & van Hook, H. J. (2003). Near theoretical microwave loss in hot isostatic pressed (hipped) polycrystalline yttrium iron garnet. Journal of Applied Physics, 94(11), 7227–7234. doi:10.1063/1.1622996

Ndazi, B. S., Karlsson, S., Tesha, J. V., & Nyahumwa, C. W. (2007). Chemical and physical modifications of rice husks for use as composite panels. Composites. Part A, Applied Science and Manufacturing, 38(3), 925–935.

Negi, K., Kumar, M., Singh, G., Chauhan, S., & Chauhan, M. S. (2018). Nanostructured CeO2 for selective-sensing and smart photocatalytic applications. Ceramics International, 44(13), 15281–15289. doi:10.1016/j.ceramint.2018.05.172

Nehra, M., Dilbaghi, N., Hassan, A. A., & Kumar, S. (2019). Carbon-Based Nanomaterials for the development of sensitive nanosensor platforms Advances in Nanosensors for Biological and Environmental Analysis. Elsevier.

Nejati, K., & Zabihi, R. (2012). Preparation and magnetic properties of nano size nickel ferrite particles using hydrothermal method. Chemistry Central Journal, 6(1), 23–28. doi:10.1186/1752-153X-6-23 PMID:22462726

Nematollahi, R., Ghotbi, C., Khorasheh, F., & Larimi, A. (2020). Ni-Bi co-doped TiO2 as highly visible light response nano-photocatalyst for CO2 photo-reduction in a batch photo-reactor. Journal of CO2 Utilization, 41, 101289. doi:10.1016/j.jcou.2020.101289

Neupane, N. P., Kushwaha, A. K., Karn, A. K., Khalilullah, H., Uzzaman Khan, M. M., Kaushik, A., & Verma, A. (2022). Anti-bacterial efficacy of bio-fabricated silver nanoparticles of aerial part of Moringa oleifera lam: Rapid green synthesis, In-Vitro and In-Silico screening. Biocatalysis and Agricultural Biotechnology, 39, 102229.

Newton, B., Pearson, R., & Williams, J. (1975). The design and realisation of M.I.C. subassemblies on ferrite substrates. IEEE Transactions on Magnetics, 11(5), 1270–1272. doi:10.1109/TMAG.1975.1058899

Nguyen, L. T. T., Nguyen, L. T. H., Manh, N. C., Quoc, D. N., Quang, H. N., Nguyen, H. T. T., Nguyen, D. C., & Bach, L. G. (2019). A Facile Synthesis, Characterization, and Photocatalytic Activity of Magnesium Ferrite Nanoparticles via the Solution Combustion Method. Hindawi Journal of Chemistry, 2019(8), 3428681. doi:10.1155/2019/3428681

Ni, X., Cui, Z., Luo, H., Chen, H., Liu, C., Wu, Q., & Ju, A. (2021). Hollow multi-nanochannel carbon nanofibers@ MoSe2 nanosheets composite as flexible anodes for high performance lithium-ion batteries. Chemical Engineering Journal, 404, 126249.

Nie, H., Li, M., Li, Q., Liang, S., Tan, Y., Sheng, L., Shi, W., & Zhang, S. X. (2014). Carbon dots with continuously tunable full-color emission and their application in radiometric pH sensing. Chemistry of Materials, 26(10), 3104–3112. doi:10.1021/cm5003669

Niederberger, M., Pinna, N., Polleux, J., & Antonietti, M. (2004). A general soft-chemistry route to perovskites and related materials: Synthesis of BaTiO3, BaZrO3, and LiNbO3 nanoparticles. Angewandte Chemie, 43(17), 2270–2273. doi:10.1002/anie.200353300 PMID:15108141

Nikeghbal, K., Zamanian, Z., Shahidi, S., Spagnuolo, G., & Soltani, P. (2020). Designing and fabricating nano-structured and micro-structured radiation shields for protection against CBCT exposure. Materials (Basel), 13(19), 4371. doi:10.3390/ma13194371 PMID:33008078

Nisar, A., Afzulpurkar, N., Mahaisavariya, B., & Tuantranont, A. (2008). MEMS-Based Micropumps in Drug Delivery and Biomedical Applications. Sensors and Actuators. B, Chemical, 130(2), 917–942. doi:10.1016/j.snb.2007.10.064

Nishino, H., & Nishijima, Y. (1996). CdTe (111) B/Si (100) structure grown by metalorganic vapor phase epitaxy with Te adsorption and annealing. Journal of Crystal Growth, 167(3-4), 488–494. doi:10.1016/0022-0248(96)00288-6

Nißler, R., Bader, O., Dohmen, M., Walter, S. G., Noll, C., Selvaggio, G., Groß, U., & Kruss, S. (2020). Remote near infrared identification of pathogens with multiplexed nanosensors. Nature Communications, 11(1), 1–12. doi:10.103841467-020-19718-5 PMID:33239609

Noor, N. A., Rashid, M., Mustafa, G. M., Khan, M. I., Mahmood, A., & Ramay, S. M. (2020). Study of Pressure Induced Physical Properties of ZnZrO 3 Perovskite Using Density Functional Theory. Chemical *Physics Letters, 753(January),* 137601. doi:10.1016/j.cplett.2020.137601

Noor Azman, N. Z., Musa, N. F., Nik Ab Razak, N. N., Ramli, R. M., Mustafa, I. S., Abdul Rahman, A., & Yahaya, N. Z. (2016). Effect of Bi2O3 particle sizes and *addition of starch* into Bi2O3–PVA composites for X-ray shielding. Applied Physics. A, Materials Science & Processing, 122(9), 1–9. doi:10.100700339-016-0329-8

Noor Azman, N. Z., Siddiqui, S. A., Haroosh, H. J., Albetran, H. M., Johannessen, *B., Dong, Y., & Low, I.* M. (2013). Characteristics of X-ray attenuation in electrospun bismuth oxide/polylactic acid nanofibre mats. Journal of Synchrotron Radiation, 20(5), 741–748. doi:10.1107/S0909049513017871 PMID:23955038

Nozari, B., M*ontazer, M., & Mahmoudi* Rad, M. (2021). Stable ZnO/SiO2 nano coating on polyester for anti-bacterial, self-cleaning and flame retardant applications. Materials Chemistry and Physics, 267, 1246*74. doi:10.1016/j.matchemphys.2021.*124674

Nuffer, J., & Bein, T. (2006). Applications of Piezoelectric Materials in Transportation Industry. Global Symposium on Innovative Solutions for the Advancement of the Transport Industry.

Nur-E-Alam, M., Vasiliev, M., Bel*otelov, V., & Alameh, K. (2018). Properties of* ferrite garnet (Bi, Lu, Y)3(Fe, Ga)5O12 thin film materials prepared by RF magnetron sputtering. Nanomaterials (Basel, Switzerland), 8(5), 355. doi:10.3390/nano8050355 PMID:29789463

Obaidat, I. M., Issa, B., & Haik, Y. (2015). Magnetic properties of magnetic nanoparticles for efficient hyperthermia. Nanomaterials (Basel, Switzerland), 5(1), 63–89. doi:*10.3390/nano*5*010*063 PMID:28347000

Ohno, H., Endoh, T., Hanyu, T., Ando, Y., & Ikeda, S. (2014). Spin-transfer-torque magnetoresistive random access memory (STT-MRAM) technology. Advances in Non-volatile Memory and Storage Technology, 455-494; doi:10.1533/*9780857098092.3.455*

Ojemaye, M. O., Okoh, S. O., & Okoh, A. I. (2020). Silver nanoparticles (AgNPs) facilitated by plant parts of Crataegus ambigua Becker AK extracts and their antibacterial, antioxidant and antimalarial activities. Green Chemistry Letters and R*eviews, 14(1), 49–59.*

*Oj*o, A. A., & Dharmadasa, I. M. (2017). Analysis of electrodeposited CdTe thin films grown using Cadmium chloride precursor fo*r applications in solar cells. Journal* of *M*aterials Science Materials in Electronics, 28(19), 14110–14120. doi:10.100710854-017-7264-0

Okamoto, A. (2009). The Invention of Ferrites and Their Contribution to the Min*iaturiz*ati*o*n of Radios. IEEE Globecom Workshops, 1–42. doi:10.1109/GLOCOMW.2009.5360693

Olabemiwo, O. M., Akintelu, S. A., Waheed, A. A., Okunlola, D. S., Akinwale, D. R., Adeyinka, G. C., Adeniji, T. W., & Adebisi, S. A. (2021). *Green synthesis of* silver nanoparticles using stem bark extract of Annona senegalensis: Characterization and its antibacterial potency. Current Research in Green and Sustainable Chemistry, 4, 100219. doi:10.1016/j.crgsc.2021.100219

Olsson, R. T., Salazar, A. G., Hedenqvist, S. M., Gedde, U. W., Lindberg, F.*, & Savage, S. J. (2005). Controlled synthes*is *of* near stoichiometric cobalt ferrite nanoparticles. Chemistry of Materials, 17(20), 5109–5118. doi:10.1021/cm0501665

Oluyamo, S. S., Faremi, A. A., Olusola, O. I. O., & Odusote, Y. A. (2020). Tunability of *conductivity* type and energy band gap of CdTe thin film in the electrodeposition technique. Materials Today: Proceedings, 38, 558–563. doi:10.1016/j.matpr.2020.02.962

Olvera-*Felix, C., Ramirez-Bon, R., Ochoa*-Landi'N, R., Ruvalcba-Manzo, S. G., & Castillo, S. J. (2020). Optical and Structural Characterization of CdTe Nanoparticles synthesized Using Chemical Bath Deposition Technique. Journal of Electronic Materials, 49(2), 1257–1265. doi:10.100711664-019-07719-0

Omidi, E., Mahmoodi, S. N., & Shepard, W. S. Jr. (2016). Mechatronics Multi Posit*ive Feedback Control Method for Active Vibration Suppres*sion in Flexible Structures. Mechatronics, 33, 23–33. doi:10.1016/j.mechatronics.2015.12.003

Or, T., Gourley, S. W., Kaliyappan, K., Yu, A., & Chen, Z. (2020). Recycling of mixed cathode lithium-ion batteries for electric *vehicles: Current status and fu*ture outlook. Carbon Energy, 2(1), 6–43.

Oto, B., Gulebaglan, S. E., Madak, Z., & Kavaz, E. (2019). Effective atomic numbers, electron densities and gamma rays buildup factors of inorganic metal halide cubic perovskites CsBX3 (B = Sn, Ge; X = I, Br, Cl). Radiation Physics *and Chemistry, 159, 195–206. doi:10.1016/j.radphysch*em.2019.03.010

Padam, G. K., & Malhotra, G. L. (1989). preparation and study of CdTe thin films grown by the solution method. Materials Research Bulletin, 24(5), 595–601. d$_o$i$_:$10.1016/0025-5408(89)90107-4

Padmapriya, G., Manikandan, A., Krishnasamy, V., Jaganathan, S. *K., & Antony, S. A. (2016). Spi*nel NixZn1-xFe2O4 (0.0 £ x £ 1.0) nano-photocatalysts: Synthesis, characterization and photocatalytic degradation of methylene blue dye. Journal of Molecular Structure, 1119, 39–47. doi:10.1016/j.molstruc.2016.04.049

Palea$_r$i, L$_.$, *Bragaglia, M., Fab*brocino, F., & Nanni, F. (2021a). Structural monitoring of glass fiber/epoxy laminates by means of carbon nanotubes and carbon black self-monitoring plies. Nanomaterials (Basel, Switzerland), 11(6), 1543. do*i:10.3390/nano1106*1543 PMID:34208091

Pan, D. Y., Zhang, J. C., Li, Z., & Wu, M. H. (2010). Hydrothermal route for cutting graphene sheets into blue-luminescent graphene quantum dots. Advanced Materials, 22(6), 734–738. doi:10.1002/adma.200902825 PMID:20217780

Pan, D. Y., Zhang, J. *C., Li, Z., Wu, C., Yan, X. M.,* & Wu, M. H. (2010). Observation of pH- solvent-, spin-, and excitation-dependent blue photolu[mi]nescence from carbon nanoparticles. Chemical Communications, 46(21), 3681–3683. doi:10.1039/c000114g PMID:20396809

Pan, J., *Wang, P., Wang, P., Yu, Q., Wang, J., Song, C., Zhen*g, *Y.*, & Li, C. (2021). The photocatalytic overall water splitting hydrogen production of g-C3N4/CdS hollow core–shell heterojunction via the HER/OER matching of Pt/MnOx. Chemical Engineering Journal, 405, 126622. doi:10.1016/j.cej.2020.126622

Pandit, R., Sharma, *K. K., Kaur, P., & Kumar, R.* (2014). Cation distribution controlled dielectric, electrical and magnetic behavior of In3+ substituted cobalt ferrites synthes*ized via solid-state* reaction technique. Materials Chemistry and Physics, 148(3), 988–999. doi:10.1016/j.matchemphys.2014.09.009

Pang, *H., Yu, P.*, Xu, F., Zhang, W., Peng, J., Hu, H., ... Liang, Y. (2020). A general strategy for metal oxide nanoparticles embedded into heterogeneous carbon nanosheets as high-rate lithium-ion battery anodes. Journal of Materials Chemistry. A, Materials for Energy and Sustainability, 8(47), 25382–25389. doi:10.1039/D0TA08198A

Paramesh, D., Kumar, K. V., & Reddy, P. V. (2016). Influence of nickel addition on s*tructural and magnetic properties of aluminum substituted Ni*-Zn ferrite nanoparticles. Processing and Application of Ceramics, 10(3), 161–167. doi:10.2298/PAC1603161P

Parashar, M., Shukla, V. K., & Singh, R. (2020). Metal oxides nanoparticles via sol–gel method: A review on synthesis, characterization, *and applications. Journal of Materials Sci*ence Materials in Electronics, 31(5), 3729–3749. doi:10.100710854-020-02994-8

Park, J., Xu, Z.-L., & Kang, K. (2020). Sol*vated ion intercalation in graph*ite*:* Sodium and beyond. Frontiers in Chemistry, 8, 432.

Park, J.-M., Kim, D.-S., Kim, S.-J., Kim, P.-G., Yoon, D.-J., & DeVries, K. L. (2007). Inherent sensing and interfacial evaluation of carbon nanofiber and nanotube/epoxy composites u*sing electrical re*sis*t*ance measurement and micromechanical technique. Composites. Part B, Engineering, 38(7–8), 847–861. doi:10.1016/j.compositesb.2006.12.004

Park, J. M., Kim, P. G., Jang, J. H., Wang, Z., *Kim, J. W., il Lee, W., Park, J. G., & Lawrence DeVri*es, K. (2008). Self-sensing and dispersive evaluation of single carbon fiber/carbon nanotube (CNT)-epoxy co$_m$posites using electro-micromechanical technique and nondestruc*tive acoustic emission.* Composites. Part B, Engineering, 39(7–8), 1170–1182. doi:10.1016/j.compositesb.2008.03.004

Patel, M., & Karera, A. (1991). SiC whisker from rice husk: Microscopic study. Powder Metallurgy International;(Germany, FR), 23(1).

Pathania, A., Thakur, P., Trukhanov, A. V., Trukhanov, S. V., Panina, L. V., Luders, U., & Thakur, A. (2019). Development of tungsten doped Ni-Zn nano-ferrites with fast response and r*ecovery time for hydrogen* gas sensing application. Results in Physics, 15, 102531. doi:10.1016/j.rinp.2019.102531

Patil, K. C., Aruna, S. T., & Mimani, T. (2002). C$_o$mbustion synthesis: An update. Current Opinion in Solid Sta$_t$e and Materials Science, 6(6), 507–512. doi*:10.1016/S135$_9$-0286(02)00123-7*

Patil, M. R., & Shrivastava, V. S. (2015). Adsorptive removal of methylene blue from aqueous solution by polyaniline-nickel ferrite nanocomposite: A kinetic approach. Desalination and Water Treatment, 57(13), 5879–5887. doi:10.1080/19443994.2015.1004594

Patil, S., & Chandrasekaran, R. (2020). Biogenic nanoparticles*: A comprehensive perspective in synthesis, cha*racterization, application and its challenges. Journal of Genetic Engineering and Biotechnology, 18(1), 67. doi:10.118643141-020-00081-*3 PMID:33104931*

*Pawar, C. S., Gu*jar, M. P., & Mathe, V. L. (2015). Synthesis and Characterization of Nanocrystalline Nickel Zinc Ferrite for Chlorine gas sensor at room temperature. In 59th DAE solid state physics symposium 2014, AIP Conference Proceedings, 1665(1), 050060. AIP. 10.1063/1.4917701

Periodic Table of Elements: Yttrium. (2015), Los *Alamos National Laboratory.*

*Pe*rumal, D., Zulkifli, S. N., Een, L. G., Albert, E. L., Muhamad Yusop, M. A., & Che Abdullah, C. A. (2022). Green biosynthesis of metallic nanoparticles and their future biomedical applications. I$_n$ Green $_N$anomaterial$_s$ for Industrial *Applications (pp. 4*1–70). doi:10.1016/B978-0-12-823296-5.00006-X

Perveen, R., Shujaat, S., Qureshi, Z., Nawaz, S., Khan, M., & Iqbal, M. (2020). Green versus sol-gel synthesis of ZnO nanoparticles and antimicrobial activity evaluation against panel of pathogens. Journal of Mat*erials Research a*nd *T*echnology, 9(4), 7817–7827. doi:10.1016/j.jmrt.2020.05.004

Phumying, S., Labuayai, S., Swatsitang, E., Amornkitbamrung, V., & Maensiri, S. (2013). Nanocrystalline spinel ferrite (MFe2O4, M = *Ni, Co, Mn, Mg, Zn) powders prepa*red *by* a simple aloe vera plant-extracted solution hydrothermal route. Materials Research Bulletin, 48(6), 2060–2065. doi:10.1016/j.materresbull.2013.02.042

Piezoelectric, C. (2017). Author's Ac*cepted Manuscript. Nano En*ergy. doi:10.1016/j.nanoen.2017.05.029

Pinnaduwage, L., Boiadjiev, V., Hawk, J., & Thundat, T. (2003). Sensitive detection of plastic explosives with self-assembled monolayer-coated microcantilevers. Applied Physics Letters, 83(7), 1471*–1473. doi:10.1063/1.1*602156

Plouffe, B. D., Murthy, S. K., & Lewis, L. H. (2014). Fundamentals and application of magnetic particles in cell isolation and enrichment. Reports on Progress in Physics, 78(1), 016601. doi:10.1088/0034-4885/78/1/016601 PMID:25471081

Pokropivny, V. V., & *Skorokhod, V. V. (2007). Classif*ication of nanostructures by dimensionality and concept of surface forms engineering in nanomaterial science. Materials Science and Engineering C, $2_7(_5$-8), 990–993. doi:10.1016/j.msec.2006.09.023

$_Po_z$ar, D. M. (1998). Microwave Engineeri*ng (2nd ed.). John Wiley & Sons Inc.*

Prabaharan, M., & Mano, J. F. (2006). Stimuli-responsive hydrogels based on polysaccharides incorporated with thermo-responsive polymers as novel biomaterials. Macromolecular Bioscience, 6(12), 991–1008. doi:10.1002/mabi.200600164 PMID:17128423

Prabhu, S., Bubbly, S. G.*, & Gudennavar, S. B. (2021a). Therm*al, mechanical and γ-ray shielding properties of micro-and nano-Ta2O5 loaded DGEBA epoxy resin composites. Journal of Applied Polymer Science, 138(44), 51289. doi:10.1002/app.51289

Prabhu, S., Bubbly, S. G., & Gudennavar, S. B. (2022). Bismuth (III) *oxide decorated graphen*e *oxi*de filled epoxy nanocomposites: Thermo-mechanical and photon attenuation properties. Advanced Composite Materials, 1–27. doi:10.1080/09243046.2022.2128264

Prabhu, S., Bubbly, S. G., & Gudennavar, S. B. (2023). X-ray *and γ-ray shielding efficiency of p*olymer composites: Choice of fillers, effect of loading and filler size, photon energy and multifunctionality. Polymer Reviews (Philadelphia, Pa.), *63(1), 246–288. doi:10.1080/15583724.2022.2067867*

Prabhu, S., Jayaram, S., Bubbly, S. G., & Gudennavar, S. B. (2021b). A simple software for swift computation of photon and charged particle inte$_r$action p$_a$r$_{am}$eters: PAGEX. Applied Radiation and Isotopes, 176, 109903. *doi:10.1016/j.apradiso.2021.109903* PMID:34419873

PradeshM.UpadhyayB.TongyaB.GuptaC.PradeshM. (2020). Study of alternative fuels and their knock behavior. https://ssrn.com/abstract=3372828

Prajapati, C. *S., Benedict, S., & Bhat, N. (2019*). An ultralow power nanosensor array for selective detection of air pollutants. Nanotechnology, 31(2), 025301. doi:10.1088/1361-6528/ab44fd PMID:31530742

Prakash, C., & Baijal, J. S. (1985). Diele*ctric behaviour of tetravalent titanium-substituted Ni-*Zn ferrites. Journal of the Less Common Metals, 107(1), 51–57. doi:10.1016/0022-5088(85)90240-1

Prakash, J., Parveen, A., Mishra, Y. K., & Kaushik, A. K. (2020). Nanotechnology-assisted liquid crystals-based biosensors: Towards fundamental to advanced app*lications. Biosensors & Bioelectron*ics, 168, 112562. doi:10.1016/j.bios.2020.112562 PMID:32919227

Pramanik, P. K. D., Solanki, A., Debnath, A., Nayyar, A., El-Sappagh, S., & Kwak, K.-S. (2020). Advancing mode*rn healthcare with nanotechnology, nanobiosensors, an*d *in*ternet of nano things: Taxonomies, applications, architecture, and challenges. IEEE Access: Practical Innovations, Open Solutions, 8, 65230–65266. doi:*10.1109/ACCESS.2020.298*4269

Prasad, R., & Bhattacharyya, A. Nguyen, Quang D. (2017) Nanotechnology in Sustainable Agriculture: Recent Developments, Challenges, and Perspectives. Frontiers in Microbiology. doi:10 doi:.3389/fmicb.2017.01014 PMID:28676790

Pugazhendhi, A., Edison, T. N. J. I., Karuppusamy, I., & Kathirvel, B. (2018). Inorganic *nanoparticles: A potential cancer therapy for huma*n welfare. International Journal of Pharmaceutics, 539(1–2), 104–111. doi:10.1016/j.ijpharm.2018.01.034 PMID:29366941

Pullar, R. C. (2012). Hexagonal ferrites: A review of the synthesis, properties, and applications of hexaferrite ceramics. Progress in Materials Science, 57(7), 1191–1334. doi:10.1016/j.pmatsci.2012.04.001

Pyo, S., Eom, W., Kim, Y. J., Lee, S. H., Han, T. H., & Ryu, W.-H. (2020). Super-expansion of assembled reduced graphene oxide interlayers by segregation of Al nanoparticle pillars for high-capacity Na-ion battery anodes. ACS Applied Materials & Interfaces, 12(21), 23781–23788.

Qader, I. N., Kok, M., Dagdelen, F., & Aydogdu, Y. (2019). A Review of Smart Materials: Researches and Applications. ECJSE, (3), 755–788.

Qin, R., Li, F., Jiang, W., & Liu, L. (2009). Salt-assisted low temperature solid state synthesis of high surface area CoFe2O4 nanoparticles. Journal of Materials Science and Technology, 25, 69–72.

Qiu, Z., Wang, B., Zhou, X., Yang, S., Gao, Q., Cai, X., Zhang, S., & Fang, Y. (2022). CdS@Mg(OH)2 core/shell composite photocatalyst for efficient visible-light photocatalytic overall water splitting. International Journal of Hydrogen Energy, 47(14), 8729–8738. doi:10.1016/j.ijhydene.2021.12.216

Qu, Y., Yang, H., Yang, N., Fan, Y., Zhu, H., & Zou, G. (2006). The effect of reaction temperature on the particle size, structure and magnetic properties of coprecipitated CoFe2O4 nanoparticles. Materials Letters, 60(29-30), 3548–3552. doi:10.1016/j.matlet.2006.03.055

Rahman, S., Nadeem, K., Anis-ur-Rehman, M., Mumtaz, M., Naeem, S., & Letofsky-Papst, I. (2013). Structural and magnetic properties of ZnMg-ferrite nanoparticles prepared using the co-precipitation method. Ceramics International, 39(5), 5235–5239. doi:10.1016/j.ceramint.2012.12.023

Rainbow, J., Sedlackova, E., Jiang, S., Maxted, G., Moschou, D., Richter, L., & Estrela, P. (2020). Integrated electrochemical biosensors for detection of waterborne pathogens in low-resource settings. Biosensors (Basel), 10(4), 36. doi:10.3390/bios10040036 PMID:32294961

Raj, S., Trivedi, R., & Soni, V. (2021). Biogenic Synthesis of Silver Nanoparticles, Characterization and Their Applications—A Review. Surfaces, 5(1), 67–90. doi:10.3390urfaces5010003

Rakibuddin, M., & Kim, H. (2020). Samarium (III)-doped ZnO/graphitic−C3N4 composites for enhanced hydrogen generation from water under visible light photocatalysis. Journal of Alloys and Compounds, 832, 154887. doi:10.1016/j.jallcom.2020.154887

Ramakrishna, K. S., Srinivas, Ch., Prajapat, C. L., Meena, S. S., Mehar, M. V. K., Potukuchi, D. M., & Sastry, D. L. (2018). Structural and magnetic investigations: Study of magnetocrystalline anisotropy and magnetic behavior of 0.1% Cu2+ substituted Ni-Zn ferrite nanoparticles. Ceramics International, 44(1), 1193–1200. doi:10.1016/j.ceramint.2017.10.011

Ramanathan, K., Dhanraj, J., Sekaran, C., & Manohar, L. (2020). A Comprehensive Review on Rail Wheel Crack Inspection System. JMCMS. . doi:10.26782/jmcms.2020.03.00014

Ramanayaka, S., Vithanage, M., Alessi, D., & Liu, W. J. (2021). Nanobiochar: Production, properties, and Multifunctional Applications. Environmental Science. Nano, 7(11). doi:10.1039/D0EN00486C

Rana, G., Dhiman, P., Kumar, A., Vo, D. N., Sharma, G., Sharma, S., & Nausad, M. (2021). Recent advances on nickel nano-ferrite: A review on processing techniques, properties and diverse applications. Chemical Engineering Research & Design, 175, 182–208. doi:10.1016/j.cherd.2021.08.040

Randhawa, B. S., Dosanjh, H. S., & Kumar, N. (2007). Synthesis of lithium ferrite by precursor and combustion methods: A comparative study. Journal of Radioanalytical and Nuclear Chemistry, 274(3), 581–591. doi:10.100710967-006-6924-y

Rane, A. V., Kanny, K., Abitha, V. K., & Thomas, S. (2018). Methods for Synthesis of Nanoparticles and Fabrication of Nanocomposites. Synthesis of Inorganic Nanomaterials, 121–139. doi:10.1016/b978-0-08-101975-7.00005-1

Rashad, M., Tekin, H. O., Zakaly, H. M., Pyshkina, M., Issa, S. A., & Susoy, G. (2020). Physical and nuclear shielding properties of newly synthesized magnesium oxide and zinc oxide nanoparticles. Nuclear Engineering and Technology, 52(9), 2078–2084. doi:10.1016/j.net.2020.02.013

Ratan, Z. A., Haidere, M. F., Nurunnabi, M., Shahriar, S. M., Ahammad, A. J. S., Shim, Y. Y., Reaney, M. J. T., & Cho, J. Y. (2020). Green chemistry synthesis of silver nanoparticles and their potential anticancer effects. Cancers (Basel), 12(4), 1–26. doi:10.3390/cancers12040855 PMID:32244822

Ratan, Z. A., Mashrur, F. R., Chhoan, A. P., Shahriar, S. M., Haidere, M. F., Runa, N. J., Kim, S., Kweon, D.-H., Hosseinzadeh, H., & Cho, J. Y. (2021). Silver nanoparticles as potential antiviral agents. Pharmaceutics, 13(12), 1–25. doi:10.3390/pharmaceutics13122034 PMID:34959320

Rathore, D., Kurchania, R., & Pandey, R. K. (2015). Gas Sensing Properties of Size Varying CoFe2O4 Nanoparticles. IEEE Sensors Journal, 15(9), 4961–4966. doi:10.1109/JSEN.2015.2432035

Rattan, R. K. (2015). Nanotechnology in Soil Science. Soil Science: An Introduction, 775-807.

Rauf, H., Yasmin, N., Ali, G., Ashiq, M. N., Safdar, M., & Mirza, M. (2022). New insight in photocatalytic degradation of textile dyes over CeO2/Ce2S3 composite. Physica B, Condensed Matter, 632, 413760. doi:10.1016/j.physb.2022.413760

Ravishankar, T. N., Ramakrishnappa, T., Nagaraju, G., & Rajanaika, H. (2015). Synthesis and Characterization of CeO2 Nanoparticles via Solution Combustion Method for Photocatalytic and Antibacterial Activity Studies. ChemistryOpen, 4(2), 146–154. doi:10.1002/open.201402046 PMID:25969812

Reddy, B. C., Manjunatha, H. C., Vidya, Y. S., Sridhar, K. N., Pasha, U. M., Seenappa, L., Mahendrakumar, C., Sadashivamurthy, B., Dhananjaya, N., Sankarshan, B. M., Krishnaveni, S., Sathish, K. V., & Gupta, P. S. D. (2021). Synthesis and characterization of multi-functional nickel ferrite nano-particles for X-ray/gamma radiation shielding, display and antimicrobial applications. Journal of Physics and Chemistry of Solids, 159, 110260. doi:10.1016/j.jpcs.2021.110260

Reddy, G. C. V., Seela, K. K., & Manorama, S. V. (2000). Preparation of Fe2O3 by the hydrazine method: Application as an alcohol sensor. International Journal of Inorganic Materials, 2, 301–305. doi:10.1016/S1466-6049(00)00044-1

Reddy, L. H., Arias, J. L., Nicolas, J., & Couvreur, P. (2012). Magnetic nanoparticles: Design and characterization, toxicity and biocompatibility, pharmaceutical and bio-medical applications. Chemical Reviews, 112(11), 5818–5878. doi:10.1021/cr300068p PMID:23043508

Reese, M. O., Perkins, C. L., Burst, J. M., Farrell, S., Barnes, T. M., Johnston, S. W., Kuciauskas, D., Gessert, T. A., & Metzger, W. K. (2015). Intrinsic surface passivation of CdTe. Journal of Applied Physics, 118(15), 155305. doi:10.1063/1.4933186

Remington, P. J. (2014). Wheel/Rail Rolling Noise, II: Validation of the Theory. Acoustical Society of America, 1824–32.

Ren, X., Chaurasia, A. K., Oliva-Avilés, A. I., Ku-Herrera, J. J., Seidel, G. D., & Avilés, F. (2015). Modeling of mesoscale dispersion effect on the piezoresistivity of carbon nanotube-polymer nanocomposites via 3D computational multiscale micromechanics methods. Smart Materials and Structures, 24(6), 065031. doi:10.1088/0964-1726/24/6/065031

Rezlescu, N., Rezlescu, E., Tudorache, F., & Popa, P. D. (2007). Some spinel oxide compounds as reducing gas sensors. Sensors & Transducers, 78, 1134–1142.

Rezlescu, N., Rezlescu, E., Tudorache, F., & Popa, P. D. (2009). Gas sensing properties of porous Cu-, Cd- and Zn-ferrites. Romanian Reports in Physics, 61(2), 223–234.

Ricote, J., & Hungrı, T. (2007). High-Sensitivity Piezoelectric, Low-Tolerance-*Factor Perovskites by Mechanosynthes*is. Chemistry of Materials, (8), 4982–4990.

Ristic, M., Nowik, I., Popovic, S., Felner, I., & Music, S. (2003). Influence of synthesis procedure on the YIG formation. Materials Leters, 57(16-17), 2584–2590. doi:10.1016*/S0167-577X(02)01315-0*

*Rocha-San*tos, T. A. P. (2014). Sensors and biosensors based on magnetic nanoparticles. Trends in Analytical Chemistry, 62, 28–36. doi:10.1016/j.trac.2014.06.016

Rosman, N. S. R., Harun, N. A., Idris, I., & Wan Ismail, W. I. (2021). *Nanobiotechnology: Nature-i*nspired silver nanoparticles towards green synthesis. Energy & Environment, 32(7), 1183–1206. doi:10.1177/0958305X21989883

Ruan, L. F., Wang, J. W., & Ying, S. M. (2021). Research progress and application of modified silicon-based anode materials fo*r lithium-ion batteries. Paper presente*d at the Materials Science Forum.

Ruoff, R. S., Qian, D., & Liu, W. K. (2003). Mechanical properties of carbon nanotubes: Theoretical predictions and experimental measurements. Comptes Rendus Physique, 4(9), 993–1008. d*oi:10.1016/j.crhy.2003.08.001*

R*uss*o, R., Malinconico, M., & Santagata, G. (2007). Effect of cross-linking with calcium ions on the physical properties of alginate ðlms. Biomacromolecules, 8(10), 3193–3197. doi:10.1021/bm700565h PMID:17803277

Rybak, M. J. (2006). The pharmacokinetic and pharmacodynamic properties of vancomycin. Clinical Infectious Diseases, 42*(Suppl.1), S35–*S39*.* doi:10.1086/491712 PMID:16323118

Rödel, J., Webber, K. G., Dittmer, R., Jo, W., Kimura, M., & Damjanovic, D. (2015). Feature Article Transferring Lead-Free Piezoelec*tric Ceramics into Application. J*our*na*l of the European Ceramic Society, 35(6), 1659–1681. doi:10.1016/j.jeurceramsoc.2014.12.013

Saba, N., & Jawaid, M. (2018). Energy and environmental applications of graphene and its derivatives. In Polymer-based Nano*composites for Energy and E*nviro*n*mental Applications (pp. 105–129). Elsevier.

Saez, P. R., Fernandez, M. J., Gutierrez, V. B., Gomez, R., & Marquina, V. (2008). Ferrites nanoparticles MFe2O4 (M = Ni and Zn): Hydrothermal synthesis and magnetic properties. Bulletin of the Spanish Society of Ceramics and Glass, 47, 133–137.

Safaei, *M., Sodano, H. A., & Anton, S. R. (2019). A Review of*Energy Harvesting Using Piezoelectric Materials : State-of-the-Art a Decade Later (2008 – 2018). Smart Materials and Structures, 28(11), 113001. doi:10.1088/1361-665X/ab36e4

Sagadevan, S., Vennila, *S., Singh, P., Lett, J.* A., Johan, M. R., Marlinda, A. R., Muthiah, B., & Lakshmipathy, M. (2019). Facile synthesis of silver nanoparticles using Averrhoa bilimbi L and Plum extracts and investigation on the synergistic bioa*ctivity using in vitro models. Green P*rocessing and Synthesis, 8(1), 873–884. doi:10.1515/gps-2019-0058

Salayová, A., Bedlovičová, Z., Daneu, N., Baláž, M., Lukáčová Bujňáková, Z., Balážová, L., & Tkáčiková, L. (2021). Green synth*esis of silver nanoparticles* wi*th* antibacterial activity using various medicinal plant extracts: Morphology and antibacterial efficacy. Nanomaterials (Basel, Switzerland), 11(4), 1005. doi:10.3390/nano11041005 PMID:33919801

Salikhov, R. F., Makushev, Y. P., Musagitova, G. N., Volkova, L. U., & Suleymanov, *R. S. (2019). Diagnosis of Fuel* Eq*ui*pment of Diesel Engines in Oil-and-Gas Machinery and Facilities Diagnosis of Fuel Equipment of Diesel Engines in Oil-and- Gas Machinery a*nd F*acilities. AIP Conference Proceedings, 050009(August), 050009. doi:10.1063/1.5122152

Salim, M. & Abdulrazig, O. (2020). Using Smart-Piezoelectric Materials to *Generate Electricity. Bright Star Journal f*or *S*cientist Research, 1(x), 0–1.

Salim, M., Salim, D., Chandran, D., Aljibori, H. S., & Kherbeet, A. S. (2018). Review of Nano Piezoelectric Devices in Biomedicine Applications. Journal of Intelligent Material Systems and Structures, 29(10), 2105–2121. doi:10.1177/1045389X17754272

Salleh, A., Naomi, R., Utami, N. D., Mohammad, A. W., Mahmoudi, E., Mustafa, N., & Fauzi, M. B. (2020). The potential of silver nanoparticles for antiviral and antibacterial applications: A mechanism of action. Nanomaterials (Basel, Switzerland), 10(8), 1–20. doi:10.3390/nano10081566 PMID:32784939

Sam-Daliri, O., Farahani, M., Faller, L.-M., & Zangl, H. (2020). Structural health monitoring of defective single lap adhesive joints using graphene nanoplatelets. Journal of Manufacturing Processes, 55, 119–130. doi:10.1016/j.jmapro.2020.03.063

Samanta, S., Shinde, M. S., Salunke, D., & Patil, R. S. (2015). Studies on growth and characterization of CdTe thin film deposited by Chemical Bath Deposition. Applied Scientific Research, 11, 10–15.

Sanchez, D., Ramos, C. A., Rivas, J., Vaqueiro, P., & Lopez-Quintela, M. A. (2004). Ferromagnetic resonance and magnetic properties of single-domain particles of Y3Fe5O12 prepared by sol-gel method. Physica B, Condensed Matter, 354(1-4), 104–107. doi:10.1016/j.physb.2004.09.028

Sandler, J., Shaffer, M. S. P., Prasse, T., Bauhofer, W., Schulte, K., & Windle, A. H. (1999). Development of a dispersion process for carbon nanotubes in an epoxy matrix and the resulting electrical properties. Polymer, 40(21), 5967–5971. doi:10.1016/S0032-3861(99)00166-4

Sanpo, N., Berndt, C. C., & Wang, J. (2012). Microstructural and antibacterial properties of zinc-substituted cobalt ferrite nanopowders synthesized by sol-gel methods. Journal of Applied Physics, 112(8), 1–7. doi:10.1063/1.4761987

Santos, J., Calpa, L. R. S., & Souza, F. G. Jr. (2022). Main countries contributing on Nanotechnology. Qeios., doi:10.32388/4MZGZR

Sapei, L., Nöske, R., Strauch, P., & Paris, O. (2008). Isolation of mesoporous biogenic silica from the perennial plant Equisetum hyemale. Chemistry of Materials, 20(5), 2020–2025.

Saroja, A. P. V. K., Kamaraj, M., & Ramaprabhu, S. (2020). Strongly coupled sulfur nanoparticles on graphene-carbon nanotube hybrid electrode for multifunctional sodium and aluminium ion storage. Journal of Alloys and Compounds, 818, 152864.

Sasaki, H., & Iguchi, K. (2000). Driving Method of Piezoelectric Transformer and Driving Circuit for the Same. US Patent 6,133,672, no. 19.

Satayeva, A., Baimenov, A., Azat, S., Zhantikeyev, U., Seisenova, A., & Tauanov, Z. (2022). Review on coal fly ash generation and utilization for resolving mercury contamination issues in Central Asia: Kazakhstan. Environmental Reviews, (ja).

Satheesh, R., Vignesh, K., Rajarajan, M., Suganthi, A., Sreekantan, S., Kang, M., & Kwak, B. S. (2016). Removal of congo red from water using quercetin modified α-Fe2O3 nanoparticles as effective nanoadsorbent. Materials Chemistry and Physics, 180, 53–65. doi:10.1016/j.matchemphys.2016.05.029

Satsangi, N. (2020). Synthesis and Characterization of Biocompatible Silver Nanoparticles for Anticancer Application. Journal of Inorganic and Organometallic Polymers and Materials, 30(6), 1907–1914. doi:10.1007/s10904-019-01372-0

Satyanarayana, L., Reddy, K. M., & Manorama, S. V. (2003). Synthesis of nanocrystalline Ni1−xCoxMnxFe2−xO4: A material for liquefied petroleum gas sensing. Sensors and Actuators. B, Chemical, 89(1-2), 62–67. doi:10.1016/S0925-4005(02)00429-X

Saylan, Y., & Denizli, A. (2020). Virus detection using nanosensors Nanosensors for Smart Cities. Elsevier.

Schierbaum, K. D., Weimar, U., & Gopel, W. (1992). Comparison of ceramic, thick-film and thin-film chemical sensors based upon SnO2. Sensors and Actuators. B, Chemical, 7(1-3), 709–716. doi:10.1016/0925-4005(92)80390-J

Sekar, S., Preethi, V., Srivishnu, V. S., Saravanan, S., & Lee, S. (2022). Highly-efficient photocatalytic activity of TiO2-AC nanocomposites for hydrogen production from sulphide wastewater. International Journal of Hydrogen Energy, 47(95), 40275–40285. doi:10.1016/j.ijhydene.2022.02.019

Seki, K., Nakayama, Y., Chiba, S., & Murakami, K. (1994). Construction of Pressure and Photo Sensors Using Ferrite Thick Films. IEEE Translation Journal on Magnetics in Japan, 9(2), 92–96. doi:10.1109/TJMJ.1994.4565832

Sellami, H., Khan, S. A., Ahmad, I., Alarfaj, A. A., Hirad, A. H., & Al-Sabri, A. E. (2021). Green synthesis of silver nanoparticles using olea europaea leaf extract for their enhanced antibacterial, antioxidant, cytotoxic and biocompatibility applications. International Journal of Molecular Sciences, 22(22), 12562. doi:10.3390/ijms222212562 PMID:34830442

Selvan, R. K., Augustin, C. O., Berchmans, L. J., & Saraswathi, R. (2003). Combustion synthesis of CuFe2O4. Materials Research Bulletin, 38(1), 41–54. doi:10.1016/S0025-5408(02)01004-8

Semancik, S., & Cavicchi, R. E. (1991). The growth of thin, epitaxial SnO2 films for gas sensing applications. Thin Solid Films, 206(1-2), 81–87. doi:10.1016/0040-6090(91)90397-G

Semenov, A. A., Karmanenko, S. F., Demidov, V. E., Kalinikos, B. A., Srinivasan, G., Slavin, A. N., & Mantese, J. V. (2006). Ferrite-ferroelectric layered structures for electrically and magnetically tunable microwave resonators. Applied Physics Letters, 88(3), 033503. doi:10.1063/1.2166489

Senapati, K. K., Borgohain, C., & Phukan, P. (2011). Synthesis of highly stable CoFe2O4 nanoparticles and their use as magnetically separable catalyst for Knoevenagel reaction in aqueous medium. Journal of Molecular Catalysis A Chemical, 339(1-2), 24–31. doi:10.1016/j.molcata.2011.02.007

Seok, J. Y., Kim, S., Yang, I., Park, J. H., Lee, J., Kwon, S., & Woo, K. (2021). Strategically Controlled Flash Irradiation on Silicon Anode for Enhancing Cycling Stability and Rate Capability toward High-Performance Lithium-Ion Batteries. ACS Applied Materials & Interfaces, 13(13), 15205–15215. doi:10.1021/acsami.0c22983

Sethi, V. K., Pandey, M., & Shukla, P. (2011). Use of Nanotechnology in Solar PV Cell. International Journal of Chemical Engineering and Applications, 2(2), 77–80. doi:10.7763/IJCEA.2011.V2.79

Sevilla, M., & Fuertes, A. B. (2009). Chemical and structural properties of carbonaceous products obtained by hydrothermal carbonization of saccharides. Chemistry (Weinheim an der Bergstrasse, Germany), 15(16), 4195–4203. doi:10.1002/chem.200802097 PMID:19248078

Sezer, N., & Koç, M. (2020, November). Nano Energy A Comprehensive Review on the State-of-the-Art of Piezoelectric Energy Harvesting. *Nano Energy, 80*, 105567. doi:10.1016/j.nanoen.2020.105567

Sha, R., & Bhattacharyya, T. K. (2020). MoS2-based nanosensors in biomedical and environmental monitoring applications. Electrochimica Acta, 349, 136370. doi:10.1016/j.electacta.2020.136370

Shafiq, M., Anjum, S., Hano, C., Anjum, I., & Abbasi, B. H. (2020). An overview of the applications of nanomaterials and nanodevices in the food industry. Foods, 9(2), 148. doi:10.3390/foods9020148 PMID:32028580

Shah, M. A., Pirzada, B. M., Priece, G., Shibiru, A. L., & Qurashi, A. (2022). Applications of nanotechnology in smart textile industry: A critical review. Journal of Advanced Research, 38, 55–75. doi:10.1016/j.jare.2022.01.008

Shahbazi, N., & Zare-Dorabei, R. (2019). A novel "off-on" fluorescence nanosensor for sensitive determination of sulfide ions based on carbon quantum dots and gold nanoparticles: Central composite design optimization. *Microchemical Journal, 145*, 996–1002. doi:10.1016/j.microc.2018.12.022

Shahi, D., Bhattarai, E., Poudel, M., Pradhan, P., Pradhananga, R. R., & Awal, S. C. (2018). Green Synthesis of Silver Nanoparticles Using Different Plant Materials and Their Antibacterial Activity. In*ternational Journal of* Applied Sciences and Biotechnology, 6(4), 294–301. doi:10.3126/ijasbt.v6i4.22112

Shaikh, S. F., Ubaidullah, M., Rajaram, S. M., & Al-Enizi, A. M. (2020). *Types, Synthesis methods and applications of ferrites. In R. Mane & V.* Jadhav (Eds.), Spinel Ferrite Nanostructures for Energy Storage Devices (pp. 51–82). Elsevier. doi:10.1016/B978-0-12-819237-5.00004-3

Shanmuganathan, R., Karuppusamy, I., Saravanan, M., Muthukumar, H., P*onnuchamy, K., Ramkumar, V. S., & Pugazhendhi, A. (2019).* Synthesis of Silver Nanoparticles and their Biomedical Applications - A Comprehensive Review. Current Pharmaceutical Design, 25(24), 2650–2660. doi:10.2174/138161282566619*0708185506 PMID:31298154*

*Sha*riat*i*-Rad, M. S., Irandoust, M., Amri, S., Feyzi, M., & Jafari, F. (2014). Magnetic solid phase adsorption, preconcentration and determination of methyl orange in water samples using silica coated magnetic nanoparticles and central composite design. International Nano Letters, 4(4), 91–101. doi:10.100740089-014-0124-5

Sharma, *N. K., Williams, W. S., & Zang*v*i*l, A. (1984). Formation and structure of silicon carbide whiskers from rice hulls. Journal of the American Ceramic Society, 67(11), 715–720.

Sharma, P., Pandey, V., Sharma, M. M. M., Patra, A., Singh, B., Mehta, S., & Husen, A. (2021). A Review on Biosensors and Nanosensors Application in Agroecosyst*ems. Nanoscale Research Letters, 16(1),* 1–24. doi:10.118611671-021-03593-0 PMID:34460019

Sharma, R., Bansal, S., & Singhal, S. (2015). Tailoring the photo-Fenton activity of spinel ferrites (MFe2O4) by incorporating different cations (M = Cu, Zn, Ni, and Co) in the structure. RSC Advances, 5(8), 6006–6018. doi:10.1039/C4RA13692F

Sharma, R. P., Raut, S. D., *Kadam, A. S., Mulani, R.* M.*, & Ma*ne, R. S. (2020). In-vitro antibacterial and anti-biofilm efficiencies of chitosan-encapsulated zinc ferrite nanoparticles. Applied Physic*s. A, Materials Science & Processing, 126(10)*, 824. doi:10.100700339-020-04007-1

Sharma, V., Borkute, G., & Gumfekar, S. P. (2022). Biomimetic nanofiltration membranes: Critical review of materials, *structures, and applications to water purification. Ch*emic*a*l Engineering Journal, 433(3), 133823. doi:10.1016/j.cej.2021.133823

Sharma, V., Saha, J., Patnaik, S., & Kuanr, B. K. (2017). Synthesis and characterization of yttrium iron garnet (YIG) nanoparticles - Microwave material. AIP Advances, 7(5), 056405. doi:1*0.1063/1.4973199*

*Sharma, Y., Kawa*t*ra*, A., Sharma, V., Dhull, D., Kaushik, S., Yadav, J. P., & Kaushik, S. (2021). In-vitro and in-silico evaluation of the anti-chikungunya potential of Psidium guajava leaf extract and their synthesized silver nanop*articles. Virusdisease, 32(*2), 260–265. doi:10.100713337-021-00685-4 PMID:33869673

Sharmin, F., Chandra Roy, D., & Basith, M. A. (2021). Photocatalytic water splitting ability of Fe/MgO-rGO nanocomposites *towards hydrogen evolution. Inter*nat*i*onal Journal of Hydrogen Energy, 46(77), 38232–38246. doi:10.1016/j.ijhydene.2021.09.072

Shawgo, R. S., Richards Grayson, A. C., Li, Y., & Cima, M. J. (2002). BioMEMS for Drug Delivery. Current Opinion in Solid State and Materials Science, 6(4), 329–334. d*oi:10.1*01*6/*S1359-0286(02)00032-3

Shawon, Z. B. Z., Hoque, M. E., & Chowdhury, S. R. (2020). Nanosensors and nanobiosensors: Agricultural and food technology aspects. Nanofabrication for Smart Nanosensor Applications. Elsevier.

She, *Z. W., Kibsgaard, J., Dickens*, C. F., Chorkendorff, I., Nørskov, J. K., & Jaramillo, T. F. (2017). Combining theory and *experiment in electrocatalysis: Insights in*to materials design. Science, 355(6321). doi:1 doi:0.1126cience.aad4998 PMID:28082532

Shen, J. H., Zhu, Y. H., Chen, C., Yang, X. L., & Li, C. Z. (2011). Facile preparat*ion and up conversion* lum*i*nescence of graphene quantum dots. Chemical Communications, 47(9), 2580–2582. doi:10.1039/C0CC04812G PMID:21173992

Shende, R. C., Chandran, P., & Ramaprabhu, S. (2020). In situ nitrogen-doped, defect-induced carb*on nanotubes as an efficient an*ode *f*or sodium-ion batteries. Nanotechnology, 31(23), 235403.

Sheng, M., Zhang, F., Ji, B., Tong, X., & Tang, Y. (2017). A nov*el tin-graphite dua*l-ion battery based on sodium-ion electrolyte with high energy density. Advanced Energy Materials, 7(7), 1601963.

Shenouda, A. Y., Sayed, E., & Sayed, M. E. (2015). Electrodeposition, characterization and photo electrochemic*al properties of CdSe* and CdTe. Ain Shams Engineering Journal, 6(1), 341–346. doi:10.1016/j.asej.2014.07.010

Shi, Q., Zhang, Y., Chen, K., Yuan, S., Chang, T., Tian, F., ... Yang, S. (2020). Cobalt vacancies assisted ion diffusion in Co 2 A*lO 4 carbon nanofibers for enha*ncing lithium battery performance. Dalton Transactions (Cambridge, England), 49(29), 10127–10137.

Shi, W., Zhang, X., He, S., & Huang, Y. (2011). CoFe2O4 magnetic nanopar*ticles as a peroxidase mimic mediated chemiluminescence for hy*drog*en peroxide and glucose. Chemical Communications, 47(38), 10785–10787. doi:10.1039/c1cc14300j PMID:21892508

Shibata, K., Wang, R., Tou, T., & Koruza, J. (2018). Applications of Lead-Free Piezoelectric *Materials. Advanced Materials, 612–16(*8), 612–616. doi:10.1557/mrs.2018.180

Shin, D. & Hong, S. (2020). Recent Advances *in Organic Piezoelectric Biomaterials for Energy and Biomedica*l Applications. Nanomaterials, 10 (123), 1–15.

Shirane, G., Sawaguchi, E., & Takagi, Y. (1951). Dielectric Properties of Lead Zirconate. Physical *Review, 84(3), 476–481. doi:10.11*03/PhysRev.84.476

Shiravand, G., Badiei, A., Goldooz, H., Karimi, M., Ziarani, G. M., Faridbod, F., & Ganjali, M. R. (2020). A Fluorescent g-C3N4 nanosensor for detection of dichromate ions. Current Analytical Chemistry, 16(5), 593–601. doi:10.*2174/1573411014666180627150248*

*Shirkhan*loo, H., Safari, M., Amini, S. M., & Rashidi, M. (2017). Novel semisolid design based on bismuth oxide (Bi2O3) nanoparticles for radiation protection. Nanomedicine Research Journal, 2, 2*30–238.*

*Shirvanimoghaddam, M., Shirvanimoghadd*am, K., Abolhasani, M. M., Farhangi, M., Zahiri Barsari, V., Liu, H., Dohler, M., & Naebe, M. (2019). Towards a Green and Self-Powered Internet of Things Using Piezoelectric Energy Harvesting. IEEE Access: Practical Innovations, Open Solutions, 7, 94533–94556. doi:10.1109/ACCESS.2019.2928523

Shishoo, R. *(2004). The Impact of Nanotechnology in t*he *T*extile Industry: Market Potential (Paper Presentation). EuroNanoTex Conference, Barcelona.

Shivashankar, P., & Gopalakrishnan, S. (2020). Review on the Use of P*iezoelectric Materials for* Activ*e Vibration, Noise, and Flow Control. Smart Materials and Structures, 29(5), 053001. doi:10.1088/1361-665X/ab7541

Shkir, M., Kilany, M., & Yahia, I. S. (2017*). Facile microw*ave*-a*ssisted synthesis of tungsten-doped hydroxyapatite nanorods: A systematic structural, morphological, dielectric, radiation and microbial activity studies. Ceramics International, 43(17), 14923–14931. doi:10.1016/j.ceramint.2017.08.009

Shojaeiarani, J., Bajwa, D., & Holt, G. (2020). Sonication amplitude and processing time influence the cellulose nanocrystals morphology and dispersion. Nanocomposites, 6(1), 41–46. doi:10.1080/20550324.2019.1710974

Shung, K. K., Cannata, J. M., & Zhou, Q. F. (2007). Piezoelectric Materials for High Frequency Medical Imaging Applications : *A Review. Journal of Electroceramics,* 19(1), 139–145. doi:10.100710832-007-9044-3

Sik, S., Seo, H., Kawahara, M., Glinsek, S., Lee, J., Kim, Y., Kyu, C., & Angus, I. (2018, *October). Nano Energy Flexible Vibrational Energy Harvesting* Devices Using Strain-Engineered Perovskite Piezoelectric Thin Films. Nano Energy, 55, 182–192. doi:10.1016/j.nanoen.2018.10.068

Silva, F. S., Cristale, J., André, P. A., Saldiva, *P. H., &* Marchi, M. R. (2010). PM2. 5 and PM10: The influence of sugarcane burning on potential cancer risk. Atmospheric Environment, 44(39), 5133–5138.

Singh, A., Gautam, P. K., Verma, A., Singh, V., Shivapriya, *P.* M., Shivalkar, S., Sahoo, A. K., & Samanta, S. K. (2020). Green synthesis of metallic nanoparticles as effective alternatives to treat antibiotics resistant bacterial infections: A review. Biotechnology Reports (Amsterdam, Netherlands), 25, 1–11. doi:10.1016/j.btre.2020.*e00427 PMID:32055457*

Singh, H., Du, J., Singh, P., & Yi, T. H. (2018). Role of green silver nanoparticles synthesized from Symphytum officinale leaf extract in protection against UVB-induced photoaging. Journal of Nanostructure in Chemistry, 8(3), 359–368. doi:10.100740*097-018-0281-6*

Singh, I., Gupta, S., Gautam, H. K., Dhawan, G., & Kumar, P. (2021). Antimicrobial, radical scavenging, and dye degradation potential of nontoxic biogenic silver nanoparticles using Cassia fistula pods. Chemical Papers, 75(3), 979–991. doi:10.1007*11696-020-01355-3*

Singh, P., & Yadava, R. (2020). Nanosensors for health care Nanosensors for Smart Cities. Elsevier.

Singh, S., Ralhan, N. K., Kotnala, R. K., & Verma, K. C. (2012). Nanosize dependent electrical and magnetic prop*erties of NiFe2O4 ferrite. Indian Journal of Pure and A*pplied Physics, 50, 739–743.

Sirdeshmukh, L., Krishna Kumar, K., Bal Laxman, S., Rama Krishna, A., & Sathaiah, G. (1998). Dielectric properties and electrical conduction in yttrium iron garnet (YIG). Bulletin of Materials Science, 21(3), 219–0. doi:10.1007/BF02744973

Sirohi, R. S. (1*990). Optical Components, Sys*tems and Measurement Techniques. CRC Press.

Sivakumar, N., Narayanasamy, A., Greneche, J. M., Murugaraj, R., & Lee, Y. S. (2010). Electrical and magnetic behaviour of nanostructured MgFe2O4 sp*inel ferrite. Journ*al *of* Alloys and Compounds, 504(2), 395–402. doi:10.1016/j.jallcom.2010.05.125

Skotadis, E., Aslanidis, E., Kainourgiaki, M., & Tsoukalas, D. (2020). Nanoparticles Synthesised in the Gas-Phase and Their Applications in Sensors: A Review. Applied Nanoscience, 1(1), 70–86.

Snoek, J. L. (1947). Gyromagnet*ic resonance in ferrites.* Nature, 160(4055), 90. doi:10.1038/160090a0 PMID:20256149

Sohn, C. H., Park, S. P., Choi, S. H., Park, S. H., Kim, S., Xu, L., Kim, S. H., Hur, J. A., Choi, J., & Choi, *T. H. (2015). MRI molecular imaging us*ing GLUT1 antibody-Fe3O4 nanoparticles in the hemangioma animal model for differentiating infantile hemangioma from vascular malformation. Nanomedicine; Nanotechnology, Biology, and Me*dicine, 11(1), 127–135. do*i:*10.*1016/j.nano.2014.08.003 PMID:25168935

Soliman, W. E., Khan, S., Rizvi, S. M. D., Moin, A., Elsewedy, H. S., Abulila, A. S., & Shehata, T. M. (2020). Therapeutic applications of biostable silver nanoparticles synthesized using peel extract *of benincasa hispida: Antibacterial and anticance*r *act*ivities. Nanomaterials (Basel, Switzerland), 10(10), 1–13. doi:10.3390/nano10101954 PMID:33008104

Soltani, N., Bahrami, A., Pech-Canul, M. I., & $_{G}$o$_{n}$zález, L. A. (2015). Review on the physicochemical treatments of rice husk for p*roduction of* advanced materials. Chemical Engineering Journal, 264, 899–935.

Song, Z., Zhou, D., & Liu, Q. (2019). Tolerance factor and phase stability of the garnet structure. Acta Crystallographica. Section C, Crystal Stru*cture Communications, 75(10), 135*3–1358. doi:10.1107/S2053229619011975 PMID:31589151

Souriou, D., Mattei, J. L., Chevalier, A., & Queffelec, P. (2010). Influential parameters on electromagnetic properties of nickel-zinc ferrites for antenna miniaturization. Journal of Applied Physics, 107, 09A518.

Spinelli, G., La*mberti, P.,* Tuc*c*i, V., Guadagno, L., & Vertuccio, L. (2020). Damage Monitoring of Structural Resins Loaded with Carbon Fillers: Experimental and Theoretical Study. Nanomaterials (Basel, Switzerland), 10(3), 434. doi:10.3390/*nano10030434 PMID:32121350*

*Spirou, S. V., B*asini, M., Lascialfari, A., Sangregorio, C., & Innocenti, C. (2018). Magnetic hyperthermia and radiation therapy: Radiobiological principles and current *practice. Nanomaterials (Basel, Switzerland), 8(6),* 1–22. doi:10.3390/nano8060401 PMID:29865277

Srivastava, R., & Yadav, B. C. (2012). Ferrite Materials: Introduction, Synthesis Techniques, and Applications as Sensors. International Journal of Green Nanotechnology, 4(2), 141–154. doi:10.1080/19430892.2012.676918

Stafiej, A., & Pyrzynska, K. (2008). Solid phase extraction of metal ions using carbon nanotubes. Microchemical Journal, 89(1), 29–33. doi:1*0.1016/j.micro*c.2*0*07.11.001

Stergiou, C. A., & Litsardakis, G. (2016). Y-type hexagonal ferrites for microwave absorber and antenna applications. Journal of Magnetism and Magnetic Materia*ls, 405, 54–61. doi:10.10*16/j.jmmm.2015.12.027

Stitzer, S. N., Carter, P. S., & Goldie, H. (1997). A High Power X-Band Frequency Selective Passive YIG Limiter. In IEEE MTT-S Internatio*nal Microwave Symposium Diges*t. IEEE.

Strassberger, M., & Waller, H. (2000). Active Noise Reduction by Structural Control Using Piezo-Electric Actuators. Mechatronics, 10(8), 851–868. doi:10.1016/S0957-4158(99)00096-3

Su, L., Feng, J., Zhou, X., Ren, C., Li, H., & *Chen, X. (20*12). Colorimetric detection of urine glucose based ZnFe2O4 magnetic nanoparticles. Analytical Chemistry, 84(13), 5753–5758. doi:10.1021/ac300939z PMID:2270223*6*

Su, Y.-F., Han, G., *Kon*g, Z., Nantung, T., & Lu, N. (2020). Embeddable piezoelectric sensors for strength gain monitoring of cementitious materials: The influence of coating materials. *Engineered Scienc*e, 11(8), 66–75. doi:10.30919/es8d1114

Subha, V., Ernest Ravindran, R., Preethi, R., Cyrril, J., Kirubanandan, S., & Renganathan, S. (2018). Silver Nanoparticles - Green Synthesis with Aq. Extract of Stems Ipomo*ea Pes-Caprae, Characterization, Antimic*rob*i*al and Anti-Cancer Potential. International Journal of Medical Nano Research, 5(1), 1–9. doi:10.23937/2378-366*4.1410024*

*Subraman*iam, S., Kumarasamy, S., Narayanan, M., Ranganathan, M., Rathinavel, T., Chinnathambi, A., & Whangchai, K. (2022). Spectral and structure characterization of Ferula assafoetida fabricated silver nanoparticles and ev*aluation of its cytotoxic, a*nd *p*hotocatalytic competence. Environmental Research, 204(PA), 111987.

Sugimoto, M. (1999). The Past, Present, and Future of Ferrites. Journal of the American Ceramic So$_{c}$i$_{e}$ty, 82(2), 269–280. doi:10.1111/j.1551-291*6.1999.tb20058.x*

Sui, D., Xu, L., Zhang, H., Sun, Z., Kan, B., Ma, Y., & Chen, Y. (2020). A 3D cross-linked graphene-based honeycomb carbon composite with excellent confinement effect of organic cathode material for lithium-ion batteries. Carbon, 157, 656–662.

Sukhanova, A., *Devy, M., Venteo, L., Kaplan, H., Artemyev, M.,* Oleinikov, V., Klinov, D., Pluot, M., Cohen, J. H. M., & Nabiev, I. (2004). Biocompatible fluorescent nanocrystals for immunolabeling of membrane proteins and cells. Analytical Biochemistry, 324(1), 60–67. doi:10.1016/j.ab.2003.09.031 PMID:14654046

Sultanov, F., Daulbayev, C., Azat, S., Kuterbekov, K., Bekmyrza, K., Bakbolat, B., & Man*surov, Z. (2020). Influence of* Metal Oxide Particles on Bandgap of 1D Photocatalysts Based on SrTiO3/PAN Fibers. Nanomaterials (Basel, Switzerland), 10(9), 1734.

Sun, C., Pan, L., Zhang, L., Huang, J., Yao, D., Wang, C.-Z., Zhang, Y., Jian*g, N., Chen, L., & Yuan*, *C.* (2019). A biomimetic fluorescent nanosensor based on imprinted polymers modified with carbon dots for sensitive detection of alpha-fetoprotein in clinical samples. Analyst (London), 144(22), 6760–6772. doi:10.1039/C9AN01065C PMID:31617507

Sun, D., Ban, R., Zhang, P. H., Wu, *G. H., Zhang, J. R.,* & Z*h*u, J. J. (2013). Hair ðber as a precursor for synthesizing of sulfur-and nitrogen-co-doped carbon dots with tunable luminescence properties. Carbon, 64, 424–434. doi:10.1016/j.carbon.2013.07.095

Sun, J., Gua*n, Q., Liu, Y.,* & Leng, J. (2016). Morphing Aircraft Based on Smart Materials and Structures : A State-of-the-Art Review. Journal of Intelligent Material Systems and Structures, 27(2), 2289–2312. Adva*nce online publication. doi:10*.1177/1045389X16629569

Sun, K., Zhang, Z., Fan, R., Chen, M., Cheng, C., Hou, Q., Zhang, X., & Liu, Y. (2015). Random copper/yttrium iron garnet composites with tunable negative electromagnetic parameters prepared by in situ synthesis. RSC Advances, 5(75), 61155–*61160. doi:*10.*1*039/C5RA09882C

Sun, L., & Gong, K. (2001). Silicon-based materials from rice husks and their applications. Industrial & Engineering Chemistry Research, 40(25), 5861–5877. doi:10.1021/ie010284b

Sun, Y., B*ao, H.-D., Guo,* Z.-*X., & Yu*, J. (2009). Modeling of the Electrical Percolation of Mixed Carbon Fillers in Polymer-Based Composites. Macromolecules, 42(1), 459–463. doi:10.1021/ma8023188

Sun, Y., Chen, J., Li, X., Lu, Y., Zhang, S., & Cheng, Z. (2019). Nano Energy Flexible Piezoelectric Energy Ha*rvester / Sensor with High Voltage Output over* Wid*e* Temperature Range. Nano Energy, 61(March), 337–345. doi:10.1016/j.nanoen.2019.04.055

Sun, Y. P., Zhou, B., Lin, Y., Wang, W., Fernando, K. S., Pathak, P., Meziani, M. J., Harruff, B. A., Wang, X., Wang, H., Luo, P. G., Yang, H., Kose, *M. E., Chen, B., Veca, L. M., & Xie*, *S*.-Y. (2006). Quantum-sized carbon dots for bright and colorful photoluminescence. Journal of the American Chemical Society, 128(24), 7756–7757. doi:10.1021/ja062677d PMID:16771487

Surabhi, S., Anurag, K., & Kumar, S. R. (2021). Effect of anneali*ng on the struc*tura*l*, compositional and optical properties of CdTe films. Materials Today: Proc*eedings, 45, 4477–4482. doi:10.1016/j.matpr.2020.12.988*

Surabhi, S., Anurag, K., & Kumar, S. R. (2022). Structural, morphological, electrical and optical properties of Zn doped CdTe thin films. Chalcogenide *Letters, 19(2), 143–152. doi:10.15251/CL.*2022.192.143

Surabhi, S., Anurag, K., Rajpal, S., & Kumar, S. R. (2021). A new route for preparing CdTe thin films by chemical bath deposition. Materials Today: Proceedings, 44, *1463–1467. doi:10.1016/j.matpr.*2020.11.635

Sutka, A., & Gross, K. A. (2016). Spinel ferrite oxide semiconductor gas sensors. Sensors and Actuators. B, Chemical, 222, 95–105. doi:10.1016/j.snb.2015.08.027

Sutka, A., & Mezinskis, G. (2012). Sol-*gel auto-combustio*n *s*ynthesis of spinel-type ferrite nanomaterials. Frontiers of Materials *Science*, *6(*2), 128–141. doi:10.100711706-012-0167-3

Sutka, A., Gross, K. A., Mezinskis, G., & Knite, M. (2011). The effect of heating conditions on the properties of nano- and microstructured Ni-Zn ferrite. Physica Scripta, 83(2), 025601–025606. doi:10.1088/0031-8949/83/02/025601

Sutka, A., Mezinskis, G., Pludons, A., & Lagzdina, S. (2010). Ch*aracterization of sol-gel auto-combustion derived* spinel ferrite nano-materials. Energetika, 56(3-4), 254–259.

Sylvester, P., Westerhoff, P., Mooller, T., Badruzzaman, M., Boyd, O. (2007) A hybrid sorbent utilizing nanoparticles of hydrous iron oxide for arsenic removal from drinking water. Environmental Engineering Science 24 (1), 104e112.

Szperlich, P., & Toroń, B. (2019). *An ultrasonic fabrication method* for epoxy resin/SbSI nanowire composites, and their application in nanosensors and nanogenerators. Polymers, 11(3), 479. doi:10.3390/polym11030479 PMID:30960463

Tachibana, Y., Vayssieres, L., & Durrant, J. R. (2012). A*rtificial photosynthesis for sola*r water-splitting. Nature Photonics, 6(8), 511–518. doi:10.1038/nphoton.2012.175

Tahir, M. B., Iqbal, T., Rafiq*ue, M., Rafique, M. S., Nawaz, T., & Sagir, M. (2020). Nanomaterial*s *fo*r photocatalysis. Nanotechnology and Photocatalysis for Environmental Applications, 65–76. doi:10.1016/b978-0-12-821192-2.00005-x

Tahir, M. B., Nabi, G., Iqbal, T., Sagir, M., & Rafique, M. (2018). Role of MoSe2 on nanostructures WO3-*CNT performance for photocatalyt*ic hydrogen evolution. Ceramics International, 44(6), 6686–6690. doi:10.1016/j.ceramint.2018.01.081

Takada, T., & Kiyama, M. (1971). Preparation of ferrites by wet method. In Ferrites: Proc*eedings of the International Confer*ence, Y. Hoshino, S. Iida and M. Sugimoto (Eds.), University of Tokyo Press, Tokyo, pp. 69-71.

Tallman, T. N., Gungor, S., Wang, K. W., & Bakis, C. E. (2015). Damage detection via electrical impedance tomography in glass fiber/epoxy *laminates with carbon black fille*r. Structural Health Monitoring, 14(1), 100–109. doi:10.1177/1475921714554142

Tamaura, Y., Kojima, M., Hasegawa, N., Tsuji, M., Ehrensberger, K., & Steinfeld, A. (1997). Solar *Energy Conversion into H2 Energy Using Ferri*tes. Journal of Physics, IV, 673-674.

Tan, B., Han, S., Luo, W., Chao, Z., Fan, J., & Wang, M. (2020). Synthesis of RGO-supported layered MoS2 with *enhanced electrochemi*cal performance for aluminum ion batteries. Journal of Alloys and Compounds, 841, 155732.

Tang, H., Yan, M., Zhang, H., Xia, M., & Yang, D. (2005). Preparation and cha*racterization of water-soluble CdS nanocrys*tals by surface modification of ethylene diamine. Materials Letters, 59(8-9), 1024–1027. doi:10.1016/j.matlet.2004.11.049

Tang, L., Ji, R., Cao, X., Lin, J., Jia*ng, H., Li,* X., Teng, K. S., Luk, C. M., Zeng, S., Hao, J., & Lau, S. P. (2012). Deep ultraviolet photoluminescence of water-soluble self-passivated graphene quantum dots. ACS Nano, 6(6), 5102–5110. doi:10.1021/nn300760g PMID:22559247

Tatarenko, A. S., Srinivasan, G., & Bichurin, M. I. (2006). Magnetoelectric *microwave phase shifter. Applied Physics Lett*ers, 88(18), 183507. doi:10.1063/1.2198111

Tatsuoka, H., Kuwabara, H., Nakanishi, Y., & Fujiyasu, H. (1993). CdTe (111) growth on misoriented Si (100) substrates by hot-wall epitaxy. Journal of Crystal Growth, 129(3-4), 686–690. doi:10.1016/0022-0248(93)90504-P

Tauanov, Z., Azat, S., & Baibatyrova, A. (2022). A mini-review *on coal fly ash properties*, utilization and synthesis of zeolites. International Journal of Coal Preparati*on and Utilization, 42(7), 1968–1990.*

Te*ki*n, H. O., Singh, V. P., Kara, U., Manici, T., & Altunsoy, E. E. (2016). Investigation of nanoparticle effect on radiation shielding property using Monte Carlo method. CBU Journal of Science, 12(2), 196–200. doi:10.18466/cbujos.15586

Thakur, A. K., Ahmed, M. S., Oh, G., Kang, *H., Jeong,* Y., Prabakaran, R., ... Hwang, J.-Y. (2021). Advancement in graphene-based nanocomposites as high capacity anode materials for sodium-ion batteries. Journal of Materials Chemistry. A, Materials for Energy and Sustainability, 9(5), 2628–2661.

*Thant, A. A., Srimal*a, *S.*, Kaung, P., Itoh, M., Radzali, O., & Fauzi, M. N. A. (2010). Low temperature synthesis of MgFe2O4 soft ferrite nanocrystallites. Journal of the Australian. Ceramic Socie,y 46(1), 11–14.

Theron, J.*, Cloete, T.E., de* Kwa*a*dsteniet, M. (2010) Current molecular and emerging nanobiotechnology approaches for the detection of microbial pathogens. Critical Reviews in Microbiology 36 (4), 318e339.

Thiagarajan, S., Sanmugam, A., & Vikraman, D. (2017). Facile Methodology of Sol-Ge*l Synthesis for Metal Oxide Nanos*truc*t*ures. Recent Applications in Sol-Gel Synthesis.

Thostenson, E. T., & Chou, T. W. (2006). Processing-structure-multi-functional property relationship in carbon nanotube/epoxy composites. Carbon, 44(14), 3022–3029. doi:10.1016/j.carbon.2006.05.014

Thostenson, E. T., & Chou, T. W. *(2008a). Carbon* na*no*tube-based health monitoring of mechanically fastened composite joints. Composites Science and Technology, 68(12), 2557–2561. doi:10.1016/j.compscitech.2008.05.016

Thostenson, E. T., & Cho*u, T. W. (2008b). Real-time in situ sensing of damage* evo*l*ution in advanced fiber composites using carbon nanotube networks. Nanotechnology, 19(21), 215713. doi:10.1088/0957-4484/19/21/215713 PMID:21730592

Thostenson, E. T.*, Ren, Z., & Chou, T.-W. (2001). Advances* in *th*e science and technology of carbon nanotubes and their composites: A review. Composites Science and Technology, 61(13), 1899–1912. doi:10.1016/S0266-3538(01)00094-X

Tian, J., Leng, *Y., Cui, H., & Li*u, H. (2015). Hydrogenated TiO2 nanobelts as highly efficient photocatalytic organic dye degradation and hydrogen evolution photocatalyst. Journal of Hazardous Materials, 299, 165–173. doi:10.1016/j.jhazmat.2015.06.019 *PMID:26118*828

Tian, L., Ghosh, D., Chen, W., Pradhan, S., Chang, X., & Chen, S. (2009). Nanosized carbon particles from natural gas soot. Chemistry of Materials, 21(13), 2803–2809. doi:10.1021/cm900709w

T*ian, S., Cai, T., Wang, D.,* Ko*n*g, D., Ren, H., & Xing, W. (2021). A superior cathode of sodium-ion battery: A V2O5 nanosheets anchored on carbon nanofibers. Paper presented at the IOP Conference Series: Earth and Env*ironmental Science.*

T*i*an, Z., Lin, L., Meng, F., & Huang, W. (2009). Combustion synthesis and characterization of nanocrystalline Ba(Mg1/3Nb2/3)O3 powders. Materials Science and Engineering B. Solid-*State Materials for Advance*d *T*echnology, 158(1–3), 88–91. doi:10.1016/j.mseb.2009.01.023

Tiede, K., Boxall, A.B.A., Tear, S.P., Lewis, J., David, H., Hassello*v, M. (2008) Detection and charact*eri*za*tion of engineered nanoparticles in food and the environment. Food Additives and Contaminants, 25 (7), 795e821.

Tiraferri, A., Vecitis, C.D., Elimelech, M. (2011) Covalent binding of single-walle*d carbon nanotu*be*s* to polyamide membranes for antimicrobial surface properties. Acs Applied Materials & Interfaces 3(8), 2869e2877.

Tiwary, C. S., Sarkar, R., Kumbhakar, P., Mitra, *A. K. (2008). Synthesis and op*tical characterization of monodispersed Mn2+ doped CdS nanoparticles. Phys. Lett. A, 327, 5825-5830.

Tiwary, C. S., Sarkar, R., Kumbhakar, P., Mitra, A. K., & Chattopadhyay, K. (2009). Synth*esis of wu*rt*z*ite-phase ZnS nanocrystal and its optical properties. Journal of Luminescence, 129(11), 1366–1370. doi:10.1016/j.jlumin.2009.07.004

Toshtay, K., Auyezov, A., Korkembay, Z., Toktassynov, S., Seytkhan, A.*, & Nurakyshev, A.* (*20*21). Partial hydrogenation of sunflower oil on platinum catalysts: Influence of process conditions on the mass content of geometric isomers. Molecular Catalysis, 513, 111819.

Townshield, A., & Jackwerth, E. (1989). Precipitation of major constitutuents for trace preconcentration: Potential and problems. Pure and Applied Chemistry, 61(9), 1643–1656. doi:10.1351/pac198961091643

Trivedi, P., & Axe, L. (2000). Modeling Cd and Zn sorption to hydrous metal oxides. Environmental Science & Technology, 34(11), 2215–2223. doi:10.1021/es991110c

Tsao, Y.-H., Husain, R. A., Lin, Y.-J., Khan, I., Chen, S.-W., & Lin, Z.-H. (2019). A self-powered mercury ion nanosensor based on the thermoelectric effect and chemical transformation mechanism. Nano Energy, 62, 268–274. doi:10.1016/j.nanoen.2019.05.032

Tu, J., Wang, J., Zhu, H., & Jiao, S. (2020). The molten chlorides for aluminum-graphite rechargeable batteries. Journal of Alloys and Compounds, 821, 153285.

Tu, Y., Wu, J., Xu, G., Yang, X., Cai, R., Gong, Q., & Huang, W. (2021). Perovskite solar cells for space applications: Progress and challenges. Advanced Materials, 33(21), 2006545. doi:10.1002/adma.202006545 PMID:33861877

Turan, F., Guclu, M., Gurkan, K., Durmus, A., & Taskin, Y. (2022). The effect of carbon nanotubes loading and processing parameters on the electrical, mechanical, and viscoelastic properties of epoxy-based composites. Journal of the Brazilian Society of Mechanical Sciences and Engineering, 44(3), 93. doi:10.100740430-022-03393-2

Tzeng, Y., Chen, R., & He, J.-L. (2020). Silicon-based anode of lithium ion battery made of nano silicon flakes partially encapsulated by silicon dioxide. Nanomaterials (Basel, Switzerland), 10(12), 2467.

Ubale, A. U., Dhokne, R. J., Chikhlikar, P. S., Sangawar, V. S., & Kulkarni, D. K. (2006). Characterization of nanocrystalline cadmium telluride thin films grown by successive ionic layer adsorption and reaction (SILAR) method. Bulletin of Materials Science, 29(2), 165–168. doi:10.1007/BF02704610

Uchino, K. (2008). Piezoelectric Actuators 2006. Journal of Electroceramics, 20(3-4), 301–311. doi:10.100710832-007-9196-1

Uchino, K., & Uchino, K. (2015). Glory of Piezoelectric Perovskites. Science and Technology of Advanced Materials, 16(4), 46001. doi:10.1088/1468-6996/16/4/046001 PMID:27877827

Ur Rahim, H., Qaswar, M., Uddin, M., Giannini, C., Herrera, M. L., & Rea, G. (2021). Nano-enable materials promoting sustainability and resilience in modern agriculture. Nanomaterials (Basel, Switzerland), 11(8), 2068. doi:10.3390/nano11082068 PMID:34443899

Ustinov, A. B., Srinivasan, G., & Kalinikos, B. A. (2007). Ferrite-ferroelectric hybrid wave phase shifters. Applied Physics Letters, 90(3), 031913. doi:10.1063/1.2432953

Vaidyanathan, G., Sendhilnathan, S., & Arulmurugan, R. (2007). Structural and magnetic properties of Co1-xZnxFe2O4 nanoparticles by co-precipitation method. Journal of Magnetism and Magnetic Materials, 313(2), 293–299. doi:10.1016/j.jmmm.2007.01.010

Valenzuela, R. (1994). Magnetic ceramics. Cambridge Press. doi:10.1017/CBO9780511600296

Valenzuela, R. (2012). Novel Applications of Ferrites. Physics Research International, 591839.

Vallinayagam, S., Rajendran, K., & Sekar, V. (2021). Green synthesis and characterization of silver nanoparticles using Naringi crenulate leaf extract: Key challenges for anticancer activities. Journal of Molecular Structure, 1243, 130829–130831. doi:10.1016/j.molstruc.2021.130829

Vanlalveni, C., Lallianrawna, S., Biswas, A., Selvaraj, M., Changmai, B., & Rokhum, S. L. (2021). Green synthesis of silver nanoparticles using plant extracts and their antimicrobial activities: A review of recent literature. RSC Advances, 11(5), 2804–2837. doi:10.1039/D0RA09941D PMID:35424248

Varghese, O. K., Paulose, M., LaTempa, T. J., & Grimes, C. A. (2009). High-Rate Solar Photocatalytic Conversion of CO2 and Water Vapor to Hydrocarbon Fuels. Nano Letters, 9(2), 731–737. doi:10.1021/nl803258p PMID:19173633

Varma, P. C. R., Manna, R. S., Benerjee, D., Varma, M. R., Suresh, K. G., & Nigam, A. K. (2008). Magnetic properties of CoFe2O4 synthesized by solid state, *citrate precursor and polymerized complex methods: A comparative study. Journal of* Alloys and Compounds, 453(1-2), 298–303. doi:10.1016/j.jallcom.2006.11.058

Vasic, M. V., Bozin, E. S., Bessais, L., Stojanovic, G., Luburic, U. K., Abeykoon, M., Jancar, B., Meden, A., Kremenovic, A., & Antic, B. (2013). Thermal Evolution of Cation Distribution/Crystallite Size and Their Correlation with the Magnetic State of Yb-Substituted Zinc Ferrite Nanoparticles. The Journal of Physical Chemistry C, 117(23), 12358–12365. doi:10.1021/jp403459t

Vecitis, C. D., Schnoor, M. H., Rahaman, M. S., Schiffman, J. D., & Elimelech, M. (2011). *Electrochemical multiwalled carbon* nanotube filter for viral and bacterial removal and inactivation. Environmental Science & Technology, 45(8), 3672–3679. doi:10.1021/es2000062 PMID:21388183

Vecitis, C. D., Zodrow, K. R., Kang, S., & Elimelech, M. (2010). Electronic-structure-dependent bacterial cytotoxicity of single-walled carbon nanotubes. ACS Nano, 4(9), 5471–5479. doi:10.1021/nn101558x PMID:20812689

Vedrtnam, A., Lalauni, K., Dubey, S., & Kumar, A. (2020). A comprehensive study on structure, properties, synthesis and characterization *of ferrites. AIMS* Materials Science, 7(6), 800–835. doi:10.3934/matersci.2020.6.800

Veerasubramani, G. K., Park, M.-S., Choi, J.-Y., & Kim, D.-W. (2020). Ultrasmall SnS quantum dots anchored onto nitrogen-enriched carbon nanospheres as an advanced anode material for sodium-ion batteries. ACS Applied Materials & Interfaces, 12(6), 7114–7124.

Venkatadri, B., Shanparvish, E., Rameshkumar, M. R., Arasu, M. V., Al-Dhabi, N. A., Ponnusamy, V. K., & Agastian, P. (2020). Green synthesis of silver nanoparticles using aqueous rhizome extract of Zingiber officinale *and Curcuma longa: In-vitro anti-cancer* potential on human colon carcinoma HT-29 cells. Saudi Journal of Biological Sciences, 27(11), 2980–2986. doi:10.1016/j.sjbs.2020.09.021 PMID:33100856

Venkatesan, M., Veeramuthu, L., Liang, F.-C., Chen, W.-C., Cho, C.-J., Chen, C.-W., Chen, J.-Y., Yan, Y., Chang, S.-H., & Kuo, C.-C. (2020). Evolution of electrospun nanofibers fluorescent and colorimetric sensors for environmental toxicants, pH, temperature, and cancer cells–A review with insights on applications. Chemical Engineer*ing Journal, 397,* 125431. doi:10.1016/j.cej.2020.125431

Venturini, J., Piva, D. H., da Cunha, J. B. M., & Bergmann, C. P. (2016). Effect of the thermal treatment on the magnetic and structural properties of cobalt ferrite particles. *Ceramics International, 42(14),* 15183–15188. doi:10.1016/j.ceramint.2016.06.120

Verdipoor, K., Alemi, A., & Mesbahi, A. (2018). Photon mass attenuation coefficients of a silicon resin loaded with WO3, PbO, and Bi2O3 micro and nano-particles for radiation shielding. Radiation Ph*ysics and Chemistry, 147,* 85–90. doi:10.1016/j.radphyschem.2018.02.017

Viana, R. L. S., Fidelis, G. P., Medeiros, M. J. C., Morgano, M. A., Alves, M. G. C. F., Passero, L. F. D., & Rocha, H. A. O. (2020). Green synthesis of antileishmanial and antifungal silver n*anoparticles using corn cob xylan as a reducing* and stabilizing agent. Biomolecules, 10(9), 1–21. PMID:32854282

Viets, C., Kaysser, S., & Schulte, K. (2014). Damage mapping of GFRP via electrical resistance measurements using nanocomposite epoxy matrix systems. Composites. Part *B, Engineering, 65, 80–88. doi:10.1016*/j.compositesb.2013.09.049

Vikesland, P. J., & Wigginton, K. R. (2010). _ Nanomaterial enabled biosensors for pathogen monitoring e a review. Environmen tal S c ie n c e & Technology, 44(10), 3656–3669. doi:10.10*21/es903704z PMID:20405829*

*Vinay Sharma, J., Sa*ha, S., Patnaik, B., & Kuanr, K. (2017), YIG based broad band microwa*ve absorber: A pe*rspective on synthesis methods. Journal of Magnetism and Magnetic Materials, 439, 277-286.

Vincent, S., *Arokiyaraj, S., Saravanan, M.*, & Dhanraj, M. (2020). Molecular Docking Studies on the Anti-viral Effects of Compounds From Kabasura Kudineer on SARS-CoV-2 3CLpro. Frontiers in Molecular Biosciences, 7, 1–12. doi:10.3389/fmolb.2020.613*401 PMID:33425994*

Vitola, J., Pozo, F., & Tibaduiza, D. (2017). Distributed Piezoelectric Sensor System for Damage. Sensors, 1–26. . doi:10.3390/s17061252

Vo, T. N., Kim, D. S., Mun, Y. S., Lee, H. J., Ahn, S.-k., & Kim, I. T. (2020). Fast charging sodium-ion batteries based on Te-PC composites and insights to *low-frequency lim*its of four common equivalent impedance circuits. Chemical Engineering Journal, 398, 125703.

Vojna, D., Slezák, O., Yasuhara, R., Furuse, H., Lucianetti, A., & Mocek, T. (2020). Faraday Rotation of Dy2O3, CeF3 *and Y3Fe5O12 at* the Mid-Infrared Wavelengths. Materials, 13(23), 324. doi:10.3390/ma13235324

Voke, E., Pinals, R. L., Goh, N. S., & Landry, M. P. (2021). In planta nanosensors: Understandin g B iocorona formation for functional design. ACS Sensors, 6(8), 2802–2814. doi:10.1021/acssensors.1c0115*9 PMID:34279907*

Waghule, T., Sankar, S., Rapalli, V. K., Gorantla, S., Dubey, S. K., Chellappan, D. K., Dua, K., & Singhvi, G. (2020). Emerging role of nanocarriers based topical delivery of anti-fungal agents in combating growing fungal infections. Dermatologic Therapy, 33(6), 1–34. doi:10.1111/dth.13905 PMID:32588940

Wang, C., Ren, H., Zeng, M., Zhu, Q., Zhang, Q., Kan, Z., Wa*ng, Z., Shen, M., Thalavitiya Achar*ige, M. J., Ruths, M., Ryan, D. K., Li, G., Kolesov, G., Kaxiras, E., & Mazur, E. (2020). Low-cost visible-light photosynthesis of water and adsorbed carbon dioxide into long-chain hydrocarbons. Chemical Physics Letters, 739, 136985. doi:*10.1016/j.cplett.2019.136985*

Wang, G., Dou, K., Cao, H., Du, R., Liu, J., Tsidaeva, N., & Wang, W. (2022). Designing Z-scheme CdS/WS2 heterojunctions with enhanced photocatalytic degradation of organic dyes and photoreduction of Cr (VI): Experi*ments, D*FT calculations and mechanism. Separation and Purification Technology, 291, 120976. doi:10.1016/j.seppur.2022.120976

Wang, G., Yu, M., & Feng, X. (2021). Carbon materials for ion-intercalation involved rec*hargeable battery tech*nologies. Chemical Society Reviews, 50(4), 2388–2443. doi:10.1039/D0CS00187B

Wang, H., Zhang, L., Chen, Z., Hu, J., Li, S., Wang, Z., Liu, J., & Wang, X. (2014). Semiconductor heterojunction photocatalysts: Design, construction, and photocatalytic perfor*mances. Chemical Society Reviews,* 43*(1*5), 5234–5244. doi:10.1039/C4CS00126E PMID:24841176

Wang, J., Li, Q., Mu, Y., Li, S., Yang, L., Lv, P., Su, S., Liu, T., Fu, W., & Yang, H. (2015). Fabrication of CdTe thin films grown by the two-step electrodeposition technique on Ni foils. Journal of Alloys and Compounds, 636, 97–101. doi:10.1016/j.jallcom.2015.02.094

Wang, J., Ma, F., Liang, W., & Sun, M. (2017). Electrical properties and applications of graphene, hexagonal boron nitride (h-BN), and graphene/h-BN heterostructures. Materials Today Physics, 2, 6–34. doi:10.1016/j.mtphys.2017.07.001

Wang, J., Zheng, Z., Chan, J., & John, T. W. (2020, August 24). Yeow. 2020. "Capacitive Micromachined Ultrasound Transducers for Intravascular Ultrasound Imaging. Microsystems & Nanoen*gineering, 6(1), 73. doi:10.*1038*4*1378-020-0181-z PMID:34567683

Wang, L., Deng, J., Deng, J., Fei, Y., Fang, Y., & Hu, Y. H. (2020). Ultra-fast and ultra-long-life Li ion batteries with 3D surface-porous graphene anodes synthesized from CO 2. Journal *of Materials Chemistry. A,* Materials for Energy and Sustainability, 8(26), 13385–13392.

Wang, M., Zhang, M., Wang, Y., Zhao, R., & Yan, S. (2019). Fano resonance in an asymmetric MIM waveguide struc ure and its a p ication in a refractive index nanosensor. Sensors *(Basel), 19(4), 791. doi:10.3390*19040791 PMID:30781360

Wang, Q., & Domen, K. (2020). Particulate Photocatalysts for Light-Driven Water Splitting: Mechanisms, Challenges, and Design Strategies. Chemical Reviews, 120(2), 919–985. doi:10.1021/acs.chemrev.9b00201 PMID:31393702

Wang, R., Yu, J., Islam, F., Tahmasebi, A., Lee, S., & Chen, Y. (2020). *State-of-the-Art* Research and Applications of Carbon Foam Composite Materials as Electrodes for High-Capacity Lithium Batteries. Energy & Fuels, 34(7), 7935–7954. doi:10.1021/acs.energyfuels.0*c01802*

*Wang, S., Cheng, Y., Xu*e, *H*., Liu, W., Yi, Z., Chang, L., & Wang, L. (2021). Multifunctional sulfur-mediated strategy enabling fast-charging Sb 2 S 3 micro-package anode for lithium-ion stor*age. Journal of Materials Chemistry. A*, Materials for Energy and Sustainability, 9(12), 7838–7847.

Wang, S., Corredor Garcia, J. L., Davidson, J., & Nichols, A. (2020). Conductance-Based Interface Detection for Multi-Phase Pipe Flow. Sensors (Basel), 20(20), 5854. do*i:10.3390*20205854 *PMID:33081178*

*W*ang, S., Huang, S., Yao, M., Zhang, Y., & Niu, Z. (2020). Engineering active sites of polyaniline for AlCl2+ storage in an aluminum-ion battery. Angewandt*e Chemi*e International Edition, 59(29), 11800–11807.

Wang, S. J., Geng, Y., Zheng, Q., & Kim, J.-K. (2010). Fabrication of highly conducting and transparent graphene films. Carbon, *48(6), 1815–1*823. doi:10.1016/j.carbon.2010.01.027

Wang, W., Huang, H., Ke, X., Liu, X., Yang, S., Wang, K., Huang, L., Tu, C., Zheng, Z., Luo, D., & Zhang, M. (2022). Morphology- and size-dependent FAPbBr3/WO3 Z-scheme photocatalysts for the controllable photo-oxidation of benzyl alcohol. Materia*ls & Design, 215, 110502. do*i:*10.*1016/j.matdes.2022.110502

Wang, X., Qu, K., Xu, B., Ren, J., & Qu, X. (2011). Microwave assisted one-step green synthesis of cell-permeable multicolor photoluminescent carbon dots without surface passivation reagents. Journal of Materials Chemi*stry, 21(8), 2445–245*0. doi:10.1039/c0jm02963g

Wang, X., Wang, J., Chen, Z., Yang, K., Zhang, Z., & Shi, Z. (2020). Yolk-double shell Fe3O4@ *C@ C composite as high-performance anode materials for lithium-ion batteries. Academic Press.*

Wang, X., Wang, J., Chen, Z., Yang, K., Zhang, Z., Shi, Z., ... Wang, X. (2020). Yolk-double shell Fe3O4@ C@ C composite as high-performance anode materials for lithium-ion batteries. Journal of Alloys and Compounds, 822, 153656.

Wang, Y., Li, B., Zhang, C., Cui, L., Kang, S., Li, X., & Zhou, L. (2013). Ordered *mesoporous CeO2-TiO2 composit*es: Highly efficient photocatalysts for the reduction of CO2 with H2O under simulated solar irradiation. Applied Catalysis B: Environmental, 130–131, 277–284. doi:10.1016/j.apcatb.2012.11.019

Wang, Y., Zhang, Y., Jiang, Z., Jiang, G., Zhao, Z., Wu, Q., Liu, Y., Xu, Q., Duan, A., & Xu, C. (2016). *Controlled fabrication and enhanced v*isible-light photocatalytic hydrogen production of Au at CdS/MIL-101 heterostructure. Applied Catalysis B: Environmental, 185, 307–314. doi:10.1016/j.apcatb.2015.12.020

Viets, C., Kaysser, S., & Schulte, K. (2014). Damage mapping of GFRP via electrical resistance measurements using nanocomposite epoxy matrix systems. Composites. Part *B, Engineering, 65, 80–88. doi:10.1016*/j.compositesb.2013.09.049

Vikesland, P. J., & Wigginton, K. R. (2010). _ Nanomaterial enabled biosensors for pathogen monitoring e a review. Environmental Science & Technology, 44(10), 3656–3669. doi:10.102*1/es903704z PMID:20405829*

*Vinay Sharma, J., Sa*ha, S., Patnaik, B., & Kuanr, K. (2017), YIG based broad band microwa*ve absorber: A pe*rspective on synthesis methods. Journal of Magnetism and Magnetic Materials, 439, 277-286.

Vincent, S., *Arokiyaraj, S., Saravanan, M.*, & Dhanraj, M. (2020). Molecular Docking Studies on the Anti-viral Effects of Compounds From Kabasura Kudineer on SARS-CoV-2 3CLpro. Frontiers in Molecular Biosciences, 7, 1–12. doi:10.3389/fmolb.2020.613*401 PMID:33425994*

Vitola, J., Pozo, F., & Tibaduiza, D. (2017). Distributed Piezoelectric Sensor System for Damage. Sensors, 1–26. . doi:10.3390/s17061252

Vo, T. N., Kim, D. S., Mun, Y. S., Lee, H. J., Ahn, S.-k., & Kim, I. T. (2020). Fast charging sodium-ion batteries based on Te-PC composites and insights to l*ow-frequency lim*its of four common equivalent impedance circuits. Chemical Engineering Journal, 398, 125703.

Vojna, D., Slezák, O., Yasuhara, R., Furuse, H., Lucianetti, A., & Mocek, T. (2020). Faraday Rotation of Dy2O3, CeF3 *and Y3Fe5O12 at* the Mid-Infrared Wavelengths. Materials, 13(23), 324. doi:10.3390/ma13235324

Voke, E., Pinals, R. L., Goh, N. S., & Landry, M. P. (2021). In planta nanosensors: Understanding Biocorona formation for functional design. ACS Sensors, 6(8), 2802–2814. doi:10.1021/acssensors.1c011*59 PMID:34279907*

Waghule, T., Sankar, S., Rapalli, V. K., Gorantla, S., Dubey, S. K., Chellappan, D. K., Dua, K., & Singhvi, G. (2020). Emerging role of nanocarriers based topical delivery of anti-fungal agents in combating growing fungal infections. Dermatologic Therapy, 33(6), 1–34. doi:10.1111/dth.13905 PMID:32588940

Wang, C., Ren, H., Zeng, M., Zhu, Q., Zhang, Q., Kan, Z., Wa*ng, Z., Shen, M., Thalavitiya Achari*ge, M. J., Ruths, M., Ryan, D. K., Li, G., Kolesov, G., Kaxiras, E., & Mazur, E. (2020). Low-cost visible-light photosynthesis of water and adsorbed carbon dioxide into long-chain hydrocarbons. Chemical Physics Letters, 739, 136985. doi:*10.1016/j.cplett.2019.136985*

Wang, G., Dou, K., Cao, H., Du, R., Liu, J., Tsidaeva, N., & Wang, W. (2022). Designing Z-scheme CdS/WS2 heterojunctions with enhanced photocatalytic degradation of organic dyes and photoreduction of Cr (VI): Experi*ments, D*FT calculations and mechanism. Separation and Purification Technology, 291, 120976. doi:10.1016/j.seppur.2022.120976

Wang, G., Yu, M., & Feng, X. (2021). Carbon materials for ion-intercalation involved rec*hargeable battery tech*no*log*ies. Chemical Society Reviews, 50(4), 2388–2443. doi:10.1039/D0CS00187B

Wang, H., Zhang, L., Chen, Z., Hu, J., Li, S., Wang, Z., Liu, J., & Wang, X. (2014). Semiconductor heterojunction photocatalysts: Design, construction, and photocatalytic perfor*mances. Chemical Society Reviews,* 43*(15*), 5234–5244. doi:10.1039/C4CS00126E PMID:24841176

Wang, J., Li, Q., Mu, Y., Li, S., Yang, L., Lv, P., Su, S., Liu, T., Fu, W., & Yang, H. (2015). Fabrication of CdTe thin films grown by the two-step electrodeposition technique on Ni foils. Journal of Alloys and Compounds, 636, 97–101. doi:10.1016/j.jallcom.2015.02.094

Wang, J., Ma, F., Liang, W., & Sun, M. (2017). Electrical properties and applications of graphene, hexagonal boron nitride (h-BN), and graphene/h-BN heterostructures. Materials Today Physics, 2, 6–34. doi:10.1016/j.mtphys.2017.07.001

Wang, J., Zheng, Z., Chan, J., & John, T. W. (2020, August 24). Yeow. 2020. "Capacitive Micromachined Ultrasound Transducers for Intravascular Ultrasound Imaging. Microsystems & Nanoen*gineering, 6(1), 73. doi:10.*103841378-020-0181-z PMID:34567683

Wang, L., Deng, J., Deng, J., Fei, Y., Fang, Y., & Hu, Y. H. (2020). Ultra-fast and ultra-long-life Li ion batteries with 3D surface-porous graphene anodes synthesized from CO 2. Journal *of Materials Chemistry. A,* Materials for Energy and Sustainability, 8(26), 13385–13392.

Wang, M., Zhang, M., Wang, Y., Zhao, R., & Yan, S. (2019). Fano resonance in an asymmetric MIM waveguide struc$_{t}$ure and its a$_{p}$p$_{l}$ication in a refractive index nanosensor. Sensors *(Basel), 19(4), 791. doi:10.339019040*791 PMID:30781360

Wang, Q., & Domen, K. (2020). Particulate Photocatalysts for Light-Driven Water Splitting: Mechanisms, Challenges, and Design Strategies. Chemical Reviews, 120(2), 919–985. doi:10.1021/acs.chemrev.9b00201 PMID:31393702

Wang, R., Yu, J., Islam, F., Tahmasebi, A., Lee, S., & Chen, Y. (2020). *State-of-the-A*rt Research and Applications of Carbon Foam Composite Materials as Electrodes for High-Capacity Lithium Batteries. Energy & Fuels, 34(7), 7935–7954. doi:10.1021/acs.energyfuels.0*c01802*

*Wang, S., Cheng, Y., Xu*e, *H*., Liu, W., Yi, Z., Chang, L., & Wang, L. (2021). Multifunctional sulfur-mediated strategy enabling fast-charging Sb 2 S 3 micro-package anode for lithium-ion stor*age. Journal of Materials Chemistry. A*, Materials for Energy and Sustainability, 9(12), 7838–7847.

Wang, S., Corredor Garcia, J. L., Davidson, J., & Nichols, A. (2020). Conductance-Based Interface Detection for Multi-Phase Pipe Flow. Sensors (Basel), 20(20), 5854. do*i:10.339020205854 PMID:33081178*

Wang, S., Huang, S., Yao, M., Zhang, Y., & Niu, Z. (2020). Engineering active sites of polyaniline for AlCl2+ storage in an aluminum-ion battery. Angewandt*e Chemi*e International Edition, 59(29), 11800–11807.

Wang, S. J., Geng, Y., Zheng, Q., & Kim, J.-K. (2010). Fabrication of highly conducting and transparent graphene films. Carbon, *48(6), 1815–1*823. doi:10.1016/j.carbon.2010.01.027

Wang, W., Huang, H., Ke, X., Liu, X., Yang, S., Wang, K., Huang, L., Tu, C., Zheng, Z., Luo, D., & Zhang, M. (2022). Morphology- and size-dependent FAPbBr3/WO3 Z-scheme photocatalysts for the controllable photo-oxidation of benzyl alcohol. Materia*ls & Design, 215, 110502. doi:10.*1016/j.matdes.2022.110502

Wang, X., Qu, K., Xu, B., Ren, J., & Qu, X. (2011). Microwave assisted one-step green synthesis of cell-permeable multicolor photoluminescent carbon dots without surface passivation reagents. Journal of Materials Chemi*stry, 21(8), 2445–245*0. doi:10.1039/c0jm02963g

Wang, X., Wang, J., Chen, Z., Yang, K., Zhang, Z., & Shi, Z. (2020). Yolk-double shell Fe3O4@ *C@ C composite as high-performance anode materials for lithium-ion batteries. Academic Press.*

Wang, X., Wang, J., Chen, Z., Yang, K., Zhang, Z., Shi, Z., ... Wang, X. (2020). Yolk-double shell Fe3O4@ C@ C composite as high-performance anode materials for lithium-ion batteries. Journal of Alloys and Compounds, 822, 153656.

Wang, Y., Li, B., Zhang, C., Cui, L., Kang, S., Li, X., & Zhou, L. (2013). Ordered *mesoporous CeO2-TiO2 composit*es: Highly efficient photocatalysts for the reduction of CO2 with H2O under simulated solar irradiation. Applied Catalysis B: Environmental, 130–131, 277–284. doi:10.1016/j.apcatb.2012.11.019

Wang, Y., Zhang, Y., Jiang, Z., Jiang, G., Zhao, Z., Wu, Q., Liu, Y., Xu, Q., Duan, A., & Xu, C. (2016). *Controlled fabrication and enhanced visible*-light photocatalytic hydrogen production of Au at CdS/MIL-101 heterostructure. Applied Catalysis B: Environmental, 185, 307–314. doi:10.1016/j.apcatb.2015.12.020

Xiang, J., Shen, X., & Meng, X. (2009). Prep*aration of Co-substituted MnZn* ferrite fibers and their magnetic properties. Materials Chemistry and Physics, 114(1), 362–366. doi:10.1016/j.matchemphys.2008.09.037

Xiang, Y., Xin, L., Hu, J., Li, C., Qi, J., Hou, Y., & Wei, X. (2021). Advances in the applications of graphene-based nanocomposites in clean energy materials. Crystals, 11(1), 47.

Xiangfeng, C., & Chenmou, Z. (2003). *Sulfide-sensing charact*eri*st*ics of MFe2O4 (M = Zn, Cd, Mg and Cu) thick film prepared by co-precipitation method. Sensors and Actuators. B, Chemical, 96(3), 50_4–5_08. doi:10.1016/S0925-4005(03)00626-9

Xiao, Z., Wang, R., Jiang, D., Qian, Z., Li, Y., Yang, K., ... Wu, F. (2021). Recent Developments *of Two-Dimensional Anode Materi*al*s an*d Their Composites in Lithium-Ion Batteries. Academic Press.

Xie, F., Xu, Z., Guo, Z., & Titirici, M.-M. (2020). Hard carbons for sodium-ion batteries and beyond. Progress in Energy, 2(4), 042002.

Xie, G., Xi, P., Liu, H.*, Chen, F., Hu*ang, L., Shi, Y., Hou, F., Zeng, Z., Shao, C., & Wang, J. (2012). A facile chemical method to produce superparamagnetic graphene oxide–Fe3O4 hy*brid composite and its* app*l*ication in the removal of dyes from aqueous solution. Journal of Materials Chemistry, 22(3), 1033–1039. doi:10.1039/C1JM13433G

Xiong, W., Ndokoye, P., & Leung, M. K. (2020). Noble Metal–Based Nanosensors for Environme*ntal Detect*ion. Advanced Nanomaterials for Pollutant Sensing and Environmental Catalysis. Elsevier.

Xu, F., & Ma, T. (2019). Modeling and Studying Acceleration-Induced Effects of Piezoelectric Pressure Sensors Using System Identification Theory. Sensors (Basel), 19(5), 1052. Ad*vance online publication.* d*oi:10.*339019051052 PMID:30832233

Xu, H.T. (2008). Synthesis of Ni nanoparticles and their catalytic effect on the decomposition of ammonium perchlorate. J. Mater. Process. Technol., 208(1–3), 494-498.

Xu, K., Shen, J., Zhang, S., Xu, D., & Chen, X. (2022). Efficient interfacial charge transfer of BiOCl-In2O3 ste*p-scheme heterojunction for* b*oosted* photocatalytic degradation of ciprofloxacin. Journal of Materials Science and Technology, 121, 236–244. doi:10.1016/j.jmst.2021.12.070

Xu, L., Shoaie, N., Jahanpeyma, F., Zhao, J., Azimzadeh, M., & Al, K. T. (2020). Optical, electro*chemical, and electrical (nano) biosensors f*or *d*etection of exosomes: A comprehensive overview. Biosensors & Bioelectronics, 161, 112222. doi:10.1016/j.bios.2020.112222 PMID:32365010

Xu, Q., Shao, W., Li, Y., Zhang, H., Ouyang, X., Zhang, X., Nie, J., Ouyang, X., & L*iu, B. (2019). Vi*s*ual* clarity methylammonium lead trichloride perovskite single crystals for X and gamma rays protection. Journal of Alloys and Compounds, 810, 151896. doi:10.1016/j.jallcom.2019.151896

Xu, Q., Shao, W., Li, Y., Zhang, X., Ouyang, X., Liu, J., Liu, *B., Wu, Z., Ouyang, X., Tang, X., &* Jia, W. (2019b). High-performance surface barrier X-ray detector based on methylammonium lead tribromide single crystals. ACS Applied Ma$_t$erials & Interfaces, 11(10), 9679–9684. doi:10.1021/acsami.8b21605 PMID:307935*85*

Xu, Y., Jie, W., Sellin, P., Wang, T., Liu, W., Zha, G., Veeramani, P., & Mills, C. (2008). Study on temperature dependent res$_i$s$_t$ivity of indium-doped cadmium Zinc telluride. Journal of Physics. D, Appl*ied Physics, 42(3), 35105. doi:10.*1088/0022-3727/42/3/035105

Xu, Y., Sun, X., & Wei, C. (2020). Improving the electrochemical performance of lithium Si batteries by multilayer porous carbon nanosheets/multi-*walled carbon nanotubes compos*ite *in*ert nano-Ag. Ionics, 26, 1149–1158.

Xu, Y. C., Song, C., Ding, X. Y., Zhao, Y., Xu, D. G., Zhang, Q. P., & Zhou, Y. L. (2019a). Tailoring lattices of Bi2WO6 crystals via Ce doping to improve the shielding properties against low-energy gamma rays. Journal of Physics and Chemistry of Solids, 127, 76–80. doi:10.1016/j.jpcs.2018.12.007

Xu, Y. C., Zhang, Q. P., Liu, J. H., Wu, Y., Liu, L. P., Xu, D. G., & Zhou, Y. L. (2018). PbWO4 nanofibers for shielding gamma radiation: Crystal growth, morphology and performance evaluation. CrystEngComm, 20(40), 6197–6206. doi:10.1039/C8CE01224E

Xuan, X., & Qian, S. (2017). Editorial for the Special Issue on Micro/Nano-Chip Electrokinetics; MDPI. Micromechanics, 8(145), 1–2. doi:10.3390/mi8050145

Xue, H., Li, Z., Wang, X., & Fu, X. (2007). Facile synthesis of nanocrystalline zinc ferrite via self-propagating combustion method. Materials Letters, 61(2), 347–350. doi:10.1016/j.matlet.2006.04.061

Xue, Z., Xiong, Q., Zou, C., Chi, H., Hu, X., & Ji, Z. (2021). Growth of carbon nanofibers through chemical vapor deposition for enhanced sodium ion storage. Materials Research Bulletin, 133, 111049.

Yadav, R. S., Havlica, J., Hnatko, M., Sajgalik, P., Alexander, C., Palou, M., Bartonickova, E., Bohac, M., Frajkorova, F., Masilko, J., Zmrzly, M., Kalina, L., Hajduchova, M., & Enev, V. (2015). Magnetic properties of Co1-xZnxFe2O4 spinel ferrite nanoparticles synthesizedby starch-assisted sol-gel autocombustion method and its ballmilling. Journal of Magnetism and Magnetic Materials, 378, 190–199. doi:10.1016/j.jmmm.2014.11.027

Yadegari, A., Khezri, J., Esfandiari, S., Mahdavi, H., Karkhane, A. A., Rahighi, R., Heidarimoghadam, R., Tayebi, L., Hashemi, E., & Farmany, A. (2019). Bottom-up synthesis of nitrogen and oxygen co-decorated carbon quantum dots with enhanced DNA plasmid expression. Colloids and Surfaces. B, Biointerfaces, 184, 110543. doi:10.1016/j.colsurfb.2019.110543 PMID:31627102

Yahia, I. S., Shkir, M., AlFaify, S., Ganesh, V., Zahran, H. Y., & Kilany, M. (2017). Facile microwave-assisted synthesis of Te-doped hydroxyapatite nanorods and nanosheets and their characterizations for bone cement applications. Materials Science and Engineering C, 72, 472–480. doi:10.1016/j.msec.2016.11.074 PMID:28024611

Yahya, N., Zhu, T. W., Rusop, M., & Soga, T. (2009). Development of Y3Fe5O12 nano-magnetic feeder for em source of an intelligent horizontal twin dipoles. AIP Conference Proceedings, 1136, 269–276. doi:10.1063/1.3160145

Yahya, N. Akhtar, M.N., Nasir, N. Shafie, A., Jabeli, M. S., & Koziol, K. (2011). CNT Fibres /Aluminium -NiZnFe2O4 Based EM Transmitter for Improved Magnitude vs. Offset (MVO) in a Scaled Marine Environment. J. Nanosci. Nanotechnol.

Yamagata, S., Takahara, I., Wang, M., Mizoguchi, T., & Yagi, S. (2020). EQCM analysis of intercalation species into graphite positive electrodes for Al batteries. Journal of Alloys and Compounds, 846, 156469.

Yamaguchi, M., Naoe, M., Yasuda, T., & Yamanaka, S. (1974). Fabrication and Characteristics of Ferrite Films for High-Frequency Electroacoustic Transducers. Electrical Engineering in Japan, 94(7), 295–302.

Yan, J., Jin, H., Sun, H., & Qing. (2019). Active Monitoring of Fatigue Crack in the Weld Zone of Bogie Frames Using Ultrasonic Guided Waves. Sensors (Basel), 19(15), 19. doi:10.339019153372 PMID:31370343

Yan, X., Cui, X., & Li, L. S. (2010). Synthesis of large, stable colloidal graphene quantum dots with tunable size. Journal of the American Chemical Society, 132(17), 5944–5945. doi:10.1021/ja1009376 PMID:20377260

Yan, X., Liang, S., Shi, H., Hu, Y., Liu, L., & Xu, Z. (2021). Nitrogen-enriched carbon nanofibers with tunable semi-ionic CF bonds as a stable long cycle anode for sodium-ion batteries. Journal of Colloid and Interface Science, 583, 535–543.

Yan, Z., Gao, J., Li, Y., Zhang, M., & Guo, M. (2015). Hydrothermal synthesis and structure evolution of metal-doped magnesium ferrite from saprolite laterite. RSC Advances, 5(112), 92778–92787. doi:10.1039/C5RA17145H

Yan, Z., Yuan, H., Zhao, Q., Xing, L., *Zheng,* X., Wang, W., Zhao, Y., Yu, Y., Hu, L., & Yao, W. (2020). Recent developments of nanoenzyme-based colorimetric sensors for heavy metal detection and the interaction mechanism. Analyst (London), 145(9), 3173–3187. doi:10.1039/D0AN00339E PMID:32222739

Yang, C., Su, X., Wang, J., Cao, X., Wang, S., & Zhang, L. (2013). Facile microwave assisted hydrothermal syn*thesis of varied-shaped CuO nanoparticles a*nd *th*eir gas sensing properties. Sensors and Actuators. B, Chemical, 185, 159–165. doi:10.1016/j.snb.2013.04.100

Yang, G. M., Xing, X., Obi, O., Daigle, A., Liu, M., Stoute, S., Naishadham, K., & Sun, N. (2010). Loading effects of self-biased magnetic films on patch antennas with substrate/superstrate sandwich st*ructure. IET Microwaves, Antennas & Pro*pag*ati*on, 4(9), 1172–1181. doi:10.1049/iet-map.2009.0429

Yang, J., Lukashuk, L., Li, H., Föttinger, K., Rupprechter, G., & Schubert, U. (2014). High surface area ceria for CO oxidation prepared from cerium t-butoxide by combined sol-gel and solvothermal processing. Catalysis Letters, 144(3), 40*3–412. doi:10.100710562-013-1162-8* PM*ID*:24532959

Yang, T., & Duncan, T. V. (2021). Challenges and potential solutions for nanosensors intended for use with foods. Nature Nanotechnology, 16(3), 251–265. doi:10.103841565-021-00867-7 PMID:33712739

Yang, X. (20*14), Voltage Tunable Multiferro*ic *Ph*ase Shifter With YIG/PMN-PT Heterostructure. IEEE Microwave and Wireless Components Letters, 24(3), 191-193. doi:10.1109/LMWC.2013.2292924

Yang, X., Liang, H.-J., Zhao, *X.-X., Yu, H.-Y., Wang, M.-Y., Ni*e, X.-J., & Wu, X.-L. (2021). A sandwich nanocomposite composed of commercially available SnO and reduced graphene oxide as advanced anode materials for sodium-ion full batteries. Inorganic Chemistry *Frontiers, 8(2), 396–404. doi:10.1039/D*0QI01033B

Yang, X., Zhuo, Y., Zhu, S., Luo, Y., Feng, Y., & Dou, Y. (2014). Novel and green synthesis of high-ñuorescent carbon dots originated from honey for sensing and imaging. Biosensors & Bioelectronics, 60, 292–298. doi:10.1016/j.bios.2014.04.04*6 PMID:24832204*

Yang, Y., Chen, Y., Li, Z., Tang, S., Li, Y., Fu, Z., Yang, S., Yang, M., & Xie, H. (2022). Homojunction type of carbon nitride as a robust photo-catalyst for reduction conversion of CO2 in water vapor under visible light. Chemical Engineering Journal, 430, 132668. doi:10.1016/j.*cej.2021.132668*

Ya*ng, Z*., Gao, M., Wu, W., Yang, X., Sun, X. W., Zhang, J., Wang, H.-C., Liu, R.-S., Han, C.-Y., Yang, H., & Li, W. (2019). Recent advances in quantum dot-based light-emitting *devices: Challenges a*nd *p*ossible solutions. Materials Today, 24, 69–93. doi:10.1016/j.mattod.2018.09.002

Yasin, G., Arif, M., Mehtab, T., Shakeel, M., Mushtaq, M. A., Kumar, A., ... Song, H. (2020). A novel strategy for the synthesis of hard carbon spheres encapsulated with graphene networks as a low-cost and large-scalable *anode material for fast sodi*um storage with an ultralong cycle life. Inorganic Chemistry Frontiers, 7(2), 402–410.

Yasmin, A., Luo, J.-J., & Daniel, I. M. (2006). Processing of expanded graphite reinforced polymer nanocomposites. Composites Science and Technology, 66(9), 1182–1189. doi:10.1*016/j.compscitech.2005.10.01*4

*Y*ellepeddi, V., Joseph, A., & Nance, E. (2019). Pharmacokinetics of nanotechnology-Based Formulations in Pediatric Populations. Advanced Drug Delivery Reviews, 151-152, 44–45. doi:10.10*16/j.addr.2019*.08.*0*08

Yin, F., Ye, D., Zhu, C., Qiu, L., & Huang, Y. (2017). Stretchable, Highly Durable Ternary Nanocomposite Strain Sensor for Structural Health Monitoring of Flexible Aircraft. Sensors (Basel), 17(11), 2677. doi:10.339017112677 PM*ID:29156620*

*Yiyan, S. M. W. (2013). Yttri*um *I*ron Garnet Nano Films: Epitaxial Growth, Spin-Pumping Efficiency, and Pt-Capping-Caused Damping. Solid State Physics, 64, 157–191. doi:10.1016/B978-0-12-408130-7.00006-X

Yoo, D.-J., Heeney, M., Glöcklhofer, F., & Choi, J. W. (2021). Tetradiketone macrocycle for divalent aluminium ion batteries. Nature Communications, 12(1), 2386.

Yoshida, M., & Lahann, J. (2008). Smart nanomaterials. ACS Nano, 2(6), 1101–1107. doi:10.1021/nn800332g PMID:19206325

Yousaf, H., Mehmood, A., Ahmad, K. S., & Raffi, M. (2020). Green synthesis of silver nanoparticles and their applications as an alternative antibacterial and antioxidant agents. Materials Science and Engineering C, 112, 110901–110908. doi:10.1016/j.msec.2020.110901 PMID:32409057

Yttrium Iron Garnet. (2015). American Elements.

Yu, H., Liu, Y., Tian, X., Zhang, S., & Liu, J. (2020). Sensors and Actuators A : Physical A Precise Rotary Positioner Driven by Piezoelectric Bimorphs : Design, Analysis and Experimental Evaluation. *Sensors and Actuators. A, Physical*, 313, 112197. doi:10.1016/j.sna.2020.112197

Yu, H., Zhao, Y., Zhou, C., Shang, L., Peng, Y., Cao, Y., Wu, L. Z., Tung, C. H., & Zhang, T. (2014). Carbon quantum dots/TiO2 composites for efficient photocatalytic hydrogen evolution. Journal of Materials Chemistry. A, Materials for Energy and Sustainability, 2(10), 3344–3351. doi:10.1039/c3ta14108j

Yu, J., Song, N., Zhang, Y. K., Zhong, S. X., Wang, A. J., & Chen, J. (2015). Green preparation of carbon dots by Jinhua bergamot for sensitive and selective fluorescent detection of Hg2+ and Fe3+. Sensors and Actuators. B, Chemical, 214, 29–35. doi:10.1016/j.snb.2015.03.006

Yu, M., Liu, X., Wang, Y., Zheng, Y., Zhang, J., Li, M., Lan, W., & Su, Q. (2012). Gas sensing properties of p-type semiconducting vanadium oxide nanotubes. Applied Surface Science, 258(24), 9554–9558. doi:10.1016/j.apsusc.2012.05.120

Yu, S., Jing, W., Tang, M., Xu, T., Yin, W., & Kang, B. (2019). Fabrication of Nd:YAG transparent ceramics using powders synthesized by citrate sol-gel method. Journal of Alloys and Compounds, 772, 751–759. doi:10.1016/j.jallcom.2018.09.184

Yue, R., Raisi, B., Rahmatinejad, J., Ye, Z., Barbeau, B., & Rahaman, M. S. (2021). A photo-Fenton nanocomposite ultrafiltration membrane for enhanced dye removal with self-cleaning properties. Journal of Colloid and Interface Science, 604, 458–468. doi:10.1016/j.jcis.2021.06.157 PMID:34273782

Yue, Z., Zhou, J., Li, L., & Gui, Z. (2000). Synthesis of nanocrystalline NiCuZn ferrite powders by sol-gel auto-combustion method. Journal of Magnetism and Magnetic Materials, 208(1-2), 55–60. doi:10.1016/S0304-8853(99)00566-1

Yue, Z., Zhou, J., Li, L., & Gui, Z. (2001). Effect of copper on the electromagnetic properties of Mg-Zn-Cu ferrites prepared by sol-gel auto-combustion method. Materials Science and Engineering B, 86(1), 64–69. doi:10.1016/S0921-5107(01)00660-2

Zahi, S., Daud, A. R., & Hashim, M. (2007). A comparative study of nickel-zinc ferrites by sol-gel route and solid-state reaction. Materials Chemistry and Physics, 106(2), 452–456. doi:10.1016/j.matchemphys.2007.06.031

Zahn, M. (2004). Arthur Robert von Hippel (1898–2003). News & Views, 3(March), 2004.

Zahra, Z., Habib, Z., Chung, S., & Badshah, M. A. (2020). Exposure route of TiO2 NPs from industrial applications to wastewater treatment and their impacts on the agro-environment. Nanomaterials (Basel, Switzerland), 10(8), 1469. doi:10.3390/nano10081469 PMID:32727126

Zarejousheghani, M., Rahimi, P., Borsdorf, H., Zimmermann, S., & Joseph, Y. (2021). Molecularly imprinted Polymer-Based Sensors for Priority Pollutants; MDPI. Sensors (Basel), 21(7), 2406. doi:10.339021072406 PMID:33807242

Zaszczy, A., Gradys, A., & Sajkiewicz, P. (2020). *Progress in the Applications* of *Sma*rt Piezoelectric Materials for Medical Devices. Polymers, 1–19. PMID:33266424

Zeng, Y., Huang, Y., Liu, N., *Wang, X.,* Zha*ng*, Y., Guo, Y., ... Zhang, Q. (2021). N-doped porous carbon nanofibers sheathed pumpkin-like Si/C composites as free-standing anodes for lithium-ion batteries. Journal of Energy Chemistry, 54, 727–*735. doi:10.1016/j.jechem.2020.06.*022

Zhang, C. Y., Shen, X. Q., Zhou, J. X., Jing, M. X., & Cao, K. (2007). Preparation of spinel ferrite NiFe2O4 fibres by organic gel-thermal decomposition process. Journal of Sol-Gel Science and Technology, 42(1), *95–100. doi:1*0.*100*710971-006-1515-5

Zhang, H., Zhang, X., Lin, H., Wang, K., Sun, X., Xu, N., Li, C., & Ma, Y. (2015). Graphene and maghemite composites based supercapacitors delivering high volu*metric c*apacitance and extraordinary cycling stability. Electrochimica Acta, 156, 70–76. doi:10.1016/j.electacta.2015.01.041

Zhang, L., He, Y., Luo, P., Ma, L., Li, S., Nie, Y., Zhong, F., Wang, Y., & Chen, L. (2022). Photocatalytic G*O/M88A "interceptor plate"* as*se*mbled nanofibrous membrane with photo-Fenton self-cleaning performance for oil/water emulsion separation. Chemical Engineering Journal, 427, 130948. doi:10.1016/j.cej.2021.130948

Zhang, L., Wang, W., Lu, *S., & Xiang, Y. (2021). Carbon anode materials: A detail*ed comparison between Na-ion and K-ion batteries. Advanced Energy Materials, *11(11), 2003640.*

*Zhang, Q. P., Sun, N., Zheng, J., Wang, S. W., Wu, Y., Luo, D. L., & Zhou, Y. L. (2017). Fabrication of lead borate singl*e crystal nanosheets for attenuating both neutron and gamma radiations. Advanced Engineering Materials, 19(3), 1600650. doi:10.1002/adem.201600650

Zhang, X., Li, Q., Yan, S., Lei, *W., Chen, J., & Qasim, K. (2019). A novel* pho*t*otransistor device with dual active layers composited CsPbBr3 and ZnO quantum dots. Materials (Basel), 12(8), 1215. doi:10.3390/ma12081215 PMID:31013902

Zhang, X., Qin, D., Chen, L., Liang, W., & Xu, J. (2018). Research on Recognition and Location of Rail *Damage Acoustic Emissio*n Signal Based on Dual Sensors. 7th Asia- Pacific Workshop on Structural Health Monitoring, 1–6.

Zhang, Y., Yang, Z., Yin, D., Liu, Y., Fei, C. L., Xiong, R., Shi, J., & Yan, G. L. (2010). Composition and magnetic properties of cobalt ferrite nano-particles prepared by the co-precipitation *method. Journal of Magnetism and* Magnetic Materials, 322(21), 3470–3475. doi:10.1016/j.jmmm.2010.06.047

Zhang, Y., Zhang, B., Li, J., Liu, J., Huo, X., & Kang, F. (2021). SnSe nano-particles as advanced p*ositive electrode materia*ls *f*or rechargeable aluminum-ion batteries. Chemical Engineering Journal, 403, 126377.

Zhang, Y., Zhang, Y., Zhang, K., Jia, K., Liu, G., & He, X. (2020). Nitrogen-doped graphene nanosheet coated nanospherical Fe*3O4 from zeolitic imidazolate* frame*w*orks template as anode of lithium ion batteries. Academic Press.

Zhang, Z., Shan, Y., Wang, J., Ling, H., Zang, S., Gao, W., Zhao, Z., & Zhang, H. (2007). Investigation on the rapid degradation of congo red catalyzed by activated carbon powd*er under microwave irradiation. Journal of* Haz*a*rdous Materials, 147(1-2), 325–333. doi:10.1016/j.jhazmat.2006.12.083 PMID:17293037

Zhang, Z., Yao, G., Zhang, X., Ma, J., & Lin, H. (2015). Synthesis and characterization of nickel ferrite nanoparticles via planetary *ball milling assisted solid-*state reaction. Ceramics International, 41(3), 4523–4530. doi:10.1016/j.ceramint.2014.11.147

Zhao, H. X., Liu, L. Q., Liu, Z. D., Wang, Y., Zhao, X. J., & Huang, C. Z. (2011). Highly selective detection of phosphate in very *complicated matrixes with an* of*f–o*n fluorescent probe of europium adjusted carbon dots. Chemical Communications, 47(9), 2604–2606. doi:10.1039/c0cc04399k PMID:21234476

Zhao, R., Wang, M., Zhao, D., Li, H., Wang, C., & Yin, L. (2018). Molecular-Level Heterostructures Assembled *from Titanium Carbide* MXene and Ni-Co-Al Layered Double-Hydroxide Nanosheets for All-Solid-State Flexible Asymmetric High-Energy Supercapacitors. ACS Energy Letters, 3(1), 132–140. doi:10.1021/acsenergylett.7b01063

Zhao, W., Li, M., Qi, Y., Tao, Y., Shi, Z., Liu, Y., & Cheng, J. (2021). Ultrasound sonochemical synthesis of amorphous *Sb2S3-graphene* composites for sodium-ion batteries. Journal of Colloid and Interface Science, 586, 404–411. doi:10.1016/j.jcis.2020.10.104

Zhao, X., Li, H., Han, F., Dai, M., Sun, Y., Song, Z., ... Niu, L. (2020). Electrochemical exfoliation *of graphene as an anode material for ultra*-long cycle lithium ion batteries. Journal of Physics and Chemistry of Solids, 139, 109301. doi:10.1016/j.jpcs.2019.109301

Zhao, X., Wang, W., Zhang, Y., Wu, S., Li, F., & Liu, J. P. (2014). Synthesis and characterization *of gadolinium doped cobalt ferrite nanoparti*cles with enhanced adsorption capability for Congo Red. Chemical Engineering Journal, 250, 164–174. doi:10.1016/j.cej.2014.03.113

Zhao, Z., Yang, Z., Hu, Y., Li, J., & Fan, X. (2013). Multiple functionalization of multi-walled carbon nanotub*es with carboxyl and amino groups*. Applied Surface Science, 276, 476–481. doi:10.1016/j.apsusc.2013.03.119

Zheng, Q., Shi, B., Li, Z., & Lin Wang, Z. (2017). Recent Progress on Piezoelectric and Triboelectric Energy Harv*esters in Biomedical Systems*. Advanced Science News, 1–23. . doi:10.1002/advs.201700029

Zheng, W., & Wong, S.-C. (2003). Electrical conductivity and dielectric properties of PMMA/expanded graphite composites. Compos*ites Science and Tech*nology, 63(2), 225–235. doi:10.1016/S0266-3538(02)00201-4

Zheng, X., Hu, Q., Zhou, X., Nie, W., Li, C., & Yuan, N. (2021). Graphene-based fibers for the energ*y devices application: A comprehens*ive review. Materials & Design, 201, 109476. doi:10.1016/j.matdes.2021.109476

Zhou, D., Huang, H., Wang, Y., Wang, Y., Hu, Z., & Li, X. (2019). A yellow-emissive carbon nanodot-based ratiometric f*luorescent nanosens*or *for* visualization of exogenous and endogenous hydroxyl radicals in the mitochondria of live cells. Journal of Materials Chemistry. B, Materials for Biology and Medicine, 7(23), 3737–3744. doi:10.1039/C9TB00289H

Zhou, D., Zhang, Q. P., Zheng, J., Wu, Y., Zhao, Y., & Zhou, Y. L. (2017). *Co-shielding of neutron and γ-ray with bismuth borate nanoparticles f*abricated via a facile sol-gel method. Inorganic Chemistry Communications, 77, 55–58. doi:10.1016/j.inoche.2017.01.034

Zhou, Q., Lau, S., Wu, D., & Kirk Shung, K. (2011). Progress in Materials Science Piezoelectric Films for *High Frequency Ultrasonic Transduc*ers in Biomedical Applications. Progress in Materials Science, 56(2), 139–174. doi:10.1016/j.pmatsci.2010.09.001 PMID:21720451

Zhou, Q., Zheng, Y., Wang, D., Lian, Y., Ban, C., Zhao, J., & Zhang, H. (2020). Cathode ma*terials in non-aqueous alumin*um-*io*n batteries: Progress and challenges. Ceramics International, 46(17), 26454–26465.

Zhu, G., Lin, Z., Jing, Q., Bai, P., Pan, C., Yang, Y., Zhou, Y., & Wang, Z. L. (2013). Toward Large-Scale Energy Harvesting by a Nanopartic*le-Enhanced Triboelect*ric Nanogenerator. Nano Letters, 13(2), 847–853. doi:10.1021/nl4001053 PMID:23360548

Zhu, G., Tang, C., Jiang, M., Du, A., Zhang, H., & Yang, J. (2021). Regulating the interfacial beha*vior* of carbon nanotubes for fast lithium storage. Electrochimica Acta, 388, 138591. doi:10.1016/j.electacta.2021.138591

Zhu, G. L., Zhao, C. Z., Huang, J. Q., He, C., Zhang, J., Chen, S., ... Zhang, Q. (*2019). Fast* char*g*ing lithium batteries: Recent progress and future prospects. Small, 15(15), 1805389.

Zhu, H.-W., Ge, J., Zhao, H.-Y., Shi, L.-A., Huang, J., Xu, L., & Yu, S.-H. (2020). Sponge-templating synthesis of *sandwich-like reduc*ed *g*raphene oxide nanoplates with confined gold nanoparticles and their enhanced stability for solar evaporation. Science China Materials, 63(10), 1957–1965. doi:10.100740843-020-1446-5

Zhu, J., Tu, W., Pan, H., Zhang, H., Liu, B., Cheng, Y., ... Zhang, H. (2020). Self-templating synthesis of hollo*w Co3O4 nanoparticles* embe*dd*ed in N, S-dual-doped reduced graphene oxide for lithium ion batteries. ACS Nano, 14(5), 5780–5787.

Zhu, S., & Wang, D. (2017). Photocatalysis: Basic principles, diverse forms of implementations and emerging scientific opportunities. Advanced Energy Materia*ls, 7(23*), *1*700841. doi:10.1002/aenm.201700841

Zhu, S., Huang, A., Wang, Q., & Xu, Y. (2021). MOF-derived porous carbon nanofibers wrapping Sn nanoparticles as flexible anodes for lithium/sod*ium ion batte*ri*es.* Nanotechnology, 32(16), 165401.

Zhu, X., Lian, Q., Tang, K., Bai, W., & Chu, J. (2019). Epitaxial processing optimization and photoluminescence spe*c-tra of CdTe thin films* grown on highly dissimilar SrTiO3 (001) by molecular beam epitaxy. Thin Solid Films, 6691, 551–557. doi:10.1016/j.tsf.2018.11.047

Zhu, X., Zhou, J., Chen, M., Shi, M., Feng, W., & Li, F. (2012). Core-shell Fe3O4@NaLuF4:Yb, Er/Tm nanostructure for MRI, *CT and upconversion lu*minescence tri-modality imaging. Biomaterials, 33(18), 4618–4627. doi:10.1016/j.biomaterials.2012.03.007 PMID:22444645

$_{z}o_{p}$f, D., Pittner, A., Dathe, A., Grosse, N., Csáki, A., Arstila, K., & Uhlrich, G. (2019). Plasmon*ic nanosenso*r *ar*ray for multiplexed DNA-based pathogen detection. ACS Sensors, 4(2), 335–343. doi:10.1021/acssensors.8b01073 PMID:30657315

Zu, H., Wu, H., & Wang, Q. (2016). High-Temperature Piezoelectric Crystals for Acoustic Wave Sensor Applications. *IEEE Tran*sactions on Ultrasonics, Ferroelectrics, and Frequency Control, 15261(c), 1–55. doi:10.1109/TUFFC.2016.2527599 PMID:26886982

Gehrke, I.; Geiser, A.; Somborn-Schulz, A. (2015) I*nnovations in nanotechnology for water treatment; Nanotechnology, Scien*ce *and* Applications, 1-17. Dovepress. doi:10.2147/NSA.S43773

About the Contributors

Suneev Anil Bansal has more than 15 years of diverse experience in the industry, research laboratories and academia. He has served in key positions in key projects like automobile development at Hero Motor Corp., fighter aircraft development at DRDO, and nanomaterials & manufacturing development at various universities. His research interest is in the fields of micro/nano-manufacturing, materials, 2D materials, graphene, polymer/metal composite materials, sensing, etc. He has published several research papers, holds 1 patent and has served as a subject expert for a number of journals.

Virat Khanna is an Associate Professor and Head of the Department of Mechanical Engineering at Maharaja Agrasen University, HP, India, and has 12 years of teaching experience. He is Ph.D. in Mechanical Engineering from Maharaja Agrasen University, HP, India. He completed his M-Tech in Mechanical Engineering in 2013 and B Tech in Mechanical Engineering in 2008 from I. K. Gujral Punjab Technical University, Jalandhar, India. His areas of research are Aluminium Metal matrix nanocomposite (Graphene, CNT, Processing technique, optimization, and mathematical modeling techniques). He has published and presented more than 18 papers in Journals and conferences of national and international repute. He has authored one book named "Steady State Creep in Rotating Thick-Walled Composite Cylinders" with Lambert Academic Publishing, Germany (ISBN: 978-3-659-71203-6). He is also the Editor of Book Metal Matrix composites: "Properties & Applications" (ISBN 9781032048574), CRC Press, USA, Metal Matrix Composites: "Fabrication, Production and 3D Printing" (ISBN 9781032048536), CRC Press, USA, and Emerging & Futuristic Trends in Engineering & Technology" (ISBN: 978-93- 82068-14-3).

Nilanthy Balakrishnan is a Lecturer in Physics at Keele University, United Kingdom, and has more than 10 years of experience in semiconductor physics and nanofabrication. Her current research interest is fundamental science and applications of 2D materials. She has extensive experience in developing and exploiting 2D materials growth, especially for optoelectronics, ferroelectrics, and thermoelectrics. She obtained her BSc Hons degree in Physics from the University of Jaffna, Sri Lanka in 2008, MSc by Research, and PhD in Condensed Matter Physics from the University of Nottingham, United Kingdom in 2012 and 2015, respectively. She has published 16 research articles in high-profile journals and given talks at 12 major international conferences, including 2 keynote and 2 invited talks. The success of her research was recognised by the European Materials Research Society (E-MRS) Young Scientist Award in 2012. Also, She received more than £200k value of research grants and scholarships from various professional bodies and funding agencies.

Pallav Gupta is presently working as an Assistant Professor (Grade-II) in the Department of Mechanical Engineering, Amity School of Engineering and Technology, Amity University Uttar Pradesh, Noida (INDIA). He completed his B.Tech. (Honors) from Department of Mechanical Engineering, Integral University, Lucknow, INDIA in the year 2009, Qualified GATE in 2009 with AIR-3291 and then completed his M.Tech. (Honors) from I.I.T. (B.H.U.), INDIA in the year 2011 followed by Ph.D. in the year 2015 from I.I.T.(B.H.U.), INDIA. His area of research includes Material Processing; Composite Materials; Mechanical Behaviour and Corrosion. He also has a good command over teaching subjects like Material Science; Engineering Materials; Manufacturing Processes; Measurement and Metrology; Composite Materials. Dr. Gupta has over 08 years of teaching and research experience. He has published more than 100 research papers in peer reviewed international journals as well as in reputed international and national conferences in India as well as in abroad. Apart from this he has also published 12 chapters in books published by Springer, Elsevier and Taylor & Francis. Dr. Gupta has edited 03 books and has also authored a textbook on "Manufacturing Processes" (IP Innovative Publication Pvt. Ltd.). He has been granted 01 Australian Patent. A large number of students have completed their Summer Internships, B.Tech. Projects and M.Tech. Dissertations under his guidance. 03 scholars have completed and 05 are presently registered for their Ph.D. research work under his Supervision.

* * *

Enas Amdeha, Researcher at Egyptian Petroleum Research Institute (EPRI), has her B.Sc. degree in special chemistry, 2007, from the faculty of science, Al-Azhar University with a grade of Very Good with Honor Rank, and has her Master's degree in Inorganic and Analytical Chemistry, 2012, from faculty of science, Al-Azhar University, in water treatment using agricultural wastes. She got her PhD in Applied Physical Chemistry, in 2017, from the faculty of science, Al-Azhar University, in water treatment via photocatalysis using agricultural waste-derived activated carbon. Moreover, she has a patent on "Method for preparing activated carbon (AC) from dried sugar cane waste in nanosize" patent No. 29959 – Academy of Scientific Research and Technology (ASRT) – Patent office (2020), Egypt. She published many articles in Scopus indexed journals and working now in the preparation of nanomaterials specially waste-based-nanomaterials and their applications especially in water treatment via different techniques, e.g., adsorption and photocatalysis. She is an associate member of the Royal Society of Chemistry (AMRSC), and a member of the Egyptian Chemical Society (Member of the Union of Arab Chemists). She is a Mendeley advisor and acts as a reviewer for many international journals.

Kumar Anurag is currently a Senior Research fellow in the Department of Applied Sciences and Humanities Thin Film Laboratory at National Institute of Advanced Manufacturing Technology (NIAMT) formerly National Institute of Foundry and Forge Technology in Ranchi, Jharkhand, India. Kumar Anurag received his post graduate degree in Nanotechnology from Central University of Jharkhand- Ranchi in year 2018. Kumar Anurag published a number of papers in preferred journals and chapters in books and participated in a range of forums on Material Science & Nanotechnology. He also presented various academic as well as research-based papers at several national and international conferences. His research activities are currently two fold: while first research activity is to develop binary and ternary direct band gap semiconductors for photovoltaics applications; the second major research activity that he is pursuing is fabrication of 2D materials for energy storage applications.

Kydyr Askaruly is a PhD student at Satbayev University with a degree in engineering physics. For the last 6 years he has been working as a teacher in physics and also works part-time as a research assistant. He has experience in the fields of thin films, electrochemical energy storage and water treatment.

Seitkhan Azat is a PhD Doctor in Nanotechnology and Nanomaterials, Associate Professor in Chemistry, Head of the Engineering profile Laboratory. Satbayev university, Kazakhstan. At the moment, he is a co-PI of the project "Valorization of biomass waste into highly efficient materials for CBRN protection" (NATO, G5636, 2019-2021), project leader AP09058425 "Development of an innovative technology for obtaining a premix from vegetable raw materials to improve the protein value of local feed for cattle" (GF for young scientists 2021-2023). responsible executor of the project for the commercialization of the results of scientific and (or) scientific and technical activities (No. 283 "Development and production of new high-performance sorbents and filters for water purification", 2017-2019). PI project of Plasma-Assisted water treatment for degradation of persistent polymeric and organic pollutants: from micro- to Nanoparticles to molecules (2022-2024).

Kalampyr Bexseitova is a Master of Technical Sciences. PhD student in the specialty of "Nanotechnology and nanomaterials". She has experience in the synthesis of silica-carbon materials, as well as in the analysis of physico-chemical characteristics.

Sumanta Bhattacharya is a research scholar at Maulana Abul Kalam Azad University of Technology, West Bengal, India and a policy Analyst. He completed his B.Tech, M.Tech in Textile Technology and currently pursuing his Ph.D (2) in Tech, along with that he has MA in development studies, LLB, MA in security and defence law, PGDESD, MPI (Oxford University). Being a research scholar he has 190 research papers published in International and Scopus index journal, 34 Book chapters published, 32 patents International and National, 4 copyrights, 54 International conference and presented papers and won 18 awards for his excellent in education, policy making and innovation.

Zulfiqar Buzdar has graduated from the most prestigious medical university of Pakistan, the King Edward Medical University in the year 2009. He has a diploma in Medical Jurisprudence to his name in 2014 and has attained Master's Degree of Philosophy (M. Phil) in the subject of Forensic Medicine and Toxicology just after three years of his previous post grad degree from the parent Institute. He has served as a professorial staff in different medical colleges of Punjab Pakistan since last 10 years. He is author of several original research articles to his name of medico-legal and postmortem studies. Currently the author is a Ph.D Scholar in the University of Health Sciences and serving as an Associate Professor in Forensic Medicine and Toxicology.

Abir Chakravorty completed his M.Tech in Chemical Engineering from University College of Science Technology, and Agriculture, University of Calcutta, and PhD in Chemical Engineering topic from Department of Mechanical Engineering IIT Kharagpur. Thereafter he pursued Post-Doctoral research in the field of Perovskite Solar Cells with Dr. Subhasis Roy, Assistant Professor Department of Chemical Engineering, University Of Calcutta. At present he is employed as an Assistant Professor at Department of Agricultural and Food Engineering, IIT Kharagpur. Prior to his employment he worked as a Senior Project Associate at West Bengal Pollution Control Board.

Che Azurahanim Che Abdullah (Dr CACA) is currently the Head of the Nanomaterial Synthesis and Characterization Lab at Universiti Putra Malaysia and an Associate Professor of Nanobiophysics in the Faculty of Science. Her research is concentrated at the interface of nanotechnology and biology. Additionally, she possesses an exceptional ability at the interface of material science (nanotechnology), agricultural science, and biotechnology, which will be critical in the twenty-first century on both a scientific and economic level. She developed into a self-sufficient researcher. She is the primary supervisor of 11 postgraduate students and has published papers in prestigious international journals. She received more than RM500,000 in total research funding as the project's leader. She has served as an ambassador for international conferences, a member of the technical program committee, an editorial board member, and a reviewer for a number of journals. She is frequently invited to speak at seminars, workshops, and conferences on a local and international level. She has received numerous honors, including the Young Educator Award, a Visiting Fellowship at Universitas Airlangga, an ACP lectureship, and a Global Asian Women2Win fellowship. Additionally, she is involved in a variety of youth and community projects (mobility and science) on a local and international level. As she became more involved in science, she became convinced that the academic community owed it to the public to engage in research, be receptive to novel ideas, and dedicated to the advancement of others. As an accomplished researcher, she currently serves as the leader of the NANOTEDD () research group, which was formed to initiate and promote research activities.

Avijit Dhara is currently pursuing B.Tech in Chemical Engineering at University of Calcutta. He is graduated with B.Sc (Hons) in Chemistry from Raja Peary Mohan College, CU. He has worked as an intern at Berger Pvt Ltd. His topic of interest is in the field of energy and nanotechnology.

Akula Durga is presently working as a Design verification engineer at Analog devices, Bangalore. She did her B.Tech in Electronics and Communication Engineering from National Institute of Technology (NIT), Andhra Pradesh.

Mehmet Güçlü currently working as Research Assistant in the Mechanical Engineering Department at Istanbul University-Cerrahpaşa. My expertise areas are following: Fiber reinforced composite materials Structural health monitoring Nano composites and Statistical methods.

Shivappa B. Gudennavar is a Professor in the Department of Physics and Electronics, CHRIST University with over 24 years of teaching and research experience. His areas of expertise include polymer composites for radiation shielding, X-ray fluorescence, neutron star and black hole X-ray binaries, interstellar medium, galaxies. He has 73 research publications in highly reputed journals and over 848 citations with h-index of 15 and i-10 index of 29 as of January 2022. He has been serving as reviewer for several international referred journals such as The Astrophysical Journal, Spectroscopy Letters, Physica Scripta, American Journal of Physics, Canadian Journal of Physics, Journal of Applied Polymer Science, Environmental Science and Pollution Research, Biomedical Physics & Engineering Express and International Journal of Astronomy and Astrophysics.

Laxmi J. Hathiya is a Ph.D. Research Scholar at Department of Physics, Saurashtra University, Rajkot. She is awarded with a national fellowship by UGC, New Delhi. Prior to this, she was a lecturer of Physics at VVP Engineering College, Rajkot. She Received her M.Sc. (Physics) from Saurashtra University in 2014. The career objective of the author is to work in the field of nano ferrites with reference to their applications. She has worked as a JRF and SRF at the Department of Physics, Saurashtra University. She has more than 6 years of research experience. She has presented her research work at many national and international conferences and achieved the best presentation awards. She has published research papers in some conference proceedings and peer-reviewed journals.

Muhammad Jawad completed his PhD in Forensic Sciences from University of Health Sciences, Lahore. He is serving in Forensic Agency from previous few years.

Hiren H. Joshi is a Professor and Head at the Department of Physics, Saurashtra University, Rajkot. He received his Ph.D. (Physics) entitled “Investigation of Magnetic and Electrical Properties of Ferrites” from Saurashtra University in 1987. He has more than 37 years of teaching experience. He is working in the field of materials science and especially in ferrites since more than three decades. The author has published more than 100 research papers in national and international scientific journals. He has also written a book entitled “Mossbauer spectroscopy and its applications” in 1999. He has guided 10 Ph.D. and 25 M.Phil. scholars and completed 5 major research projects funded by DST, UGC, IUAC New Delhi and IUC-DAEF Mumbai.

Sumit Kashyap is presently working in Carrier Technologies India Limited as Management Trainee. He has completed his Bachelors of Technology in Electronics and Communication Engineering from National Institute of Technology, Andhra Pradesh. He has worked on Synthesis and Characterization of Quantum Dots from Organic material. Apart from this, the industrial project that he has worked on is Virtualization of COTS (Commercial Off the Shelf) hardware like Raspberry-pi, Beagle bone etc. using some of the popular virtualizing tools. He has also written some application programs for microprocessors. The area of his expertise is embedded software.

Bakht Mand Khan obtained his BS degree in Physics (2021) from the Department of Applied Physical and Materials Science, University of Swat. His research interest is the evolution of graphene/ CNTs based anode material for Li Ion batteries. Specializing in Li-ion batteries or energy storage devices has likely focused his studies on the design, production, and application of these technologies. He has gained expertise in the chemistry and physics of battery systems, including the properties and behavior of lithium-ion cells. He has also studied energy storage systems, including advanced batteries, supercapacitors, and fuel cells. He has the skills to develop and evaluate the performance, safety, and sustainability of these technologies.

Fawad Khan obtained his BS degree in Physics (2021) from the Department of Applied Physical and Materials Science, University of Swat. His research interest is development of anode material for Li Ion batteries. Specializing in anodes of batteries and environmental applications has likely focused his studies on the development and optimization of battery anode materials for use in energy storage systems. He has gained knowledge in the properties and behavior of different anode materials, including their impact on battery performance, safety, and sustainability. Additionally, he has studied the environmental impact of battery production and usage, and has developed skills in the design and implementation of sustainable energy storage solutions.

S. R. Kumar holds the position of Professor in the Department of Applied Sciences and Humanities at National Institute of Advanced Manufacturing Technology(NIAMT) formerly National Institute of Foundry and Forge Technology in Ranchi, Jharkhand, India. In the form of research and teaching, he has encouraged multidisciplinary activities that are centered on nanotechnology. Dr. Kumar works in an area related to nanomaterials and thin films. Dr. Kumar frequently oversees a small team of research scientists who develop thin films and energy devices. His earlier work focused on photovoltaic and photoelectrochemical solar cells. His team claimed a 12% efficiency for the CdSe, CdTe, CuInSe2, Cu(In, Ga)Se2, and CdS solar cells they had fabricated. Their group is unique in that they synthesis semiconductors in non-aqueous medium since non-aqueous baths allow for more deposition source options, higher operating temperatures, etc. His team is currently focusing on 2D materials for energy applications, non-aqueous deposition of semiconducting materials like ZnTe, ZnSe, CdZnTe, CdZnS, and ZnSe, and quantum dots.

Prashant Kumar is working as Assistant Professor in Department of Electronics, Keshav Mahavidyalaya, University of Delhi. He is currently pursuing Ph.D from Department of Electronic Science, University of Delhi. Research interests include thin films, Nanoscience and Nanotechnology along with his current research in spintronics.

Ridipt Mishra is currently a final year bachelor of technology student of Department of Chemical Engineering. His key area of interests include nano-materials and nanotechnology, piezoelectric materials, perovskites, sustainable energy development, microalgae, and growth. His aim is to harness such materials and systems which will be beneficial for environment and reduce pollution too.

Nur Farahah Mohd Khairuddin received her PhD from UTM and currently work as research assistance to Dr CACA. Her project focuses on nanomaterial synthesis for wastewater application.

Anam Munawar is currently working as an Assistant Professor in the Department of Forensic Sciences at University of Health Sciences Lahore, Pakistan. Her area of interest if the synthesis of nanomaterials and their application in the area of sensing.

Won Oh is a Full Professor in the Department of Advanced materials and engineering at Hanseo University in Korea and the School of Materials Science and Engineering at Anhui University of Science and Technology in China. He is a guest professor in some of the Universities in China, Thailand, and Indonesia. He obtained the 'Research Front' award from Korean Carbon Society and obtained the 'Yangsong' award from Korea Ceramic Society, the "Excellent Paper Award" from Korea Journal of Material Research, and the Best Paper Award" from Journal of Industrial and Engineering Chemistry for his pioneering work, and "Award of appreciation" from ICMMA. He is an ICMMA committee board member and is appointed as one of the "Conference Chairman and Vice-Chairman" from 2007 to the present year. He is the author and a co-author of 890 papers published in domestic and international journals.

Dharshini Perumal is currently a Master student working on synthesis, characterization and application of metal oxide nanoparticles prepared via green approach for biomedical applications.

Srilakshmi Prabhu is an active researcher in the field of Materials Science and Nuclear Physics who received her PhD (Physics) from CHRIST University in 2023. She has over 13 publications including book chapters and 60 citations with an h-index of 5 and i-10 index of 3 as of January 2023. Her areas of interest include polymer composites and nanomaterials, biopolymers, photon and EMI shielding materials and computational physics.

Sunil Kumar Pradhan is presently working as Associate Professor at School of Electronics Engineering at Vellore Institute of Technology(VIT), Chennai. His research areas include Nano-technology, VLSI Technology, Nano-Electronics.

Ahsan Riaz is currently enrolled in MPhil Forensic Sciences at University of Health Sciences, Lahore, pakistan. His research is based on the synthesis and characterization of molybdenum based composite materials and their application in explosives detection.

Subhasis Roy is working as Assistant Professor at Department of Chemical Engineering, University of Calcutta, India. Dr. Roy received Ph.D. from Indian Institute of Technology, Kharagpur, India in 2012. Before joining as a faculty member at Calcutta University, he worked as a Postdoctoral Researcher at Sungkyunkwan University, South Korea, Research Scientist at IISER, Kolkata, and worked as a visiting researcher Raman Fellow, Ohio University. Dr. Roy has published more than 55 scientific research papers in peer-reviewed journals. He has 10 book chapters to his credit and filed several patents. Dr. Roy was awarded DST-Fast Track Young Scientist Start-Up Research Grant from SERB –DST, Mission innovation MI IC#5' DBT-DST joint funding grant, Teachers Associateship for Research Excellence (TARE) SERB –DST. He is a member of many Professional Societies and reviewers as well as editorial board member of many international journals. His expertise is in perovskite solar cells, multiferroic composite thin films, photocatalysts for water splitting, fuel cells, and hydrogen energy.

Bubbly S. G. is a Professor in the Department of Physics and Electronics, CHRIST University with over 26 years of teaching and research experience. Her areas of expertise include polymer composites and nanomaterials, radiation shielding materials, crystallography, X-ray fluorescence and X-ray binaries. She has 54 research publications and over 564 citations with an h-index of 13 and an i-10 index of 22 as of January 2023. She is a reviewer for several international refereed journals such as Physica Scripta, RSC Advances, Advanced Materials Interfaces, Scientific Reports, Journal of Applied Polymer Science, Luminescence: The Journal of Biological and Chemical, Luminescence and Advances in Materials Physics and Chemistry.

Rahul Saha is currently pursuing B.Tech in Chemical Engineering from University Of Calcutta. He obtained B.Sc (Hons) in Chemistry from Maulana Azad College, CU. He has successful internship engagement in Indian Farmers Fertilizer Cooperative (IFFCO), Paradip unit. He is now working as an undergraduate project intern in the nano-science and technology field under the supervision of Dr. Subhasis Roy. His research interest is concentrated in the area of nano-science and technology, solar cell and energy field.

Yerbolat Sailaukhanuly is a PhD in chemistry and leading researcher at engineering profile laboratory Satbaev University. His research interest is in the field of analysis of persistent organic pollutants in environmental objects.

Arkadyuti Seth is currently pursuing his Masters in Development studies from IGNOU, he has completed his B.Sc. in Chemistry from Calcutta University and M.Sc. in Forensic Science from BHU. He has 11 research papers published, 6 Book chapters and attended 6 conference and presented paper.

Neha Sharma is working as Professor in Electronics in Department of Electronics, Keshav Mahavidyalaya, University of Delhi. Research interests include Nano science and Nanotechnology and Optoelectronics.

Sudeshna Surabhi is currently a Senior Research fellow in the Department of Applied Sciences and Humanities Thin Film Laboratory at National Institute of Advanced Manufacturing Technology (NIAMT) formerly National Institute of Foundry and Forge Technology in Ranchi, Jharkhand, India. Sudeshna Surabhi received his post graduate degree in Electronics and Instrumentation from Bharath University- Chennai in year 2014. Sudeshna Surabhi published a number of papers in preferred journals and chapters in books and participated in a range of forums on Material Science & Nanotechnology. She also presented various academic as well as research-based papers at several national and international conferences. Her research activities are currently twofold: while first research activity is to develop binary and ternary direct band gap semiconductors for photovoltaics applications; the second major research activity that she is pursuing is fabrication of 2D materials for energy storage applications.

Zhandos Tauanov, Ph.D. in Materials Science and Engineering (Nazarbayev University, Kazakhstan), M.Sc. in Chemical Engineering (Loughborough University, UK), currently working as an acting associate professor and postdoc for al-Farabi Kazakh National University. Science tutor in physical sciences at King Abdullah University of Science and Technology (SRSI Summer Program, KAUST, Saudi Arabia, 2021, 2022), winner of the Korea Research Fellowship (South Korea, 2019-2020), winner of the state stipend for talented young scholars of Kazakhstan (2021-2022). Author of more than 16 scientific papers in leading journals and international proceedings. His h-index is 9 on Scopus with 230+ total citations.

Kainaubek Toshtay is a Ph.D. in Materials Science and Engineering, M.Sc. in Chemical Engineering, currently working as acting associate professor and postdoc for al-Farabi Kazakh National University, senior researcher of LLP "NPTC "Zhalyn". Worked as a postdoc scholar at Chungnam National University (South Korea, 2019-2020), senior researcher of the Environmental Science and Technology Group (ESTg) of Nazarbayev University (2016-2019). PI of the research grant for Young Scholars Grant No. AP09058298 (2021-2023) and co-PI of the project Grant No. AP08857007 (2020-2022) funded by the Science Committee of the Ministry of Education and Science of the Republic of Kazakhstan. Science tutor in physical sciences at King Abdullah University of Science and Technology (SRSI Program, KAUST, Saudi Arabia, 2021).

Fatih Turan is currently working as an Associate Professor in the Mechanical Engineering Department at Eskisehir Technical University. My expertise areas are following: Fiber reinforced composite materials Structural health monitoring Nano composites.

Kefayat Ullah completed his PhD degree in graphene-based nanomaterials form Hanseo University South Korea in the year 2015. He Joined University of Swat in January 2016 as Assistant Professor. Kefayat currently works at the Department of Applied Physical and Materials Science, University of Swat.

His research interests are materials science, graphene-based nanomaterials, solar cell and development of anode/cathode for Li Ion Batteries. During the study course he specialized in the development of novel nanocomposite materials for energy storage and environmental applications. He has gained extensive knowledge in the areas of materials synthesis, characterization, and electrochemistry. Additionally, He has gained valuable experience working in a multidisciplinary team and collaborating with industry and partner organizations.

Jostin Vara has completed his Bachelors of Technology in Electronics and Communication Engineering from National Institute of Technology, Andhra Pradesh. He has worked on Synthesis and Characterization of Quantum Dots from Organic material. Apart from this, he has worked for an Non-Governmental Organization (Endnow Foundation) on Digital Wellbeing and Awareness and also written a research paper on that.

Jie Hui Wong is currently an undergraduate student at Faculty of Health and Medical Sciences UPM and previously attended internship program with Nanotedd research group led by Dr CACA.

Darkhan Ybyraiymkul is a Master of Materials Science and Technology. He is a lecturer at Satbayev University in Materials Science. He worked as a researcher on a commercialization project. He has experience in solar cells, the production of carbon nanotubes.

Ulan Zhantikeyev is a Master of Technical Sciences. PhD student in the specialty of "chemical technology of explosives and pyrotechnics". Junior researcher of the Laboratory of Engineering Profile. He has experience in the synthesis of activated carbon, as well as in the analysis of physico-chemical characteristics.

Index

G

H

L

M

N

O

P

Q

R

S

T

U

V

W

X

Y

Z

www.ingramcontent.com/pod-product-compliance
Lightning Source LLC
LaVergne TN
LVHW062112300926
849161LV00010B/2440
9781668473580